# Childhood Trauma in Mental Disorders

Gianfranco Spalletta • Delfina Janiri
Federica Piras • Gabriele Sani
Editors

# Childhood Trauma in Mental Disorders

## A Comprehensive Approach

Springer

*Editors*
Gianfranco Spalletta
Laboratory of Neuropsychiatry
IRCCS Fondazione Santa Lucia
Rome
Italy

Menninger Department of Psychiatry
and Behavioral Sciences, Baylor College
of Medicine
Houston, TX
USA

Federica Piras
Laboratory of Neuropsychiatry
IRCCS Fondazione Santa Lucia
Rome
Italy

Delfina Janiri
Department of Psychiatry and Neurology
Sapienza University of Rome
Rome
Italy

Fondazione Policlinico Universitario
"Agostino Gemelli" IRCCS
Rome
Italy

Centro Lucio Bini
Rome
Italy

Gabriele Sani
Department of Neuroscience, Section
of Psychiatry, Università Cattolica del
Sacro Cuore
Rome
Italy

Fondazione Policlinico Universitario
"Agostino Gemelli" IRCCS
Rome
Italy

Centro Lucio Bini
Rome
Italy

ISBN 978-3-030-49413-1 ISBN 978-3-030-49414-8 (eBook)
https://doi.org/10.1007/978-3-030-49414-8



This Springer imprint is published by the registered company Springer Nature Switzerland AG
The registered company address is: Gewerbestrasse 11, 6330 Cham, Switzerland

# Preface

This book reports the current achievements in the understanding of traumatic spectrum symptomatology across mental disorders and tries to clarify some of the many controversies about the underpinning neurobiological mechanisms. Editors and authors focused on the definition of symptoms of childhood trauma and their persistence and/or development into adolescence and adulthood. Our goal was to emphasize the mental health risk related to childhood trauma and its enduring psychobiological effects. In light of this, the way to understand trauma-related disorders needs to be revisited and perhaps, to be provocative, even reconceptualized. We think that improving our knowledge on this topic may be a challenge of extreme importance, given that early life trauma/adversities affect about half of the population of children and young adults. Apart from the immediate effects of trauma, early adversities also increase the risk for developing mental and physical problems later in life. We hope that a closer consideration of early traumatic experiences may help, in the near future, to revolutionize diagnostic and therapeutic approaches to mental disorders.

Readers will be informed about how traumatizing events, including events perceived as such, evoke defense states, and thus determine responses aimed at ensuring biological survival. The persistent deregulation of brain–body psycho-physiological integration prolongs symptoms even after the danger has passed, leading to what we call trauma-related disorders. Although sensitization and/or conditional learning processes may account for symptom duration, we must try to understand whether the prolongation of defensive responses, once the threat has ceased, might have some biological value. Clinicians and researchers should examine this potential mechanistic link consistently with a developmental perspective, since traumatizing nurturing environment predicts unfavorable outcomes, such as chronic overwhelming stress, and determines psychobiological and psychiatric effects on child development. On the other hand, we may hypothesize that what is commonly regarded as a deviation from "optimal" development could conversely reflect an initial adaptation, even if the final developmental outcome is unfavorable. Examining trauma-related disorders through the prism of biological survival, we should consider that psychiatric phenomenology related to childhood trauma may represent "facultative" responses (i.e., responses selected as biologically valuable in the behavioral repertoire) in specific environmental contexts. This wide range of possibilities could allow us to even reconceptualize the relationship between childhood trauma and mental disorders and to consider some psychiatric symptoms as adaptive responses. In this case, we may assume that natural selection enhances developmental strategies and promotes immediate survival, even if negative consequences may occur later in life.

In the book, a panel of experts in the field thoroughly reviewed, from a psychobiological perspective, the harmful effects of early adversities on the developing brain and analyzed biological mechanisms underpinning subsequent emergence of psychiatric symptoms. An alternative view has also been considered and involves the neuroplastic adaptive variation in individuals with heightened neurobiological susceptibility. In the section of the book dedicated to the potential pathogenic link between childhood trauma and mental disorders, authors addressed the individual differential susceptibility to (eventually unfavorable) environmental conditions. In order to promote biological survival, individuals with specific neurobiological susceptibility may change their structural/functional neuro-circuitries, and their immune/neuroendocrine defense systems in response to peculiar nurturing environments. Later in life, this mechanism results in developmental trajectories culminating in maladaptive psychopathology, which is likely determined by the interaction between neurobiological individual susceptibility and negative environmental factors. This new perspective may have crucial clinical implications. If we consider childhood trauma-related manifestations as attempts to adapt homeostasis in response to environmental conditions, we can treat symptoms by promoting supportive contexts and alternative strategies for maintaining physiological integrity, enhance individual adaptive traits, and potentially divert negative developmental outcomes.

Research and treatment in modern psychiatry cannot ignore that: (i) childhood trauma has a role in shaping subsequent clinical pictures, and (ii) psycho/biological-developmental approaches should be applied to the individual prediction of outcomes and to tailor intervention strategies in the framework of personilized medicine. Experts in social and phenomenological fields suggest to further focus on the reliability of biological indices of neurobiological susceptibility, in order to apply multidimensional metrics in the design and the evaluation of interventions, particularly at the prevention level.

We hope that this book will provide the knowledge and awareness necessary to embrace an advantageous integrative research approach to prevent, treat, and rehabilitate childhood trauma-related mental disorders. An extended hope is that new treatment avenues will derive from the consideration of the adaptive strategies and traits expressed by traumatized children, in the struggle to maintain their physiological and psychological integrity.

Rome, Italy Gianfranco Spalletta
Rome, Italy Delfina Janiri
Rome, Italy Federica Piras
Rome, Italy Gabriele Sani

# Contents

# Part I

# General

# Introduction on Childhood Trauma in Mental Disorders: A Comprehensive Approach

# 1

Delfina Janiri, Gabriele Sani, Federica Piras, and Gianfranco Spalletta

The exposure to negative events emotionally overwhelming can lead to traumatic memories, altered sense of self, emotional dysregulation, and avoidance. This is a picture consistently described since Homer's poems [1]. In book 19 of the Odyssey, Odysseus, which has at last come home to Ithaca, is recognized by a scar on his thigh. Odysseus describes the origin of the scar, a hunting accident occurred in his boyhood, at a boar hunt during the time of his visit to his grandfather Autolycus. The scar was already in his name: his grandfather called him "Odysseus" meaning "the hated one" because Autolycus had many enemies. The second great Homeric hero, Achilles, was killed by an arrow hitting his heel. To prevent his predicted early

D. Janiri
Department of Psychiatry and Neurology, Sapienza University of Rome, Rome, Italy

Fondazione Policlinico Universitario "Agostino Gemelli" IRCCS, Rome, Italy

Centro Lucio Bini, Rome, Italy
e-mail: delfina.janiri@gmail.com

G. Sani
Fondazione Policlinico Universitario "Agostino Gemelli" IRCCS, Rome, Italy

Centro Lucio Bini, Rome, Italy

Department of Neuroscience, Section of Psychiatry, Università Cattolica del Sacro Cuore, Rome, Italy
e-mail: gabriele.sani@unicatt.it

F. Piras
Laboratory of Neuropsychiatry, IRCCS Fondazione Santa Lucia, Rome, Italy
e-mail: federica.piras@hsantalucia.it

G. Spalletta (✉)
Laboratory of Neuropsychiatry, IRCCS Fondazione Santa Lucia, Rome, Italy

Menninger Department of Psychiatry and Behavioral Sciences, Baylor College of Medicine, Houston, TX, USA
e-mail: g.spalletta@hsantalucia.it



G. Spalletta et al. (eds.), *Childhood Trauma in Mental Disorders*,
https://doi.org/10.1007/978-3-030-49414-8_1

death, Achilles's mother, Thetis, took him to the River Styx in the Underworld, which was supposed to offer powers of invulnerability, and completely dipped his body into the infernal water. However, as Thetis held Achilles by the heel, his heel was not washed over by the river and became vulnerable. Thus, early negative events seem to shape hero's psychological traits, leading to two different pictures: strike first; withdraw and isolate oneself from others (e.g., Achilles); or create deceptions, distractions, false identities, and narratives to spoil the aim of what is expected (e.g., Odysseus).

In the last centuries, early traumatic events have been specifically linked with the genesis of psychopathology. In 1859, Briquet [2] described, for the first time, the relationship between trauma and hysteria, highlighting that out of 501 patients, 381 presented with past traumatic experiences. Later, Charcot [3], Janet [4] and particularly Freud [5], proposed trauma models to better understand mental disorders. These models were founded on the clinical observation that patients often reported severe and repeated childhood traumatic experiences influencing subsequent psychopathological behaviors. This book aims to further enlighten this observation and to frame it in the context of a substantial modern body of literature corroborating the link between childhood trauma and mental disorders.

Childhood trauma could be defined as "the experience of an event by a child that is emotionally distressful." Traumatic events deal great damage, not just because of the immediate harm they cause, but because of the lingering need to reevaluate one's view of oneself and the world. Among the consequences of childhood trauma, many are biologically based; this is sustained by a variety of studies linking early adverse events with disrupted neurodevelopment. Specifically, childhood maltreatment has been associated with long-term structural and functional brain abnormalities [6–8], alterations in neurochemistry [9], and in other biological targets. Different forms of childhood trauma have been described; they include physical, sexual and emotional abuse and a different type of traumatic experience, particularly pertaining to carelessness, that has been defined "neglect" [10]. Childhood trauma can also be divided into a single traumatic experience or into "complex trauma," indicating traumatic events that occurred in combination or cumulatively [11]. Recent studies confirmed the association between early life stress and mental disorders [12–17]. Childhood traumatic events are more frequently reported in patients than in healthy controls, with a high prevalence rate (e.g., childhood abuse and neglect have been reported by 50% of patients diagnosed with bipolar disorders [18]). Even though early maltreatment has been found to trans-diagnostically increase vulnerability to psychopathology, it should not be considered as unspecific risk factor. Distinct types of traumatic experiences differently impact on multiple mental disorders and they can independently modulate the clinical expression [12–17].

This book, divided into 21 chapters, all of which disentangling different aspects of the relationship between early maltreatment and mental disorders, addresses the complicated and multilayered intersection between childhood trauma and vulnerability to mental disorders. It deals with the complexity of events that impact directly on personal development and interact with individual susceptibility to mental illness.

After an exhaustive overview of the concept of childhood trauma and its historical perspective, the first part of the book is dedicated to identifying and examining biological mechanisms underpinning alterations that result from early adverse events. In these chapters, experts in the field review the latest evidence about the effects of traumatic experiences on brain development resulting in cognitive dysfunctions and neuroimaging and electroencephalographic changes. In the same section, a specific chapter also provides an overview of the candidate genes involved in mental disorders, potentially interacting with childhood trauma. This part of the book is crucial to understand how the neurobiological implications of early adverse events intersect with mechanisms underlying psychopathology. Relating to this, a fundamental question to explain is when and how childhood trauma *per se* causes biological changes and increases the risk for mental disorders, and how biological vulnerabilities to mental disorders early interact with adverse events at the neurodevelopmental level. Furthermore, a specific chapter also traces some pieces of the increasing evidence that maternal trauma can be transmitted to the offspring. In this case, early adverse events experienced by the mother seem to influence the child's vulnerability to both trauma exposure and mental disorders, possibly trough epigenetic pathways. An evolutionary understanding of the physical and mental effects of childhood trauma is also summarized in this first part of the volume that clarifies some intriguing points of view of the Darwinian psychiatry.

The second part of the book is concerned with the impact that childhood trauma has on each mental disorder. Chapters in this section synthesize the latest evidence on the relationship between early adverse events and specific clinical pictures including psychosis, mood and anxiety disorders, substance use, and eating problems. Each chapter provides information about the epidemiology of early trauma in each specific disorder, systematically describes the risk associated with different types of childhood maltreatment and focuses on their impact on the clinical presentation and outcome. These chapters also present a specific framework defining neurobiological mechanisms of interaction between diagnosis and early negative events. Childhood trauma emerges as a crucial risk factor for multiple mental disorders and authors make the effort to tackle its intersectionality with other variables that increase the likelihood to develop mental illness. Special cases are the two chapters on post-traumatic stress and dissociative disorders. In this case, indeed, traumatic events represent *per se* putative causes of symptomatology. Therefore, what becomes crucial in these different perspectives is to evaluate the cumulative effect of traumatic events from childhood to adult life. Other two peculiar situations considered in this section are the occurrence of trauma in children with neurodevelopmental disorders and the association between childhood trauma and the risk of neurological deficits later in life. In the first case, trauma and mental disorders occur in the same time framework; thus, authors investigate the diagnostic complexities of co-occurring trauma-related symptoms and neurodevelopmental disorders. In the second case, a specific chapter faces the innovative topic of neurobiological sequelae of early trauma later in life, considering the potential mediating role of cognitive functions.

The third part of the book addresses treatment strategies in the context of the presence of childhood traumatic memories in patients with mental disorders. Psychotherapy treatments are considered across different developmental periods and a separate chapter specifically reviews pharmacological approaches as integrated treatments in subjects with childhood trauma. This is particularly relevant because recent studies demonstrated that patients reporting childhood adverse events often show inadequate response to different classes of mental health medications [19]. In the same section of the book, a specific chapter is also dedicated to stigma as peculiar consequence of early maltreatment. The chapter highlights the importance of tailoring therapeutic interventions on decreasing stigma, in order to minimize long-term negative effects of traumatic events.

By tracing psychopathological and neurobiological consequences of childhood trauma, this book proposes methodological tools to define and evaluate resilience. The concept of resilience in psychiatric research is based on studies on children exposed to social adversity, including family dysfunction, economic deprivation, and institutionalization. The British psychiatrist Sir Michael Rutter defined resilience as "*an interactive concept that is concerned with the combination of serious risk experiences and a relatively positive psychological outcome despite those experiences*" [20]. Resilience can be viewed as a dynamic construct, which is not the reverse of risk but is associated with biological changes indicative of adaptive functions. Resilience can be termed as "*ordinary magic*" [21], suggesting that it arises from the normative functions of human adaptation systems. Understanding the multifaceted nature of childhood trauma, by bringing together its many perspectives, can help to explain how people cope with traumatic life experiences giving the means to enhance resilience processes.

It is our hope that the ideas in this volume might make a contribution to identifying new venues for preventing and treating mental disorders. In the light of its prominent contribution to neuroplasticity, early trauma should be included in the clinical assessment of all patients with mental disorders. Physicians should have a specific framework to properly address a pharmacological/psychotherapeutic/rehabilitative treatment in patients reporting a history of early adverse events. We do believe that a wider contextualization of childhood trauma can really make the difference between caring for symptoms and treating mental disorders.

## References


1. Harrison JA, Jordan RH, editors. Homer. Iliad. Bristol cl. Bristol, U.K.; 1991.
2. Briquet P. Traité clinique et thérapeutique de l'hystérie. Paris: Ballière; 1959.
3. Charcot JM. Leçons sur les maladies du système nerveux, faites a la Salpêtrière, vol. 3. Paris: Progres medical en A. Delahaye & E. Lecrosnier; 1887.
4. Janet P. L'Amnésie et la dissociation des souvenirs par l'émotion. J Psycologie. 1904;1:417–53.
5. Freud S. Zur Ätiologie der Hysterie. Wien; 1896.
6. Dannlowski U, Stuhrmann A, Beutelmann V, Zwanzger P, Lenzen T, Grotegerd D, et al. Limbic scars: long-term consequences of childhood maltreatment revealed by functional and structural magnetic resonance imaging. Biol Psychiatry. 2012;71:286–93.

7. Janiri D, Sani G, De Rossi P, Piras F, Iorio M, Banaj N, et al. Amygdala and hippocampus volumes are differently affected by childhood trauma in patients with bipolar disorders and healthy controls. Bipolar Disord. 2017;19:353–62.
8. Janiri D, Sani G, De Rossi P, Piras F, Banaj N, Ciullo V, et al. Hippocampal subfield volumes and childhood trauma in bipolar disorders. J Affect Disord. 2019;253:35–43.
9. Treadway MT, Grant MM, Ding Z, Hollon SD, Gore JC, Shelton RC. Early adverse events, HPA activity and rostral anterior cingulate volume in MDD. PLoS One. 2009;4(3):e4887.
10. Bernstein DP, Fink L, Handelsman L, Foote J, Lovejoy M, Wenzel K, et al. Initial reliability and validity of a new retrospective measure of child abuse and neglect. Am J Psychiatry. 1994;151:1132–6.
11. Terr LC. Childhood trauma: an outline and overview. Am J Psychiatry. 1991;148:10–6.
12. Janiri D, Sani G, Danese E, Simonetti A, Ambrosi E, Angeletti G, et al. Childhood traumatic experiences of patients with bipolar disorder type I and type II. J Affect Disord. 2014;175:92–7.
13. Üçok A, Bikmaz S. The effects of childhood trauma in patients with first-episode schizophrenia. Acta Psychiatr Scand. 2007;116:371–7.
14. Carpenter L, Chung MC. Childhood trauma in obsessive compulsive disorder: the roles of alexithymia and attachment. Psychol Psychother Theory Res Pract. 2011;84:367–88.
15. Martinotti G, Carli V, Tedeschi D, Di Giannantonio M, Roy A, Janiri L, et al. Mono- and polysubstance dependent subjects differ on social factors, childhood trauma, personality, suicidal behaviour, and comorbid Axis I diagnoses. Addict Behav. 2009;34:790–3.
16. Hovens JGFM, Giltay EJ, Wiersma JE, Spinhoven P, Penninx BWJH, Zitman FG. Impact of childhood life events and trauma on the course of depressive and anxiety disorders. Acta Psychiatr Scand. 2012;126:198–207.
17. Herman JL, Perry JC, van der Kolk B. A. Childhood trauma in borderline personality disorder. Am J Psychiatry. 1989;146:490–5.
18. Garno J, Goldberg J, Ramirez P, Ritzler B. Impact of childhood abuse on the clinical course of bipolar disorder. Br J Psychiatry. 2005;186:121–5.
19. Etain B, Lajnef M, Brichant-Petitjean C, Geoffroy PA, Henry C, Gard S, et al. Childhood trauma and mixed episodes are associated with poor response to lithium in bipolar disorders. Acta Psychiatr Scand. 2017;135:319–27.
20. Rutter M. Implications of resilience concepts for scientific understanding. Annals of the New York Academy of Sciences; 2006. p. 1–12.
21. Masten AS. Ordinary magic: resilience processes in development. Am Psychol. 2001;56: 227–38.

# The Concept of Childhood Trauma in Psychopathology: Definitions and Historical Perspectives

# 2

Gretchen Buchanan, Abigail H. Gewirtz, Cara Lucke, and Monica R. Wambach

## 2.1 Introduction

In defining childhood trauma, it is critical to distinguish exposure to traumatic events from a child's reaction to those events (e.g., psychopathological symptoms). We use the word "trauma" to refer to a traumatic event or events. We use the definitions of traumatic events from the ICD-11 and DSM-5, as outlined in the diagnostic criteria (Criterion A in the DSM) for trauma-related disorders. ICD characterizes a traumatic event as "*an extremely threatening or horrific event or series of events (e.g., natural or human-made disasters, combat, serious accidents, sexual violence, assault).*" The DSM provides the following definition: "*The person was exposed to death, threatened death, actual or threatened serious injury, or actual or threatened sexual violence* as follows: (one required)

1. Direct exposure.
2. Witnessing, in person.
3. Indirectly, by learning that a close relative or close friend was exposed to trauma. If the event involved actual or threatened death, it must have been violent or accidental.
4. (Not generally relevant to children.) Repeated or extreme indirect exposure to aversive details of the event(s), usually in the course of professional duties (for example, first responders, those collecting body parts or professionals repeatedly exposed to details of child abuse). This does not include indirect non-professional exposure through electronic media, television, movies or pictures."

---

All authors contributed equally to the manuscript. Order of authors is presented alphabetically.

---

G. Buchanan · A. H. Gewirtz (✉) · C. Lucke · M. R. Wambach
University of Minnesota, St. Paul, MN, USA
e-mail: bucha245@umn.edu; agewirtz@umn.edu; lucke124@umn.edu; corre053@umn.edu



G. Spalletta et al. (eds.), *Childhood Trauma in Mental Disorders*,
https://doi.org/10.1007/978-3-030-49414-8_2

In childhood, trauma exposure typically occurs as a result of violence, disasters, unintentional accidents, or illness. Arguably, the most common source of exposure to violence is within the family (i.e., via domestic violence, childhood maltreatment) though large numbers of children in specific countries and regions across the world are exposed to violence outside the home (i.e., community violence, political violence, terrorism, and/or war). Natural and man-made disasters include weather events such as hurricanes, typhoons, tornados, and flooding, as well as fires and earthquakes. Unintentional accidents include motor vehicle accidents, accidental drownings, falls, electrical and other accidents within and outside the home. Finally, less attention has been paid to medical trauma exposure, which refers to the experience of life-threatening or chronic illness resulting in painful and sometimes frequent medical procedures and treatments.

It should be noted that though the above definitions of trauma include those put forth by the World Health Organization (i.e., ICD-11), there are national and regional variations regarding the degree to which traumatic events are recognized as such. For example, only quite recently has attention been paid to the extensive trauma exposure suffered by child soldiers, e.g., [1]. Female genital mutilation is still widely practiced—legally— in countries around the world where it is not viewed as traumatic; in 2016, an estimated 200 million girls worldwide were victims of genital cutting [2]. Worldwide, few countries have surveillance systems to monitor children's trauma exposure. Moreover, many countries do not have legislation aimed at limiting or preventing child abuse or domestic violence; legal system responses vary widely across nations.

## 2.2 Epidemiology of Childhood Trauma Worldwide

Childhood trauma exposure is prevalent worldwide, though we could find no recent worldwide estimates. In the United States, large, nationally representative studies indicate that approximately two-thirds of children have been exposed to at least one traumatic event, and 20% report exposure to multiple traumas by early adulthood [3–5]. The epidemiologic pattern of childhood trauma varies worldwide, and is often perpetuated by sociocultural factors [6, 7]. For example, in parts of Africa, the Middle East, and Asia, female genital mutilation (FGM) is considered an unquestioned social norm and is often viewed as a precondition for marriage [2]. Internationally, FGM is recognized as traumatic, and a human rights violation, but it is estimated that over 200 million girls, across 30 countries, have undergone FGM before the age of 18 [2].

Violence against children exists in every country across the world, and both international and smaller scale studies elucidate the magnitude and pervasive scope of this problem [6]. Since many countries lack both formal systems to investigate violence against children, as well as operationalized, culturally sensitive definitions of child maltreatment, only a proportion of childhood trauma is reported worldwide [8]. In addition, children who are victims of violence are often fearful, or may not have the capacity to report such trauma, and similarly, parents of children are less likely to report incidents of violence against their children if the act was conducted by a person of power in their community [8]. Despite these limitations, as well as varying methodology among the studies that attempt to ascertain children's exposure to trauma, numerous studies have investigated the epidemiology of children's exposure to trauma that provide critical insight on estimates of its prevalence throughout the world.

### 2.2.1 Physical and Sexual Abuse, Neglect, Witnessing Violence

Across a series of meta-analyses that examined 244 studies, the self-reported prevalence of physical violence to children ranged from approximately 22.8% of children in Africa, 16.7% in Asia, 14.3% in Australia, 22.9% in Europe, 24.0% in North America, and 54.8% in South America [9]. Although the prevalence of physical abuse varies by region and countries, more than 50% of youth report having been abused as children [10].

In a prospective longitudinal study of adolescents and young adults in Germany, 7.5% of youth had experienced physical assault, while 9–19% of American youth have experienced physical abuse by a caregiver [4, 11]. In the Kurdistan Province of the Islamic Republic of Iran, 38.5% of children aged 11–17 reported physical violence at home which caused mild to severe physical injury [8, 12]. Across Africa, the prevalence of child physical abuse ranged from 7.6% to 45%, and was greatly influenced by study methodology and child abuse definitions [13]. A national population-based survey in Zimbabwe estimated that 63.9% of girls and 76% of boys experienced physical violence before the age of 18 [10]. Across 13.830 individuals from 171 countries, over 50% of the perpetrators of physical and emotional violence were household members [14].

Worldwide, it is estimated that 150 million girls and 73 million boys have experienced sexual violence [8, 15]. Across 21 high- and middle-income countries, approximately 7–36% of girls and 3–29% of boys reported sexual abuse [16]. The prevalence of sexual abuse varied widely across 17 studies based in Africa (1.6% to 77.7%) and was greatly influenced by study methodology [13]. In the United States, 8–12% of children experienced at least one sexual assault, and girls were approximately 1.5 times more likely to be a victim of sexual assault than boys [4, 17]. In addition to females, younger children are at higher risk for child sexual abuse [13]. Approximately 19% of youth in Namibia and 30% of youth in Zimbabwe were physically coerced to have sex [18], and among victims of sexual abuse, reports indicate that family members often constitute a high proportion of the perpetrators. Across a multicountry study, approximately 14% to 56% of girls and 25% of boys were sexually abused by a relative or step-parent [8, 16]. Among Peruvian women who reported childhood sexual abuse, the perpetrator was a family member in approximately half of the cases [15]. Sexual abuse is also prevalent in the school setting, and findings from UNICEF indicate that 9% of children in Nepal were sexually abused, and that among these individuals, the teacher was the perpetrator in 18% of these cases [19].

Epidemiological data on neglect is difficult to estimate worldwide. In developing countries, many caregivers and children reside in communities with poor public health infrastructures and a severe lack of resources [20]. In a review of 22 studies on childhood maltreatment in humanitarian contexts over the past 20 years, neglect was only measured in three studies, and significantly overlapped with measures of poverty [21]. Therefore, it is difficult to distinguish deliberate neglect compared to an inability to provide sufficient care to children [8, 20]. In contrast, formal reporting systems for child maltreatment are more likely to exist in industrialized countries, and estimates suggest that neglect often constitutes the largest proportion of child maltreatment cases reported to the authorities in those countries [8, 20].

Children may also be traumatized through witnessing violence, and it is estimated that between 133 and 275 million children worldwide witness domestic violence annually [22]. Approximately 38–70% of American youth have witnessed community violence, and among these, 10% witnessed severe violence between caregivers [4, 23]. Although boys were more likely to be exposed to community violence, girls were more likely to witness sexual assault [4.4% vs. 2%; 17].

### 2.2.2 Large-Scale Natural and Manmade Disasters: Children in Specific Regions of the World and Marginalized Children Are Disproportionately at Risk

Child exposure to trauma is not equally distributed across the world, and there is strong evidence to suggest that children residing in low-income or developing countries are disproportionately exposed to trauma on a large scale. For example, between 1998 and 2017, natural disasters killed 1.3 million people worldwide, but on average, the rate of mortality was seven times greater in low-income countries [24]. Over this 20-year period, an additional 4.4 billion individuals were left injured, homeless, displaced, or in need of emergency assistance [24], but individuals in low-income countries were on average six times more likely to suffer from such effects [24]. It is estimated that 175 million children each year will be affected by natural disasters, and this number is expected to rise due to climate change [25, 26]. Children exposed to natural disasters are at risk for depression, anxiety, and post-traumatic stress disorder, but children in poverty who have little access to resources, are at an even greater risk for adverse outcomes [25, 27].

Children in war-torn countries are at significant risk for mortality and exposure to violence. Across the world, it is estimated that over one billion children reside in countries affected by armed conflict [28]. In over 65 countries, children are recruited into armed forces either legally as volunteers, but more frequently, through force or deception [8, 28]. Child soldiers are at significant risk for exposure to violence and sexual and physical abuse by military groups [8, 28]. In recent years, the nature of warfare has shifted, and armed groups systematically perpetrate violence on the civilian population [28]. These strategies dismantle protective infrastructures such as health, education, and justice, and children in the community often become targets of terrorist attacks [28]. Children in war-torn communities are also vulnerable to abduction. For example, during the war in Uganda, it is estimated that upwards of 25,000 children were abducted and subject to abuse by military forces [28].

More than half of the world's 60 million refugees are children [29, 30]. Although the majority of refugee children attempt to re-locate with family members, a significant number are unaccompanied and are at heightened risk for trauma exposure, including physical and sexual abuse and gender-based violence [6, 8, 31]. As children seek asylum, they are often held in centers that are unequipped to meet their needs. In Australia, estimates suggest that youth seeking asylum are held in refugee detention centers for an average of 20 months, which can adversely impact both physical and mental health [8, 32]. Among migrant children from Northern Africa seeking asylum in Spain, many were victims of theft and physical abuse by older children [8, 32]. A significant number of unaccompanied children run away from these centers due to the threat of violence from adults and peers, but a significant proportion are also victim to child trafficking while residing in these centers [8, 33].

The United States has one of the world's largest immigration detention systems in the world [34, 35]. In 2017, 41.435 unaccompanied children and 75.622 family units were detained by the US Border Patrol [34, 36] and in April of 2018, the Trump administration authorized a zero-tolerance strategy that prosecuted parents for illegal immigration and separated over 2,600 children from their parents [37]. This policy was halted after 6 weeks, but the traumatic effects on children and families were devastating and enduring. In addition to the trauma of forced separation, children were detained in housing facilities with a history of poor conditions and abusive treatment [37]. Approximately 1,000 of these unaccompanied children were under the age of 10 [38], and upwards of 500 children were still detained in these temporary facilities several months after the separation [38].

In addition to child refugees, other marginalized groups of children are especially vulnerable to traumatic experiences. These include youth who identify with a racial, ethnic, or sexual minority, have a disability, are in conflict with the law or in forced labor, or are otherwise displaced [6]. Across the world, over 200 million children and adolescents are involved in child labor, and estimates suggest that approximately 126 million of these youth endure hazardous work [8, 39]. Children constitute 26% of all forced labor victims [40], and it is estimated that 5.5 million were in forced or bonded labor, 1.8 million in prostitution and pornography, and 1.2 million were victims of trafficking [6, 40, 41]. In a study conducted in the Philippines and Peru, almost all child domestic workers reported that they had suffered maltreatment that consisted of physical punishment and sexual harassment [8, 42], and across a 13-country study, findings suggest that child sexual exploitation is increasing worldwide [8, 43]. In El Salvador, 66% of girls in domestic service reported abuse and that they always felt the threat of sexual advances from their employer [8, 41].

Institutionalized children are also disproportionately exposed to trauma, and are vulnerable to physical, physiological, or sexual abuse violence from staff [8].

Children in residential care are at risk for similar abuse since they are often placed in facilities that are overcrowded, unsanitary, and lack trained staff [8]. Displaced children, such as those living on the street, are subject to stigmatization and police harassment. In Kenya, children living on the street are frequently charged with petty offenses, without a proper trial, and are consequently held in corrective institutions or incarcerated [8, 44, 45]. In Yemen, approximately 33% of detained children in detention centers reported physical abuse and over 50% had been sexually abused by guards or teachers [8, 46].

### 2.2.3 Unintentional Injuries

In additional to intentional acts of violence, children also experience trauma through unintentional injuries. Upwards of 950,000 children die from injury each year, and road traffic collisions, drowning, burns, falls or poisoning account for approximately 60% of these fatalities [47]. An additional 23% of these unintentional injuries were due to smothering, asphyxiation, choking, animal or snakebites, hypothermia, and hyperthermia [47]. Among adolescents aged 15–19, motor vehicle accidents accounted for 72.3% of unintentional injury deaths [48].

Although unintentional fatal injuries are prevalent worldwide, death is not the most common outcome of unintentional injuries among children. In the United States alone, it is estimated that for every child who was fatally injured, 12 required hospitalization, and 641 children were treated in an emergency department [49]. Across a 28-country study, falls were the leading cause of nonfatal injuries, and similarly in the United States, falls and being struck by a person or object were the most common type of injuries documented in emergency departments among children [4, 48]. Injuries to the head are the most common outcome of unintentional injuries, and a significant proportion of children who survive unintentional injuries endure lifelong disability, especially among children injured in traffic accidents or fires [47].

Children in low- and middle-income countries are disproportionately affected by unintentional injuries [47, 50, 51]. More than 95% of injury deaths among children occur in low- and middle-income countries, and it is estimated that children residing in these countries are 11 times more likely to die from fires, six times more likely to die from drowning, four times more likely to die from poisoning, and six times more likely to die from falls than children in high-income countries [47]. Children in poverty are less likely to have adequate supervision, and regulations for safety equipment, such as smoke alarms, are less stringent in low- and middle-income countries [47]. Furthermore, children residing in low- and middle-income countries are more likely to be exposed to hazardous environments, such as a lack of space for safe play, fast-moving traffic, and cramped living conditions [47]. Reports suggest that boys and younger children are more likely to have more severe unintentional injuries than girls and older children [47, 52–56].

## 2.3 Historical Perspectives on Childhood Trauma

Over the past 100 years in particular, the conceptualization of and response to childhood trauma has changed significantly, in parallel with changing conceptualizations of childhood. Prior to the nineteenth century, childhood was not considered to be a distinct developmental period [57]. Once children were no longer dependent on their mother to meet their basic needs, they were integrated into adult society and expected to contribute to the family. Children were expected to help raise younger siblings, engage in strenuous or even dangerous household chores, and contribute economically through paid work, if possible. Adolescents often got married and had children of their own once they reached reproductive maturity [57].

Views of childhood changed in the late nineteenth and early twentieth centuries. Philosophers such as John Locke proposed children to be "blank slates," introducing the idea that childhood might be an important socialization period [58]. In the USA in the nineteenth century, temperance groups of mostly women concerned about widespread domestic violence, thought to occur as a result of alcohol use, successfully lobbied for mandatory kindergarten in order to teach children the moral importance of temperance. This marked a shift in Western thought and policy toward childhood as a malleable period: an investment in the future of a moral society [57]. In both the United States and Europe, this shift led to a publicly funded education system, child labor laws, and child welfare laws [59]. As the conceptualization of childhood changed, so did the understanding of and responses to childhood trauma. In the United States, the case of Mary Ellen Wilson, an abused child, brought attention to child abuse and neglect as a social problem. Prior to this case, children were considered the property of their fathers or caretakers, and harsh physical punishment was an acceptable form of discipline. The case, brought by the American Society for the Prevention of Cruelty of Animals, was built on the legal precedent that animals could be removed from the home if they were subjected to torture or intense physical harm. Her lawyer argued that if Mary Ellen was not removed from the home, she would face "irreparable harm." The case resulted in legal reform in the interest of protecting children from abuse inflicted by family members or other adults and led to the development of the first ever child protection agency, The New York Society for the Prevention of Cruelty to Children. The case also influenced a similar movement and set of legal policies overseas in Great Britain [59, 60].

At around the same time, the child welfare field in the United States began to take shape. This field was initially developed by various religious and charitable organizations motivated to "save" children from what were characterized as the deplorable living conditions of European immigrants who brought different cultural and religious practices to their new lives in the USA. As a result, the first of these early child welfare organizations would convince urban dwelling immigrant parents to give up their children; the children would then be transported to Protestant families in rural areas [59]. While there still exists racial and ethnic disparities in the removal of children from home, the family began to be seen as a crucial social institution for the wellbeing of children. The United States Children's Bureau, founded in 1912, was the first department of the US government solely focused on the welfare of children, and it still exists, as does the tension within the child welfare system between the desire to protect children and the need to keep families together [59].

These changes are important to consider with regard to a broader concept of childhood trauma. Cultural notions began to shift dramatically during this time, and children were increasingly viewed as more vulnerable not only to the harm caused by child maltreatment, but to other events that could also pose a threat to their socialization and development, such as domestic violence, parental death or separation, parental substance abuse or criminal behavior, medical trauma, institutionalization, and natural disaster and war. In particular, the world wars of the twentieth century contributed to a more nuanced view of childhood trauma, as well as to a broader view of what events were considered traumatic to children.

The world wars of the twentieth century raised public awareness of traumatic stress as a medical condition of concern in both the United States and in Europe. Some scholars estimate that "mental breakdowns" may have accounted for approximately 40% of British battle casualties during the First World War [61]. Many soldiers returned home "shell shocked" and traumatized from their experiences on the battlefield and in particular, the horrific conditions of trench warfare. Traumatic stress had become a cause for concern, with important implications for the conceptualization of war trauma more broadly, and later on, for childhood trauma [62].

Large numbers of children were orphaned as a result of World Wars I and II, and were subsequently homeless or institutionalized in conditions that we now understand as inherently traumatic and developmentally adverse. John Bowlby, the British psychiatrist, psychologist, and psychoanalyst who developed *Attachment Theory*, was impacted by these postwar circumstances. Some of his first publications noted that many of the children he worked with who were considered juvenile delinquents, had experienced prolonged separation from their parents earlier in childhood. He also collaborated with filmmaker James Robertson to create documentaries including "A Two Year Old Goes to the Hospital" [63] and "John" [64], which showcased the adverse effects of separation of children from their parents for the hospitalization of a mother for the birth of a subsequent child, or a childhood hospitalization. These films garnered mass public attention, particularly in light of the postwar orphan crisis [65].

Bowlby authored a World Health Organization report "*Maternal Care and Mental Health*" [66] to highlight the alarming rate of homelessness and institutionalization of children in postwar Europe. In this report, Bowlby boldly stated that a child must have an enduring, warm, and sensitive relationship with his or her mother in order to grow up psychologically healthy [67]. He also emphasized the traumatic and harmful impact of parental separation, medical trauma, and homelessness on children. This marked a dramatic shift in multiple fields, including psychology and social work as well as psychoanalytic and clinical practice [68]. Bowlby's contributions changed

Western notions of children's mental health, creating a foundation for almost seven decades of subsequent research on attachment, loss, and trauma.

As psychologists and psychiatrists such as Bowlby advanced understanding of childhood trauma, medical professionals also began to consider the issue of child maltreatment by discussing cases in which parents were the cause of numerous children's injuries [69]. Kempe's landmark paper *The Battered Child Syndrome* published in 1962 included a specific call to action for doctors to report suspected child abuse to police. In the USA in 1974, the Child Abuse Prevention and Treatment Act mandated states to set up reporting and investigative agencies in order to continue receiving federal funding [59].

Following Kempe's landmark paper, social institutions and child welfare organizations in both the United States and Europe developed and defined a different approach and role for social workers in responding to child trauma [59] which continue to influence practice. In the scientific literature and across disciplines, ideas such as traumatic stress as a response to childhood maltreatment, intergenerational transmission of trauma, and common psychological attributes of abusive caretakers were raised, e.g., [70, 71].

The feminist movements of this era also brought increased attention to the social problem of family violence. Feminist scholars argued that post-traumatic stress disorder could not only be applied to soldiers returning from battle but also to survivors of rape, incest, and intimate partner violence [72]. Some feminist scholars criticized the scientific community's overemphasis on mothers as the main perpetrators of abuse toward children [68].

The increased interest generated in the study of social and cultural factors associated with child maltreatment and family violence, as well as the idea that society itself could pose harm toward children determined broader policies and cultural practices [73]. Concurrently, multiple other fields such as social work, sociology, and public health began to examine the causes and consequences of societal-level childhood traumas such as child marriage, sex trafficking, and lack of economic and educational safety [72]. This confluence of events set the stage for the emergence of a new field uniquely situated to study childhood trauma and its impact on children's mental health: developmental psychopathology.

Developmental psychopathology emerged as a subfield of developmental psychology in the 1980s and sought to generate research utilizing a multidisciplinary developmental approach in order to understand mental health [74]. One core tenet of this subfield is that typical and atypical development are fundamentally intertwined, and that to understand one, you must understand the other as psychopathology is ultimately the result of an inability to achieve developmental competency in one or more domains. This naturally implicated survivors of childhood trauma and specifically child maltreatment, as key populations of importance in research design. By utilizing these populations as study participants, alongside typically developing samples, the etiology of many psychological disorders and the role of trauma in their development could be further elucidated.

## 2.4 The Relationship Between Childhood Trauma and Developmental Psychopathology

Developmental psychopathology is "*the study of the origins and course of individual patterns of behavioral maladaptation, whatever the age of onset, whatever the causes, whatever the transformations in behavioral manifestation, and however complex the course of the developmental pattern may be*" [75]. Masten [76] describes developmental psychopathology simply as "*the study of behavioral health and adaptation in a developmental context*" [76]. The uniqueness of the developmental psychopathology perspective is that it is fundamentally interested in the emergence and *developmental process* of psychopathology.

According to Sroufe [77], the goals of the developmental psychopathology field are twofold: (1) to identify and define premorbid developmental patterns in order to treat them preemptively, and (2) to develop a classification system based on empirical studies of development rather than as an extension of an adult classification system. The field of developmental psychopathology, therefore, is not limited to one age group, though it is often associated with childhood, or even only with psychopathology; the study of adaptive development is also of interest to developmental psychopathologists as a key to understanding deviations in development [78]. Concepts often discussed in the developmental psychopathology literature include the following:

- Developmental cascades and developmental trajectories
- Equifinality and multifinality
- Resilience
- Risk and protective factors

Each will be discussed in further detail in this section.

### 2.4.1 Childhood Trauma, Childhood Mental Disorders, and Adult Mental Disorders

Much more is known about adult mental disorders than child mental disorders, due to easier access to research participants, and a more extensive history of the study of adult mental disorders compared with those of children [79, 80]. Several diagnostic paradigms including DC:0-5 (Diagnostic Classification of Mental Health and Developmental Disorders of Infancy and Early Childhood), and the Research Diagnostic Criteria-Preschool Age/RDC-PA; [79, 81, 82] now provide a developmental approach to assessing young children for mental disorders, but diagnosis of childhood disorders did not become prevalent until the 1980s. Only relatively recently have clinicians argued for the inclusion of additional categories in the standard diagnostic manuals (i.e., the International Classification of Diseases and the Diagnostic & Statistical Manual of Mental Disorders [83]) related to the developmental impact of interpersonal trauma, e.g., [84]. This is likely due to a historical minimization of the impact of exposure to trauma on children's development, with misconceptions that children could not remember traumatic events, or would naturally recover from them [80].

### 2.4.2 Developmental Trajectories and Developmental Cascades

Despite the historical minimization of the impact of trauma on children, the field of developmental psychopathology has encompassed substantial research on the development of children who experience trauma. It may be helpful here to differentiate between the concepts of *developmental trajectories* and *developmental cascades*. *Developmental trajectories* can be defined as "*the processes by which function in one domain or level or system influences another system or level of function over time to shape the course of ontogenesis and epigenesis*" [85]. Developmental trajectories are typically studied within the context of normative or typical development in the field of developmental psychology, e.g., [86]; they may examine variation but not within the context of psychopathology. *Developmental cascades*, on the other hand, are defined as "*the cumulative consequences for development of the many interactions and transactions occurring in developing systems that result in spreading effects across levels, among domains at the same level, and across different systems or generations*" [85].

An example of a developmental cascade may be that of a child with an unidentified and therefore untreated history of sexual abuse. This child may experience difficulty with self-regulation and concentration due to trauma symptoms, and perhaps struggles with reading. He starts off in school behind other children who are able to concentrate and engage with the classroom materials and the teacher. He becomes more frustrated over time as he falls further behind, and perhaps as his teachers and parents begin to label him as lazy, unintelligent, or uncooperative. He stops trying to learn to read well. When math problems involve extensive word problems, as they often do, he does not engage with them; therefore, he falls behind in math as well, continuing a growing experience of anger, shame, and other negative emotions around school. In middle or high school, he may find solace in spending time with other adolescents who perform poorly in school, which reinforces his beliefs that school is punitive, that he is not "made for" school, and that his future lies elsewhere. He may stop attending school altogether, spending time with other dropouts. They begin experimenting with drinking and using drugs. His parents do not know what to do, so they attempt to provide consequences, at which he gets angry, and the parent–child relationship deteriorates. The adolescent is arrested for using substances, continues to spend time with his peer group, and becomes more heavily involved in a drug and criminal culture.

Childhood trauma has a fundamental influence on development across multiple domains, leading to various poor outcomes. These include post-traumatic stress disorder, as well as major depression [87], psychosis [88], cardiovascular disease [89], learning disabilities [90], and the perpetuation of violence in the next generation [91].

### 2.4.3 Equifinality and Multifinality

Equifinality, a key developmental psychopathology concept of relevance to child trauma, refers to the observation that different conditions (in this case, trauma experiences) can lead to the same outcome (in this case, psychopathology). Multifinality is the obverse, i.e., "*the concept that people can experience the same life events or have similar histories yet their developmental outcomes can vary widely*" [92]. Cicchetti and Rogosch [93] emphasize that multifinality allows for the observation that the same component, trait, or experience can result in different outcomes due to the context in which they occur. Equifinality and multifinality together are key concepts to understand both how two people might experience the same situation and respond very differently, and how two people might present with the same symptoms (for example) and have had very different life experiences. These concepts account for complex childhood experiences that can result in a vast array of complex adult experiences, varying amounts of success and struggle, which can ebb and flow over the course of an individual's development. Developmental psychopathology researchers have increasingly incorporated this perspective into their research questions. Instead of asking cause-effect questions such as "what are the early indicators of depression?" or "what type of abuse is most likely to lead to post-traumatic stress disorder?" they instead ask questions about the factors that may contribute to initiating and maintaining certain pathways or disrupting them or diverging from them [94].

### 2.4.4 Typical Development, Resilient Development, and Pathological Development

When a child has experienced trauma, development may maintain a typical course with or without an initial "bump" or it may go awry (i.e., resulting in psychopathology). The reality, of course, is that there are various degrees of deviation from typical developmental trajectories. *Resilience* refers to "*The capacity of a dynamic system to withstand or recover from significant challenges that threaten its stability, viability, or development*" [95], which Masten also has termed "*ordinary magic*" by virtue of the fact that resilience harnesses typical intra- and interpersonal processes [96]. Resilience by definition then requires that some significant challenge has been faced and either withstood or overcome, as reflected in a child's "OK" functioning across developmental domains (e.g., emotional, cognitive, social).

*Pathological development* or "patterns of maladaptation" [97], may be precipitated by childhood trauma, which may take the form of a discrete, singular event (e.g., car accident), or multiple events unfolding over time (e.g., war, abuse, domestic violence), with cascading impact. Of note, even ostensibly single event traumas, such as a natural disaster, can have cascading effects over time—for example, a flood dispossesses families, who lose their homes and their livelihoods, resulting in disruptions in children's education, homelessness, and increased vulnerability to violence. The emergence of patterns of maladaptation also vary depending upon the developmental stage in which the traumatic event occurred, and the severity of the trauma(s) [97–99].

## 2.5 Risk and Protective Factors

How typically a child develops from infancy through adulthood depends on neutral, positive, and negative internal and external factors and processes [100]. The concept of risk and protection is integral to the study of developmental psychopathology. Risk factors are "*variables within the child or the surrounding environment that correlate with an increased probability of the child experiencing negative outcomes*" [101]. Protective factors "*facilitate the attainment of positive outcomes*" [102]. Whether a trait, experience, or context is considered risky or protective may sometimes also depend on the outcome under study, cf. [103]. Though childhood trauma clearly increases the risk of many negative biopsychosocial outcomes, as well as having immediate impacts [84], the latter can be buffered by various protective factors or processes [104].

## 2.6 Theories of Vulnerability

Several competing theories explain a child's vulnerability to stress and trauma. *Diathesis-stress theory,* for many decades the leading theory, reflected the view that children with various behavioral/temperamental (e.g., difficult temperament), physiological, endophenotypic (e.g., highly physiologically reactive), and/or genetic (e.g., 5-HTTLPR short alleles) characteristics were "*disproportionately or even exclusively likely to be affected adversely by an environmental stressor*" [100, 105]. The child's vulnerabilities interact with the trauma s/he experiences, increasing the risk for psychopathology, e.g., [95]. Evidence indicates that a number of individual differences may be attributed to this "dual risk," such as the complex interaction between baseline plasma cortisol levels (the primary active stress hormone), the attachment relationship the child is involved in, and the development of problematic externalizing behaviors, or the interaction between temperament, family environment and the adolescent onset of substance abuse [106, 107].

More recently, Belsky & Pluess [100] proposed a differential susceptibility theory. Rather than focusing solely on vulnerability, differential susceptibility theory reframes vulnerability as susceptibility to both positive and negative environmental influences. More susceptible individuals (sometimes known as "orchids") do worse in highly stressful or traumatic environments, but do better than their less susceptible peers (also known as "dandelions") in enriched and supportive environments lacking adversity. Scholars have recently called for conceptualizing a continuum of sensitivity, with at least three categories of low (dandelions), moderate (tulips), and high sensitivity (orchids) [108]. More research is needed to understand whether and how differential susceptibility theory and diathesis-stress models apply to large populations of children.

Much of the rest of this book will focus more specifically on individual disorders and their relationship to childhood trauma. Factors at multiple levels of analysis: genetic, biological, behavioral, environmental, interpersonal, and others, influence a child's vulnerability to psychopathology after exposure to trauma [109]. The development from childhood trauma to mental disorders in childhood and adulthood is a complex process that involves ongoing, sometimes cascading processes at multiple levels within the individual, and across the child's ecosystem (family, school, community, etc.). The complexities of these processes are just starting to be understood.

## References

1. Betancourt TS, Borisova I, Williams TP, Meyers-Ohki SE, Rubin-Smith JE, Annan J, Kohrt BA. Research review: psychosocial adjustment and mental health in former child soldiers--a systematic review of the literature and recommendations for future research. J Child Psychol Psychiatry. 2013;54:17–36.
2. UNICEF. Female genital mutilation/cutting: a statistical overview and exploration of the dynamics of change. New York; 2016, 2013.
3. Breslau N. Epidemiologic studies of trauma, posttraumatic stress disorder, and other psychiatric disorders. Can J Psychiatr. 2002;47:923–9.
4. Saunders BE, Adams ZW. Epidemiology of traumatic experiences in childhood. Child Adolesc Psychiatr Clin N Am. 2014;23(167–84):vii.
5. Saunders BE. Understanding children exposed to violence: toward an integration of overlapping fields. J Interpers Violence. 2003;18:356–76.
6. Pinheiro PS (2006) Rights of the child. report of the independent expert for the United Nations study on violence against children.
7. Hahm HC, Guterman NB. The emerging problem of physical child abuse in South Korea. Child Maltreat. 2001;6:169–79.
8. Pinheiro PS (2006) World report on violence against children.
9. Stoltenborgh M, Bakermans-Kranenburg MJ, Alink LRA, van Ijzendoorn MH. The prevalence of child maltreatment across the globe: review of a series of meta-analyses. Child Abuse Rev. 2015;24:37–50.
10. Chigiji H, Fry D, Mwadiwa TE, Elizalde A, Izumi N, Baago-Rasmussen L, Maternowska MC. Risk factors and health consequences of physical and emotional violence against children in Zimbabwe: a nationally representative survey. BMJ Glob Health. 2018;3:e000533.
11. Perkonigg A, Kessler RC, Storz S, Wittchen H-U. Traumatic events and post-traumatic stress disorder in the community: prevalence, risk factors and comorbidity. Acta Psychiatr Scand. 2000;101:46–59.
12. Sheikhattari P, Stephenson R, Assasi N, Eftekhar H, Zamani Q, Maleki B, Kiabayan H. Child maltreatment among school children in the Kurdistan Province, Iran. Child Abuse Negl. 2006;30:231–45.
13. Meinck F, Cluver LD, Boyes ME, Ndhlovu LD. Risk and protective factors for physical and emotional abuse victimisation amongst vulnerable children in South Africa. Child Abuse Rev. 2015;24:182–97.
14. Devries K, Knight L, Petzold M, et al. Who perpetrates violence against children? A systematic analysis of age-specific and sex-specific data. BMJ Paediatr Open. 2018;2:e000180.
15. World Health Organization. Global estimates of health consequences due to violence against children. Geneva: World Health Organization; 2006.
16. Finkelhor D. The international epidemiology of child sexual abuse. Child Abuse Negl. 1994;18:409–17.

17. Finkelhor D, Turner H, Ormrod R, Hamby SL. Violence, abuse, and crime exposure in a national sample of children and youth. Pediatrics. 2009;124:1411–23.
18. Andrews G, Corry J, Slade T, Issakidis C, Swanston H, et al. Child sexual abuse. In: Ezzati M, Lopez AD, Rodgers A, Murray CJL, editors. Comparative quantification of health risks: global and regional burden of disease attributable to selected major risk factors. Geneva: World Health Organization; 2004. p. 1851–940.
19. United Nations Secretary-General's Study on Violence against Children (2005) Regional desk review. North America.
20. Dong M, Anda RF, Felitti VJ, Dube SR, Williamson DF, Thompson TJ, Loo CM, Giles WH. The interrelatedness of multiple forms of childhood abuse, neglect, and household dysfunction. Child Abuse Negl. 2004;28:771–84.
21. Stark L, Landis D. Violence against children in humanitarian settings: a literature review of population-based approaches. Soc Sci Med. 2016;152:125–37.
22. UNICEF. Behind closed doors: the impact of domestic violence on children. New York: UNICEF; 2006.
23. Zinzow HM, Ruggiero KJ, Resnick H, Hanson R, Smith D, Saunders B, Kilpatrick D. Prevalence and mental health correlates of witnessed parental and community violence in a national sample of adolescents. J Child Psychol Psychiatry. 2009;50:441–50.
24. Centre for Research on the Epidemiology of Disasters CRED, & United Nations Office for Disaster Risk Reduction (2018) Economic Losses, Poverty & Disasters, 1998–2017.
25. Dyregrov A, Yule W, Olff M. Children and natural disasters. Eur J Psychotraumatol. 2018;9:1500823.
26. Codreanu TA, Celenza A, Jacobs I. Does disaster education of teenagers translate into better survival knowledge, knowledge of skills, and adaptive behavioral change? A systematic literature review. Prehosp Disaster Med. 2014;29:629–42.
27. Philipsborn RP, Chan K. Climate change and global child health. 2018; https://doi.org/10.1542/peds.2017-3774.
28. United Nations. Machel study 10-year strategic review: children and conflict in a changing world. UNICEF; 2009.
29. Domonoske C. Refugees, displaced people surpass 60 million for first time. NPR: UNHCR Says; 2016.
30. Children's Rights Violations during Armed Conflicts on Rise despite National Action Plans to End Abuse, Security Council Told in Day-long Debate | Meetings Coverage and Press Releases. https://www.un.org/press/en/2016/sc12470.doc.htm. Accessed 30 Nov 2018.
31. Human Rights Watch. Darfur: Women Raped Even After Seeking Refuge; 2005.
32. A last resort? A summary guide to the National Inquiry into Children in Immigration Detention. Human Rights and Equal Opportunity Commission Australia, Australia; 2004.
33. Separated Children in Europe Programme Newsletter No. 22; 2005. http://scep.sitespirit.nl/images/12/110.pdf. Accessed 26 Nov 2018.
34. Flores G. An urgent call to action: building a better America and world by prioritizing Children's health, health care, and well-being. Acad Pediatr. 2018;18:493–5.
35. Freedom for Immigrants. In: Freedom for immigrants; 2018. https://www.freedomforimmigrants.org/. Accessed 28 Nov 2018.
36. U.S. Department of Homeland Security. U.S. Border Patrol Southwest Border Apprehensions by Sector FY2017 | U.S. Customs and Border Protection; 2017. https://www.cbp.gov/newsroom/stats/usbp-sw-border-apprehensions-fy2017. Accessed 26 Mar 2018.
37. Domonoske C, Gonzales R. What we know: family separation and "Zero Tolerance" at the border. National Public Radio; 2018.
38. Sacchetti M. Still separated: Nearly 500 migrant children taken from their parents remain in U.S. custody. The Washington Post; 2018.
39. International Labour Conference, 95th Session. The end of child labour: within reach. Geneva: International Labour Office; 2006.
40. International Labour Organization. ILO global estimate of forced labour: results and methodology. Geneva: ILO Publications; 2012.

41. International Programme on The Elimination of Child Labour. Helping Hands or Shackled Lives?: Understanding Child Domestic Labour and Responses to it. Geneva: International Labour Organization; 2004.
42. Black M. Child domestic workers: a handbook on good practice in programme interventions. London: Anti-Slavery International; 2005.
43. Save the Children. 10 Essential Learning Points: Listen and Speak out against Sexual Abuse of Girls and Boys: Based on Country Reports from Save the Children in Canada, Columbia, Brazil, Nicaragua, Syria, South Africa, Mozambique, Rwanda, Uganda, Bangladesh, Nepal, Spain and Romania, The International Save the Children Alliance, Oslo; 2005.
44. United Nations Secretary-General's Study on Violence against Children. Regional Desk Review: South Asia; 2005.
45. United Nations Secretary-General's Study on Violence against Children. Regional Desk Review: Eastern and Southern Africa; 2005.
46. United Nations Secretary-General's Study on Violence against Children. Regional Desk Review: Middle East and North Africa; 2005.
47. Branche C, Oyebite K, Hyder AA, Ozanne-Smith J, Bartolomeos K, Rivara F. World report on child injury prevention. World Health Organization; 2008.
48. Sleet DA, Ballesteros MF, Borse NN. A review of unintentional injuries in adolescents. Annu Rev Public Health. 2010;31:195–212.
49. Centers for Disease Control (2009) Web-based Injury Statistics Query and Reporting System (WISQARS).
50. Hyder AA, Peden M. Inequality and road-traffic injuries: Call for action. Lancet. 2003;362:2034–5.
51. Hulme D. Chronic poverty and development policy: an introduction. World Dev. 2003;31:399–402.
52. Baker SP, Ginsburg MJ, Li G, O'Neill B. The injury fact book. 2nd ed. Lexington, MA: Lexington Books; 1992.
53. Bartlett SN. The problem of children's injuries in low-income countries: a review. Health Policy Plan. 2002;17:1–13.
54. Rivara FP, Bergman AB, LoGerfo JP, Weiss NS. Epidemiology of childhood injuries. II. Sex differences in injury rates. Am J Dis Child. 1982;136:502–6.
55. Spady DW, Saunders DL, Schopflocher DP, Svenson LW. Patterns of injury in children: A population-based approach. Pediatrics. 2004;113:522–9.
56. Lansdown G. The evolving capacities of the child. Firenze: UNICEF Innocenti Research Centre; 2005.
57. Veerman PE. The rights of the child and the changing image of childhood. Dordrecht: Martinus Nijhoff Publishers; 1992.
58. Ezell MJM. John Locke's Images of Childhood: Early Eighteenth Century Response to Some Thoughts Concerning Education. Eighteenth Century Stud. 1983;17:139–55.
59. Parton N. The natural history of child abuse: a study in social problem definition. Br J Soc Work. 1979.
60. Courtney ME. Child Welfare: History and Policy Framework. Encyclopedia of Social Work. 2013; https://doi.org/10.1093/acrefore/9780199975839.013.530.
61. Showalter E. Female malady: women, madness and English culture. New York: Virago Press; 1985.
62. Crocq MA, Crocq L. From shell shock and war neurosis to posttraumatic stress disorder: a history of psychotraumatology. Dialogues Clin Neurosci. 2000;2:47–55.
63. Robertson J, Robertson JA. two year old goes to the hospital. PsycEXTRA Dataset. https://doi.org/10.1037/e528272004-001.
64. Robertson J, Robertson J. John, Seventeen months: for nine days in a residential nursery (film). London: Tavistock Institute of Human Relations, Tavistock Clinic; 1969.
65. Bretherton I. The origins of attachment theory: John Bowlby and Mary Ainsworth. Dev Psychol. 1992;28:759.
66. Bowlby J. Maternal care and mental health. Bull World Health Organ. 1951;3:355–533.

67. Bowlby J. Maternal care and mental health: a report prepared on behalf of the World Health Organization as a contribution to the United Nations programme for the welfare of homeless children. Geneva: World Health Organization; 1952.
68. Vicedo M. The social nature of the mother's tie to her child: John Bowlby's theory of attachment in post-war America. Br J Hist Sci. 2011;44:401–26.
69. Kempe CH, Silverman FN, Steele BF, Droegemueller W, Silver HK. The Battered-child Syndrome. In: Krugman RD, Korbin JE, editors. C. Henry Kempe: a 50 year legacy to the field of child abuse and neglect. Dordrecht: Springer; 2013. p. 23–38.
70. Caffey J. The parent-infant traumatic stress syndrome. Am J Roentgenol Radium Therapy, Nucl Med. 1972;114:218–29.
71. Spinetta JJ, Rigler D. The child-abusing parent: a psychological review. Psychol Bull. 1972;77:296–304.
72. Korbin JE. The cultural context of child abuse and neglect. Child Abuse Negl. 1980;4:3–13.
73. Helfer RE. A review of the literature on the prevention of child abuse and neglect. Child Abuse Negl. 1982;6:251–61.
74. Cicchetti D. The emergence of developmental psychopathology. Child Dev. 1984;55:1–7.
75. Sroufe LA, Rutter M. The domain of developmental psychopathology. Child Dev. 1984;55:17–29.
76. Masten AS. Developmental psychopathology: pathways to the future. Int J Behav Dev. 2006;30:47–54.
77. Sroufe LA. The concept of development in developmental psychopathology. Child Dev Perspect. 2009;3:178–83.
78. Cicchetti D. Development and psychopathology. In: Cicchetti D, Cohen DJ, editors. Developmental psychopathology, theory and method, 2nd ed, vol. 1. New York: Wiley; 2006. p. 1–23.
79. Angold A, Costello EJ. Nosology and measurement in child and adolescent psychiatry. J Child Psychol Psychiatry. 2009;50:9–15.
80. Osofsky JD, Stepka PT, King LS. Introduction: recognizing the impact of trauma exposure on young children. In: Osofsky JD, Stepka PT, King LS, editors. Treating infants and young children impacted by trauma: interventions that promote healthy development. Washington: American Psychological Association; 2017. p. 3–13.
81. Emde RN. RDC-PA: a major step forward and some issues. J Am Acad Child Adolesc Psychiatry. 2003;42:1513–6.
82. Zeanah CH, Carter AS, Cohen J, Egger H, Gleason MM, Keren M, Lieberman A, Mulrooney K, Oser C. Diagnostic classification of mental health and developmental disorders of infancy and early childhood DC:0-5: selective reviews from a new nosology for early childhood psychopathology. Infant Ment Health J. 2016;37:471–5.
83. American Psychiatric Association. DSM history; 2016. https://www.psychiatry.org/psychiatrists/practice/dsm/history-of-the-dsm. Accessed 26 Nov 2018.
84. D'Andrea W, Ford J, Stolbach B, Spinazzola J, van der Kolk BA. Understanding interpersonal trauma in children: why we need a developmentally appropriate trauma diagnosis. Am J Orthopsychiatry. 2012;82:187–200.
85. Masten AS, Cicchetti D. Developmental cascades. Dev Psychopathol. 2010;22:491–5.
86. Taylor SJ, Barker LA, Heavey L, McHale S. The typical developmental trajectory of social and executive functions in late adolescence and early adulthood. Dev Psychol. 2013;49:1253–65.
87. Mandelli L, Petrelli C, Serretti A. The role of specific early trauma in adult depression: a meta-analysis of published literature. Childhood trauma and adult depression. Eur Psychiatry. 2015;30:665–80.
88. Varese F, Smeets F, Drukker M, Lieverse R, Lataster T, Viechtbauer W, Read J, van Os J, Bentall RP. Childhood adversities increase the risk of psychosis: a meta-analysis of patient-control, prospective- and cross-sectional cohort studies. Schizophr Bull. 2012;38:661–71.
89. Min MO, Minnes S, Kim H, Singer LT. Pathways linking childhood maltreatment and adult physical health. Child Abuse Negl. 2013;37:361–73.

90. Eigsti I-M, Cicchetti D. The impact of child maltreatment on expressive syntax at 60 months. Dev Sci. 2004;7:88–102.
91. Evans SE, Davies C, DiLillo D. Exposure to domestic violence: a meta-analysis of child and adolescent outcomes. Aggress Violent Behav. 2008;13:131–40.
92. Howe TR. Multifinality. In: Goldstein S, Naglieri JA, editors. Encyclopedia of child behavior and development. Boston, MA: Springer US; 2011. p. 982.
93. Cicchetti D, Rogosch FA. Equifinality and multifinality in developmental psychopathology. Dev Psychopathol. 1996;8:597–600.
94. Cicchetti D, Toth SL. The past achievements and future promises of developmental psychopathology: the coming of age of a discipline. J Child Psychol Psychiatry. 2009;50:16–25.
95. Masten AS. Resilience in children threatened by extreme adversity: frameworks for research, practice, and translational synergy. Dev Psychopathol. 2011;23:493–506.
96. Masten AS. Ordinary magic. Resilience processes in development. Am Psychol. 2001;56:227–38.
97. Sroufe LA. Considering normal and abnormal together: the essence of developmental psychopathology. Dev Psychopathol. 1990;2:335–47.
98. Kaplow JB, Widom CS. Age of onset of child maltreatment predicts long-term mental health outcomes. J Abnorm Psychol. 2007;116:176–87.
99. Sroufe LA, Egeland B, Kreutzer T. The fate of early experience following developmental change: Longitudinal approaches to individual adaptation in childhood. Child Dev. 1990;61:1363–73.
100. Belsky J, Pluess M. Beyond diathesis stress: differential susceptibility to environmental influences. Psychol Bull. 2009;135:885–908.
101. Schantz AR, Pham AV, Carlson JS. Risk factors. In: Goldstein S, Naglieri JA, editors. Encyclopedia of child behavior and development. Boston, MA: Springer US; 2011. p. 1273–4.
102. Bell C, Pham AV, Carlson JS. Protective factors. In: Goldstein S, Naglieri JA, editors. Encyclopedia of child behavior and development. Boston, MA: Springer US; 2011. p. 1168–9.
103. Ogbu JU. Origins of human competence: a cultural-ecological perspective. Child Dev. 1981;52:413–29.
104. Pynoos RS, Steinberg AM, Piacentini JC. A developmental psychopathology model of childhood traumatic stress and intersection with anxiety disorders. Biol Psychiatry. 1999;46:1542–54.
105. Meehl PE. Schizotaxia, schizotypy, schizophrenia. Am Psychol. 1962;17:827–38.
106. Fong MC, Measelle J, Conradt E, Ablow JC. Links between early baseline cortisol, attachment classification, and problem behaviors: a test of differential susceptibility versus diathesis-stress. Infant Behav Dev. 2017;46:158–68.
107. Rioux C, Castellanos-Ryan N, Parent S, Séguin JR. The interaction between temperament and the family environment in adolescent substance use and externalizing behaviors: support for diathesis–stress or differential susceptibility? Dev Rev. 2016;40:117–50.
108. Lionetti F, Aron A, Aron EN, Burns GL, Jagiellowicz J, Pluess M. Dandelions, tulips and orchids: Evidence for the existence of low-sensitive, medium-sensitive and high-sensitive individuals. Transl Psychiatry. 2018;8:24.
109. Cicchetti D, Dawson G. Multiple levels of analysis. Dev Psychopathol. 2002;14:417–20.

# Part II

# Neurobiology

# Neuroimaging and Cognition of Early Traumatic Experiences

# 3

Michela Sala, Giuseppe Delvecchio, and Paolo Brambilla

## 3.1 Introduction

In the last decades, several neuroimaging studies have been carried out for elucidating the mechanisms underpinning brain development and for identifying how early traumatic experiences affect this developmental process. Early adversities in the life of a child have been shown to have detrimental effects on mental health, ultimately leading to behavioral and cognitive problems [1]. Furthermore, childhood maltreatment produces a cascade of physiological, neurochemical, and hormonal changes, which can lead to enduring alterations in brain structure and function [2].

In this chapter, we will first describe how the human brain develops. Then we will provide a brief description of biological mechanism of stress response, focusing on hypothalamus-pituitary-adrenal (HPA) axis functioning trying to explain how this process can be strongly influenced by childhood experiences. Finally, we will provide an overview of the existing evidence exploring the association between child maltreatment and cognitive or neuroimaging abnormalities. In this regard, human and animal studies showed that childhood adversity can have a long-lasting impact on brain structure and function. Particularly, in animal studies, childhood adversity was found to be associated with changes in selective brain structures

M. Sala
Department of Mental Health, Azienda Sanitaria Locale Alessandria, Alessandria, Italy
e-mail: michelasala@hotmail.com; msala@aslal.it

G. Delvecchio
Department of Pathophysiology and Transplantation, University of Milan, Milan, Italy
e-mail: giuseppe.delvecchio@unimi.it

P. Brambilla (✉)
Department of Pathophysiology and Transplantation, University of Milan, Milan, Italy

Department of Neurosciences and Mental Health, Fondazione IRCCS Ca' Granda, Ospedale Maggiore Policlinico, Milan, Italy
e-mail: paolo.brambilla1@unimi.it



G. Spalletta et al. (eds.), *Childhood Trauma in Mental Disorders*,
https://doi.org/10.1007/978-3-030-49414-8_3

involved in stress and emotion regulation, including the hippocampus and prefrontal regions [3]. Interestingly, structural and functional alterations in emotion- and stress-regulating brain structures were also reported in human adults with histories of childhood abuse or emotional maltreatment [4]. Moreover, animal studies showed that compensatory fostering partially reversed several changes in the brain, whereas treatment with some antidepressant was reported to have similar effect in both animal and human studies [5].

In this chapter, we will also try to explain how many different factors like type and time of trauma, gender differences, and sensitive periods of different brain structures can influence structural and functional changes in subjects with traumatic history.

In the last part of this chapter, we will discuss the recent functional magnetic resonance imaging (fMRI) studies that support the theory of latent vulnerability [6]. According to this theory, maltreatment results in measurable alterations in a number of neurobiological systems that reflect calibration to neglectful and/or abusive early environments. These changes are often beneficial within the early maladaptive context, but they suggest that the individual is poorly optimized to negotiate the demands of more normative environments, thus increasing vulnerability to future stressors [6]. Moreover, childhood maltreatment may have a great impact on brain development, and it has been an important confounding factor for the identification of the etiology of psychiatric illnesses for many years. Indeed, maltreated and non-maltreated individuals with the same primary psychiatric diagnosis are clinically, neurobiologically, and genetically distinct. Therefore, the maltreated subgroup must be considered a unique ecophenotype, as proposed by Teicher and Samson [7]. The relationship between brain changes in maltreated subjects and psychopathology is complex as such changes are present both in susceptible and resilient individuals with maltreatment histories. Mechanism fostering resilience are still largely unknown; however, increasing evidence on functional changes in maltreated children who have not yet developed any manifest psychiatric condition seem to suggest the possibility that early psychological treatment might represent an effective preventive approach. Elucidating mechanisms fostering resilience will be the focus of future studies.

### 3.1.1 Trauma and Normal Brain Development

Human brain development is a protracted process that begins in the third gestational week, with the differentiation of neural progenitor cells, and extends at least through late adolescence, arguably through the lifespan [8]. Neural production in humans begins on embryonic day 42 and is largely complete by midgestation [9]. The mature brain is composed of more than 100 billion neurons [10]. As they are produced, neurons migrate to different brain areas where they begin to make connections with other neurons establishing rudimentary neural networks. The point of connection between two neurons is called a synapse [11]. Populations of neurons are connected to one another by two type of fibers that extend from cell bodies of individual neurons: dendrites that are short fibers that receive the electrochemical input signals from other neurons and axons, and long connecting fibers wrapped in a fatty substance called myelin. Brain development involves different processes like neuron production, neuron migration and differentiation, synaptogenesis, and myelination. Beside those events that involve the proliferation of neural elements, two important processes involve substantial loss of neural elements: apoptosis and pruning. Apoptosis is a naturally occurring cell death, which involves the normal loss of 50% or more of neurons within a brain region. This event is typical in prenatal period, even though apoptosis in glial cell populations occurs in postnatal period.

During the early postnatal period, there is also a massive excess production of connections (synaptic exuberance) followed by the systematic elimination of up to 50% of them (pruning). Both apoptosis and pruning reflect non-pathological events that play an essential role in generating the complex network of the developing brain. Particularly, the events during the prenatal period serve to establish the core compartments of the developing nervous system from the spinal cord and hindbrain, to the cortical structures of the telencephalon. These early events also provide initial patterning within each of the major subdivisions of the brain, but this early patterning, especially in the neocortex, is both unspecified and malleable. The mature organization of the neocortex emerges over a protracted time during the postnatal period and requires diverse forms of input. Furthermore, the development of normal brain organization requires input from the major sensory systems.

When specific aspects of inputs are lacking, alternative patterns of brain organization can emerge [12]. These alternative patterns of organization reflect the effects of altered profiles of neural competition, and capture a fundamental property of mammalian brain development: the capacity for plastic adaptation.

In animal studies, two simple ways to alter the environmental experience are enrichment and deprivation. Both have dramatic effects on the structural and functional organization of the developing brain. Greenough [13] has shown that simply rearing animals in either impoverished (standard laboratory cage) or enriched environments (large enclosures with interesting and changing landmarks and multiple littermates) affects the development of a wide range of brain structures and functions [14]. Animal reared in complex environments show greater density of cortical synapses, increased number of brain support cells and even augmented complexity of the brain vascular systems. Sensory deprivation has more selective effects that target particular cortical sensory systems. The seminal studies of Hubel and Weisel [15] showed that monocular visual deprivation in the early postnatal period can substantially alter basic patterns of organization within primary visual cortex (PVC). Within the typical primary visual pathway, inputs from the two eyes remain segregated from the retina to the thalamus to PVC. In PVC, the inputs from the two eyes form a distinctive banded pattern, called ocular dominance columns (ODC), which give the input layer of PVC a striped appearance. When patterned input from one eye is blocked by suturing the eyelid closed, the effect of this altered experience on ODC organization is striking. The bands representing the active eye widen and expand into the territory of the deprived eye, while the bands representing the deprived eye shrink to thin stripes. The monocular reduction in activity introduced by the suturing procedure alters the competitive balance of the input from the two eyes. The input from the active eye invades and subsumes territory that would normally receive input from the deprived eye.

Considering the above mechanisms of brain development, childhood trauma can be regarded as an experience that induce a complex developmental disorder, a condition that strongly affects the sensorial inputs essential for a normal brain development. In children, motor vehicle accidents, bullying, terrorism, exposure to war, child maltreatment (physical, sexual, and emotional abuse or neglect) and exposure to domestic and community violence are common types of childhood traumas that cause an overwhelming activation of the body's biological stress response systems, which may lead to posttraumatic stress disorder (PTSD), posttraumatic stress symptoms (PTSS), depression, anxiety, antisocial behaviors, and greater risk for alcohol and substance abuse disorders [16–19]. The biological stress response system is formed by different and interacting systems working together to direct the body's attention toward protecting the individual against environmental threats to life and health as to shift metabolic resources away from homeostasis and toward a "fight or flight" reaction [20]. Specifically, the stressors associated with the traumatic event are processed by the body's sensory systems through the thalamus, a small structure within the brain, which then activates the amygdala, a central component of the brain's fear detection and anxiety circuits. Through amygdala activation, fear signals reach neurons in the prefrontal cortex, hypothalamus, and hippocampus. At the same time, the activation of the main biological stress response system, the limbic-hypothalamic-pituitary-adrenal (LHPA) axis, leads to increased levels of stress hormones cortisol and catecholamine with consequent changes in heart rate, metabolic rate, blood pressure and alertness, ultimately activating other biological stress systems. Activation of the LHPA axis triggers the hypothalamus to secrete the corticotrophin-releasing hormone (CRH). CRH stimulates the release of the adrenocorticotrophic hormone (ACTH) by binding to CRH receptors in the anterior pituitary gland. ACTH, in turn, binds to transmembrane receptors in the adrenal cortex and stimulates the secretion of cortisol, a glucocorticoid hormone that plays an important role throughout the central nervous system. Cortisol activates glucocorticoids and mineralocorticoid receptors, which are located and expressed throughout the brain. Through negative feedback, cortisol controls its own secretion, inhibiting the hypothalamus' release of CRH and the pituitary's release of ACTH, thereby bringing the body back to a state of homeostasis rather than arousal [20]. Glucocorticoids are important for normal brain maturation including initiation of terminal maturation, for remodeling axons and dendrites and by affecting cell survival [21]. Both suppressed and elevated glucocorticoids levels can impair brain development and function [21]. During brain maturation, stress and elevated levels of stress hormones and neurotransmitters may lead to adverse brain development through apoptosis [22], delays in myelination [23], abnormalities in developmentally appropriate pruning [24], the inhibition of neurogenesis [25], or stress-induced decrease in brain growth factors [26]. Furthermore, glucocorticoids acting via glucocorticoid receptors can impair neural plasticity [27]. This explains why brain regions with a particular high density of glucocorticoid receptors and characterized by prolonged phases of postnatal development (e.g., prefrontal cortex, hippocampus) are more susceptible to disturbances [2]. Interestingly, there are several mediating and moderating mechanisms involved in the LHPA axis functioning. Specifically, they include genetic and epigenetic factors [28] as well as priming effects (or sensitization, defined as enhanced neuroendocrine, autonomic, and behavioral responsiveness to stress) [29] and the downregulation of pituitary CHR receptors [21]. A detailed examination of these mechanisms is important to explain individual differences in stress response (vulnerability or resilience) and to identify different endophenotypes. Although LHPA axis is the most studied stress response system, an individual's biological stress response is formed by different integrating systems, which are strictly intertwined and show regulatory mechanism that can be disrupted by early traumatic experiences [28]. These integrated systems are the locus coeruleus (LC)-norepinephrine/sympathetic nervous system (SNS)/catecholamine system, the serotonin system, the oxytocin system, and the immune system. Briefly, the LPHA axis, through the release of CHR from the hypothalamus due to stressors, indirectly activates the LC-norepinephrine/SNS and immune system through the amygdala, which in turn, causes an increase of norepinephrine, ultimately determining anxiety symptoms. Similarly, the LC also increases the SNS activity, which controls the "fight of flight" response [28]. However, how early traumatic experiences affect cognitive functioning and emotional well-being via specific neurobiological pathways is still poorly understood. Therefore, the following section will provide insight on how particular cognitive and affective functions may be affected by disruptions in brain development following early traumatic experiences.

### 3.1.2 The Effects of Childhood Trauma on Neuropsychological Function and Cognitive Development

Several strands of evidence have focused their attention toward the investigation of neuropsychological functioning in children exposed to traumatic events with or without current/resolved acute/ chronic post-traumatic stress reactions [30]. Impaired intellectual ability, worse academic performance, and greater needs for individualized education programs have been observed in children who experienced early traumatic events, including early institutionalization, neglect, or various forms of maltreatment [1]. Additionally, exposure to trauma in childhood has also been associated with executive deficits [28]. Moreover, in addition to the impact of trauma exposure, the development of PTSD may probably place a child at increased risk of adverse neurodevelopmental outcomes [30]. Specifically, PTSD is a psychiatric disorder that may result from a single traumatic event, although a dose–response relationship or "building-block effect" is reported in the literature, with cumulative trauma exposure posing an individual at increased risk of developing the disorder [31]. This disorder is characterized by the onset of trauma-related symptoms, including intrusive memories, flashbacks, or nightmares with trauma-related content, post-trauma alterations in arousal and reactivity, as well as post-trauma deficits in cognition and mood. Importantly, sequelae of early traumatic experiences depend on type of adversity, number of recurrences, and, in particular age and time of occurrence [1].

With regard to the type of adversity, a recent meta-analysis carried out by Malarbi et al. (2017) [30] showed that familial trauma exposure is more detrimental to neurodevelopment than nonfamilial trauma, especially because the young child depends on his or her parent/caregiver to survive and to learn how to regulate his/her emotions [27]. In contrast, children exposed to nonfamilial trauma may have the opportunity to seek comfort from their parent/caregiver and feel a sense of safety and security from this relationship, thus limiting the impact of the experienced trauma [32]. Indeed, children exposed to nonfamilial trauma show deficits only in the presence of PTSD symptoms. On the other hand, poorer cognitive functioning in children exposed to familial trauma is more likely to result not only from PTSD but also from other psychopathologies, probably associated with additional neurodysfunctions (including additional somatic, cognitive, affective, behavioral, relational, and personality disorders, such as chronic affect dysregulation, poor self-esteem, interpersonal distrust, dissociation, emotional numbing, self-injury, identity confusion, and impulsivity) [33]. Finally, while it is probable that familial trauma has a stronger impact on cognition than nonfamilial trauma, it is also possible that children exposed to familial trauma are at greater risk of preexisting cognitive deficits and more likely to experience trauma [34]. In line with this hypothesis, a recent study examined two population-representative birth cohorts [35]. Authors suggested that the association between childhood violence victimization and later cognition is largely noncausal. Specifically, they found that adolescents and adults with a history of childhood victimization had pervasive impairments in clinically relevant cognitive functions including general intelligence, executive functions, processing speed, memory, perceptual reasoning, and verbal comprehension. However, these cognitive deficits were largely explained by alterations already present before the observational period for childhood victimization and by nonspecific effects of childhood socioeconomic disadvantage. Therefore, authors suggested that cognitive deficits should be conceptualized as individual risk factors for victimization, as well as potential complicating features during treatment. Furthermore, type of adversity should also be considered as the pathways that process and convey the aversive experience (auditory, visual, and somatosensory cortices) are specifically involved in adaptive changes occurring during brain development. Interestingly, childhood trauma exposure and PTSD seem to be associated to altered right hemisphere development [36]. This is not surprising especially because the right hemisphere is central in perceptual/visuospatial functioning [37], and plays a dominant role in the modulation of HPA axis and sympathetic-adrenomedullary activity [38], as well as in attachment and emotional regulation processes, which are often found disrupted as a result of early familial trauma exposure [36, 39].

With regard to time of exposure, Teicher et al. [40] conducted extensive studies on "sensitive periods" emphasizing that the timing of early traumatic experiences may especially affect those brain regions undergoing specific growth spurts at that time, while brain regions with extended postnatal development are particularly vulnerable to the long-term effect of stress [2]. The last brain regions reaching fully maturity during development, which include the Dorsolateral Prefrontal Cortex (DLPFC), Orbitofrontal Cortex (OFC), temporal lobe and superior parietal lobe, are linked to higher order, complex skills, and are at greater risk of impairment following early life adversities. Finally, gender-related differences have been observed in the brain, with males having a proportionally larger mean cerebral volume, despite reaching gray matter peak volume 1–3 years later than females [41]. Moreover, gender-specific differences have also been observed in selective brain regions and primarily in those containing sex steroid receptors (e.g., hypothalamus) or regions with strong connections to areas with high sex steroid receptors density (e.g., amygdala, parts of the nucleus of the stria terminalis) [42]. As developmental trajectories differ in males and females, early adverse experiences occurring at the same time may lead to different, sex-specific outcomes [1].

In conclusion, the abovementioned evidence suggests that chronic trauma exposure is likely to result in the ongoing activation of the physiological stress response, while PTSD probably originates from a maladaptive stress response that remains activated in the absence of actual trauma.

The neural structures and networks involved in both cases may well be very similar, if not the same [30]. Furthermore, the evidence seems also to suggest that cognitive deficits may represent a preexisting risk factor for children that will undergo childhood victimization. In contrast, there is evidence from literature showing that different high-order cognitive-affective functions, like reward processing, emotional perception, and regulation, which will be discussed in the last section, may be affected in adults reporting early traumatic events and presenting common psychiatric disorders (like PTSD, depression, and anxiety). Although alterations in these processes are considered to carry adaptive value in early adverse caregiving environments, they also confer long-term risk.

### 3.1.3 The Effects of Childhood Trauma on Brain Development and Morphology

As mentioned above, abusive experiences induce a cascade of stress-mediated effects on hormones and neurotransmitters, which may affect the development of vulnerable brain regions [43, 44] (Table 3.1). However, the brain structures that are more vulnerable to the effects of childhood abuse are those with protracted postnatal development, high density of glucocorticoid receptors, and some degree of postnatal neurogenesis [2].

In this paragraph we will review magnetic resonance imaging (MRI) studies assessing structural differences in specific brain regions, including hippocampus, amygdala, prefrontal cortices, sensory cortex, corpus callosum, and cerebellum (Tables 3.2, 3.3, 3.4, and 3.5).

**Table 3.1** Review studies on structural magnetic resonance imaging investigations on the association between early adversities (EA: either institutionalization or maltreatment) and alterations in limbic structures

| Study | Type of study | Structures investigated | Number of studies reported | Results |
|---|---|---|---|---|
| Smith (2005) | Meta-analysis | Hippocampus | 13 papers reporting hippocampal findings in patients with PTSD versus matched controls | PTSD patients had 6.9% smaller left hippocampal volume and 6.6% smaller right hippocampal volume |
| Teicher and Samson (2016) | Review study | Hippocampus Amygdala | 37 papers reporting hippocampal findings in adults with childhood maltreatment<br>27 studies reporting amygdala volumes findings in subjects with maltreatment histories | 30 papers reported one or more significant difference between groups with and groups without trauma exposure or an inverse correlation between severity of trauma exposure and volume<br>3 papers reporting nearly significant reductions or correlations<br>4 studies failed to find any difference<br>8 studies reported a significant amygdala volume reduction in maltreated versus non-maltreated subjects<br>13 studies reported no differences<br>4 studies reported a significant increase |
| Rinne-Albers 2013 | Review study on children and adolescent | Hippocampus Amygdala | 9 papers reporting hippocampal findings in traumatized children<br>3 studies reporting amygdala findings in traumatized children | 6 studies on children (prepubertal children and postpubertal adolescents) found no difference between groups with and groups without trauma exposure<br>2 studies found a significant decrease in hippocampal volume in children with trauma exposure<br>1 study found an increase in hippocampal white matter volume in exposed children<br>1 study reported increased amygdala volume in severe early deprived Romanian adoptees<br>2 studies reported no differences |

**Table 3.2** Magnetic Resonance Imaging studies on amygdala and hippocampal morphological alterations in subjects with early life adversities

| Study | Mean age | Sample size | Structures investigated | Type of study | Type of trauma | Results |
|---|---|---|---|---|---|---|
| Mehta (2009) | 12.5 | 14 adopted (6 males and 8 females) adolescents who had experienced severe early institutional deprivation<br>11 noninstitutionalized controls | Amygdala<br>Hippocampus<br>Corpus callosum | MRI study | Institutional deprivation | PI subjects had greater right amygdala volume<br>No difference was observed in hippocampal volume or corpus callosum mid-sagittal area<br>The left amygdala volume was inversely correlated to the time spent in the institutions |
| Tottenham (2010) | 24 | 34 PI subjects: 17 early adopted, before 15 months of age (15 girls, 2 boys), and 17 late adopted, after 15 months of age (10 girls and 7 boys)<br>28 non-PI matched subjects | Amygdala<br>Hippocampus<br>Caudate | MRI study | Institutional deprivation | Later adopted PI children had significantly larger amygdala volume than the early adopted group and the comparison group<br>No differences were found between hippocampal and caudate volumes among PI and non-PI subjects |
| Lupien (2011) | 10 | 17 children (7 boys and 10 girls) exposed to maternal depressive symptomatology since birth<br>21 children not exposed to maternal depressive symptomatology | Amygdala<br>Hippocampus | MRI study | Continuative exposure to maternal depressive symptomatology | No group difference in hippocampal volume<br>Larger left and right amygdala volumes in children exposed to maternal depressive symptomatology since birth |
| Driessen (2000) | 29.5 | 21 female patients with BPD<br>21 female control subjects | Hippocampus<br>Amygdala | MRI study | Physical and emotional abuse, sexual abuse, emotional and physical neglect | BPD had significantly smaller bilateral hippocampal and amygdala volumes |

(continued)

**Table 3.2** (continued)

| Study | Mean age | Sample size | Structures investigated | Type of study | Type of trauma | Results |
|---|---|---|---|---|---|---|
| Schmahl (2003) | 31 | 10 female patients with BPD<br>23 female patients without BPD | Hippocampus<br>Amygdala | MRI study | Childhood sexual and/or physical abuse | Patients with BPD had significantly smaller hippocampal and amygdala volumes |
| Vermetten (2006) | 38 | 15 patients with dissociative identity disorder<br>23 healthy comparison subjects | Hippocampus<br>Amygdala | MRI study | Childhood sexual and/or physical abuse | Patients with dissociative identity disorder had significantly smaller bilateral hippocampal and amygdala volumes |
| Yehuda et al. 2007 | 62.5 | 17 veterans with chronic PTSD and 16 veterans without chronic PTSD | Hippocampus | Case–control MRI study | Combat veterans with early trauma exposure | Hippocampal volume did not differ between subjects with and without PTSD. Presence and severity of early trauma exposure was a risk factors for PTSD, but not associated with hippocampal volume |
| Kuo (2012) | 53 | 42 combat veterans who developed PTSD<br>45 combat veterans who did not develop PTSD | Amygdala | MRI study | Combat veterans with a history of early life trauma | Combat veterans who developed PTSD exhibited larger total amygdala volumes compared with non-PTSD veterans. Greater severity of combat exposure and presence of early life trauma were significantly associated with smaller amygdala volume |
| Whittle (2013) Longitudinal study | 13 | 117 adolescents | Hippocampus<br>Amygdala | MRI study | Emotional abuse, physical abuse, emotional neglect, physical neglect, sexual abuse | Childhood maltreatment was associated with larger baseline left hippocampal volumes and retarded growth of the left amygdala over time |

*PI* previously institutionalized/previous institutionalization, *PTSD* posttraumatic stress disorder, *BPD* borderline personality disorder, *MRI* magnetic resonance imaging

**Table 3.3** Magnetic resonance imaging studies on corpus callosum alterations in subject with early life adversities

| Study | Mean age | Sample size | Structures investigated | Type of study | Type of trauma | Results |
|---|---|---|---|---|---|---|
| De Bellis (1999) | 14 | 44 maltreated children with PTSD<br>61 healthy controls | Corpus callosum | MRI study | Neglect, physical abuse, sexual abuse, emotional maltreatment | The total midsagittal area, and middle and posterior regions of corpus callosum were significantly smaller in PTSD patients than controls<br>A significant gender by diagnosis effect revealed greater corpus callosum area reduction in maltreated males with PTSD |
| De Bellis (2002) | 14 | 28 maltreated children and adolescent with PTSD<br>66 healthy controls | Corpus callosum | MRI study | Neglect, physical abuse, sexual abuse, emotional maltreatment | The total midsagittal area of corpus callosum and middle and posterior regions were significantly smaller in PTSD patients than controls |
| De Bellis (2003) | 15 | 61 maltreated children (31 males and 30 females) and adolescent with PTSD<br>122 healthy controls (62 males and 60 females) | Corpus callosum | MRI study | Physical and sexual abuse | Significant smaller midsagittal area of the corpus callosum subregion (splenium) in PTSD patients compared to gender matched comparison subjects<br>Significant sex by group effects demonstrated smaller corpus callosum regions (rostrum and isthmus) in maltreated males with PTSD compared to females with PTSD |
| Teicher (2004) | | 51 children and adolescent with psychiatric diagnosis including 28 maltreated children and adolescent<br>115 healthy controls | Corpus callosum | MRI study | Physical and sexual abuse<br>Neglect | Total corpus callosum area of the abused/neglected patients was significantly smaller than in control subjects.<br>Total corpus callosum area of the abused/neglected patients was significantly smaller than in psychiatric patients who had not been abused or neglected |
| Jackowski (2008) | 11 | 17 maltreted children (7 males and 10 females)<br>15 matched healthy controls (7 males and 8 females) | Corpus callosum | DTI study | Intrafamilial abuse | Maltreated children with PTSD had significantly reduced fractional anisotropy (microstructural alterations) in anterior and posterior midbody of the corpus callosum |

(continued)

**Table 3.3** (continued)

| Study | Mean age | Sample size | Structures investigated | Type of study | Type of trauma | Results |
|---|---|---|---|---|---|---|
| Galinowski (2015) | 14 | 2224 healthy community adolescent<br>55 resilient with low risk of mental disorder despite high exposure to lifetime stress<br>68 at risk of mental disorder exposed to the same level of stress<br>123 controls | Corpus callosum | DTI study | Negative life events | Higher fractional anisotropy values (suggesting decreases in axonal diameter, packing density, and branching or strengthened connectivity) were detected in the anterior corpus callosum of resilient compared to both nonresilient and control adolescents |

*DTI* diffusion tensor imaging, *MRI* magnetic resonance imaging, *PTSD* posttraumatic stress disorder

**Table 3.4** Studies investigating sensitive periods in the effect of childhood abuse on brain development

| Study | Mean age | Sample size | Structures investigated | Type of study | Type of trauma | Results |
|---|---|---|---|---|---|---|
| Andersen (2008) | 20 | 26 women with reported episode of childhood abuse<br>17 healthy female controls | Hippocampus<br>Corpus callosum<br>Frontal cortex | Case–control MRI study | Sexual abuse | Hippocampal volume was reduced in association with childhood abuse at 3–5 years and 11–13 years<br>Corpus callosum was reduced in association with childhood abuse at 9–10 years<br>Frontal cortex was reduced in association with childhood abuse at 14–16 years |
| Pechtel (2014) | 25 | 18 participants form a longitudinal cohort with early and continued life stress | Amygdala<br>Hippocampus | Case–control MRI study | Childhood maltreatment | Hippocampal volume was reduced in association with childhood maltreatment at 7–14 years<br>Amygdala volume was reduced in association with childhood maltreatment at 10–11 years |

*MRI* magnetic resonance imaging

#### 3.1.3.1 Hippocampus

The hippocampus is involved in the formation and retrieval of memories, particularly explicit memories. It is densely populated with glucocorticoid receptors, and preclinical studies showed that excessive exposure to glucocorticoids led to reversible atrophy of dendritic processes on pyramidal cells in the cornu ammonis (one of the hippocampus subfields) and to the suppression of neurogenesis in the dentate gyrus (a separate structure wrapped around the hippocampus) [45]. Interestingly, in females, the volumetric reduction in the hippocampus is less evident compared to males mainly because they show increased resilience due to a potential

**Table 3.5** Functional magnetic resonance imaging studies investigating the association between early adversity and alterations across four neurocognitive domains in children and adolescents

| Study | Neurocognitive domain explored | Mean age | Sample | Task | Maltreatment subtype | Results |
|---|---|---|---|---|---|---|
| Maheu (2010) | Threat processing | 13.5 | 11 youths with history of emotional neglect<br>19 healthy controls | Emotion discrimination task | Caregiver deprivation and emotional neglect | Higher activation of left amygdala and left anterior hippocampus in MS versus healthy controls |
| McCrory (2011) | Threat processing | 12.5 | 20 children exposed to family violence<br>23 children not exposed | Emotion discrimination task | Family violence | Higher amygdala and insula activation to angry faces in MS versus healthy controls |
| Tottenham (2011) | Threat processing | 10.1 | 22 children who experienced orphanage rearing<br>22 nonexposed children | Emotion discrimination task | Children who experienced orphanage rearing | Higher amygdala and insula activation to angry faces in MS versus healthy controls |
| White (2012) | Threat processing | 13.5 | 139 healthy youths | Emotion discrimination task | Emotional neglect | Higher amygdala activation to angry faces in carrier of FKBP5 polymorphism and in the context of emotional neglect |
| McCrory (2013) | Threat processing | 12.5 | 18 maltreated subjects<br>23 non-maltreated subjects | Emotion discrimination task | Physical, sexual, emotional abuse, and neglect | Higher amygdala activation to angry and happy faces in MS versus healthy controls |
| McLaughlin (2015) | Passive viewing of negative and positive emotional stimuli and effortful attempts to regulate emotional responses using cognitive reappraisal | 16.6 | 21 maltreated subjects<br>21 non-maltreated subjects | Event-related task | Physical or sexual abuse | Higher response of amygdala, putamen, and anterior insula to negative relative to neutral stimuli in maltreated subjects<br>Higher DLPFC, mPFC, and dACC activation during effortful attempt to decrease emotional response to negative stimuli in MS versus healthy controls |
| Hart (2018) | Threat processing | 17 | 20 maltreated subjects<br>20 adolescents with psychiatric diagnosis but no maltreatment history<br>27 healthy controls | Emotion discrimination task | Severe physical abuse prior to age 12 | Higher activation of vmPFC and ACC in MS versus healthy controls |
| Dillon (2009) | Reward processing | 30 | 13 maltreated subjects<br>31 non-maltreated controls | Monetary incentive delay task | Physical, sexual, emotional abuse | Lower left basal ganglia activation to reward cues in MS versus healthy controls |

| | | | | | | |
|---|---|---|---|---|---|---|
| Mehta (2010) | Reward processing | 16 | 12 adolescents who experienced global deprivation prior to adoption<br>11 nonadopted comparison | Monetary reward anticipation task | Neglect | Lower ventral striatum activation to reward cues in MS versus healthy controls |
| Goff (2013) | Reward processing | 10 | 38 PI Children<br>11 never institutionalized children | Passive viewing of emotional stimuli | PI | Lower nucleus accumbens activation to positive stimuli in MS versus healthy controls |
| Hanson (2015) Longitudinal study. | Reward processing | 14 | 106 children | Operant conditioning task | Neglect | Greater levels of emotional neglect were associated with blunted development of reward-related vs. activity between the first and second assessments<br>Decreases in this reward-related vs. activity were related to greater depressive symptomatology and partially mediated the association between emotional neglect and subsequent depressive symptomatology |
| Boecker (2014) | Reward processing | 24 | 162 healthy young adults | Monetary reward anticipation task | Self-reported early familial adversities | Decreased activation during reward anticipation in ventral striatum, putamen, thalamus when early familial adversities increased. In contrast, during reward delivery, activation of the bilateral insula, right pallidum, and bilateral putamen increased with early familial adversity |
| Dennison (2016) | Reward processing | 17 | 21 adolescents with a history of maltreatment<br>38 adolescents without a history of maltreatment | Passive viewing of emotional stimuli | Physical, sexual, emotional abuse | Higher basal ganglia (pallidum and putamen) activation to positive stimuli cross-sectionally predicted lower depression symptoms longitudinally |
| Gee (2013) | Emotion regulation | 12 | 41 PI children<br>48 non-PI controls | Face processing | PI | Children with history of early adversity evidence a mature connectivity pattern (resembling the adolescent phenotype) between amygdala and medial prefrontal cortex. This pattern was not present in comparison subjects |

(continued)

**Table 3.5** (continued)

| Study | Neurocognitive domain explored | Mean age | Sample | Task | Maltreatment subtype | Results |
|---|---|---|---|---|---|---|
| Marusak (2015) | Emotion regulation | 12 | 14 adolescents with a history of maltreatment<br>16 adolescents without a history of maltreatment | Emotional conflict task | Physical abuse<br>Neglect<br>Domestic violence | Increased amygdala-ACC connectivity and higher DLPFC activation during emotional conflict in MS versus healthy controls |
| Puetz (2014) | Emotion regulation | 11 | 25 adolescents with a history of maltreatment<br>26 adolescents with a history of maltreatment | Psychological stress | Physical abuse<br>Neglect<br>Domestic violence | Reduced dACC/DLPFC connectivity and lower dACC/DLPFC activation during social rejection in MS versus HC |
| Lee (2015) | Emotion regulation | 16 | 31 healthy subjects | Face processing | Verbal abuse | Reduced amygdala/ACC connectivity during implicit affect processing |
| Elsey (2015) | Emotion regulation | 15 | 31 adolescents with a history of maltreatment<br>33 adolescents without a history of maltreatment | Personalized imagery | Physical abuse<br>Neglect<br>Sexual abuse | Higher mPFC, lPFC, dACC, PCC, and insula activation during personalized stress cues in MS versus healthy controls |
| Puetz (2016) | Emotion regulation | 13 | 21 adolescents with a history of maltreatment<br>19 adolescents without a history of maltreatment | Emotional conflict task | Physical abuse<br>Neglect<br>Sexual abuse<br>Domestic violence | Lower vlPFC, insula, amygdala, and STS activation during emotional conflict for rejection-themed words in MS versus controls |
| Mueller (2010) | Executive control | 13 | 12 PI<br>21 non-PI | Stop signal task | PI | Higher dACC/vlPFC, basal ganglia, and insula activation during error monitoring and cognitive control functions in MS versus healthy controls |
| Lim (2015) | Executive control | 17 | 22 adolescents with a history of maltreatment<br>17 matched psychiatric patients<br>27 adolescents without a history of maltreatment | Stop signal task | Physical abuse<br>Neglect<br>Emotional abuse | Higher dACC/mCC and dorsomedial frontal regions activation during error monitoring in MS versus healthy controls |

*MS* maltreated subjects, *DLPFC* dorsolateral prefrontal cortex, *mPFC* medial Prefrontal Cortex, *vlPFC* vetrolateral Prefrontal Cortex, *vmPFC* ventromedial Prefrontal Cortex, *ACC* anterior cingulate cortex, *dACC* dorsal anterior cingulate cortex, *mCC* mid Cingulate Cortex, *PCC* Posterior Cingulate Cortex, *PI* previously Institutionalized/Previous Institutionalization, *STS* Superior Temporal Sulcus neuroprotective effect of estrogens, as observed in translational studies [46]. Moreover, there is compelling evidence that adults with histories of maltreatment have smaller hippocampi than non-maltreated comparison subjects as reported in previous reviews [47, 48] and a meta-analysis [49]. However, the relationship between childhood maltreatment and hippocampal volume is less clear in studies involving children or adolescents [2, 49]. It seems that the volumetric reduction in hippocampus appears over time and depends on the type, duration, and time of trauma. This fits with the hypothesis that there may be a silent period between exposure to maltreatment and discernible neurobiological differences [50]. Indeed, two retrospective studies reported data on potential sensitive exposure period. Specifically, Andersen et al. (2008) [51] found that, in a study on young female adults, decreased bilateral hippocampal volume was associated with sexual abuse at age 3–5 years and 11–13 years. Similarly, Pechtel (2014) [52] also reported, in a cross-sectional analysis of a mix-gender longitudinal sample with disturbed attachment and exposure to emotional abuse and neglect, that right hippocampal volume appeared to be most sensitive to maltreatment at 7 and 14 years of age.

Another interesting theory suggests that smaller hippocampal volume could be a sign of vulnerability instead of a consequence of psychological trauma [29, 53]. Gilbertson et al. (2002) [53] reported the association between hippocampal volumes in twins with PTSD and their nontrauma-exposed co-twin, while Yehuda et al. (2007) [29] found smaller left hippocampal volumes in veterans who developed PTSD in response to their first reported traumatic exposure, compared to veterans who had first experienced a traumatic event after which they did not develop PTSD. Actually, both theories might be true: a smaller hippocampus may be a vulnerability factor for PTSD, and after trauma exposure PTSD can result in a smaller hippocampus [48]. Indeed, eight different psychiatric disorders have been associated with reduced hippocampal volume [52], and maltreatment is a major risk factor for all of them. Maltreatment appears to exert a predominant influence on hippocampal development regardless of the presence or absence of psychiatric disorders. This is important because the confounding influence of childhood maltreatment needs to be taken into account before considering the role of hippocampal abnormalities in the pathogenesis of psychiatric disorder [47] (Tables 3.1, 3.2, and 3.4).

#### 3.1.3.2 Amygdala

The amygdala is another key limbic structure that is critically involved in the encoding of implicit emotional memories [54] and in detecting and responding to salient stimuli, such as facial expressions and potential threats [55]. Similarly to the hippocampus, the amygdala has a high density of glucocorticoid receptors on stress-susceptible pyramidal cells [56] and a postnatal development that reaches its peak at 11 years, with gradual pruning thereafter [57]. However, differently from hippocampus, in the amygdala psychological stressors and stress hormones stimulate dendritic arborization and new spine formation on pyramidal cell, which ultimately lead to an increase in volume that endures long after cessation of the stressor [58]. Studies exploring amygdala volumes in association with childhood trauma are highly heterogeneous. There are several studies reporting a significant increase in amygdala volume in subjects with early exposure to emotional or physical neglect [52, 59–61]. Conversely, other studies on adults with exposure to multiple forms of maltreatment across development reported significant reductions in amygdala volume [62–64]. However, it is important to point out that studies showing significant reductions in amygdala volume had, on average, much older participants, greater degrees of psychopathology and exposure to multiple types of abuse during childhood. Moreover, it has also been reported that early exposure to maltreatment or neglect may result in an initial increase in amygdala volume, particularly noticeable during childhood. However, early exposure may also sensitize the amygdala to further stress, which may result in a substantial reduction of amygdala volume in late adolescence or adulthood [7]. The latter hypothesis is further supported by two fairly recent longitudinal studies [65, 66]. Specifically, the study by Kuo et al. (2012) [65] on a group of 87 combat veterans exposed to traumatic events before 13 years of age showed that childhood trauma was associated with a non-significant increase in amygdala volume and a highly significant interactive effect with the degree of combat exposure. Early trauma exposure appeared to sensitize the amygdala resulting in volume reductions with subsequent exposure to stress.

On the other hand, the longitudinal study carried out by Whittle et al. (2013) [66] on 139 children/adolescents provided evidence that early exposure to maltreatment may increase amygdala volume, whereas later exposure to maltreatment resulted in decreased gray matter volume (Tables 3.2 and 3.4).

#### 3.1.3.3 Prefrontal Cortex

Prefrontal regions have a protracted development that lasts till early adulthood [67, 68]. Studies on individuals with histories of childhood maltreatment found an overall reduction in brain size [69, 70] with a reduction more pronounced in prefrontal regions. Morphometric studies on individuals with histories of childhood maltreatment, neglect, or early deprivation showed volumetric reduction of anterior cingulate cortex (ACC), DLPFC, and OFC. These three portions of the prefrontal cortex seem to play an important role in decision-making and emotional regulation. Moreover, neuroplastic changes in function and connectivity of these structures appear to be a critical factor in addiction [71]. Indeed, both animal and human studies consistently reported that these structures are part of a widely distributed network underpinning key elements of the addiction cycle, including craving and inhibitory control, and, therefore, in the development of drug addictions. This is not surprising especially because these regions play a key role in the regulation of cognitive and emotional processes that, if disrupted, might alter the response to drugs and enhance the sensitivity to stressors [71]. Interestingly, maltreatment-related alterations in these structures is one of the ways by which maltreatments affect the brain to enhance risk of addiction [47]. Moreover, atrophy in these three prefrontal regions has been consistently reported in adults with histories of childhood maltreatment without, though, a history of psychiatric illnesses [71, 72], ultimately suggesting that prefrontal cortices deficits seem to be present in maltreated subject also in the absence of psychopathology.

#### 3.1.3.4 Sensory Cortex and Fiber Pathways

Brain regions and fiber tracts that process and convey the adverse sensory input of the abuse may be specifically modified by this experience, and these changes represent modifications or adaptations, rather than a nonspecific damage. Studies using diffusion tensor imaging (DTI) and tract-based spatial statistic (TBSS) techniques identified three fiber tracts that significantly differed between verbally abused subjects and healthy controls [73]: the left arcuate fasciculum, which links superior temporal gyrus to the frontal cortex, interconnecting Broca's and Wernicke's areas and critically involved in human language [74], the left cingulum bundle, which supports prefrontal, parietal, and temporal lobe interactions, while its reduced integrity correlates with depressive and dissociative symptoms, and the left fornix, which is associated with symptoms of anxiety and somatization [73]. Moreover, other studies, using voxel-based morphometry (VBM), found gray matter volume alterations in the auditory cortex in subjects exposed to severe parental verbal abuse [75]. Also, subjects witnessing domestic violence showed alterations in the left inferior longitudinal fasciculus [76], which is a key component of the visual limbic pathway and is involved in emotional, learning, and memory functions. Similarly, in a VBM study carried out by Tomoda et al. (2012) [77] in a group of subjects who visually witnessed multiple episodes of domestic violence overlapping with the previously reported sample, authors found a robust reduction in gray matter density in the right lingual gyrus, a component of the visual system involved in visual memory for shapes and in the non-conscious processing of visual material. Finally, brain regions and fiber tracts that process and convey the adverse sensory input of the abuse may be specifically modified by this experience, particularly in subjects exposed to a single type of maltreatment. Conversely, exposure to multiple types of maltreatment may more commonly produce alterations in corticolimbic regions involved in emotional processing and stress response [76, 77]. Heim et al. [78] suggested that neuroplastic cortical adaptations might protectively shield a child from the sensory processing of the specific abusive experience. However, thinning of the somatosensory cortex may lead to the development of behavioral problems, and to impairments in verbal comprehension, visual recall, and emotional responses to witnessed events.

#### 3.1.3.5 Corpus Callosum

The corpus callosum (CC) is the largest white matter tract and plays a critically important role in interhemispheric communication. Interestingly, reduced integrity of this brain region has been one of the most consistent findings in maltreated children and adults [69, 70, 79, 80].

In the first MRI study carried out in children with PTSD, De Bellis et al. (1999) [69] found a reduction in size of the midsagittal, middle, and posterior regions of the CC. Symptoms of PTSD and dissociation negatively correlated with total and regional CC volumetric measurements. Interestingly, subsequent studies from the same research group also confirmed these findings [70, 79]. Moreover, Teicher et al. [80] found that a group of 28 abused or neglected psychiatric inpatients children had a significant reduction in a cross-sectional area of the CC compared to healthy controls and, to a lesser degree, in comparison to psychiatric inpatients with other diagnoses. In this study, this CC alteration (in the rostral CC) was associated with sexual abuse in girls, whereas in boys the CC reduction (in its caudal part) was associated with neglect. Authors suggest that this can be explained by the fact that myelinization in the CC follows a rostral–caudal pattern, and neglect usually takes place in an earlier phase of development. Moreover, preclinical studies have identified prominent sex differences in CC in response to early traumatic experiences [81]. Specifically, a reduction in CC volume consequent to childhood maltreatment appears greater in boys than girls, probably because males have an earlier sensitive period [82]. Additionally, DTI studies showed that children with maltreatment-related PTSD had reduced fractional anisotropy (FA), a quantitative indicator of white matter integrity, in the medial and posterior corpus of CC [83]. Notably, this region receives interhemispheric projections from several regions that have connections with prefrontal areas involved in emotional regulation and memory functions. Therefore, disruptions in the medial and posterior corpus of CC might explain the core deficits often observed in individuals with a trauma history, including emotional dysregulation and memory, which may in turn explain the dissociative symptoms, repetitive compulsions, startle reactions, and overreactions often observed in these subjects. Similarly, Galinowski et al. [84] assessed the relationship between exposure to negative life events and FA in a large sample of adolescents. They reported an increase in FA in CC segments II and III in resilient subjects (i.e., individuals with low risk of mental disorder despite high exposure to lifetime stress), relative to control subjects and susceptible subjects. Moreover, they also observed a linear trend for CC FA to be greater in resilient subject than controls, and greater in controls than in susceptible subjects, ultimately suggesting that increased FA in CC could be a preexisting protective factor for the development of psychiatric disorder after exposure to early stress.

The development of DTI and TBSS techniques allowed to identify two additional pathways correlated to childhood maltreatment: the uncinate fasciculus that interconnects limbic regions with the OFC, and the superior longitudinal fasciculus that conveys information regarding the spatial location of body parts, awareness of visual space, and somatosensory information to portions of the prefrontal cortex involved in motor planning, working memory, and speech. Indeed, two studies reported significant reductions in FA in the uncinate fasciculus of orphans exposed to early deprivation, and this reduction inversely correlated with duration and time of orphanage [85, 86]. Finally, reduced integrity of the superior longitudinal fasciculus has been reported in orphans with early deprivation [86], in adolescent exposed to physical abuse, sexual abuse, or witnessing domestic violence [87] and in hospitalized bipolar adults with childhood adversity [88] (Table 3.3).

#### 3.1.3.6 Cerebellum

The cerebellum, as the hippocampus and amygdala, has a high density of glucocorticoid receptors [89] and shows postnatal neurogenesis [90]. Several studies with heterogeneous samples of children exposed to different traumatic experiences (physical abuse, sexual abuse, witnessing domestic violence with PTSD symptoms or diagnosis, adoptees from Rumania, Russia, China, and other Eastern countries, adults with histories of childhood sexual abuse, PTSD resilient and susceptible adults with post childhood trauma and varying degrees of early life trauma) reported significant alteration in the vermis and significantly lower volume in the cerebellar hemisphere [43, 72, 91–96].

## 3.2 Functional Magnetic Resonance Imaging Studies in Maltreated Subjects

In recent years, fMRI studies have driven their attention toward the study of specific cognitive and emotional functions in maltreated individuals through the use of specific activation tasks. Notably, these studies demonstrated that childhood maltreatment is associated with altered functioning in a range of cognitive and emotional systems including threat processing, reward processing, emotion regulation, and executive control. Alterations of these cognitive systems do not *per se* constitute symptoms of any psychiatric disorder. Rather, these alterations reflect a complex phenotype characterized by a maladaptive regulation of higher order systems important for socioemotional and cognitive functioning, which may be indicators of future risk for developing a psychiatric disorder. These indicators are referred as "latent vulnerability." Such vulnerability could theoretically be present for months or years, but clinical symptoms may only manifest under certain conditions, e.g., stress or a developmental challenge, or conversely, may never manifest, given adequate intrinsic and extrinsic protective factors (e.g., social support, resilient genotypes).

In this section, we will summarize the available fMRI evidence on maltreated subjects exploring the four neurocognitive and emotional systems mentioned above (Table 3.5).

### 3.2.1 Threat Processing

Human and animal studies demonstrate that significant neurobiological and cognitive resources are dedicated to threat detection and response, mainly because these abilities are a necessary condition for survival.

Brain regions and pathways involved in regulating response to threatening stimuli include thalamus, visual cortex, ACC, ventromedial prefrontal cortex, amygdala, and hippocampus [97]. Within this circuit, the amygdala plays a critical role in the detection of salient stimuli [98], especially in situations when there is ambiguity [99]. In such situations, the gathering of additional environmental information is beneficial and possibly, critical. Therefore, the amygdala would be a key component of an "information gathering system." Furthermore, amygdala activation redirects mental and bodily resources, especially by engaging the hypothalamus and other brain stem structures allowing the significance of an item to be ascertained [100]. Therefore, alterations in this system may potentially place an individual at a greater risk of developing maladaptive behaviors. For instance, active avoidance of threats, enhanced threat detection, and a more rapid recognition of fearful stimuli can be considered adaptive modifications in a context of adversity; however, these functional adaptations may reduce available resources for other important functions and for exploratory behaviors [101]. The majority of fMRI studies reporting functional changes in amygdala focused on threat detection and response by using emotional face-processing paradigms. The majority of these studies explored amygdala activation during processing of negatively valenced emotions, in order to provide evidence of the role of this structure in threat detection and response [102].

Overall, these studies reported increased amygdala activation in orphans who experienced caregiver deprivation [103–105]. The degree of reactivity appears to partly relate to the severity of early adversity [104, 106–108] and possibly, to genetic differences [108]. These findings are consistent with the view that altered threat reactivity, as indexed by amygdala (and related structures) neural response to biological threat stimuli, represents a candidate neurocognitive impaired process conferring latent vulnerability in children exposed to early adversity [109]. Higher baseline levels of amygdala activation prior to trauma exposure may be present in individuals who have genetic vulnerability, or those who have both genetic vulnerability and experience of adversities [110]. Interestingly, in a recent whole-brain fMRI investigation exploring the effect of child abuse on functional activation and connectivity during dynamic emotion processing in adolescents, Hart et al. [111] demonstrated that maltreated adolescents responded faster to fear than healthy controls, showing increased activation in bilateral ventromedial prefrontal cortex (vmPFC) and ACC. Furthermore, maltreated individuals had reduced functional connectivity between left vmPFC and insula. The insula is implicated in fear processing and it is thought to convey cortical representations of fear to the amygdala [112]. Decreased vmPFC–insula connectivity could result in a weakened top-down control of vmPFC over the insula, leading to a fear regulation deficit, and increased fear sensitivity. Reduced functional connectivity parallels structural connectivity findings: two studies in maltreated individuals reported reduced integrity in the left uncinate fasciculus, a white matter tract connecting prefrontal to limbic regions including amygdala and insula [85] and in the cingulum bundle, which connects limbic structures including the insula, to cortical regions including the cingulate gyrus [74]. Importantly, amygdala volume peaks at around age 10, while the prefrontal cortex (PFC) matures relatively late in life with progressive volumetric changes into adulthood and a sharp growth between 8 and 14 years of age [113]. Because of its protracted development, the PFC is conceivably more vulnerable to environmental stressors such as child abuse, than limbic structures [111]. Finally, McLaughlin et al. [114] reported that maltreated adolescents modulate amygdala hyper-responsiveness by increasing the activation of regions involved in effortful control (superior frontal gyrus, dorsal anterior cingulate cortex, and frontal pole). Although child maltreatment appears to heighten the salience of negative emotional stimuli, maltreated adolescents seem to be able to modulate amygdala activation to a similar degree as non-maltreated youth when taught specific emotion regulation strategies. This is because they use the PFC regions involved in effortful control of emotion to a greater degree. Indeed, greater engagement of these regions may reflect the fact that maltreated youth must devote greater cognitive resources to modulating emotional responses than non-maltreated children, and also that training in cognitive reappraisal strategies is likely to be an effective tool for reducing emotional reactivity to negative emotional stimuli in these patients.

### 3.2.2 Reward Processing

Reward processing plays a central role in our ability to successfully adapt to the environment by motivating and reinforcing goal-directed behavior [115]. Reward processing can be conceptualized as comprising three main components: "liking," "wanting," and "learning" [110]. Reward anticipation or "wanting" is indexed, in particular, by neural responses in the ventral striatum, with response to reward-related cues mediated by dopamine signaling. In contrast, "liking" (reflecting a hedonic response signaling receipt of reward) is believed to be mediated by opioid and endocannabinoid systems [116]. Several fMRI studies reported decreased activity in the ventral striatum of maltreated individuals to anticipated reward in a monetary incentive delay task. These results were reported in both children and adults and include the following: orphans experiencing early deprivation [117], maltreated children in a community setting [118], adults who experienced early family adversity [119], and adults reporting exposure to physical, sexual, or emotional abuse [120]. Reduced activity to anticipated reward magnitude in the ventral striatum, alongside with heightened threat reactivity, may reflect an adaptive calibration in an approach-avoidance conflict situation to tip the balance toward avoidance. This shift may increase the likelihood of survival in an environment characterized by danger [47]. The downside is that this adaptive mechanism may also lead to symptoms of depression/anhedonia [121], anxiety [122], and enhanced risk for addiction [123]. In contrast to the abovementioned studies, Dennison et al. [124] reported a greater response during passive viewing of positive, relative to neutral, social stimuli in the left nucleus accumbens and left putamen in a group of older adolescents who had experienced at least moderate physical and/or sexual abuse compared with a control group matched on age and IQ. These discrepant findings could reflect differences in the task used to elicit neural reward reactivity. Particularly, Dennison et al. [124] measured passive reactivity to positive stimuli, whereas the instrumental reward task used by Hanson et al. [5] is likely to have captured activations related to reward expectancy and anticipation [115]. Neurobiological evidence supports the divergence of these reward-processing phases, with "liking" stages being more associated with activations in the pallidum, and "wanting" with activation in the ventral striatum [115]. Childhood adversity appears to differentially influence neural response during these discrete reward processes. Indeed, Boecker et al. [119] reported that early exposure to adversity (indexed by poverty and social disadvantage) was associated with reduced neural reactivity in the ventral and dorsal striatum during anticipation of reward, as well as higher reactivity during reward delivery in the putamen, right pallidum, and insula. This suggests that childhood adversity might be associated with lower expectations of positive outcomes and greater surprise or pleasure when positive events occur [125].

Importantly, the study by Dennison et al. [124] revealed that greater reward reactivity across behavioral and neurobiological measures moderates the association between maltreatment and baseline depression. Specifically, faster reaction times to cues paired with monetary reward relative to those unpaired with reward, and greater activity in the left pallidum were associated with lower depression symptoms in maltreated youth. Longitudinally, greater activity in the left putamen moderated changes in depression scores over time, such that higher levels of reward response were associated with reduced worsening of depressive symptoms over time in maltreated youth.

In conclusion, elevated responsiveness to (rather than anticipation of) reward cues might confer resilience to depression [110].

### 3.2.3 Emotion Regulation

Affect or emotion regulation is regarded as a dynamic and multifaceted process that can operate within or outside our conscious awareness [126], and it comprises various mechanisms and strategies including cognitive reappraisal, suppression of emotions, and attention modulation [127]. Neuroimaging and lesion studies have identified a functionally and structurally interconnected circuit involved in emotion regulation. In particular, subcortical/limbic regions involved in the evaluation of threat, reward, and appraisal of internal physiological states (such as striatum, amygdala, and insula) have been found to be strongly interconnected with frontal association cortices (such as the ACC and the lateral prefrontal regions (mPFC and lPFC)) [128]. Within this circuit, prefrontal regions exert a top-down inhibitory effect over subcortical brain structures. Although several fMRI studies investigated brain activity during emotional processing in children exposed to maltreatment by employing different paradigms, results are highly heterogeneous. Specifically, two studies used a face-processing paradigm to examine functional connectivity between amygdala and frontal regulatory regions (ventral ACC and mPFC). Both studies showed a pattern of aberrant functional connectivity in children and adolescents previously institutionalized [129] and in adolescents reporting histories of verbal abuse [130] compared to age-matched control subjects.

Research in normal human development has shown that regulatory connections between the amygdala and mPFC manifest as positive coupling in childhood with a shift to a mature pattern of negative coupling during adolescence. Gee et al. (2013) [129] suggested that children with a history of maternal deprivation display the mature pattern of connectivity (negatively coupled amygdala–mPFC activity), whereas comparison children showed an immature pattern of connectivity (positively coupled amygdala–mPFC activity) and hypothesized that in a context of early adversities, amygdala hyperactivity speeded up the development of mature connections with the mPFC.

A further study [131] examined a group of trauma-exposed children and a group of matched control children during an emotional conflict task that involved categorizing facial affect while ignoring an overlying emotion word. This study did not report the typical pattern of negative connectivity between the amygdala and the ventral ACC. In contrast, authors reported that trauma-exposed youths were less able to regulate emotional conflict and engaged the amygdala-pregenual cingulate inhibitory circuit during the regulation of emotional conflict. These findings may be explained considering the complex pattern of altered connectivity that characterizes individuals exposed to early adversity. Indeed, this pattern is influenced by the developmental stage, the nature of the adversity experienced, and the specific computational demands of any given emotional processing task.

Finally, four other fMRI studies investigated brain activity in children exposed to maltreatment during a number of paradigms requiring emotion regulation. Two studies [114, 132] found an overall pattern of increased activity in the DLPFC and also in the dorsal and ventral ACC, while the others [133, 134] found reduced activity in these brain regions. Interestingly, these contrasting findings may be related to differences across the paradigms used. Indeed, in the first two studies, participants were required to explicitly modulate their emotional response to visually or auditory presented stimuli, with increased activation in frontal regulatory regions reflecting greater effort, whereas in the latter, participants processed aversive stimuli more incidentally, such as during color naming of negative stimuli, with decreased activation in frontal regulatory regions reflecting greater avoidance [110].

### 3.2.4 Executive Control

There are three basic cognitive functions underlying executive control: updating, inhibiting, and task shifting [135]. These functions interact together to carry out effective decision-making and adaptive behaviors, including self-regulation and the ability to monitor performance and detect errors (error processing).

Interestingly, only two fMRI studies explored executive functions in children and adolescents who have been exposed to institutionalization or maltreatment [136, 137] showing impaired executive control and increased activity during error monitoring and inhibition in medial and lateral frontal regions, such as the dorsal ACC and frontal motor regions.

Impaired executive control is associated with emotion regulation difficulties, rumination, and reduced social skills, which are all predictors of psychopathology [138].

However, the existence of a causal relationship between childhood victimization and impairments in cognitive functioning has recently been confuted by Danese et al. [35] in a longitudinal study using two population-representative birth cohorts comprising more than 3000 individuals. They reported that either adult and adolescent individuals exposed to childhood victimization had clinically relevant pervasive impairments in cognitive functions, including general intelligence, executive functions, processing speed, memory, perceptual reasoning, and verbal comprehension. However, the observed cognitive deficits in victimized individuals were largely explained by cognitive deficits that predated childhood victimization and by confounding genetic and environmental risks.

## 3.3 Conclusions

Childhood abuse has been consistently found to be associated with alterations in brain structure and functions. Specific type of abuse appears to selectively target sensory systems and pathways that convey and process the aversive experience and evidence for relatively brief sensitive periods (i.e., limited periods of times during which the effects of experience on brain and behavior are particularly potent) has emerged in some studies. However, many of the reviewed studies provided only partial or incomplete descriptions of their sample in terms of kinds of maltreatment experiences to which children were exposed. In addition, it has been difficult to make any claims about sensitive periods. This is because there is a great heterogeneity in timing of exposure of abuse among victimized children in community settings. Also, in almost all of the reviewed studies psychopathology was measured concurrently with neurocognitive functioning. Longitudinal prospective studies are required to establish whether alterations in neurocognitive functioning can predict levels of future psychiatric symptomatology. These limitations preclude any strong inference regarding causality between maltreatment exposure and altered neurocognitive function, and suggest the need for further studies as to increase findings validity and to provide a full characterization of all maltreatment domains.

However, some evidence should be discussed. Many of the maltreatment-related findings might be interpreted as neuroplastic adaptive responses. For instance, enhanced amygdala response to emotional faces, and diminished striatal response to anticipated reward make sense as adaptations that would tip the balance in approach/avoidance situations toward avoidance. Thus, neurocognitive alterations may hold functional value for the child in the context of an early neglectful or abusive environment. However, they may also contribute to the pathogenesis of future mental health problems, ceasing to be adaptive or beneficial during later stages of development, or in more normative environments not characterized by threat, deprivation, or unpredictability [6].

The relationship between childhood abuse, brain changes, and psychiatric illness is still unclear. There is relevant evidence suggesting that some of the key neuroimaging findings correlated to histories of childhood maltreatment, such as reduced hippocampal, ACC, and CC volumes, as well as enhanced amygdala response to fearful faces, may be restricted to maltreated individuals (maltreated ecophenotype), independently from the presence of any specific psychiatric condition or diagnosis.

Importantly, the neuroimaging evidence suggesting that these alterations can also be observed in resilient individuals with histories of maltreatment, but without a psychiatric illness, raises concerns about the relationship between these findings and the onset of psychiatric disorders.

An important outstanding question for clinicians and researchers relates to the malleability of neurocognitive systems in "recalibrating" responses to threat and reward cues in more normal environmental contingencies. Particularly, Fonagy and Allison [139] proposed a psychological intervention that takes into account developmental psychological processes, focusing on the child's relationship with a sensitive and warm caregiver who understands the child to be an intentional agent capable of representing mental states. Such an understanding is thought to be critical for the development of epistemic trust, a necessary building block for using social referents to acquire new knowledge about the world. In the case of maltreatment, where the caregiver shows an absence of mentalization and contingent and marked mirroring, the development of epistemic trust is compromised. Enhancing such trust (and consequent learning) in typical adults, even in those who are ordinarily capable of sensitive caregiving, may be insufficient to meet the needs of those children with altered social information processing and marked behavioral problems. Thus, developing an effective set of therapeutic tools to enhance and promote the development of epistemic trust becomes a challenge, which could be seen as an important factor in promoting a resilient outcome.

The evidence from the abovementioned neuroimaging studies might have important clinical implications as they provide both the motivation and the rationale to pursue a much more explicit preventative psychiatric approach in helping those children exposed to maltreatment before they present with a frank psychiatric disorder.

## References

1. Pechtel P, Pizzagalli D. Effects of early life stress on cognitive and affective function: an integrated review of human literature. Psychopharmacol. 2011;214:55–70.
2. Teicher MH, Andersen SL, Polcari A, Anderson CM, Navalta CP, Kim DM. The neurobiological consequences of early stress and childhood maltreatment. Neurosci Biobehav Rev. 2003;27:33–44.
3. McCrory EJ, De Brito SA, Viding E. Research review: the neurobiology and genetics of maltreatment and adversity. J Child Psychol Psychiatry. 2010;51:1079–95.
4. Bremner JD, Vermetten E. Stress and development: behavioral and biological consequences. Dev Psychopathol. 2001;13:473–89.
5. Vermetten E, Vythilingam M, Southwick SM, Charney DS, Bremner JD. Long-term treatment with paroxetine increases verbal declarative memory and hippocampal volume in posttraumatic stress disorder. Biol Psychiatry. 2003;54:693–702.
6. McCrory EJ, Viding E. The theory of latent vulnerability: reconceptualizing the link between childhood maltreatment and psychiatric disorder. Dev Psychopathol. 2015;27:493–505.
7. Teicher MH, Samson JA. Childhood maltreatment and psychopathology: a case for ecophenotypic variants as clinically and neurobiologically distinct subtypes. Am J Psychiatry. 2013;170:1114–33.
8. Stiles J, Jernigan TL. The basics of brain development. Neuropsychol Rev. 2010;20:327–48.
9. Bystron L, Blakemore C, Rakic P. Development of the human cerebral cortex: boulder committee revisited. Nat Rev Neurosci. 2008;9:110–22.
10. Pakkenberg B, Gundersen HJ. Neocortical neuron number in humans: effects of sex and age. J Comp Neurol. 1997;384:312–20.

11. Stiles J. The fundamentals of brain development: integrating nature and nurture. Cambridge, MA: Harvard University Press; 2008.
12. Von Melchner L, Pallas SL. Visual behavior mediated by retinal projections directed by auditory pathway. Nature. 2000;404:871–6.
13. Greenough WT, Black JE, Wallace CS. Experience and brain development. Child Dev. 1987;58:539–59.
14. Black JE, Sirevaag AM, Greenough WT. Complex experience promotes capillary formation in young rat visual cortex. Neurosci Lett. 1987;83:351–5.
15. Hubel DH, Wiesel TN. Ferrier lectures: functional architecture of macaque monkey visual cortex. Proc Royal Soc Lond Ser B. 1977;198:1–59.
16. Widom CS. Posttraumatic stress disorder in abused and neglected children grown up. Am J Psychiatry. 1999;156:1223–9.
17. Widom CS, DuMont K, Czaja SJ. A prospective investigation of major depressive disorder and comorbidity in abused and neglected children grown up. Arch Gen Psychiatry. 2007;64:49–56.
18. Copeland W, Keeler G, Angold A, Costello E. Traumatic events and posttraumatic stress in childhood. Arch Gen Psychiatry. 2007;64:577–84.
19. Ford JD, Stockton P, Kaltman S, Green BL. Disorders of extreme stress (DESNOS) symptoms are associated with type and severity of interpersonal trauma exposure in a sample of healthy Young women. J Interpers Violence. 2006;21:1399–416.
20. Chrousos GP, Gold PW. The concepts of stress system disorders: overview of physical an behavioral homeostasis. JAMA. 1992;267:1244–52.
21. McEwen BS. Physiology and neurobiology of stress and adaptation: central role of the brain. Physiol Rev. 2007;87:873–904.
22. Simantov R, Blinder E, Ratovitski T, Tauber M, Gabbay M, Porat S. Dopamine induced apoptosis in human neuronal cells: inhibition by nucleic acid antisense to the dopamine transporter. Neurosci. 1996;74:39–50.
23. Dunlop SA, Archer MA, Quinlivan JA, Beazley LD, Newnham JP. Repeated prenatal corticosteroids delay myelination in the ovine central nervous system. J Matern Fetal Med. 1997;6:309–13.
24. Todd RD. Neural development is regulated by classical neuro-transmitters: dopamine D2 receptor stimulation enhances neurite outgrowth. Biol Psychiatry. 1992;31:794–807.
25. Gould E, McEwen BS, Tanapat P, Galea LA, Fuchs E. Neurogenesis in the dentate gyrus of the adult tree shrew is regulated by psychosocial stress and NMDA receptor activation. J Neurosci. 1997;17:2492–8.
26. Pizarro JM, Lumley LA, Medina W, Robison CL, Changa WE, Alagappa A, et al. Acute social defeat reduces neurotrophin expression in brain cortical and subcortical areas in mice. Brain Res. 2004;1025:10–20.
27. Gunnar M, Quevedo K. The neurobiology of stress and development. Annu Rev Psychol. 2007;58:145–73.
28. De Bellis M, Abigail Zisk AB. The biological effects of childhood trauma. Child Adolesc Psychiatr Clin N Am. 2014;23:185–222.
29. Yehuda R, Golier JA, Tischler L, Harvey PD, Newmark R, Yang RK, et al. Hippocampal volume in aging combat veterans with and without post-traumatic stress disorder: relation to risk and resilience factors. J Psychiatr Res. 2007;41:435–45.
30. Malarbi S, Abu Rayya HM, Muscara F, Stargatt R. Neuropsychological functioning of childhood trauma and post-traumatic stress disorder: a meta-analysis. Neurosci Biobehav Rev. 2017;72:68–86.
31. Wilker S, Pfeiffer A, Kolassa S, Koslowski D, Elbert T, Kolassa IT. How to quantify exposure to traumatic stress? Reliability and predictive validity of measures for cumulative trauma exposure in a post- conflict population. Eur J Psychotraumatol. 2015;6:1–10.
32. Frewen P, Brown M, DePierro J, D'Andrea W, Schore A. Assessing the family dynamics of childhood maltreatment history with the childhood attachment and relational trauma screen (CARTS). Eur J Psychotraumatol. 2015;6:27792.

33. Cloitre M, Stolbach BC, Herman JL, van der Kolk B, Pynoos R, Wang J, et al. A developmental approach to complex PTSD: childhood and adult cumulative trauma as predictors of symptom complexity. J Trauma Stress. 2009;22:399–408.
34. Yehuda R, Hoge CW, McFarlane AC, Vermetten E, Lanius RA, Nievergelt CM, et al. Posttraumatic stress disorder. Nat Rev Dis Primers. 2015;1:1–22.
35. Danese A, Moffit TE, Arseneault L, Bleiberg BA, Dinardo PB, Gandelman SB, et al. The origins of cognitive deficits in victimized children: implications for neuroscientists and clinicians. Am J Psychiatry. 2017;174:349–61.
36. Schore AN. Dysregulation of the right brain: a fundamental mechanism of traumatic attachment and the psychopathogenesis of posttraumatic stress disorder. Aust N Z J Psychiatry. 2002;36:9–30.
37. Casey JE. A model to guide the conceptualization, assessment and diagnosis of nonverbal learning disorder. Can J School Psychol. 2012;27:35–57.
38. Sullivan RM, Gratton A. Prefrontal cortical regulation of hypothalamic-pituitary-adrenal function in the rat and implications for psychopathology: side matters. Psychoneuroendocrinology. 2002;27:99–114.
39. Schore AN. Attachment and the regulation of the right brain. Attachment Hum Dev. 2000;2:23–47.
40. Teicher MH, Tomoda A, Andersen SL. Neurobiological consequences of early stress and childhood maltreatment: are results from human and animal studies comparable? Ann N Y Acad Sci. 2006;1071:313–23.
41. Giedd JN, Lalonde FM, Celano MJ, White SL, Wallace GL, Lee NR, et al. Anatomical brain magnetic resonance imaging of typically developing children. J Am Acad Child Adolesc Psychiatry. 2009;48:465–70.
42. Lenroot RK, Giedd JN. Sex differences in the adolescent brain. Brain Cogn. 2010;72:46–55.
43. Anderson CM, Teicher MH, Polcari A, Renshw PF. Abnormal T2 relaxation time in the cerebellar vermis of adults sexually abused in childhood: potential role of the vermis in stress-enhanced risk for drug abuse. Psychoneuroendocrinology. 2002;27:231–44.
44. Teicher MH, Sl A, Dumont NL, Ito Y, Glod CA, Vaituzis C, et al. Childhood neglect attenuates development of the corpus callosum. Soc Neurosci Abstr. 2000;26:549.
45. Sapolsky RM. Stress, glucocorticoids and damage to the nervous system: the current state of confusion. Stress. 1996;1:1–19.
46. McEwen BS, Gianaros PJ. Central role in the brain in stress and adaptation: links to socioeconomic status, health and disease. Ann N Y Acad Sci. 2010;1186:190–222.
47. Teicher MH, Samson JA. Annual research review: enduring neurobiological effects of childhood abuse and neglect. J Child Psychol Psychiatry. 2016;57:241–66.
48. Rinne-Albers MA, van der Wee NJ, Lamers-Winkelman F, Vermeiren RR. Neuroimaging in children, adolescent and young adults with psychological trauma. Eur Chil Adolesc Psychiatry. 2013;22:745–55.
49. Smith M. Bilateral hippocampal volume reduction in adults with post traumatic stress disorder: a meta-analysis of structural MRI studies. Hippocampus. 2005;15:798–807.
50. Andersen SL, Teicher MH. Delayed effects of early stress on hippocampal development. Neuropsychopharmacology. 2004;29:1988–93.
51. Andersen SL, Teicher MH. Stress, sensitive periods and maturational events in adolescent depression. Trends Neurosci. 2008;31:183–91.
52. Pechtel P, Lyons-Ruth K, Anderson CM, Teicher MH. Sensitive periods of amygdala development: the role of maltreatment in preadolescence. NeuroImage. 2014;97:236–44.
53. Gilbertson MW, Shenton ME, Ciszewski A, Kasai K, Lasko NB, Orr SP, et al. Smaller hippocampal volume predicts pathologic vulnerability to psychological trauma. Nat Neurosci. 2002;5:1242–7.
54. LeDoux JE. Emotional memory systems in the brain. Behav Brain Res. 1993;58:69–79.
55. Derntl B, Habel U, Windischberber C, Robinson S, Kryspin-Exner I, Gur RC, et al. General and specific responsiveness of the amygdala during explicit emotion recognition in females and males. BMC Neurosci. 2009;10:91.

56. Sarrieau A, Dussaillant M, Agid F, Philibert D, Agid Y, Rostene W. Autoradiographic localization of glucocorticosteroid and progesterone binding sites in the human post-mortem brain. J Steroid Biochem. 1986;25:717–21.
57. Uematsu A, Matsui M, Tanaka C, Takahashi T, Noguchi K, Suzuki M, et al. Developmental trajectories of amygdala and hippocampus from infancy to early adulthood in healthy individuals. PLoS One. 2012;7(10):e46970.
58. Mitra R, Jadhav S, McEwen BS, Vyas A, Chattarji S. Stress duration modulates the spatiotemporal patterns of spine formation in the basolateral amygdala. Proc Natl Acad Sci U S A. 2005;102:9371–6.
59. Mehta MA, Golembo NI, Nosarti C, Colvert E, Mota A, Williams SC, et al. Amygdala, hippocampal and corpus callosum size following severe early institutional deprivation: the English and Romanian adoptees study pilot. J Child Psychol Psychiatry. 2009;50:943–51.
60. Tottenham N, Hare TA, Quinn BT, McCarry TW, Nurse M, Gilhooly T, et al. Prolonged institutional rearing is associated with atypically large amygdala volume and difficulties in emotion regulation. Dev Sci. 2010;13:46–61.
61. Lupien SJ, Parent S, Evans AC, Tremblay RE, Zelazo PD, Corbo V, et al. Larger amygdala but no change in hippocampal volume in 10-year-old children exposed to maternal depressive symptomatology since birth. Proc Natl Acad Sci U S A. 2011;108:14324–9.
62. Driessen M, Herrmann J, Stahl K, Zwaan M, Meier S, Hill A, et al. Magnetic resonance imaging volumes of the hippocampus and the amygdala in women with borderline personality disorder and early traumatization. Arch Gen Psychiatry. 2000;57:1115–22.
63. Schmahl CG, Vermetten E, Elzinga BM, Douglas BJ. Magnetic resonance imaging of hippocampal and amygdala volume in women with childhood abuse and borderline personality disorder. Psychiatry Res. 2003;122:193–8.
64. Vermetten E, Schmahl C, Lindner S, Loewenstein RJ, Bremner JD. Hippocampal and amygdalar volumes in dissociative identity disorder. Am J Psychiatry. 2006;163:630–6.
65. Kuo JR, Kaloupek DG, Woodward SH. Amygdala volume in combat-exposed veterans with and without posttraumatic stress disorder: a cross-sectional study. Arch Gen Psychiatry. 2012;69:1080–6.
66. Whittle S, Dennison M, Vijayakumar N, Simmons JG, Yucel M, Lubman DI, et al. Childhood maltreatment and psychopathology affect brain development during adolescence. J Am Acad Child Adolesc Psychiatry. 2013;52:940–52.
67. Fuster JM. Frontal lobe and cognitive development. J Neurocytol. 2002;31:373–85.
68. Carballedo A, Morris D, Zill P, Fahey C, Reinhold E, Meisenzahl E, et al. Brain-derived neurotrophic factor Val66Met polymorphism and early life adversity affect hippocampal volume. Neuropsychiatr Genet. 2013;162B:183–90.
69. De Bellis MD, Keshavan MS, Clark DB, Casey BJ, Giedd JN, Boring AM, et al. Developmental traumatology. Part II: Brain Development Biol Psychiatry. 1999;45:1271–84.
70. De Bellis MD, Keshavan MS, Shifflett H, Iyengar S, Beers SR, Hall J, et al. Brain structures in pediatric maltreatment-related posttraumatic stress disorder: a sociodemographically matched study. Biol Psychiatry. 2002;52:1066–78.
71. Koob GF, Volkow ND. Neurocircuitry of addiction. Neuropsychopharmacology. 2010;35:217–38.
72. Edmiston EE, Wang F, Mazure CM, Guiney J, Sinha R, Mayes LC, et al. Corticostriatal-lymbic gray matter morphology in adolescents with self-reported exposure to childhood maltreatment. Arch Pediatr Adolesc Med. 2011;165:1069–77.
73. Daniels JK, Lamke JP, Gaebler M, Walter H, Scheel M. White matter integrity and its relationship to PTSD and childhood trauma: a systematic review and meta-analysis. Depress Anxiety. 2013;30(3):207–16.
74. Choi J, Jeong B, Rohan ML, Polcari AM, Teicher MH. Preliminary evidence for white matter tract abnormalities in young adults exposed to parental verbal abuse. Biol Psychiatry. 2009;65:227–34.
75. Tomoda A, Sheu YS, Rabi K, Suzuki H, Navalta CP, Polcari A, et al. Exposure to parental verbal abuse is associated with increased gray matter volume in superior temporal gyrus. NeuroImage. 2011;54(Suppl 1):S280–6.

76. Choi J, Jeong B, Polcari AM, Rohan ML, Teicher MH. Reduced fractional anisotropy in the visual limbic pathway of young adults witnessing domestic violence in childhood. NeuroImage. 2012;59:1071–9.
77. Tomoda A, Polcari A, Anderson CM, Teicher MH. Reduced visual cortex gray matter volume and thickness in young adults who witnessed domestic violence during childhood. PLoS One. 2012;7(12):e52528.
78. Heim CM, Mayberg HS, Mletzko T, Nemeroff CB, Pruessner JC. Decreased cortical representation of genital somatosensory field after childhood sexual abuse. Am J Psychiatry. 2013;170:616–23.
79. De Bellis MD, Keshavan MS. Sex differences in brain maturation in maltreatment-related pediatric posttraumatic stress disorder. Neurosci Biobehav Rev. 2003;27:103–17.
80. Teicher MH, Dumont NL, Ito Y, Vaituzis C, Giedd JN, Andersen SL. Childhood neglect is associated with reduced corpus callosum area. Biol Psychiatry. 2004;56:80–5.
81. Juraska JM, Kopcik JR. Sex and environmental influences on the size and ultrastructure of the rat corpus callosum. Brain Res. 1988;450:1–8.
82. Teicher MH, Parigger A. The "Maltreatment and Abuse Chronology of Exposure" (MACE) scale for the retrospective assessment of abuse and neglect during development. PLoS One. 2015;10(2):e0117423.
83. Jackowski AP, Douglas Palumberi H, Jackowski M, Win L, Schutz RT, Staib LW, et al. Corpus callosum in maltreated children with posttraumatic stress disorder: a diffusion tensor imaging study. Psychiatry Res. 2008;15:256–61.
84. Galinowski A, Miranda R, Lemaitre H, Paillere Martinot ML, Artiges E, Vulser H, et al. Resilience and corpus callosum microstructure in adolescence. Psychol Med. 2015;45:2285–94.
85. Eluvathingal TJ, Chugani HT, Behen ME, Juhasz C, Muzik O, Maqbool M, et al. Abnormal brain connectivity in children after early severe socioemotional deprivation: a diffusion tensor imaging study. Pediatrics. 2006;117:2093–100.
86. Govindan RM, Behen ME, Helder E, Makki MI, Chugani HT. Altered water diffusivity in cortical association tracts in children with early deprivation identified with Tract-Based Spatial Statistics (TBSS). Cereb Cortex. 2010;20:561–9.
87. Huang H, Gundapuneedi T, Rao U. White matter disruptions in adolescents exposed to childhood maltreatment and vulnerability to psychopathology. Neuropsychopharmacology. 2012;37:2693–701.
88. Benedetti F, Bollettini I, Redaelli D, Poletti S, Locatelli C, Falini A. Adverse childhood experiences influence white matter microstructure in patients with bipolar disorder. Psychol Med. 2014;44:3069–82.
89. Sanchez MM, Young LJ, Plotsky PM, Insel TR. Distribution of corticosteroid receptors in the rhesus brain: relative absence of glucocorticoid receptors in the hippocampal formation. J Neurosci. 2000;20:4657–68.
90. Walton RM. Postnatal neurogenesis: of mice, men and macaques. Vet Pathol. 2012;49:155–65.
91. Baldacara L, Jackowski AP, Schoedl A, Pupo M, Andreoli SB, Mello MF, et al. Reduced cerebellar left hemisphere and vermal volume in adults with PTSD from a community sample. J Psichiatry Res. 2011;45:1627–33.
92. Carrion VG, Weems CF, Eliez S, Patwardhan A, Brown W, Ray RD, et al. Attenuation of frontal asymmetry in pediatric posttraumatic stress disorder. Biol Psychiatry. 2001;50:943–51.
93. Hanson JL, Chung MK, Avants BB, Shirtcliff EA, Gee JC, Davidson RJ, et al. Early stress is associated with alterations in the orbitofrontal cortex: a tensor-based morphometry investigation of brain structure and behavioural risk. J Neurosci. 2010;30:7466–72.
94. Walsh ND, Dalgleish T, Lombardo MV, Dunn VJ, Van Harmelen AL, Ban M, et al. General and specific effects of early-life psychosocial adversities on adolescent grey matter volume. Neuroimage Clin. 2014;4:308–18.
95. Bauer PM, Hanson JL, Pierson RK, Davidson RJ, Pollak SD. Cerebellar volume and cognitive functioning in children who experienced early deprivation. Biol Psychiatry. 2009;66:1100–6.
96. De Bellis MD, Kuchibhatla M. Cerebellar volumes in pediatric maltreatment-related posttraumatic stress disorder. Biol Psychiatry. 2006;60:697–703.

97. Shin LM, Liberzon I. The neurocircuitry of fear, stress and anxiety disorders. Neuropsychopharmacology. 2010;35:169–91.
98. Phelps EA, LeDoux JE. Contributions of amygdala to emotion processing: from animal models to human behavior. Neuron. 2005;48:175–87.
99. Whalen PJ, Rauch SL, Etcoff NL, McInerney SC, Lee MB, Jenike MA. Masked presentation of emotional facial expressions modulate amygdala activity without explicit knowledge. J Neurosci. 1998;18:411–8.
100. Pessoa L, Adolphs R. Emotion processing and the amygdala: from a "low road" to "many roads" of evaluating biological significance. Nat Rev Neurosci. 2010;11:773–83.
101. Rogosch FA, Dackis MN, Cicchetti D. Child mal treatment and allostatic load: consequences for physical and mental health in children from low-income families. Dev Psychopathol. 2011;23:1107–24.
102. Suslow T, Ohrmann P, Bauer J, Rauch AV, Schwindt W, Arolt V, et al. Amygdala activation during masked presentation of emotional faces predicts conscious detection of threat-related faces. Brain Cogn. 2006;61:243–8.
103. Goff B, Gee DG, Telzer EH, Humphreys KL, Gabard-Durnam L, Flannery J, et al. Reduced nucleus accumbens reactivity and adolescent depression following early-life stress. Neurosci. 2013;249:129–38.
104. Maheu FS, Dozier M, Guyer AE, Mandell D, Peloso E, Poeth K, et al. A preliminary study of medial temporal lobe function in youths with a history of caregiver deprivation and emotional neglect. Cogn Affect Behav Neurosci. 2010;10:34–49.
105. Tottenham N, Hare TA, Millner A, Gilhooly T, Zevin JD, Casey BJ. Elevated amygdala response to faces following early deprivation. Dev Sci. 2011;14:190–204.
106. McCrory EJ, De Brito SA, Sebastian CL, Mechelli A, Bird G, Kelly PA, et al. Heightened neural reactivity to threat in child victims of family violence. Curr Biol. 2011;21:947–8.
107. McCrory EJ, De Brito SA, Kelly PA, Bird G, Sebastian CL, Mechelli A, et al. Amygdala activation in maltreated children during pre-attentive emotional processing. Br J Psychiatry. 2013;202:269–76.
108. White MG, Bogdan R, Fisher PM, Munoz KE, Williamson DE, Hariri AR. FKBP5 and emotional neglect interact to predict individual differences in amygdala reactivity. Genes Brain Behav. 2012;11:869–78.
109. Swartz JR, Knodt AR, Radtke SR, Hariri AR. A neural biomarker of psychological vulnerability to future life stress. Neuron. 2015;85:505–11.
110. McCrory EJ, Gerin MI, Viding E. Annual research review: childhood maltreatment, latent vulnerability and the shift to preventative psychiatry- the contribution of functional brain imaging. J Child Psychol Psychiatry. 2017;58:338–57.
111. Hart H, Lim L, Mehta MA, Simmons A, Mirza KAH, Rubia K. Altered fear processing in adolescents with history of severe childhood maltreatment: an fMRI study. Psychol Med. 2018;48:1092–101.
112. Phelps EA, O'Connor KJ, Gatenby JC, Gore JC, Grillon C, Davis M. Activation of the left amygdala to a cognitive representation of fear. Nat Neurosci. 2001;4:437–41.
113. Lupien SJ, McEwen BS, Gunnar MR, Heim C. Effects of stress throughout the lifespan on the brain, behavior and cognition. Nat Rev Neurosci. 2009;10:434–45.
114. McLaughlin KA, Peverill M, Gold AL, Alves S, Sheridan MA. Child maltreatment and neural systems underlying emotion regulation. J Am Acad Child Adolesc Psychiatry. 2015;54:753–62.
115. Berridge KC, Robinson TE, Aldridge JW. Dissecting components of reward: "liking", "wanting", and learning. Curr Opin Pharmacol. 2009;9:65–73.
116. Luking KR, Pagliaccio D, Luby JL, Barch DM. Reward processing and risk for depression across development. Trends Cogn Sci. 2016;20:456–68.
117. Mehta MA, Gore-Langton E, Golembo N, Colvert E, Williams SCR, Sonuga-Barke E. Hyporesponsive reward anticipation in the basal ganglia following severe institutional deprivation early in life. J Cogn Neurosci. 2010;22:2316–25.

118. Hanson JL, Hariri AR, Williamson DE. Blunted ventral striatum development in adolescent reflects emotional neglect and predicts depressive symptoms. Biol Psychol. 2015;78:598–605.
119. Boecker R, Holz NE, Buchmann AF, Blomeyer D, Plichta MM, Wolf I, et al. Impact of early life adversity on reward processing in young adults: EEG-fMRI results from a prospective study over 25 years. PLoS One. 2014;9(8):e104185.
120. Dillon DG, Holmes AJ, Birk JL, Brooks N, Lyons-Ruth K, Pizzagalli DA. Childhood adversity is associated with left basal ganglia dysfunction during reward anticipation in adulthood. Biol Psychiatry. 2009;66:206–13.
121. Pizzagalli DA, Holmes AJ, Dillon DG, Goetz EL, Birk JL, Bogdan R, et al. Reduced caudate and nucleus accumbens response to rewards in unmedicated individuals with major depressive disorder. Am J Psychiatry. 2009;166:702–10.
122. Redlich R, Grotegerd D, Opel N, Kaufmann C, Zwitserlood P, Kugel H, et al. Are you gonna leave me? Separation anxiety is associated with increased amygdala responsiveness and volume. Soc Cogn Affect Neurosci. 2015;10:278–84.
123. Balodis IM, Potenza MN. Anticipatory reward processing in addicted populations: a focus on the monetary incentive delay task. Biol Psychiatry. 2015;77:434–44.
124. Dennison MJ, Sheridan MA, Busso DS, Jenness JL, Everill M, Rosen ML, et al. Neurobehavioural markers of resilience to depression amongst adolescents exposed to child abuse. J Abnor Psychol. 2016;125:1201–12.
125. Mannella F, Gurney K, Baldassarre G. The nucleus accumbens as a nexus between values and goals in goal-directed behavior: a review and a new hypothesis. Front Behav Neurosci. 2013;7:135.
126. Williams LE, Bargh JA, Nocera CC, Gray JR. The unconscious regulation of emotion: nonconscious reappraisal goals modulate emotional reactivity. Emotion. 2009;9:847–54.
127. Koenisberg HW, Fan J, Ochsner KN, Liu X, Guise K, Pizzarello S, et al. Neural correlates of using distancing to regulate emotional responses to social situations. Neuropsychol. 2010;48:1813–22.
128. Ochsner KN, Silvers JA, Buhle JT. Functional imaging studies of emotion regulation: a synthetic review and evolving model of the cognitive control emotion. Ann N Y Acad Sci. 2012;1251:E1–24.
129. Gee DG, Gabard-Durnam LJ, Flannery J, Goff B, Humphreys KL, Telzer EH. Early developmental emergence of human amygdala-prefrontal connectivity after maternal deprivation. Proc Natl Acad Sci U S A. 2013;110:15638–43.
130. Lee SW, Yoo JH, Kim KW, Lee JS, Kim D, Park H, et al. Aberrant function of frontoamygdala circuits in adolescents with previous verbal abuse experiences. Neuropsychol. 2015;79:76–85.
131. Marusak HA, Etkin A, Thomason ME. Disrupted insula-based neural circuit organization and conflict interference in trauma-exposed youth. Neuroimage Clin. 2015;8:516–25.
132. Elsey J, Coates A, Lacadie CM, McCrory EJ, Sinha R, Mayes LC, et al. Childhood trauma and neural responses to personalized stress, favorite-food and neutral relaxing cues in adolescent. Neuropsychopharmacology. 2015;40:1580–9.
133. Puetz VB, Viding E, Palmer A, Kelly PA, Lickley R, Koutoufa I, et al. Altered neural response to rejection-related words in children exposed to maltreatment. J Child Psychol Psychiatry. 2016;57:1165–73.
134. Puetz VB, Kohn N, Dahmen B, Zvyagintsev M, Schuppen A, Schultz RT, et al. Neural response to social rejection in children with early separation experiences. J Am Acad Child Adolesc Psychiatry. 2014;53:1328–37.
135. Miyake A, Friedman NP, Emerson MJ, Witzki AH, Howerter A, Wager TD. The unity and diversity of executive functions and their contributions to complex "frontal lobe" task: a latent variable analysis. Cogn Psychol. 2000;41:49–100.
136. Mueller SC, Maheu FS, Dozier M, Peloso E, Mandell D, Leibenluft E, et al. Early life stress is associated with impairment in cognitive control in adolescence: an fMRI study. Neuropsychol. 2010;48:3037–44.

137. Lim L, Hart H, Mehta MA, Simmons A, Mirza K, Rubia K. Neural correlates of error processing in young people with a history of severe childhood abuse: an fMRI study. Am J Psychiatry. 2015;172:892–900.
138. Snyder HR, Miyake A, Hankin BL. Advancing understanding of executive function impairments and psychopathology: bridging the gap between clinical and cognitive approaches. Front Psychol. 2015;6:328.
139. Fonagy P, Allison E. The role of mentalizing and epistemic trust in the therapeutic relationship. Psychotherapy. 2015;51:372–80.

# Perinatal Mental Health and Childhood Trauma

# 4

Jessica Trinetti, Rebecca Bind, Kristi Sawyer,
and Carmine M. Pariante

## 4.1 Introduction

Adverse childhood experiences are defined as a set of exposures to personal abuse, neglect and household dysfunction prior to 18 years of age [1]. There is substantial evidence showing that childhood maltreatment is associated with a higher risk of developing depression or displaying antisocial behaviour in adulthood [2]. For example, retrospective reporting of childhood maltreatment (psychological abuse, sexual abuse, physical abuse and neglect) has been shown to be associated with a significant increase in lifetime depression for both men and women.

Studies have revealed that exposure to maternal depression in utero increases the risk of exposure to childhood maltreatment [2], and also the risk of offspring psychopathology in later life [3]. Furthermore, maternal experiences of childhood maltreatment are also associated with offspring exposure to maltreatment [4] and offspring psychopathology [5].

In this chapter, we will highlight the clinical evidence related to the intergenerational transmission of exposure to stress, and in particular, of exposure to childhood maltreatment. We will focus in particular on the mechanism by which exposure to stress and maltreatment in the childhood of women translates into exposure to stress and maltreatment in their offspring, and how pregnancy and the in utero environment

J. Trinetti (✉) · R. Bind · K. Sawyer
Department of Psychological Medicine, Institute of Psychiatry, Psychology and Neuroscience, King's College London, London, UK
e-mail: jessica.trinetti@kcl.ac.uk; Jessicatrinetti25@gmail.com; rebecca.bind@kcl.ac.uk; kristi.sawyer@kcl.ac.uk

C. M. Pariante
National Institute for Health and Research Biomedical Research Centre at South London, London, UK

Maudsley NHS Foundation Trust and King's College London, London, UK
e-mail: carmine.pariante@kcl.ac.uk



G. Spalletta et al. (eds.), *Childhood Trauma in Mental Disorders*,
https://doi.org/10.1007/978-3-030-49414-8_4

are a crucial time and biological setting where these mechanisms may operate. In particular, we will focus on the evidence that the experiences of childhood maltreatment in mothers induce persistent behavioural and biological changes in the regulation of the maternal stress response. This in turn, alters the biology of the in utero environment during pregnancy, inducing further changes in both mothers and offspring, which may contribute to the transmission of stress exposure. Indeed, mothers will be less capable in their interactions with offspring, and may be predisposed to fewer vigilant and protective behaviours. As a result, the offspring will show an altered stress response and a disturbed behavioural trajectory [6].

## 4.2 The Transmission of Exposure to Maltreatment from Mother to Child

Recent studies have shown that women who experience childhood stress tend to have children who also experience stress. Furthermore, a history of childhood maltreatment exposes the mother to a greater risk of mental disorders during the perinatal period, thus increasing the risk of mental disorders in her offspring. A study published in 2011, using 120 women recruited into the South London Child Development Study [2], explored how antenatal depression can influence offspring psychopathology, focusing in particular on the role played by childhood maltreatment. The study looked for (1) an association between antenatal depression and offspring childhood maltreatment, and (2) how this maltreatment in the offspring exacerbates the effect of mother antenatal depression on offspring psychopathology in adolescence.

The study found that a child's exposure to maternal antenatal depression and childhood maltreatment combined (but not independently) led to a 12-fold increased risk of that child developing psychopathology, most commonly either conduct or depressive disorder.

In addition, neither postnatal depression nor maternal depression at any other time during the child's first 11 years of life was associated with the child's experience of abuse. This study was the first to demonstrate that antenatal depression is associated with an increased risk of the offspring being exposed to childhood maltreatment.

Surely, mothers with a history of abuse have an increased risk of having depression throughout their life, especially in the perinatal period. These data are associated both with a reduced ability of the mother to take care of her child and to watch over him/her, exposing it to an increased risk of maltreatment, especially in the family, and to the development of an altered model of attachment that could increase this risk also in the social environment [7].

However, the mechanisms underlying transmission of trauma are less clear.

The most relevant studies on the subject are summarised in Table 4.1.

**Table 4.1** A summary of major studies investigating the association between maternal childhood maltreatment (MCM) and the risk of maltreatment in the offspring

| Author | State | Design setting | Sample | Maternal child maltreatment | Child psychopathology | Relevant findings |
|---|---|---|---|---|---|---|
| Plant (2017) [6] | United Kingdom | Cohort, community | 14,451 pregnant women followed at 12 and 32 weeks of gestation to child age 2 years 9 months | Physical abuse (7.1%), sexual abuse (6.2%), emotional abuse (7.5%) and neglect (21.9%); self-report questionnaire | Maternal ratings of child DSM-IV depressive symptoms, ($M$ = 0.3, SD = 1.0, $n$ = 6.207 at 10 years, $M$ = 0.3, SD = 1.0, $n$ = 5.591 at 13 years), ADHD, conduct disorders, ODD were summed at 10 ($M$ = 5.6, SD = 8.4, n = 6.599) and 13 years ($M$ = 5.1, SD = 8.1, $n$ = 6.057); internalising and externalising at 11 years; SDQ | (+) MCM directly and indirectly associated with internalising ($\beta$ = 0.014) and externalising problems ($\beta$ = 0.012). Increased risk of the children being exposed to maltreatment and developing psychopathology for mothers with depression during antenatal period ($\beta$ = 0.001) |
| Myhre (2014) [50] | Norway | Cohort, community | 24,452 pregnant women followed to child age 3 years | Emotional abuse (8%), physical abuse, sexual abuse (9% physical/sexual, 18% any); questions based on the NorAQ | Internalising and externalising problems at 36 months; SDQ | (+) MCM associated with externalising behaviour even after adjusting for maternal age, education, single motherhood, gender and maternal adult abuse. |

(continued)

**Table 4.1** (continued)

| Author | State | Design setting | Sample | Maternal child maltreatment | Child psychopathology | Relevant findings |
|---|---|---|---|---|---|---|
| Rijlaarsdam (2014) [32] | Netherlands | Cohort, community | 4,438 pregnant women followed to child age 6 years | Physical abuse, emotional abuse, sexual abuse, emotional neglect, physical neglect (6% physical abuse/ neglect, 8% emotional abuse/ neglect, 6% sexual abuse); CTQ | Internalising (11%) and externalising (7%) problems at 6 years; CBCL (parental reports) and BPI (child reports) | (+) MCM associated with internalising problem and BPI externalising, and predicted CBCL internalising, adjusted for maternal age, ethnicity, marital status, education, parity, family income, child prior total problems, age and gender ($\beta = 0.4$) |
| Min (2012) [51] | United States | Cohort, community | 231 mother–child dyads identified as high risk for pregnancy drug use followed from childbirth to age 9 years | Physical abuse, emotional abuse, sexual abuse, emotional neglect, physical neglect (50% any); CTQ | Total problems (internalising, externalising, attention) and DSM-IV disorder symptoms (internalising, externalising, attention; $M = 3.9$, SD = 3.6, EXT; $M = 6.6$, SD = 4.3, ATT) at 9 years; CBCL (parental reports, problems), DI (child reports, DSM-IV symptoms) | (+) MCM predicted maternal-rated problems (latent factor: INT, EXT, ATT) adjusted for maternal age, education and ethnicity ($b = 15$). Effect partially mediated by maternal psychological distress. Indirect effect of MCM on DSM-IV disorder symptoms (latent factor: INT, EXT, ATT) through maternal low social support |
| Miranda (2013) [52] | Spain | Cross-sectional, clinic | 327 outpatient adolescents (8–17 years) and their mothers | Sexual abuse, physical abuse, emotional abuse (14% any); structured interview | Internalising (36%) and externalising (33%) problems at 8–17 years; CBCL (parental reports) | (+) MCM associated with externalising problems ($\beta = 0.11$). Effect fully mediated by maternal psychological distress, adjusted for child age and gender |

| | | | | | | |
|---|---|---|---|---|---|---|
| Plant (2013) [31] | United Kingdom | Cohort, community | 125 pregnant women followed to child age 16 years | Physical abuse, sexual abuse, emotional neglect, physical neglect (18% any 2+); semi-structured interview | DSM-IV DBDs ($M = 2.8$, SD = 4.3 11 years; $M = 2.7$, SD = 4.1, 16 years) and depression ($M = 0.6$, SD = 0.9, 11 years; $M = 1.2$, SD = 1.3, 16 years) symptoms at 11 and 16 years; CAPA (combined parental and child reports) | (+) MCM associated with DBDs symptoms (latent factor: 11, 16 years; $\beta = 0.3$) effect fully mediated by child maltreatment and moderated by maternal depression, adjusted for maternal age, education, child ethnicity and gender. |
| Miranda (2011) [12] | Spain | Cross-sectional, clinic | 547 outpatient adolescents (8–17 years) and their parents | Emotional abuse, sexual abuse, physical abuse (12% any); structured interview | Internalising, externalising problems and DSM-IV disorder diagnoses (DBD, mood, anxiety, eating, elimination disorders) at 8–17 years; CBCL (parental reports problems), DICA-IV (combined parental and child reports, DSM-IV diagnoses) | (+) MCM associated with externalising problems, eating disorders and total DSM-IV diagnoses. MCM predicted DBDs diagnoses, partner violence, child age, gender, comorbid DSM-IV diagnoses and physical punishment (OR = 1.9, 95% CI [1.0, 3.6]) |
| Lang (2010) [33] | United States | Cohort, community | 31 pregnant women follow to child age 1 year | Emotional abuse, physical abuse, sexual abuse; CTQ subscales | Affect at 1 year; IBQ-R (parental reports) | (#) Maternal physical abuse predicted negative affect ($\beta = 0.5$) but maternal emotional abuse negatively predicted negative affect ($\beta = -0.6$), adjusted for maternal postnatal depression and PTSD |

(continued)

**Table 4.1** (continued)

| Author | State | Design setting | Sample | Maternal child maltreatment | Child psychopathology | Relevant findings |
|---|---|---|---|---|---|---|
| Collishaw (2007) [11] | United Kingdom | Cohort, community | 5,619 pregnant women followed to child age 7 years | Sexual abuse (11%), physical abuse (3%), emotional abuse (8%; 17% any); questionnaire | Total problems (internalising, externalising) at 4 ($M = 8.8$, SD = 4.5) and 7 years ($M = 7.7$, SD = 4.9, parent; $M = 6.6$, SD = 6.1, teacher); SDQ (parental and teacher reports-7 years only) | (+) MCM associated with total problems at 4 and 7 (parent and teacher reports) and predicted poorer adjustment trajectories 4–7 years (parent reports). Effect fully mediated by maternal psychological distress, parenting hostility and life events. |
| Thompson (2007) [53] | United States | Cohort, community | 197 mother–child dyads identified as high risk for child maltreatment followed from <2 years to 4 years | Physical abuse (26%); semi-structured interview | Total problems (internalising, externalising) at 4 years ($M = 51.1$, SD = 9.3, total problems); CBCL (parental reports) | (+) Maternal physical abuse predicted total problems, adjusted for family income, maternal education, age, marital status, depression, alcohol abuse, child temperament, gender and ethnicity ($\beta = 0.2$). Effect partially mediated by maternal parental hostility. |
| Roberts (2004) [54] | United Kingdom | Cohort, community | 8,292 pregnant women followed to child age 4 years | Sexual abuse (4%); questionnaire | Total problems (internalising, externalising) at 4 years; SDQ (parental reports) | (+) Maternal sexual abuse predicted total problems, adjusted for maternal psychological distress ($\beta = 0.4$) and childhood cruelty ($\beta = 0.4$). Effect partially mediated by maternal psychological distress and parenting confidence |

| | | | | | | |
|---|---|---|---|---|---|---|
| Dubowitz (2001) [55] | United States | Cohort, community | 419 mothers–child dyads. Identified as high risk for child health problems and/or child maltreatment, followed from <2–5 years to 6–7 years | Physical abuse, sexual abuse; questionnaire | Internalising and externalising problems at 6–7 years; CBCL (parental reports) | (+) MCM associated with internalising problems, adjusted for maternal education, child age and recruitment site |

*ADHD* attention deficit hyperactivity disorders, *ATT* attention problems, *β* Beta coefficients, *BPI* Berkeley Puppet Interview, *CAPA* Child and Adolescent Psychiatric Assessment, *CBCL* Child Behavior Checklist, *CI* Confidence Intervals, *CTQ* Childhood Trauma Questionnaire, *DBDs* disruptive behavior disorders, *DI* Dominic Interactive, *DICA-IV* Diagnostic Interview for Children and Adolescents-IV, *DSM-IV* Diagnostic and Statistical Manual of Mental Illness, *EXT* externalising problems, *IBQ-R* Infant Behavior Questionnaire-Revised, *INT* internalising problems, *MCM* maternal childhood maltreatment , *NorAQ* NorVold Abuse Questionnaire, *ODD* oppositional defiant disorder, *OR* Odds Ratio, *PTSD* Post Traumatic Stress Disorder, *SD* standard deviation, *SDQ* Strengths and Difficulties Questionnaire

## 4.3 Mechanisms and Risk Factors for the Association Between Maternal and Offspring Maltreatment

It has been identified that the transmission of exposure to maltreatment between mother and child is more likely in cases which include maternal psychopathology, single parenthood, unemployment and social isolation [4, 8]. Indeed, children whose mothers have both depression and antisocial disorders are more likely to experience physical and emotional abuse, and exposure to domestic violence [9]. Moreover, in a study of twin pairs, it was found that if one twin receives more maternal positivity and warmth, he/she is less likely to show antisocial behaviour, thus implying that antisocial behaviour can be underpinned by maternal emotional attitudes [10]. Furthermore, there is a correlation between maternal childhood exposure to stress and offspring behavioural problems, which may render them more susceptible to exposure to violence. Indeed, Berlin et al. [8] suggest that this could be due to maternal exposure to maltreatment altering her cognitive processing and interpretation of hostile or aggressive situations, in a way that is passed onto her offspring, and might impact their subsequent exposure to maltreatment.

It has been shown in more than one study using a large sample of over 5,000 British mother–child dyads that children of mothers with a history of childhood abuse were more likely to develop emotional and behavioural adjustment problems in early childhood [11]. In a similar Spanish study, it is reported that these problems can persist into adolescence [12].

From our 2018 study, as shown in the figure below (see Fig. 4.1), it is evident that the history of maternal infantile maltreatment correlates with the development of

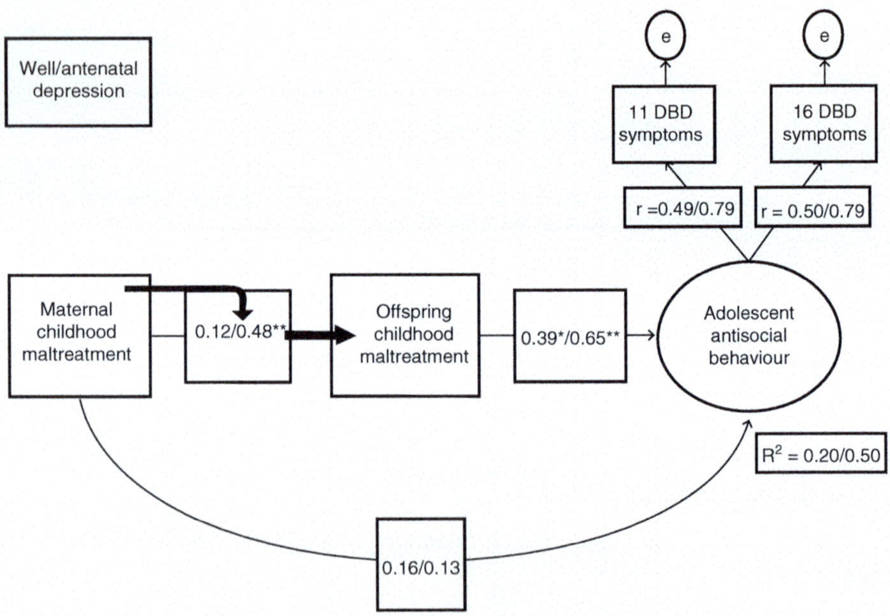


**Fig. 4.1** Illustrating mechanism of transmission of maternal maltreatment to child

depression during pregnancy. In turn, it is important to point out that depression in pregnancy is correlated with the increased risk of child maltreatment of the offspring, which in turn, increases the risk of developing behavioural disorders in adolescence (especially in the form of antisocial behaviour) [13].

## 4.4 Biological Perspective: Transmission of Biological Abnormalities from Mothers with Childhood Maltreatment to Offspring

Several studies show that childhood maltreatment increases a woman's vulnerability to develop disorders such as depression, post-traumatic stress disorder, substance dependence and immune and metabolic alterations [14–17]. What is less known, though, is whether these experiences of maternal abuse can directly influence the development of the offspring.

Certainly, intrauterine life represents a particularly delicate moment for the development of a new individual. It is therefore possible to hypothesise that the effects of maternal maltreatment can be transmitted to the offspring by means of biological mechanisms that modify foetal programming through the mother–placenta pathway [18]. In fact, several studies show that children born to mothers with a history of child maltreatment have a greater prevalence of behavioural problems, obesity, autism and poor overall heath [11, 19, 20]. Various hypotheses to explain biological mechanisms of transmission have been explored. For example, direct alterations in maternal gene expression have been identified, as a result of maltreatment suffered in childhood, that could be transmitted during foetal development [18].

It is known that mechanisms of methylation and acetylation of histone proteins that pack and order the DNA into structural units, induce modifications of gene expression. Animal studies have shown that infantile maltreatment induces epigenetic modifications in the germline (in particular, these changes happen by histone methylation [21–23]). Despite the gene remodelling that occurs between the male and female gamete, the maternal histone hypermethylation signature appears to still be transmitted to the offspring, making it more susceptible to the development of pathology during extrauterine life.

It has also been shown in animal studies that methylation of the NR3C1 gene, which encodes the glucocorticoid receptor, determines resistance to cortisone stimulation, leading to intrauterine alterations during gestation [24]. Another putative mechanism might be that maternal maltreatment can be transmitted to the offspring through modification of the cytoplasmic environment of the oocytes, although there is no empirical evidence to date [25].

In addition, we must also consider that the intrauterine environment in which the embryo develops may modify its development [26, 27]. Neuronal plasticity is usually elevated during intrauterine life and thus, any presence of stress can cause both structural and physiological alterations.

Studies show that mothers with a history of childhood maltreatment have alterations in the regulation of their hypothalamic-pituitary-adrenal (HPA) axis that can induce a proinflammatory state (due to alterations of the basal response to cortisol [28]), which can in turn alter the fetoplacental environment [29]. The increased cytokine production linked to this proinflammatory state can also lead to changes in the processes of myelination and neuronal cell proliferation, as well as in the production of neuronal growth factors. Such alterations may therefore cause a greater susceptibility to the development of different forms of mental disorder in the offspring [14, 30]. Although the prenatal environment appears to be a key player in the transmission of maternal trauma to the offspring, the effect of the postnatal environment cannot be ignored. In fact, it has been reported that women who were abused in childhood had alterations in the production of oxytocin in the postpartum period, which could alter their ability to look after, and care for their offspring [11, 31–33]. This could therefore lead to the development of a non-secure attachment style, an identified risk factor for the later development of psychopathology in the offspring [34]. Studies conducted on animal models have also shown that the high level of cortisol and the proinflammatory state of mothers with a history of infant maltreatment during pregnancy could lead to an alteration in the composition of breast milk. This alteration could contribute to the psychopathological deviation of the offspring, although this is a hypothesis still to be validated [35].

## 4.5 Psychosocial Point of View

As outlined above, scientific evidence clearly shows that mothers with a history of childhood maltreatment are at greater risk of developing depression and psychopathological disorders, with particular vulnerability in the peripartum period. It has been found that up to 58.3% of women with child abuse in the form of psychological abuse and up to 69.9% of women with a history of sexual abuse will be diagnosed with different forms of DSM-III disorders [36]. Such disorders inevitably affect the development of new mothers' children through biological transmission. In addition to biological mechanisms as to explain the risk of transmission of maltreatment from mother to child, there may also be psychological and social factors at play, too.

For example, according to Bowlby's attachment theory, the development of a child's secure attachment style is dependent on the caregiver's sensitive care and adequate response to needs. It is possible to evaluate the attachment style of a child starting from the age of one by observing his/her responses to being separated from the mother, and his/her subsequent reunion behaviour [37]. Children with a secure attachment style use their parents as a support for exploring the environment and appear stressed by the separation and relieved by the reunion [38]. On the other hand, children presenting with a non-secure attachment style (of the anxious-avoidant, ambivalent or disorganised type) do not use the caregiver as a secure base and show less adaptive ability both when separated and when reunited [38, 39]. A non-secure style of attachment determines the development of an altered model of representation of the internal self that affects the attachment capacities and the relationships of the future life [40].

It has been shown that the offspring attachment style most highly correlated with maternal history of maltreatment is a disorganised/disoriented attachment style [41]. In particular, it has been found that sexual abuse is more related to abandonment anxiety towards close relationships in both men and women, while psychological abuse correlates more with a style of attacking anxiety/avoidance especially in women [42]. There is evidence showing a different trajectory in a psychopathological sense between male and female with a history of child abuse [42]: males with a history of physical abuse seem to be more predisposed to the development of an avoidant attachment style, while women with a history of psychological maltreatment develop more separation anxiety. In a study conducted on a sample of 107 women with a history of child maltreatment, it emerged that childhood physical abuse, emotional abuse and emotional neglect are closely connected with the development of internalisation problems. Instead emotional abuse appears particularly correlated with externalising and other psychiatric disorders in general [43], in the form of depression, eating disorders, low self-esteem and sexual difficulties [44]. Studies have also brought to light how the history of child maltreatment predisposes one to being more at risk of experiencing community violence and the subsequent increased risk of developing post-traumatic stress disorder (PTSD) [45].

Additionally, another study conducted on 57 mother–child dyads showed that 83% of babies born to mothers with childhood abuse and neglect show a non-secure style of attachment at 20 months, and of these, 44% can be classified in a disorganised attachment style [46]. The study focused in particular, on the aspect of the mentalisation of maternal trauma. Mothers with childhood abuse and neglect would present a poor capacity to elaborate traumatic experiences that would predispose them to aggressiveness and poorly modulated emotional responses to offspring. They would also manifest a poor capacity for containing anguish [47] and mirroring [25], which is the cause of the development of the altered model of attachment and the psychopathological manifestations of children in adolescence and adulthood. Finally, Bifulco et al. [48] found that there was a correlation between a fearful attachment style and the development of social anxiety disorder as well as depression. Additionally, they found a correlation between an angry style of attachment and generalised anxiety disorder. They hypothesised that an angry style of attachment may have served as a protective factor against depression.

Taken together, the importance of intervention for mothers with a history of maltreatment is evident, not only for the health of the women but also as an early intervention and safeguard for the mental health of the offspring. Social interventions, trauma-focused cognitive behavioural therapy and joint sessions between mother and child were the best means of prevention of transmission found [49].

## 4.6 Conclusion

Findings suggest a need for the early identification of, and provision of support to, mothers with traumatic childhoods as a means to protect their own and their children's psychological well-being. Interventions for pregnant women with a history of maltreatment and/or depression could include high-quality social support, improved access to psychological therapies, as well as parenting programmes aimed at promoting sensitive caregiving practices. It is particularly important that vulnerable women are identified as early as possible, even before pregnancy, or if not, during pregnancy when they routinely come into contact with healthcare services, and that support and interventions are offered on an ongoing and regular basis going forward. At the same time, preventative actions such as psychotherapeutic support for both the children themselves who are born to maltreated mothers and their families where possible, could prevent the onset of psychiatric disorders.

In addition, it is also important that further resources are allocated for research to identify and better understand the mechanisms by which maternal exposure to childhood maltreatment might negatively impact her child's development.

## References

1. Letourneau N, Dewey D, Kaplan BJ, Ntanda H, Novick J, Thomas JC, et al. Intergenerational transmission of adverse childhood experiences via maternal depression and anxiety and moderation by child sex. J Dev Orig Health Dis. 2019;10(1):88–99.
2. Pawlby S, Hay D, Sharp D, Cerith SW, Pariante CM. Antenatal depression and offspring psychopathology: the influence of childhood maltreatment. Br J Psychiatry. 2011;199(2):106–12.
3. Seckl JR, Holmes MC. Mechanisms of disease: glucocorticoids, their placental metabolism and fetal 'programming' of adult pathophysiology. Nat Clin Pract Endocrinol Metab. 2007;3:479.
4. Sidebotham P, Heron J. Child maltreatment in the "children of the nineties": a cohort study of risk factors. Child Abuse Negl. 2006;30(5):497–522.
5. Collishaw S, Hammerton G, Mahedy L, Sellers R, Owen MJ, Craddock N, et al. Mental health resilience in the adolescent offspring of parents with depression: a prospective longitudinal study. Lancet Psychiatry. 2016;3(1):49–57.
6. Plant DT, Jones FW, Pariante CM, Pawlby S. Association between maternal childhood trauma and offspring childhood psychopathology: mediation analysis from the ALSPAC cohort. Br J Psychiatry. 2017;211(3):144–50.
7. Oliver JE. Successive generations of child maltreatment: social and medical disorders in the parents. Br J Psychiatry. 1985;147(5):484–90.
8. Berlin LJ, Appleyard K, Dodge KA. Mechanisms and implications for prevention. 2012;82(1):162–76.
9. Kim-Cohen J, Caspi A, Taylor A, Williams B, Newcombe R, Craig IW, et al. MAOA, maltreatment, and gene–environment interaction predicting children's mental health: new evidence and a meta-analysis. Mol Psychiatry. 2006;11:903.
10. Caspi A, Mofitti TE, Morgan J, Rutter M, Taylor A, Arsenault L, Tully L, Jacobs C, Kim-cohen J, Polo-Tomas M. Maternal expressed emotion predicts children's antisocial behavior problems: using monozygotic-twin differences to identify environmental effects on behavioral development. Dev Psychol. 2004;40(2):49–161.
11. Collishaw S, Dunn J, O'Connor TG, Golding J. Team Talsopacs. Maternal childhood abuse and offspring adjustment over time. Dev Psychopathol. 2007;19(02):367–83.
12. Miranda JK, de la Osa N, Granero R, Ezpeleta L. Maternal experiences of childhood abuse and intimate partner violence: psychopathology and functional impairment in clinical children and adolescents. Child Abuse Negl. 2011;35(9):700–11.

13. Choi KW, Houts R, Arseneault L, Pariante C, Sikkema KJ, Moffitt TE. Maternal depression in the intergenerational transmission of childhood maltreatment and its sequelae: testing postpartum effects in a longitudinal birth cohort. Dev Psychopathol. 2018;31(1):143–56.
14. Heim C, Binder EB. Current research trends in early life stress and depression: review of human studies on sensitive periods, gene–environment interactions, and epigenetics. Exp Neurol. 2012;233(1):102–11.
15. Anda RF, Felitti VJ, Bremner JD, Walker JD, Whitfield C, Perry BD, et al. The enduring effects of abuse and related adverse experiences in childhood. A convergence of evidence from neurobiology and epidemiology. Eur Arch Psychiatry Clin Neurosci. 2006;256(3):174–86.
16. MacMillan HL, Fleming JE, Streiner DL, Lin E, Boyle MH, Jamieson E, et al. Childhood abuse and lifetime psychopathology in a community sample. Am J Psychiatry. 2001;158(11):1878–83.
17. Heim C, Nemeroff CB. The role of childhood trauma in the neurobiology of mood and anxiety disorders: preclinical and clinical studies. Biol Psychiatry. 2001;49(12):1023–39. https://doi.org/10.1016/S0006-3223(01)01157-X.
18. Buss C, Entringer S, Moog NK, Toepfer P, Fair DA, Simhan HN, et al. Intergenerational transmission of maternal childhood maltreatment exposure: implications for fetal brain development. J Am Acad Child Adolesc Psychiatry. 2017;56(5):373–82.
19. Bouvette-Turcot A-A, Fleming AS, Wazana A, Sokolowski MB, Gaudreau H, Gonzalez A, et al. Maternal childhood adversity and child temperament: an association moderated by child 5-HTTLPR genotype. Genes Brain Behav. 2015;14(3):229–37. https://doi.org/10.1111/gbb.12205.
20. Myhre MC, Dyb GA, Wentzel-Larsen T, Grøgaard JB, Thoresen S. Maternal childhood abuse predicts externalizing behaviour in toddlers: a prospective cohort study. Scand J Public Health. 2013;42(3):263–9. https://doi.org/10.1177/1403494813510983.
21. Smith ZD, Chan MM, Mikkelsen TS, Gu H, Gnirke A, Regev A, et al. A unique regulatory phase of DNA methylation in the early mammalian embryo. Nature. 2012;484(7394):339–44.
22. Smallwood SA, Tomizawa S-I, Krueger F, Ruf N, Carli N, Segonds-Pichon A, et al. Dynamic CpG island methylation landscape in oocytes and preimplantation embryos. Nat Genet. 2011;43(8):811–4.
23. Sanchez-Delgado M, Court F, Vidal E, Medrano J, Monteagudo-Sánchez A, Martin-Trujillo A, et al. Human oocyte-derived methylation differences persist in the placenta revealing widespread transient imprinting. PLoS Genet. 2016;12(11):e1006427.
24. Palma-Gudiel H, Córdova-Palomera A, Eixarch E, Deuschle M, Fañanás L. Maternal psychosocial stress during pregnancy alters the epigenetic signature of the glucocorticoid receptor gene promoter in their offspring: a meta-analysis. Epigenetics. 2015;10(10):893–902.
25. Kovalchuk I. Transgenerational epigenetic inheritance in animals. Front Genet. 2012;3:76.
26. Gluckman PD, Hanson MA. Developmental origins of disease paradigm: a mechanistic and evolutionary perspective. Pediatr Res. 2004;56:311. https://doi.org/10.1203/01.PDR.0000135998.08025.FB.
27. Hanson M, Godfrey KM, Lillycrop KA, Burdge GC, Gluckman PD. Developmental plasticity and developmental origins of non-communicable disease: theoretical considerations and epigenetic mechanisms. Prog Biophys Mol Biol. 2011;106(1):272–80.
28. Heim C, Newport DJ, Bonsall R, Miller AH, Nemeroff CB. Altered pituitary-adrenal Axis responses to provocative challenge tests in adult survivors of childhood abuse. Am J Psychiatry. 2001;158(4):575–81.
29. Danese A, Pariante CM, Caspi A, Taylor A, Poulton R. Childhood maltreatment predicts adult inflammation in a life-course study. Proc Natl Acad Sci U S A. 2007;104(4):1319–24.
30. Heim C, Newport DJ, Mletzko T, Miller AH, Nemeroff CB. The link between childhood trauma and depression: Insights from HPA axis studies in humans. Psychoneuroendocrinology. 2008;33(6):693–710.
31. Plant DT, Barker ED, Waters CS, Pawlby S, Pariante CM. Intergenerational transmission of maltreatment and psychopathology: the role of antenatal depression. Psychol Med. 2013;43(3):519–28.

32. Rijlaarsdam J, Stevens GWJM, Jansen PW, Ringoot AP, Jaddoe VWV, Hofman A, et al. Maternal childhood maltreatment and offspring emotional and behavioral problems: maternal and paternal mechanisms of risk transmission. Child Maltreat. 2014;19(2):67–78. https://doi.org/10.1177/1077559514527639.
33. Lang AJ, Gartstein MA, Rodgers CS, Lebeck MM. The impact of maternal childhood abuse on parenting and infant temperament. J Child Adolesc Psychiatr Nurs. 2010;23(2):100–10. https://doi.org/10.1111/j.1744-6171.2010.00229.x.
34. Davis EP, Glynn LM, Schetter CD, Hobel C, Chicz-Demet A, Sandman CA. Prenatal exposure to maternal depression and cortisol influences infant temperament. J Am Acad Child Adolesc Psychiatry. 2007;46(6):737–46. https://doi.org/10.1097/chi.0b013e318047b775.
35. Hinde K, Skibiel AL, Foster AB, Del Rosso L, Mendoza SP, Capitanio JP. Cortisol in mother's milk across lactation reflects maternal life history and predicts infant temperament. Behav Ecol. 2015;26(1):269–81.
36. Silverman AB, Reinherz HZ, Giaconia RM. The long-term sequelae of child and adolescent abuse: a longitudinal community study. Child Abuse Negl. 1996;20(8):709–23.
37. Ainsworth MDS. An ethological approach to personality development. Am Psychol. 1991;46(April):333–41.
38. Main M, Solomon J. Discovery of an insicure-disorganized/disoriented attachment pattern. Affective development in infancy. Wetport, CT, US: Ablex publishing; 1986. p. 95–124.
39. Lamb ME. Patterns of attachment: a psychological study of the strange situation. Mary D. salter Ainsworth, Mary C. Blehar, Everett Waters, and Sally Wall. Hillsdale, N.J., Erlbaum, 1978, Halsted(Wiley), New York.
40. Thomas PM. Dissociation and internal models of protection: Psycoteraphy with child abuse survivors. Psycotherapy. 2005;42(1):20–36.
41. Lawrence AJ, Allen J, Carlson V, Cicchetti D. Child maltreatment. New York: Cambridge University Press; 1989. p. 579–619.
42. Godbout N, Sabourin S, Lussier Y. Child sexual abuse and adult romantic adjustment: comparison of single- and multiple-indicator measures. J Interpers Violence. 2008;24(4):693–705. https://doi.org/10.1177/0886260508317179.
43. Lowell A, Renk K, Adgate AH. The role of attachment in the relationship between child maltreatment and later emotional and behavioral functioning. Child Abus Negl. 2014;38(9):1436–49. https://doi.org/10.1016/j.chiabu.2014.02.006.
44. Mullen PE, Martin JL, Anderson JC, Romans SE, Herbison GP. The long-term impact of the physical, emotional, and sexual abuse of children: a community study. Child Abuse Negl. 1996;20(1):7–21.
45. London MJ, Lilly MM, Pittman L. Attachment as a mediator between community violence and posttraumatic stress symptoms among adolescents with a history of maltreatment. Child Abus Negl. 2015;42:1–9. https://doi.org/10.1016/j.chiabu.2014.11.002.
46. Ensink K, Normandin L, Target M, Fonagy P, Sabourin S, Berthelot N. Mentalization in children and mothers in the context of trauma: an initial study of the validity of the child reflective functioning scale. Br J Dev Psychol. 2015;33(2):203–17.
47. Bion WR. A theory of thinking. Int J Psycho-Analysis. 1962;43:308–15.
48. Bifulco A, Kwon J, Jacobs C, Moran PM, Bunn A, Beer N. Adult attachment style as mediator between childhood neglect/abuse and adult depression and anxiety. Soc Psychiatry Psychiatr Epidemiol. 2006;41(10):796–805.
49. Cohen J, Mannarino AP. Disseminating and implementing trauma-focused CBT in community settings. Trauma Violence Abus. 2008;9(4):214–26. https://doi.org/10.1177/1524838008324336.
50. Myhre MC, Dyb GA, Wentzel-Larsen T, Grøgaard JB, Thoresen S. Maternal childhood abuse predicts externalizing behaviour in toddlers: A prospective cohort study. Scand J Public Health. 2014;42(3):263–9.
51. Min MO, Singer LT, Minnes S, Kim H, Short E. Mediating Links Between Maternal Childhood Trauma and Preadolescent Behavioral Adjustment. Journal of Interpersonal Violence. 2012;28(4):831–51.

52. Miranda JK, de la Osa N, Granero R, Ezpeleta L. Maternal childhood abuse, intimate partner violence, and child psychopathology. Violence Against Women. 2013;19(1):50–68.
53. Thompson R. Mothers' violence victimization and child behavior problems: Examining the link. Am J Orthopsychiat. 2007;77(2):306–15.
54. Roberts R, O'Connor T, Dunn J, Golding J. The effects of child sexual abuse in later family life; mental health, parenting and adjustment of offspring. Child Abuse & Neglect. 2004;28(5):525–45.
55. Dubowitz H, Black MM, Kerr MA, Hussey JM, Morrel TM, Everson MD, Starr RH. Type and timing of mothers' victimization: effects on mothers and children. Pediatrics. 2001;107(4):728–35.

# Electroencephalography and Childhood Trauma

5

Alessio Simonetti

## 5.1 Introduction

The deleterious effects of childhood abuse are widely established. Exposure to one or more maltreatment-related childhood experiences accounts for 54% of the population attributable risk (PAR) for depression [1], 67% of the PAR for suicide attempts [1], and 64% of the PAR for illicit drugs dependence [2]. Moreover, exposure to one or more of such experiences is associated with up to a 17-fold increase in the risk for receiving a prescription of antidepressant, antipsychotic mood-stabilizing or anxiolytic medications [3]: subjects with past exposure to 6 or more traumatic childhood experiences were found to have a lifespan reduction of 20 years [4]. Therefore in the last decades many efforts have been made to understand the neurobiology underlying childhood maltreatment. To this extent literature regarding magnetic resonance imaging (MRI) techniques has provided consistent findings of altered networks subserving emotional control, arousal, inhibition, and attention in subjects exposed to traumatic childhood experiences [5]. However, even though MRI with high spatial resolution techniques can identify elements of large-scale distributed neural networks, their low temporal resolution impedes obtaining information on temporal dynamics of neural activity within these networks. Moreover, the high cost of such analyses further limits their application. Conversely, electroencephalography (EEG) is a relatively low cost, noninvasive technique with a temporal resolution in the order of milliseconds which can integrate findings from the aforementioned high spatial low temporal resolution techniques in order to give an exhaustive framework of the structural and functional alterations related to childhood trauma.

A. Simonetti (✉)
Menninger Department of Psychiatry and Behavioral Sciences, Baylor College of Medicine, Houston, TX, USA

Department of Neurology and Psychiatry, Sapienza University of Rome, Rome, Italy

Centro Lucio Bini, Rome, Italy
e-mail: alessio.simo@gmail.com



G. Spalletta et al. (eds.), *Childhood Trauma in Mental Disorders*,
https://doi.org/10.1007/978-3-030-49414-8_5

This chapter provides a review of the available studies on the electrophysiological characteristics of children with childhood trauma and adults with past exposure to maltreatment. In addition, this chapter provides an introduction to the basics of EEG. Since the field of neurophysiology is composed by a broad range of different techniques and analyses, their description will be limited to those which are useful to the comprehension of this chapter.

## 5.2 A Brief Introduction to the EEG Technique

EEG is a graphic representation of electrical activity produced by the brain over time, and more specifically, the difference in voltage between electrodes at two different locations in the brain [6]. Even though EEG techniques were most often applied to the evaluation of persons with suspected seizures or encephalopathies, digital recording methods and advanced software analyses broaden the application of such techniques to the evaluation and study of psychiatric conditions [7, 8].

### 5.2.1 Physical Bases of Electrophysiologic Activity

The electrical activity measured at the scalp by EEG is generated by cortical neurons underneath the scalp where the recording electrodes are placed; specifically, the scalp-recorded EEG reflects the synchronous electrical activity of groups of neurons that are oriented in parallel to one another and radially with respect of the scalp surface. The majority of EEG activity is generated by pyramidal neurons whose bodies are located mainly in the layer three and five of the cortex. The type of neuronal activity generating the scalp-recorded electrical signal is not constituted by action potentials, the duration of which are too short to be recorded. Instead, the sum of the inhibitory or excitatory postsynaptic potentials occurring within large groups of pyramidal cells creates the electrical signal recorded at scalp electrodes (see Fig. 5.1). Even though the brain produces activity at a wide range of frequencies, clinical electroencephalographers typically divide these frequencies in four major bands, beta, alpha, theta, and delta, from higher to lower frequencies (See Table 5.1). During waking state, the reticular formation activates the cortex and therefore neurons activate and fire independently at high frequencies. This activation is reflected in an asynchronous cortical activity on EEG as shown by fast (alpha and beta) activity. At sleep onset, reticular activity decreases and even neurons in the cortical areas start to be more synchronized and produce slower rhythms. Therefore, the EEG records slower rhythms such as delta and theta. It has been suggested that there are brain "pacemakers" that determine the frequency of these slower and more synchronous rhythms, and the thalamus may be one of these pacemakers [9].

Below, a detailed explanation of such bands is provided, together with the description of the gamma band, which was initially ignored and subsequently discovered:

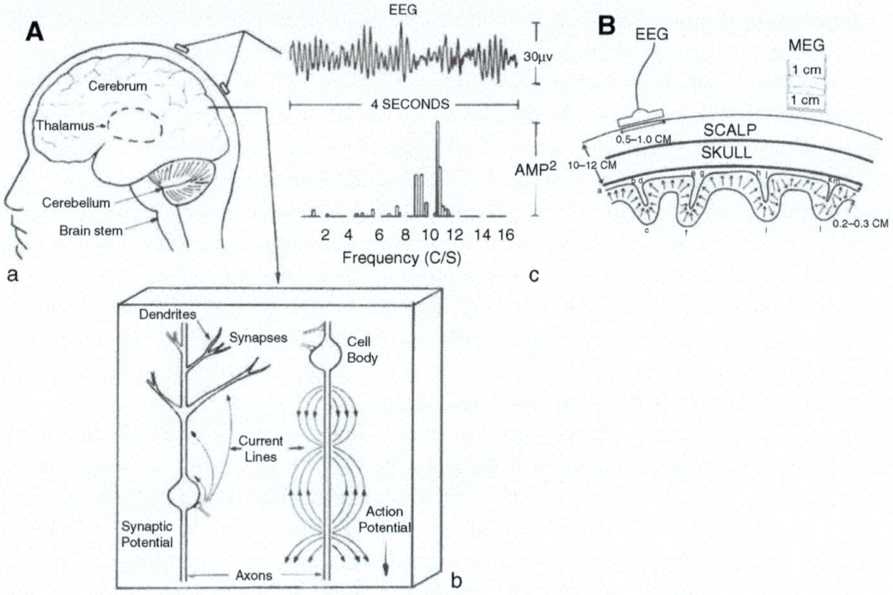


**Fig. 5.1** (**Aa**) The human brain. (**Ab**) Section of the cerebral cortex showing microcurrents. Sources due to synaptic and action potentials. (**Ac**) Each scalp EEG electrode records space averages over many square centimeters of cortical sources. A 4-s epoch of alpha rhythm and its corresponding power spectrum are shown. (**B**) Neocortical sources can be generally pictured as dipole layers (or "dipole sheets," in and out of cortical fissures and sulci) with strength varying as a function of cortical location. EEG is most sensitive to correlated dipole layer in gyri (regions ab, de, gh), less sensitive to correlated dipole layer in sulcus (region hi). And insensitive to opposing dipole layer in sulci (regions bcd, efg) and random layer (region ijldrn). This image was reprinted from [6]

**Table 5.1** Major EEG frequency bands and their characteristic location in conventional EEG

| Band | Frequency | Characteristic location in scalp recordings |
|---|---|---|
| Gamma | >30 Hz | Frontal |
| Beta | 13 to 30 Hz | Frontal (in awake records of adults) |
| Alpha | 8 to 13 Hz | Posterior (with eyes closed) |
| Theta | 4 to <8 Hz | Central |
| Delta | <4 Hz | Frontal/central (most prominent during deep sleep) |

*Alpha:* The alpha rhythm is the prominent EEG wave pattern of an adult who is awake, relaxed, and with eyes closed. Each region of the brain has a characteristic alpha rhythm, but alpha waves with the greatest amplitude are recorded at the occipital and parietal regions. In general, amplitude of alpha waves diminishes when subjects open their eyes and pay attention to external stimuli, although in some experimental settings (e.g., subjects trained in relaxation techniques) alpha amplitude can be also present in subjects with eyes open.

*Beta:* Beta rhythms occur in individuals who are alert and attentive to external stimuli or exert specific mental effort, or paradoxically beta rhythms also occur during deep sleep and REM (rapid eye movement) sleep. Their lower amplitude does not indicate that there is less electrical activity rather, that the "positive" and "negative" activities are starting to be counterbalanced so that the sum of the electrical activity is less. The beta wave has been associated with cortical arousal toward higher states of alertness or tension [10], but it may also be associated with "remembering." Beta activity can be divided into three subcomponents: low beta waves (12.5–16 Hz), beta waves (16.5–20 Hz), and high beta waves (20.5–28 Hz).

*Delta and Theta:* Delta and theta rhythms are low-frequency EEG patterns that increase during sleep in the normal adult. As people move from lighter to deeper sleep stages (prior to REM sleep), the occurrence of alpha waves diminishes and is gradually replaced by theta and then delta frequency rhythms. Although delta and theta rhythms are generally prominent during sleep, there are cases when such waves are also present during wakefulness. In fact, theta waves increase during presentation of images of threats [11] or negative arousing stimuli [12], behavioral paradigms designed to elicit perception of risk [13] or during priming involving negative words [14]. Studies using implanted electrodes in animal models also highlighted the involvement of theta-like rhythms, which are observable in the hippocampus, during fear memory retrieval [15–17]. Such "hippocampal theta" seems to be different from the band observed with classical noninvasive EEG recording; however, since theta range frequency content is strongly represented in temporal lobe EEG epileptiform activity [18], this suggests that some theta range cortical EEG activity may represent a limbic-allocortical resonant mode. Conversely, delta waves may increase during difficult mental activities requiring concentration (Figs. 5.2, and 5.3).

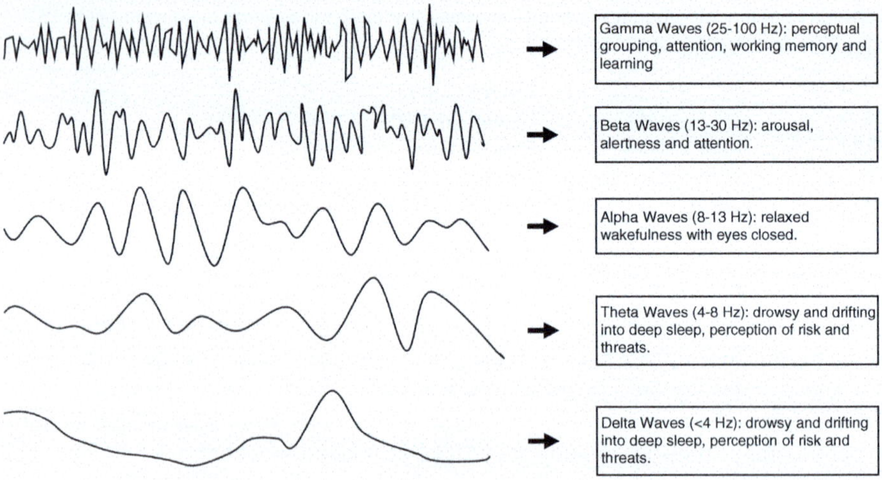


**Fig. 5.2** Morphology of the major EEG frequency bands

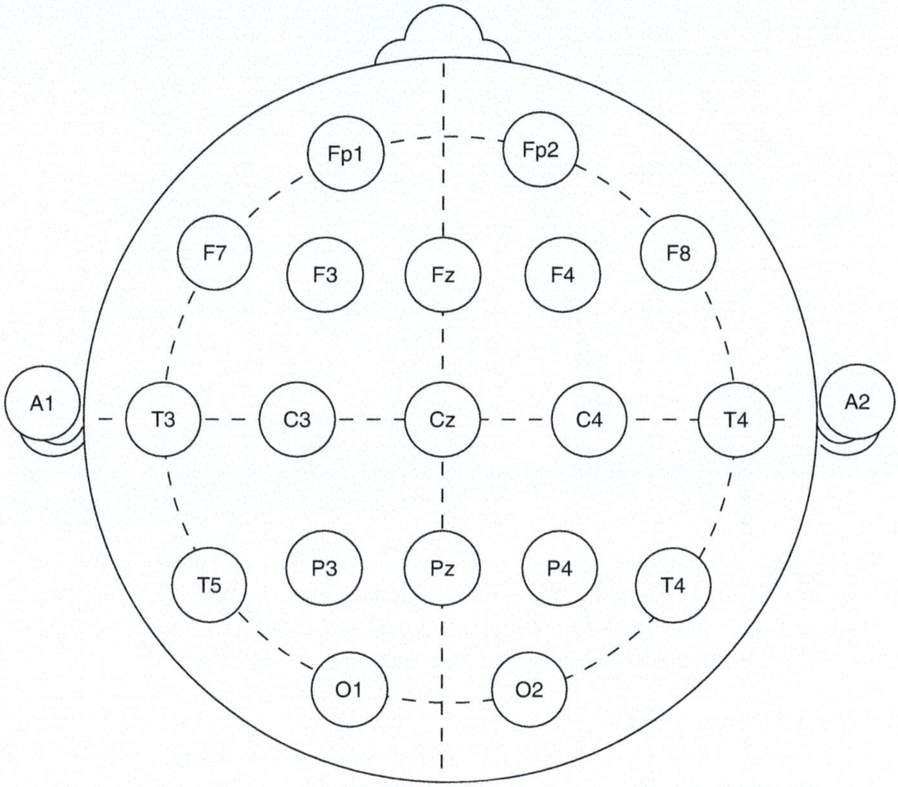


**Fig. 5.3** The 10–20 International System of Electrode Placement with labeled electrodes, including the A1 and A2 reference electrodes

*Gamma:* A gamma wave is a pattern of neural oscillation in humans with a frequency between 25 and 100 Hz [19], though 40 Hz is typical [20]. Gamma waves were initially ignored before the development of digital electroencephalography as analog EEG is restricted to recording and measuring rhythms that are usually less than 25 Hz [20]. It is related to a number of cortical areas, as well as subcortical structures in numerous species. Gamma activity is associated to a broad range of cognitive phenomena, including perceptual grouping [21] and attention [22], working memory [23] and learning [24]. In general, gamma may influence the communication between neuronal populations [25, 26] and act as a temporal reference frame so that neurons can encode stimulus orientation in "phase-of-firing" relative to gamma [27].

## 5.2.2 Types of EEG Recordings

### 5.2.2.1 Conventional EEG

The conventional or routine EEG is a 20–30 min recording of ongoing cerebral electric activity. Routine EEGs can offer useful diagnostic information about

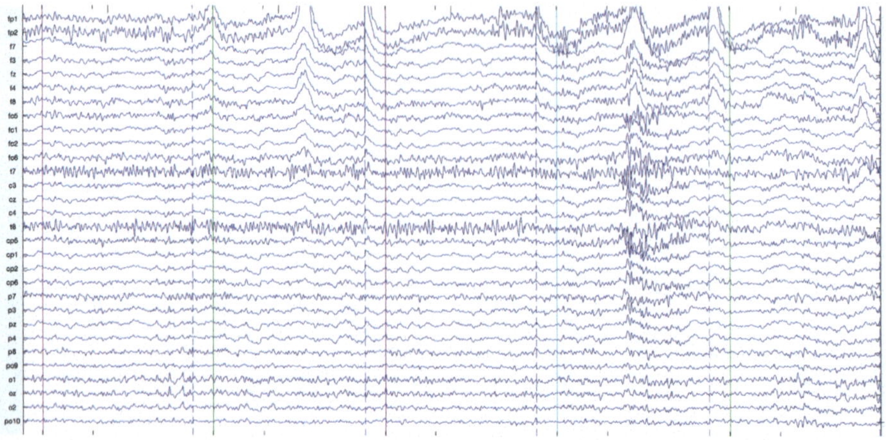

**Fig. 5.4** An example of the EEG recorded during wakefulness in a 15-year-old woman undergoing a passive-viewing task

epilepsy type, evaluate patients with delirium and dementia, and determine whether seizure activity may be causing an altered mental state. Each conventional EEG consist of three standardized conditions: relaxed wakefulness with eyes closed, a series of provocation maneuvers, and a quiet period for sleep (Fig. 5.4).

#### 5.2.2.2 Continuous EEG

Continuous EEG refers to the collection of ongoing cerebral electrical activity on a long-term continuous basis, ranging from hours up to days or weeks. Such technique can include concurrent video recordings of patient's behaviors. The primary advantage of such type of EEG is that it offers a much longer sampling of ongoing activity, and the possibility to link the brain's electrical activity to the observed behavior from the video recording.

#### 5.2.2.3 Quantitative EEG

Quantitative EEG refers to a secondary mathematical processing of EEG signals. Examples of quantitative EEG include spectral, brain asymmetry, and coherence analyses.

### 5.2.3 Types of EEG Analyses

#### 5.2.3.1 Spectral Analysis

Spectral analyses of the EEG signal are motivated by the fact that the observable frequency bands can be predominant in the EEG under certain conditions. Even though it is possible to read and analyze each of these frequency bands without computer assistance, i.e., simply by counting the cycles in the EEG and also trying to ignore the noise from the other frequency components, such methodology is not always easy to apply, mainly because the frequency content of the EEG is not invariably readily apparent in the time domain. In fact, the frequency content of the EEG is a mix of the various bands, making it difficult to determine by eye how to count the relevant cycles. To ease such procedures, each band can be isolated through computerized transformation to a frequency representation. Such transformation is usually accomplished using the Fourier transform (FT), or other variants, such as faster Fourier transform (FFT). Such analyses on multiple sensor locations often reveal clear spatial patterns which are indicative of different mental states (Table 5.1).

#### 5.2.3.2 Frontal Asymmetry

Frontal asymmetry is a measure evaluating hemispheric lateralization of functional brain activity. It is a measure of frontal alpha activity, thus reflecting the relative amount of EEG power in the alpha frequency band in the left and right frontal cortical regions. Frontal alpha asymmetry is typically obtained by subtracting the log-transformed left frontal alpha power from the log-transformed right frontal alpha power, with positive values indicating a relative left frontal asymmetry (i.e., decreased left frontal alpha power) and negative values indicating a relative right frontal asymmetry (i.e., decreased right frontal alpha power). Power in the alpha band is inversely related to activity; thus, a positive asymmetry score represents lower log-transformed power density on the left side as opposed to the right side of the brain. Hence, for the alpha band, positive asymmetry scores reflect greater left side activity. Recent empirical evidence revealed that measures of hemisphereic asymmetry offer a direct biologically based indicator of emotion regulation. In this context, high levels of left frontal activity are related to the experience and expression of positive states of approach-related emotions. Positive emotions and good emotion regulatory abilities have consistently been associated with resilient adaptation, which in turn, has been related to reduced environmental stress, thus conferring protection to the development of psychopathology [28]. Moreover, high left frontal activity is related to mood control since it has been considered to express the inhibitory processes of the prefrontal cortex on the amygdala for controlling the expression of conditioned responses [29, 30]. Conversely, high levels of right frontal activity are associated with the experience and expression of negative withdrawal-related emotions [31], which were associated with the development of depression and anxiety disorders [32–36].

#### 5.2.3.3 Coherence

Coherence analyses are essentially cross correlational analyses between the spectra of two neuroelectrical signals. The possible values of EEG coherence range from 0 to 1. Greater values (more phase synchrony of oscillations) indicate increased neuronal coupling between brain regions, and decreased values (less phase synchrony) indicate more cortical activity differentiation. Longitudinal changes in coherence probably reflect multiple neurophysiological mechanisms, including axonal sprouting, synaptogenesis, synaptic pruning, myelination, as well as changes in pre- and postsynaptic neurotransmitter dynamics [37]. Unfortunately, EEG coherence

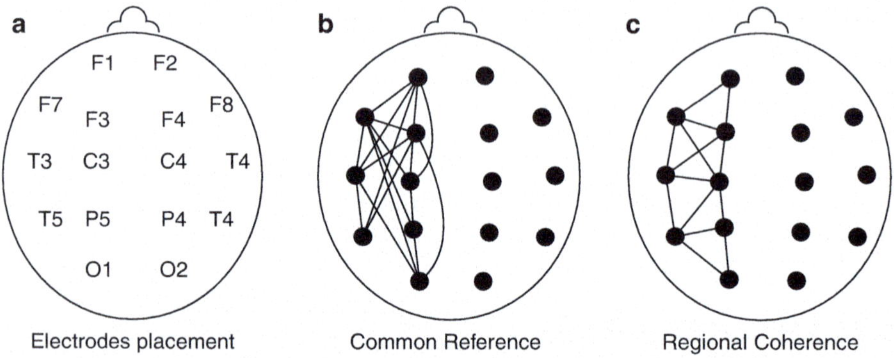


**Fig. 5.5** Representation of the electrode configurations and pairings used for calculation of alpha EEG coherence. (**a**) Alphanumeric designation for the 16 leads in the 10–20 international system. (**b**) The 28 pairings of each between each pairing (continuous line), and averaged to yield a global measure. (**c**) The coherence pairings used to calculate regional EEG coherence. EEG is collected for each lead (black circle), and local coherence is calculated by comparison with each adjacent lead (continuous line)

cannot discriminate between these distinct physical processes. Coherence can provide information regarding cortico-cortical connections within (intrahemispheric EEG coherence) and between (interhemispheric EEG coherence) cerebral hemispheres (see Fig. 5.5). Intrahemispheric cortico-cortical connections are subserved by white matter fibers that exhibit marked variations in length, whereas interhemispheric connections are subserved mostly, by corpus callosum fibers [38, 39]. High EEG coherence values are not necessarily reflective of greater maturation or optimal cortical organization. Thatcher and colleagues [40] have argued that decreased EEG coherence is associated with more complex cortical networks that support greater speed and efficiency of information processing.

#### 5.2.3.4 Event-Related Potentials (ERPs)

Event-related potentials (ERPs) are very small voltages generated in the brain structures in response to specific events or stimuli [41]. They are EEG changes that are time locked to sensory, motor, or cognitive events that provide a safe and noninvasive approach to study psychophysiological correlates of mental processes. ERPs can be elicited by a wide variety of sensory, cognitive, or motor events. They reflect the summed activity of postsynaptic potentials produced when a large number of similarly oriented cortical pyramidal neurons fire in synchrony while processing information (see Fig. 5.6). ERPs in humans can be divided into two categories: the early waves, i.e., components peaking within the first 100–200 ms expressing perceptual and attention orienting processes, and late components, which are related to the processing of the incoming information (see Fig. 5.7). In the following paragraph, description of the most relevant ERPs is provided.

*P100, N170, and N200*: The P100 is an early occipital component peaking around 100 ms, reflecting basic visual processing [42]. The P100 is thought to be

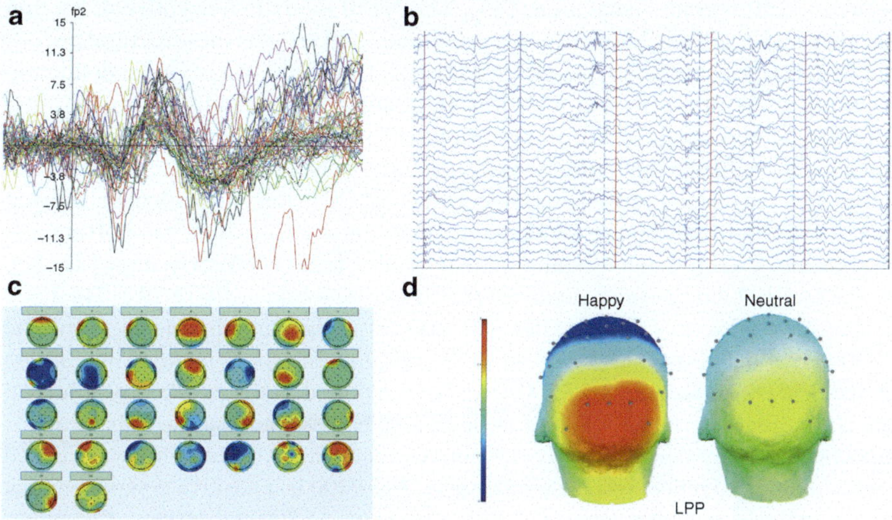


**Fig. 5.6** (**a**) ERPs at right frontopolar electrodes during emotional Go/NoGo task. Each color indicates ERPs for Go or NoGo conditions for each emotional expression. (**b**) Typical EEG during an emotional cognitive task. Color lines indicate the appearance on the screen of an emotional face. (**c**) Graphical representation of different EEG components. (**d**) Different scalp amplitude of late positive potential (LPP), averaged between 400 and 1000 ms, in subjects with pediatric bipolar disorder

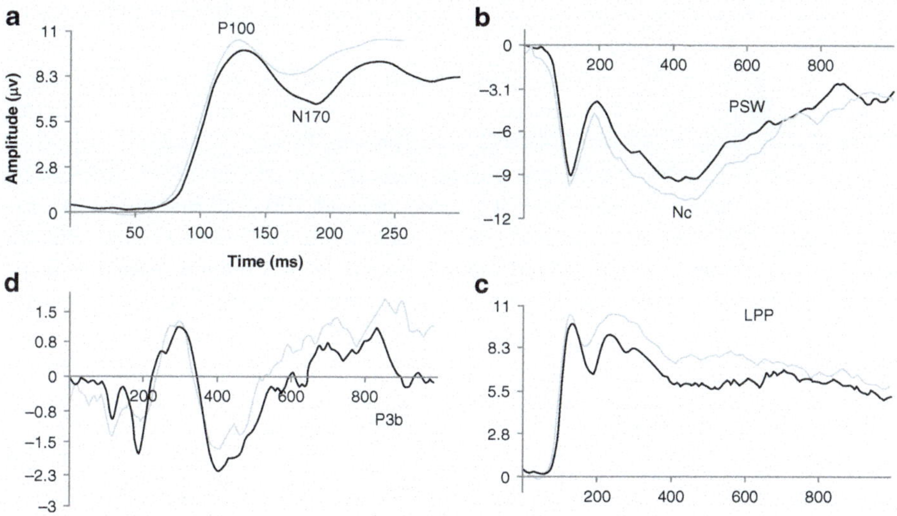


**Fig. 5.7** Event-related potentials (ERPs). (**a**) P100, N170 at occipital electrodes; (**b**) negative central (Nc), positive slow wave (PSW) at frontocentral electrodes, (**c**) P3b at parietal electrodes, and (**d**) late positive potential (LPP) at occipital electrodes

related to extrastriate visual areas [43]. The N170 refers to a negative deflection occurring 130–200 ms after stimulus presentation. It reflects the structural encoding of faces prior to face recognition [44–46]. The N200 is a negative deflection resulting from a deviation in the form, or context of a prevailing stimulus. It is sensitive to perceptual features, attention and novelty/mismatch [47], and conflict detection during the regulation of successful behavior [48]. The N200 modulation involves the anterior cingulate cortex (ACC). Recent evidences showed that these components are sensitive to emotional expression, supporting the hypothesis of a top-down modulation from limbic areas at an early stage [49, 50].

*P300, P400, late positive potential (LPP), positive slow wave (PSW)*: The P300 is a positive wave that occurs after 300–600 ms at the central-parietal scalp. The P300 may reflect processes involved in the updating of representations in working memory [51]. Three types of P300 have been reported in the psychophysiological literature. The P3a, that is elicited during passive attention, and the P3b, which is associated with participant's active cognitive processing. An additional wave called P3d is thought to reflect response inhibition. *The* P400 is a positive component that is maximal over occipital electrodes and peaks between 390 and 450 ms after stimulus onset. P400 has mainly been related to face processing [52–55] and novelty detection [56, 57]. The LPP is a positive deflection reflecting automatic allocation of attentional resources to emotionally salient stimuli, including emotion expressing faces [58]. The LPP is associated with activation of the amygdala, ACC, anterior insula, prefrontal, and parietal areas [59–61], and appears to be useful as a marker of emotion regulation processes. The PSW is a positive deflection arising 500 ms after stimulus presentation with different scalp distribution. It is thought to reflect the updating of working memory for the decision processes following target detection [62].

*Negative central (Nc), error-related negativity (ERN), and correct-response negativity (CRN):* The Nc is a large frontal negative ERP component occurring about 400–800 ms after stimulus. Nc waveform is thought to reflect allocation of attention to salient or relevant stimuli during working memory tasks [62]. The ERN is a negative deflection that peaks around 100–150 ms and is associated with a general executive control mechanism that detects and directs responses to errors [63]. Many individuals also exhibit a response-locked frontocentral negativity associated with the execution of correct responses. This component is known as the correct-(response)-related negativity or CRN [64, 65].

#### 5.2.3.5 Source Analyses

As stated above, the spatial resolution of surface EEG is low, and does not enable fine spatial discrimination, limiting the possible understanding of complex brain dynamics. Therefore, in recent years research has mainly focused on improving such gap. This aim was achieved through the study of the so-called source space, i.e., the study of linear combinations of data with the purpose of extracting latent variables hidden in the EEG. In this context, recent research capitalized on knowledge from source localization and source separation techniques developed in other fields of research, which provided reasonably accurate electromagnetic tomographies, i.e., true EEG based on 3D volumetric functional images of the brain. Low-resolution electromagnetic tomography (LORETA) is one possible approach, which computes an instantaneous three-dimensional discrete linear solution consisting of the smoothest of all possible neural current density distributions [66] (see Fig. 5.8). LORETA is a well-established method which has gained widespread popularity and has been used on hundreds of scientific publications [67]. However, other source analyses methods have been developed with similar, though not identical, approaches already used for LORETA [68].

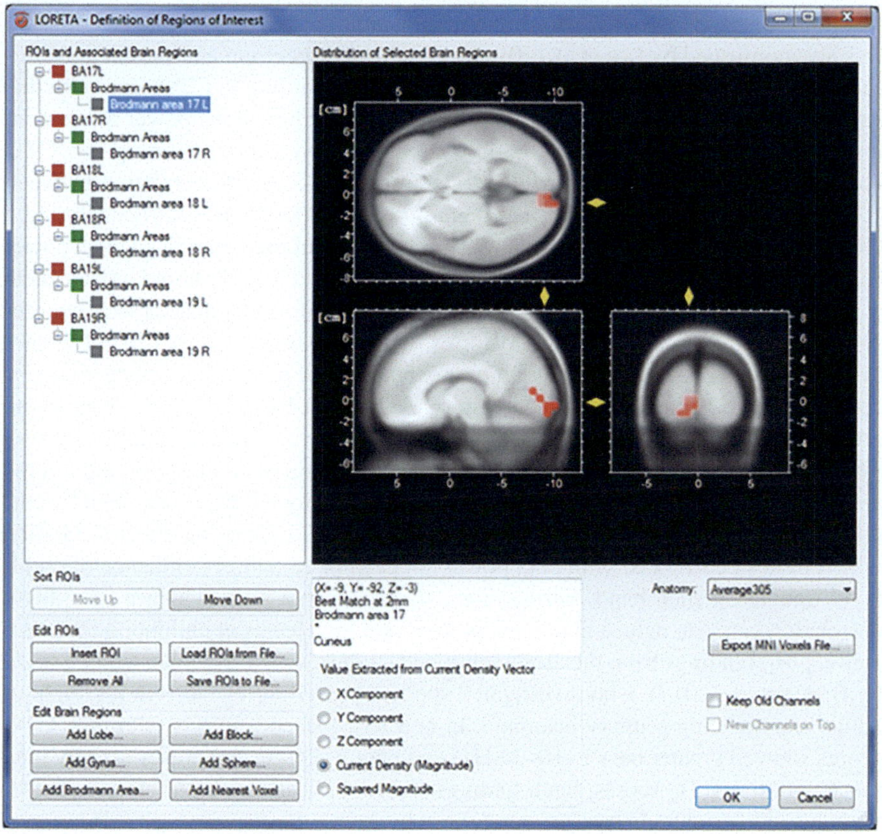


**Fig. 5.8** LORETA tree—ROIs located in the visual cortex. Dialog of the LORETA transform. Six ROIs in primary visual cortex (Brodmann area 17 right and left) and the extrastriate visual areas (Brodmann areas 18 and 19 right and left) are inserted in the ROI tree. The ROI BA17L (Brodmann area 17 left) is partially depicted in the MNI brain images. Available at https://www.pressrelease.brainproducts.com/loreta/, with permission of by ©Brain Products GmbH, Gilching, Germany

## 5.3 EEG Evidences in Childhood Trauma

### 5.3.1 Spectral Analysis

The large majority of spectral analyses revealed an inverse relationship between childhood abuse and alpha power, whereas the relationship with other band powers appears to be direct. To this extent, Howells and colleagues [69] performed correlation analyses between frequency band powers and different subscales of the childhood trauma questionnaire (CTQ) in three conditions: resting eyes open, resting eyes closed, and during a visual Go/NoGo task assessing sustained attention, response inhibition, and the ability to delay a response. They found negative correlations between physical and emotional neglect (as reported on the CTQ) and alpha power for both resting EEG and Go/NoGo conditions at left parietal electrodes, and with emotional abuse at frontal electrodes in resting conditions. Such inverse relationship was also confirmed by Lee et al. [70], who performed spectrum analyses in nonclinical volunteers, who were divided into two groups according to CTQ scores. In this study, subjects with high CTQ scores showed smaller low-alpha power at anterior, middle, and posterior regions of brain as compared with those showing low CTQ scores. On the other hand, Howells et al. [69] found a direct correlation between CTQ scores of emotional abuse and theta power at frontal left electrodes during a Go/NoGo task. The finding of a relationship between referred abuse and theta power is also confirmed by two further studies: Ben-Amitay et al. [71] found greater theta activity in frontal areas of abused women as compared to non-abused women after passive viewing of allusive sexual pictures, whereas Alper et al. [72] found greater theta activity in cocaine abusers with past experience of physical or sexual abuse, as compared with those without such history. As to beta band [70], Lee et al. found greater widespread beta power in adults with high vs. low scores of childhood abuse, whereas Jin et al. [73] did not find any direct relationship between childhood trauma and this, or any other band. However, in their work the total CTQ scores directly correlated with scores from the Beck Depression Inventory, expressing affective lability, which in turn, correlated with beta power suggesting an indirect influence of childhood trauma on such band. Furthermore, there is also a little evidence of a direct correlation between delta and gamma powers and high levels of childhood traumatization [70]. Findings from the aforementioned studies were also replicated in a study by Bader et al. (2013), who investigated spectral power during sleep in a sample of adults experiencing primary insomnia. In non-REM sleep, subjects with high CTQ scores showed greater beta 1 (16–24 Hz) and beta 2 (24–32 Hz) power activity than those with low CTQ scores. Such findings where replicated in REM sleep, together with the additional finding of greater delta activity in subjects with high CTQ scores.

### 5.3.2 Cortical Asymmetry

Curtis and Cicchetti [74] were the first authors who analyzed asymmetry in maltreated children and found greater negative asymmetry scores (reflecting greater right activity) in parietal electrodes only in males. Such findings were further confirmed by Miskovith et al. [75], who found a greater right asymmetry in young maltreated females (10–16 years), but at variance with the former study, the difference was found in frontal rather than parietal electrodes. Findings of greater right hemisphere activation were further confirmed by Tang et al. [76] who evaluated a sample of maltreated young females for 2 years. More specifically, such greater right asymmetry was found to belong to one of two possible longitudinal trajectories: the 60.5% of the sample showed a stable right (negative) alpha frontal asymmetry, that gradually became less relative, whereas 39.5% exhibited a stable left (positive) frontal alpha symmetry. Interestingly, regression analyses showed that maltreated females with positive frontal asymmetry (greater left activation) were less likely to present a current post-traumatic stress disorder (PTSD) or a past major depressive disorder (MDD), as compared to those showing right frontal asymmetry even though such differences were present only for lower CTQ scores. The relationship between right hemisphere asymmetry and childhood trauma was not replicated by Hostinar et al. [77] who found no differences in frontal asymmetry in mid-aged subjects with or without a history of childhood abuse. However, in this study right asymmetry showed a direct relationship with levels of inflammation. Moreover, the right asymmetry moderated the relationship between maltreatment and levels of inflammation. Authors concluded that such findings corroborate the hypothesis linking right frontal asymmetry to vulnerability to emotionally challenging major environmental stressors. Conversely, Popkirov et al. [78] measured the relationship between childhood trauma and asymmetrical brain activity in a sample of patients affected by bipolar disorder (BD), before and after the administration of a passive-viewing emotional task. They found that in BD a greater asymmetry on the left hemisphere directly correlated with CTQ total score, while the emotional abuse sub-scores showed a positive correlation with asymmetry at fronto-lateral electrodes. Moreover, greater asymmetry on the left hemisphere correlated with scores of the physical abuse subscale. Regression analyses also revealed that dissociative symptoms and CTQ scores predicted 51% of frontal asymmetry. This study suggests that the effect of psychopathology, BD in this case, can interact with childhood trauma, adding complexity to the neural alterations underlying frontal asymmetry, and possibly shifting such asymmetry from the right to the left.

### 5.3.3 Coherence

Only two studies evaluated coherence and its relationship with childhood trauma. Ito and colleagues [79] used intrahemispheric alpha band EEG coherence to measure cortical development in hospitalized children with severe physical or sexual abuse and their age- and gender-matched non-abused peers. Compared with non-abused children, those experiencing maltreatments showed a greater overall left hemisphere EEG coherence and a reversed hemispheric asymmetry, with left hemisphere coherence exceeding that of the right hemisphere. By contrast, normal children evidenced a greater right than left hemisphere coherence. Such alterations were confirmed by Miskovic et al. [80], who investigated alpha intra- and interhemispheric coherence in young females with and without exposure to child maltreatment. In this study, females with a history of childhood trauma showed more coherence at the left hemisphere than those without childhood trauma, and such coherence was greater also compared to right hemisphere coherence. The two groups also demonstrated different patterns of interhemispheric coherence: maltreated females showed less interhemispheric coherence in anterior areas, whereas females without a history of maltreatment had less coherence in centroparietal areas. In maltreated females, left intrahemispheric coherence and centroparietal hemispheric coherence were positively correlated with physical abuse and neglect. The observed left hemispheric coherence was also found to be the mediator of the child maltreatment effect on the risk of developing mood and anxiety symptoms.

### 5.3.4 ERPs

The vast majority of studies evaluating the relationships between ERPs and childhood trauma uses faces as emotional trigger. Some of them [81–83] required subjects to passively view faces expressing different emotions, whereas others [84–86] required subjects to perform a Go/NoGo task with faces as Go or NoGo stimuli (e.g., when a subject sees a happy face, action is needed; otherwise, when a subject sees an angry faces, the action has to be withheld, or vice-versa).

In this view, Cicchetti and Curtis [82] and Curtis and Cicchetti [83] found generalized reduced N170 amplitude in a sample of toddlers with a history of abuse and neglect, and also a different pattern of reactivity to face processing compared to healthy controls (HC). Specifically, in maltreated toddlers, angry faces elicited greater N170 amplitude than neutral and happy faces, whereas in controls, a greater N170 amplitude was related to happy faces. The aforementioned differences were partially confirmed in adult subjects by Chu et al. [81] who found a greater N170 peak amplitude after the unconscious view of angry faces as compared with happy faces, but only for patients with low-level interpersonal trauma. Such differential patterns are also present for other components: maltreated children showed greater P400 amplitude for angry as compared with happy and neutral faces, whereas no differences in P400 amplitude were seen in non-maltreated subjects. On the other hand, this latter group exclusively showed a greater Nc amplitude for happy faces, as compared with Nc amplitude for angry and neutral faces. Studies investigating passive viewing of faces also show that subjects exposed to childhood trauma exhibit alterations in late components: as compared with HC, maltreated subjects showed a greater amplitude after passive viewing of angry faces in P400 and two additional waves, namely P240 and P260, which resemble the P3b, and a smaller amplitude for PSW. Differences in late waves were also confirmed by other studies using Go/NoGo emotional paradigms. Pollack and colleagues [84, 85] used an emotional Go/NoGo task with two counterbalanced target conditions, i.e., happy and angry, and found greater p3b in maltreated children than HC when subjects attended to target angry faces. Even in this case, p3b amplitude seems greater for angry targets than happy targets in maltreated children only, whereas non-maltreated children showed no differences. Such findings were further confirmed by Shackman and colleagues [86], who used a Go/No go task with conflicting auditory and emotional cues and found greater p3b for angry faces as compared with happy and sad, in physically abused children only. Moreover, in this study maltreated children showed greater p3b than HC after viewing pictures of their mother's angry face. Alterations in late components are also present when the administered tasks do not embed emotional faces. In this perspective Howell et al. [69] used a go task with letters and found a greater correlation between P300 amplitude and latency and levels of physical and sexual abuse, whereas in patients with borderline personality disorder [87] the levels of trauma positively correlated with a positive deflection occurring after 250–300 ms, and the LPP waveform after ratings of physical and psychological pain conditions. Childhood trauma was related also to impairments in hibitory processes; adult subjects previously exposed to high levels of childhood trauma showed greater p3d than those exposed to low levels [88], whereas physically abused children showed greater N2 than HC, after hearing their angry mother's voices [86]. Finally, Pechtel and Pizzagalli [89] highlighted alterations of reinforcement learning in women with a past history of sexual abuse and remitted MDD. Using a probabilistic stimulus selection task in which subjects have to choose the most rewarded, and avoid the most punished stimulus, subjects showed greater CRN for rewarding trials as compared to HC.

### 5.3.5 Source Analyses

The aforementioned work of Pechtel and Pizzagalli [89], using a task tapping into reinforcement learning showed that women with remitted MDD and past exposure to childhood abuse demonstrated greater activation of subgenual ACC than both women without childhood trauma, and HC. Kim et al. [88] using a Go/NoGo task with numbers showed that for P3 subjects with high childhood trauma have reduced source activity in NoGo conditions, in the right ACC, bilateral medial frontal cortex (MFC), bilateral superior frontal gyrus (SFG), and right precentral gyrus (PG) than those with low level of childhood trauma. The same group [90] also performed a combined source localization and spectral analyses on a time interval between 300 and 700 ms and found, for the low beta band, a lower activation in primary somatosensory cortex (PMC) and middle occipital gyrus (MOG) in adults with high vs. low levels of trauma. For the gamma band, subjects with high levels of trauma showed lower activation in the superior temporal gyrus (STG). Moreover, the activation in the left primary somatosensory cortex (PSC) significantly correlated with the level of trauma for the low beta band, whereas the activation in the left MOG (low beta band) and superior temporal gyrus (gamma band) were significantly correlated with the level of emotional neglect. Alper et al. [72] evaluated differences in cocaine abusers during resting-state conditions and found a greater theta activity for those with past childhood abuse, mainly for the theta band at 3.9 Hz. Source analysis revealed that cocaine abusers with a past history of child maltreatment showed a greater parahippocampal (PC), FG, lingual (LG), posterior cingulate (PCC), and insular gyri (IG) activity, than those without abuse. Such alterations are mainly present on the left hemisphere, with the exception of the PG.

Trentini et al. [91] evaluated children with different forms of abuse before and after 8–10 weeks of Eye Movement Desensitization and Reprocessing therapy (EMDR) in a time window between 100 and 400 ms. As compared with the evaluation at the end of treatment, greater activation in the Inferior Frontal Gyrus (IFG) and OFC was found. Conversely, activation at the end of treatment, as compared with the evaluation before treatment, was localized mainly in the left temporal pole, the left IFG, and bilateral inferior temporal gyrus.

## 5.4 Neurobiology of Childhood Trauma: A Neurophysiological Perspective

Findings from neurophysiological studies confirm the evidence from preclinical and MRI data which demonstrate that childhood trauma is associated with impairments in a broad neuronal network involved in emotional expression and control, threat detection, arousal, information processing, and sensory gating [5]. Such network includes amygdala, the primary and secondary sensory cortices, the arousal system, the working memory circuit, the ventral striatopallidum, the hippocampus, the fornix, mammillary bodies, the anterior and posterior cingulate cortices, dorsolateral prefrontal cortex, and the corpus callosum. Specifically, findings from MRI studies consistently showed alterations in sensory systems (i.e., the secondary visual cortex and LG) [92, 93], heightened amygdala activation for emotional faces [94] especially for threatening stimuli [95, 96], disruption in prefrontal areas regulating emotional expression and action monitoring, as the ACC [97–99], the posterior cingulate cortex (PCC), the OFC [100], the dorsolateral prefrontal cortex (DLPFC) [101], and hemispheric asymmetry due to callosal alterations [102]. To this extent, neurophysiological studies corroborates such model. Specifically, studies using EEG techniques found alterations in early ERP components (P100 and N170) which are related to early visual processing and to secondary visual cortices [43, 103]. Accordingly, source analyses showed greater alterations in these areas, namely the fusiform gyrus (FG), LG, occipital gyrus (OG), thus corroborating the findings from neuroimaging techniques. Moreover, subjects exposed to childhood trauma showed greater amplitude in late ERP components such as P300, P400, and LPP for arousing, emotive stimuli, suggesting alterations in limbic and cortical areas, and giving an indirect evidence of impairments in processing of emotive information and mood control. Late ERP components reflect different aspect of cognitive processing such as updating of working memory [51], enhanced sustained attention toward emotional stimuli [104], encoding of emotional information [62], and emotional regulation [58]. As stated above, such functions are provided by interactions between subcortical and limbic structures and prefrontal areas. Specifically, sustained attention is mediated by the amygdala as a general relevance detector, through connections to sensory cortices, basal forebrain, and prefrontal areas [105], whereas encoding of emotional stimuli relies on amygdala-hippocampal reciprocal connections [106] and on connections to prefrontal areas subserving high-order functions. Working memory is the product of the hippocampal–medial prefrontal cortex interaction [107]. Finally, emotion regulation is provided by lateral prefrontal cortices inhibiting the amygdala through medial prefrontal cortex connections [108]. As previously stated, MRI studies showed that the presence of childhood maltreatment was associated with heightened amygdala activation for emotional stimuli [94], and disruption in prefrontal areas regulating emotional expression and action monitoring. Concurrently, neuropsychological studies showed alterations in memory functions [109], sustained attention [110], and emotional dysregulation [111] related to emotive stimuli. Therefore, neuroimaging and EEG studies appear to converge on the hypothesis that childhood trauma is related to a greater emotionally driven hyperactivation of limbic systems at the expenses of prefrontal control. Furthermore, greater amplitude of late ERP components in maltreated children for angry faces further support the hypothesis from preclinical and MRI studies, that amygdala activation and poor prefrontal control is to some extent selective to threatening stimuli.

Interestingly, these selective impairments are also present in early components, suggesting that this selective alteration is also present at a pre-cognitive, perceptual level. The hypothesized neuroanatomical counterpart of such alterations relies on an early emotional processing system based on a subcortical route providing low spatial frequency, rapid, coarse visual information through extra geniculate projections involving retino-collicular and pulvinar pathways. Such information reaches the amygdala which in turn projects to visual areas, providing early enhancement to emotional visual stimuli [112, 113].

Heightened amygdala transmission and reduced cortical inhibition are also confirmed by spectral analyses showing an inverse relationship between alpha activity and trauma severity, and direct correlations with beta and theta activities. Since alpha activity is related to inhibition and deactivation [114, 115], while beta activity is associated with alertness, attentive processes, and greater arousal [116, 117], such findings suggest heightened general activation and alertness as the impact of the traumatic experiences increases. Even though the spatial resolution impedes a clear localization on the anatomical structures causing such alterations, the direct correlation with theta activity, which is hypothesized to be related to functional coupling of hippocampus, amygdala, and cortex [17, 118] and to reflect emotional arousal, confirms the hypothesis of some limbic driver of such hyperactivation. Finally, a further, though partial, evidence of the aforementioned alteration comes from source analyses showing reduced activation in middle and lateral cortices, even though these studies did not involved emotional cues. Poor prefrontal control may also be manifested by the alteration in frontal symmetry. Greater activation in the right cortex at the expenses of the left, was related with poor prefrontal maturation and poor inhibitory control [29, 30] and interestingly, would represent the neurophysiological marker of a shift in emotional valence-related behavior [119]. In this view, greater activation of the right hemisphere in cases of childhood abuse may suggest a preferential development of negative emotions processing, and a tendency to produce avoidant responses at the expenses of positive emotions/approach behavior. Such preferential shift was found to be related to the further development of mood and anxiety disorders [32–36]. This shift is also confirmed by coherence studies, which demonstrate an imbalance between right/left interhemispheric coherence, together with grater interhemispheric coherence in frontal areas of abused children. This would suggest a poor development of the left hemisphere, as compared to the right [40], possibly due to alterations in the corpus callosum (CC). This view is corroborated by MRI studies which frequently found alterations in such structure, specifically in forebrain regions [120–124]. Alterations in left-right hemisphere balance, greater focus on negative emotions, and poor prefrontal control could also explain ERPs alterations related to inhibitory processes, such as in P3d and N2 in maltreated subjects, which corroborates clinical findings of the high rates of impulsive-related behaviors, namely suicide [125] and substance use [126]. Regarding different types of childhood abuse, neurophysiologic data are still insufficient to give a clear pattern for each subtype of maltreatment. However, as compared to other forms of abuse, physical abuse seems to have greater evidence of an association with left asymmetry, as well as with alterations in late component such as P3b and P300. This would suggest a greater impairment in frontal activities, thus corroborating the clinical and neuropsychological findings of a greater impulse dyscontrol, emotional dysregulation [127] and worse executive function, working memory and information processing [128] in the group of patients physically abused (See Fig. 5.9).

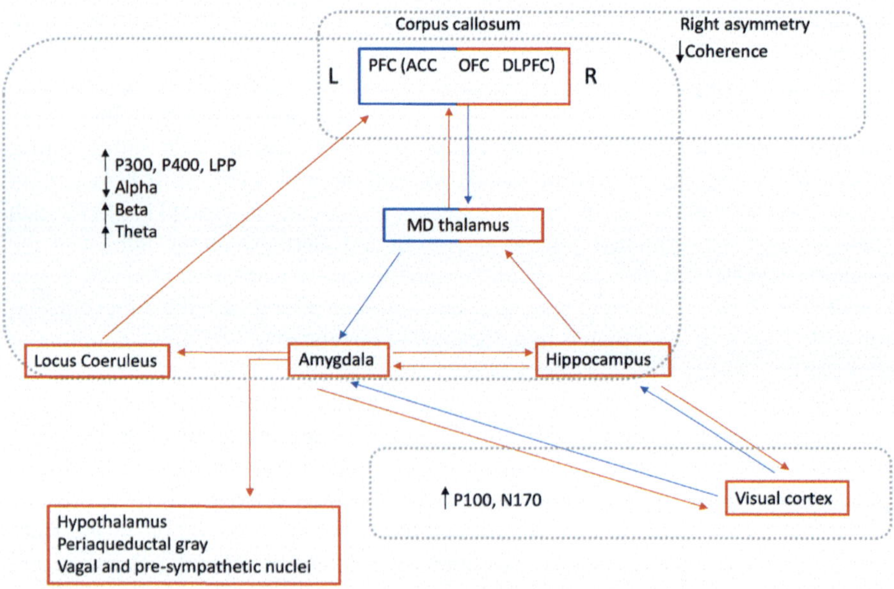


**Fig. 5.9** A schematic representation of the structures involved with childhood trauma and their functional relationships. Blue arrows indicate inhibitory processes, red arrows indicate excitatory processes. Blue squares indicated impaired function/activation, red squares indicated increased function/activation. Dashed squares define brain areas related to neurophysiological measures; *OFC* orbitofrontal cortex, *PFC* prefrontal cortex, *DLPFC* dorsolateral prefrontal cortex [129]

## 5.5 Conclusions

Childhood trauma is still an unknown devastating condition leading to high levels of impairment in social and affective domains. Neurophysiology confirm models highlighting multiple dysfunctions in sensory gating, mood control, threat perception, inhibition. Such alterations seem to start at the sensory level, and reflect multiple dysfunctions, primarily localized in limbic and prefrontal areas. As a consequence, the main underlying mechanism of the abovementioned disorder seems to be a hypersensibility to emotional stimuli and the influence that such hyperactivation has on the prefrontal regulatory cortex. The aforementioned neurofunctional alterations appear to be present since early childhood, and are still present either in the short-term period and also in adulthood. However, research on the neurobiological underpinnings of childhood trauma is still at its infancy, and more studies are needed to clarify the complex nature of this disorder.

## References


1. Dube SR, Felitti VJ, Dong M, Giles WH, Anda RF. The impact of adverse childhood experiences on health problems: evidence from four birth cohorts dating back to 1900. Prev Med. 2003;37(3):268–77.
2. Dube SR, Felitti VJ, Dong M, Chapman DP, Giles WH, Anda RF. Childhood abuse, neglect, and household dysfunction and the risk of illicit drug use: the adverse childhood experiences study. Pediatrics. 2003;111(3):564–72. https://doi.org/10.1542/peds.111.3.564.
3. Anda RF, Brown DW, Felitti VJ, Bremner JD, Dube SR, Giles WH. Adverse childhood experiences and prescribed psychotropic medications in adults. Am J Prev Med. 2007.
4. Brown DW, Anda RF, Tiemeier H, Felitti VJ, Edwards VJ, Croft JB, et al. Adverse childhood experiences and the risk of premature mortality. Am J Prev Med. 2009.
5. Teicher MH, Samson JA, Anderson CM, Ohashi K. The effects of childhood maltreatment on brain structure, function and connectivity. Nat Rev Neurosci Nature Publishing Group. 2016;17(10):652–66. https://doi.org/10.1038/nrn.2016.111.
6. Nunez PL, Srinivasan R. A theoretical basis for standing and traveling brain waves measured with human EEG with implications for an integrated consciousness. Clin Neurophysiol. 2006;117(11):2424–35.
7. Tatum WO. Handbook of EEG interpretation. Medicine 2014.
8. Berger H. Ueber das Electroencephalogram des Menschen. Arch fuer Psychatrie 1929.
9. Arciniegas DB, Anderson CA, Filley CM, Garcia TA. Behavioral neurology & neuropsychiatry. Behav Neurol Neuropsych. 2010:1–700.
10. Dustman RE, Boswell RS, Porter PB. Beta brain waves as an index of alertness. Science (80). 1962;137(3529):533–4.
11. Maratos FA, Mogg K, Bradley BP, Rippon G, Senior C. Coarse threat images reveal theta oscillations in the amygdala: a magnetoencephalography study. Cogn Affect Behav Neurosci 2009.
12. Balconi M, Brambilla E, Falbo L. BIS/BAS, cortical oscillations and coherence in response to emotional cues. Brain Res Bull 2009.
13. Qin J, Lee TMC, Han S. Theta and alpha oscillations linked to risk identifications. Brain Res. 2009.
14. Garolera M, Coppola R, Muñoz KE, Elvevåg B, Carver FW, Weinberger DR, et al. Amygdala activation in affective priming: a magnetoencephalogram study. Neuroreport 2007.

15. Narayanan RT, Seidenbecher T, Sangha S, Stork O, Pape HC. Theta resynchronization during reconsolidation of remote contextual fear memory. Neuroreport. 2007.
16. Pape HC, Narayanan RT, Smid J, Stork O, Seidenbecher T. Theta activity in neurons and networks of the amygdala related to long-term fear memory. Hippocampus. 2005.
17. Seidenbecher T, Laxmi TR, Stork O, Pape HC. Amygdalar and hippocampal theta rhythm synchronization during fear memory retrieval. Science (80). 2003.
18. Ebersole JS, Pacia S V. Localization of temporal lobe foci by ictal EEG patterns. Epilepsia. 1996.
19. Hughes JR. Gamma, fast, and ultrafast waves of the brain: their relationships with epilepsy and behavior. Epilepsy Behav 2008.
20. Gold I. Does 40-Hz oscillation play a role in visual consciousness? Conscious Cogn. 1999.
21. Tallon-Baudry C, Bertrand O. Oscillatory gamma activity in humans and its role in object representation. Trends Cogn Sci 1999.
22. Fries P, Reynolds JH, Rorie AE, Desimone R. Modulation of oscillatory neuronal synchronization by selective visual attention. Science (80). 2001;291(5508):1560–3.
23. Pesaran B, Pezaris JS, Sahani M, Mitra PP, Andersen RA. Temporal structure in neuronal activity during working memory in macaque parietal cortex. Nat Neurosci. 2002;5(8):805–11.
24. Bauer EP, Paz R, Pare D. Gamma oscillations coordinate Amygdalo-Rhinal interactions during learning. J Neurosci. 2007;27(35):9369–79. https://doi.org/10.1523/JNEUROSCI.2153-07.2007.
25. Fries P. Neuronal gamma-band synchronization as a fundamental process in cortical computation. Annu Rev Neurosci. 2009;32(1):209–24. https://doi.org/10.1146/annurev.neuro.051508.135603.
26. Womelsdorf T, Schoffelen JM, Oostenveld R, Singer W, Desimone R, Engel AK, et al. Modulation of neuronal interactions through neuronal synchronization. Science (80). 2007;316(5831):1609–12.
27. Masquelier T, Hugues E, Deco G, Thorpe SJ. Oscillations, phase-of-firing coding, and spike timing-dependent plasticity: an efficient learning scheme. J Neurosci. 2009;29(43):13484–93. https://doi.org/10.1523/JNEUROSCI.2207-09.2009.
28. Davidson RJ. Cerebral asymmetry and emotion: conceptual and methodological conundrums. Cogn Emot. 1993;7(1):115–38.
29. Ochsner KN, Bunge SA, Gross JJ, Gabrieli JDE. Rethinking feelings: an FMRI study of the cognitive regulation of emotion. In: Social neuroscience: Key readings; 2013. p. 253–70.
30. Quirk GJ, Likhtik E, Pelletier JG, Paré D. Stimulation of medial prefrontal cortex decreases the responsiveness of central amygdala output neurons. J Neurosci. 2003;23(25):8800–7.
31. Davidson R. What does the prefrontal cortex "do" in affect. Perspect Front EEG asymmetry Res. 2004;67(1–2):219–33.
32. Davidson RJ. Affective style and affective disorders: perspectives from affective neuroscience. Cogn Emot. 1998;12(3):307–30.
33. Fingelkurts AA, Fingelkurts AA. Altered structure of dynamic electroencephalogram oscillatory pattern in major depression. Biol Psychiatry. 2015:1050–60.
34. Jesulola E, Sharpley CF, Bitsika V, Agnew LL, Wilson P. Frontal alpha asymmetry as a pathway to behavioural withdrawal in depression: research findings and issues. Behav Brain Res. 2015:56–67.
35. Nusslock R, Miller GE. Early-life adversity and physical and emotional health across the lifespan: a neuroimmune network hypothesis. Biol Psychiatry. 2016:23–32.
36. Thibodeau R, Jorgensen RS, Kim S. Depression, anxiety, and resting frontal EEG asymmetry: a meta-analytic review. J Abnorm Psychol. 2006:715–29.
37. Thatcher RW. Cyclic cortical reorganization during early childhood. Brain Cogn. 1992;20(1):24–50.
38. Pogarell O, Teipel SJ, Juckel G, Gootjes L, Möller T, Bürger K, et al. EEG coherence reflects regional corpus callosum area in Alzheimer's disease. J Neurol Neurosurg Psychiatry. 2005;76(1):109–11.

39. Thatcher RW, Krause PJ, Hrybyk M. Cortico-cortical associations and EEG coherence: a two-compartmental model. Electroencephalogr Clin Neurophysiol. 1986;64(2):123–43.
40. Thatcher RW, North DM, Biver CJ. Development of cortical connections as measured by EEG coherence and phase delays. Hum Brain Mapp. 2008;29(12):1400–15.
41. Blackwood DHR, Muir WJ. Cognitive brain potentials and their application. Br J Psychiatry. 1990:96–101.
42. Allison T, Puce A, Spencer DD, McCarthy G. Electrophysiological studies of human face perception. I: potentials generated in occipitotemporal cortex by face and non-face stimuli. Cereb Cortex. 1999;9(5):415–30.
43. Di Russo F, Martínez A, Sereno MI, Pitzalis S, Hillyard SA. Cortical sources of the early components of the visual evoked potential. Hum Brain Mapp. 2002;15(2):95–111.
44. Bentin S, Allison T, Puce A, Perez E, McCarthy G. Electrophysiological studies of face perception in humans. J Cogn Neurosci 1996.
45. Eimer M. Event-related brain potentials distinguish processing stages involved in face perception and recognition. Clin Neurophysiol 2000.
46. Eimer M, Holmes A. An ERP study on the time course of emotional face processing. Neuroreport. 2002.
47. Ibanez A, Melloni M, Huepe D, Helgiu E, Rivera-Rei A, Canales-Johnson A, et al. What event-related potentials (ERPs) bring to social neuroscience? Soc Neurosci. 2012;7(6):632–49.
48. Nieuwenhuis S, Yeung N, Van Den Wildenberg W, Ridderinkhof KR. Electrophysiological correlates of anterior cingulate function in a go/no-go task: effects of response conflict and trial type frequency. Cogn Affect Behav Neurosci. 2003;3(1):17–26.
49. Herman JP, Ostrander MM, Mueller NK, Figueiredo H. Limbic system mechanisms of stress regulation: Hypothalamo-pituitary-adrenocortical axis. Prog Neuro-Psychopharmacol Biol Psychiatry. 2005;29:1201–13.
50. Wronka E, Walentowska W. Attention modulates emotional expression processing. Psychophysiology. 2011;48(8):1047–56.
51. Donchin E, Karis D, Bashore TR. Cognitive psychophysiology and human information processing. Psychophysiol Syst Proces Appl. 1986;
52. de Haan M, Nelson CA. Brain activity differentiates face and object processing in 6-month-old infants. Dev Psychol. 1999;35(4):1113–21.
53. McCleery JP, Akshoomoff N, Dobkins KR, Carver LJ. Atypical face versus object processing and hemispheric asymmetries in 10-month-old infants at risk for autism. Biol Psychiatry. 2009;66(10):950–7.
54. Halit H, Csibra G, Volein Á, Johnson MH. Face-sensitive cortical processing in early infancy. J Child Psychol Psychiatry Allied Discip. 2004;45(7):1228–34.
55. Halit H, De Haan M, Johnson MH. Cortical specialisation for face processing: face-sensitive event-related potential components in 3- and 12-month-old infants. NeuroImage. 2003;19(3):1180–93.
56. Key APF, Stone W, Williams SM. What do infants see in faces? ERP evidence of different roles of eyes and mouth for face perception in 9-month-old infants. Infant Child Dev. 2009;18(2):149–62.
57. Scott LS, Shannon RW, Nelson CA. Neural correlates of human and monkey face processing in 9-month-old infants. Infancy. 2006;10(2):171–86.
58. Hajcak G, MacNamara A, Olvet DM. Event-related potentials, emotion, and emotion regulation: an integrative review. Dev Neuropsychol. 2010;35(2):129–55.
59. Liu Y, Huang H, McGinnis-Deweese M, Keil A, Ding M. Neural substrate of the late positive potential in emotional processing. J Neurosci. 2012;32(42):14563–72.
60. Sabatinelli D, Keil A, Frank DW, Lang PJ. Emotional perception: correspondence of early and late event-related potentials with cortical and subcortical functional MRI. Biol Psychol. 2013;92(3):513–9.
61. Sabatinelli D, Lang PJ, Keil A, Bradley MM. Emotional perception: correlation of functional MRI and event-related potentials. Cereb Cortex. 2007;17(5):1085–91.

62. García-Larrea L, Cézanne-Bert G. P3, positive slow wave and working memory load: a study on the functional correlates of slow wave activity. Electroencephalogr Clin Neurophysiol Evoked Potentials. 1998;108(3):260–73.
63. Bates AT, Kiehl KA, Laurens KR, Liddle PF. Error-related negativity and correct response negativity in schizophrenia. Clin Neurophysiol. 2002.
64. Ford JM. Schizophrenia: the broken P300 and beyond. In: Psychophysiology; 1999.
65. Mathalon DH, Fedor M, Faustman WO, Gray M, Askari N, Ford JM. Response-monitoring dysfunction in schizophrenia: an event-related brain potential study. J Abnorm Psychol 2002.
66. Pascual-Marqui RD. Low resolution brain electromagnetic tomography (LORETA). Electroencephalogr Clin Neurophysiol. 1997;1(103):25–6.
67. Pascual-Marqui RD, Esslen M, Kochi K, Lehmann D. Functional imaging with low resolution electromagnetic tomography ( LORETA ): review, comparisons, and new validation brain new. Japanese J Clin Neurophysiol. 2002;30:81–94.
68. Litvak V, Mattout J, Kiebel S, Phillips C, Henson R, Kilner J, et al. EEG and MEG data analysis in SPM8. Comput Intell Neurosci. 2011;2011
69. Howells FM, Stein DJ, Russell VA. Childhood trauma is associated with altered cortical arousal: insights from an EEG study. Front Integr Neurosci. 2012;6(December):1–19. http://journal.frontiersin.org/article/10.3389/fnint.2012.00120/abstract
70. Lee S-H, Park Y, Jin MJ, Lee YJ, Hahn SW. Childhood trauma associated with enhanced high frequency band powers and induced subjective inattention of adults. Front Behav Neurosci. 2017;11(August):1–12.
71. Ben-Amitay G, Kimchi N, Wolmer L, Toren P. Psychophysiological reactivity in child sexual abuse. J Child Sex Abus Routledge. 2016;25(2):185–200. https://doi.org/10.1080/10538712.2016.1124309.
72. Alper K, Shah J, Howard B, Roy John E, Prichep LS. Childhood abuse and EEG source localization in crack cocaine dependence. Psychiatry Res. 2013;213(1):63–70. https://doi.org/10.1016/j.pscychresns.2013.01.008.
73. Jin MJ, Kim JS, Kim S, Hyun MH, Lee SH. An integrated model of emotional problems, beta power of electroencephalography, and low frequency of heart rate variability after childhood trauma in a non-clinical sample: a path analysis study. Front Psych. 2018;8.
74. Curtis WJ, Cicchetti D. Emotion and resilience: a multilevel investigation of hemispheric electroencephalogram asymmetry and emotion regulation in maltreated and nonmaltreated children. Dev Psychopathol. 2007;19(3):811–40.
75. Miskovic V, Schmidt LA, Georgiades K, Boyle M, MacMillan HL. Stability of resting frontal electroencephalogram (EEG) asymmetry and cardiac vagal tone in adolescent females exposed to child maltreatment. Dev Psychobiol. 2009;51(6):474–87.
76. Tang A, Miskovic V, Lahat A, Tanaka M, MacMillan H, Van Lieshout RJ, et al. Trajectories of resting frontal brain activity and psychopathology in female adolescents exposed to child maltreatment. Dev Psychobiol. 2018;60(1):67–77.
77. Hostinar CE, Davidson RJ, Graham EK, Mroczek DK, Lachman ME, Seeman TE, et al. Frontal brain asymmetry, childhood maltreatment, and low-grade inflammation at midlife. Psychoneuroendocrinology. 2017;75:152–63.
78. Popkirov S, Flasbeck V, Schlegel U, Juckel G, Brüne M. Childhood trauma and dissociative symptoms predict frontal EEG asymmetry in borderline personality disorder. J Trauma Dissociation Routledge. 2018;00(00):1–16. https://doi.org/10.1080/15299732.2018.1451808.
79. Ito Y, Teicher MH, Glod CA, Ackerman E. Preliminary evidence for aberrant cortical} development in abused Children. A quantitative EEG study. J Neuropsychiatry Clin Neurosci. 1998;10(3):298–307.
80. Miskovic V, Schmidt LA, Georgiades K, Boyle M, Macmillan HL. Adolescent females exposed to child maltreatment exhibit atypical EEG coherence and psychiatric impairment: linking early adversity, the brain, and psychopathology. Dev Psychopathol. 2010;22(2):419–32.

81. Chu DA, Bryant RA, Gatt JM, AWF Harris. Failure to differentiate between threat-related and positive emotion cues in healthy adults with childhood interpersonal or adult trauma. J. Psychiatr. Res. 2016;78: 31–41. https://doi.org/ 10.1016/j.jpsychires.2016.03.006.
82. Cicchetti D, Curtis WJ. An event-related potential study of the processing of affective facial expressions in young children who experienced maltreatment during the first year of life. Dev Psychopathol. 2005;17:641–77.
83. Curtis WJ, Cicchetti D. Affective facial expression processing in young children who have experienced maltreatment during the first year of life: an event-related potential study. Dev Psychopathol. 2011;23(2):373–95.
84. Pollak SD, Cicchetti D, Klorman R, Brumaghim JT. Cognitive brain event-related potentials and emotion processing in maltreated children. Child Dev. 1997;68(5):773–87.
85. Pollak SD, Klorman R, Thatcher JE, Cicchetti D. P3b reflects maltreated children's reactions to facial displays of emotion. Psychophysiology HAM-TMC Library. 2001;38(2): 267–74.
86. Shackman JE, Shackman AJ, Pollak SD. Physical abuse amplifies attention to threat and increases anxiety in children. Emotion. 2007;7(4):838–52.
87. Flasbeck V, Enzi B, Brüne M. Childhood trauma affects processing of social interactions in borderline personality disorder: an event-related potential study investigating empathy for pain. World J Biol Psychiatry. 2017:1–11. https://doi.org/10.1080/15622975.2017.1333147.
88. Kim S, Kim JS, Jin MJ, Im CH, Lee SH. Dysfunctional frontal lobe activity during inhibitory tasks in individuals with childhood trauma: an event-related potential study. NeuroImage Clin. 2018;17:935–42.
89. Pechtel P, Pizzagalli DA. Disrupted reinforcement learning and maladaptive behavior in women with a history of childhood sexual abuse: a high-density event-related potential study. JAMA Psychiat. 2013;70(5):499–507.
90. Kim S, Kim JS, Shim M, Im CH, Lee SH. Altered cortical functional network during behavioral inhibition in individuals with childhood trauma. Sci Rep. 2018;8(1):1–10.
91. Trentini C, Pagani M, Fania P, Speranza AM, Nicolais G, Sibilia A, et al. Neural processing of emotions in traumatized children treated with eye movement desensitization and reprocessing therapy: a hdEEG study. Front Psychol. 2015;6(NOV):1–12.
92. Tomoda A, Navalta CP, Polcari A, Sadato N, Teicher MH. Childhood sexual abuse is associated with reduced Gray matter volume in visual cortex of young women. Biol Psychiatry. 2009;66(7):642–8.
93. Tomoda A, Polcari A, Anderson CM, Teicher MH. Reduced visual cortex Gray matter volume and thickness in young adults who witnessed domestic violence during childhood. PLoS One. 2012;7(12)
94. Teicher MH, Samson JA. Annual research review: enduring neurobiological effects of childhood abuse and neglect. J Child Psychol Psychiatry Allied Discip. 2016:241–66.
95. Öhman A. The role of the amygdala in human fear: automatic detection of threat. Psychoneuroendocrinology. 2005:953–8.
96. Suslow T, Ohrmann P, Bauer J, Rauch AV, Schwindt W, Arolt V, et al. Amygdala activation during masked presentation of emotional faces predicts conscious detection of threat-related faces. Brain Cogn. 2006;61(3):243–8.
97. Baker LM, Williams LM, Korgaonkar MS, Cohen RA, Heaps JM, Paul RH. Impact of early vs. late childhood early life stress on brain morphometrics. Brain Imaging Behav. 2013;7(2):196–203.
98. Cohen RA, Grieve S, Hoth KF, Paul RH, Sweet L, Tate D, et al. Early life stress and Morphometry of the adult anterior cingulate cortex and caudate nuclei. Biol Psychiatry. 2006;59(10):975–82.
99. Thomaes K, Dorrepaal E, Draijer N, De Ruiter MB, Van Balkom AJ, Smit JH, et al. Reduced anterior cingulate and orbitofrontal volumes in child abuse-related complex PTSD. J Clin Psychiatry. 2010;71(12):1636–44.
100. Hanson JL, Chung MK, Avants BB, Shirtcliff EA, Gee JC, Davidson RJ, et al. Early stress is associated with alterations in the orbitofrontal cortex: a tensor-based Morphometry inves-

tigation of brain structure and behavioral risk. J Neurosci. 2010;30(22):7466–72. https://doi.org/10.1523/JNEUROSCI.0859-10.2010.

101. Carballedo A, Lisiecka D, Fagan A, Saleh K, Ferguson Y, Connolly G, et al. Early life adversity is associated with brain changes in subjects at family risk for depression. World J Biol Psychiatry. 2012;13(8):569–78.
102. Benedetti F, Bollettini I, Radaelli D, Poletti S, Locatelli C, Falini A, et al. Adverse childhood experiences influence white matter microstructure in patients with bipolar disorder. Psychol Med. 2014;44(14):3069–82.
103. Eimer M. The face-sensitive N170 component of the event-related brain potential. In: Oxford handbook of face perception; 2012.
104. Schupp HT, Junghöfer M, Weike AI, Hamm AO. The selective processing of briefly presented affective pictures: an ERP analysis. Psychophysiology. 2004;41(3):441–9.
105. Sarter M, Givens B, Bruno JP. The cognitive neuroscience of sustained attention: where top-down meets bottom-up. Brain Res Rev. 2001:146–60.
106. Richardson MP, Strange BA, Dolan RJ. Encoding of emotional memories depends on amygdala and hippocampus and their interactions. Nat Neurosci. 2004;7(3):278–85.
107. Godsil BP, Kiss JP, Spedding M, Jay TM. The hippocampal-prefrontal pathway: the weak link in psychiatric disorders? Eur Neuropsychopharmacol. 2013;23(10):1165–81.
108. Phillips ML, Swartz HA. A critical appraisal of neuroimaging studies of bipolar disorder: toward a new conceptualization of underlying neural circuitry and a road map for future research of neuroimaging studies of bipolar disorder: toward a new conceptualization of underlying ne. Am J Psychiatry. 2014:829–43.
109. Cromheeke S, Herpoel LA, Mueller SC. Childhood abuse is related to working memory impairment for positive emotion in female university students. Child Maltreat. 2014;19(1):38–48.
110. Lim M, Lee S, Park JI. Differences between impulsive and non-impulsive suicide attempts among individuals treated in emergency rooms of South Korea. Psychiatry Investig. 2016;13(4):389–96.
111. Weiss NH, Tull MT, Lavender J, Gratz KL. Role of emotion dysregulation in the relationship between childhood abuse and probable PTSD in a sample of substance abusers. Child Abuse Negl. 2013;37(11):944–54. https://www-sciencedirect-com.e.bibl.liu.se/science/article/pii/S0145213413000860?via%3Dihub
112. Vuilleumier P, Armony JL, Driver J, Dolan RJ. Distinct spatial frequency sensitivities for processing faces and emotional expressions. Nat Neurosci. 2003;6(6):624–31.
113. Pourtois G, de Gelder B, Bol A, Crommelinck M. Perception of facial expressions and voices and of their combination in the human brain. Cortex. 2005;41(1):49–59.
114. Cooper NR, Burgess AP, Croft RJ, Gruzelier JH. Investigating evoked and induced electroencephalogram activity in task-related alpha power increases during an internally directed attention task. Neuroreport. 2006;17(2):205–8.
115. Uusberg A, Uibo H, Kreegipuu K, Allik J. EEG alpha and cortical inhibition in affective attention. Int J Psychophysiol. 2013;89(1):26–36.
116. Ray WJ, Cole HW. EEG alpha activity reflects attentional demands, and beta activity reflects emotional and cognitive processes. Science (80). 1985;228(4700):750–2.
117. Steriade M, Gloor P, Llinás RR. Lopes da Silva FH, Mesulam MM. Basic mechanisms of cerebral rhythmic activities. Electroencephalogr Clin Neurophysiol. 1990;76(6):481–508.
118. Miller R. Cortico-hippocampal interplay: self-organizing phase-locked loops for indexing memory. Psychobiology 1989.
119. Fetterman AK, Ode S, Robinson MD. For which side the bell tolls: the laterality of approach-avoidance associative networks. Motiv Emot. 2013;37(1):33–8.
120. De Bellis MD, Keshavan MS, Shifflett H, Iyengar S, Beers SR, Hall J, et al. Brain structures in pediatric maltreatment-related posttraumatic stress disorder: a sociodemographically matched study. Biol Psychiatry. 2002;52(11):1066–78.
121. Andersen SL, Tomada A, Vincow ES, Valente E, Polcari A, Teicher MH. Preliminary evidence for sensitive periods in the effect of childhood sexual abuse on regional brain devel-

opment. J Neuropsychiatry Clin Neurosci. 2008;20(3):292–301.https://doi.org/10.1176/jnp.2008.20.3.292.
122. Mehta MA, Gore-Langton E, Golembo N, Colvert E, Williams SCR, Sonuga-Barke E. Hyporesponsive reward anticipation in the basal ganglia following severe institutional deprivation early in life. J Cogn Neurosci. 2010;22(10):2316–25.
123. Teicher MH, Dumont NL, Ito Y, Vaituzis C, Giedd JN, Andersen SL. Childhood neglect is associated with reduced corpus callosum area. Biol Psychiatry. 2004;56(2):80–5.
124. Teicher MH, Samson JA, Sheu YS, Polcari A, McGreenery CE. Hurtful words: association of exposure to peer verbal abuse with elevated psychiatric symptom scores and corpus callosum abnormalities. Am J Psychiatry. 2010;167(12):1464–71.
125. Zatti C, Rosa V, Barros A, Valdivia L, Calegaro VC, Freitas LH, et al. Childhood trauma and suicide attempt: a meta-analysis of longitudinal studies from the last decade. Psychiatry Res. 2017:353–8.
126. Edalati H, Krank MD. Childhood maltreatment and development of substance use disorders: a review and a model of cognitive pathways. Trauma, Violence, Abus. 2016;17(5):454–67.
127. Sugaya L, Hasin DS, Olfson M, Lin KH, Grant BF, Blanco C. Child physical abuse and adult mental health: a national study. J Trauma Stress. 2012;25(4):384–92.
128. Lysaker PH, Meyer P, Evans JD, Marks KA. Neurocognitive and symptom correlates of self-reported childhood sexual abuse in schizophrenia spectrum disorders. Ann Clin Psychiatry. 2001;13(2):89–92.
129. Nunez PL, Srinivasan R. Electric fields of the brain: the neurophysics of EEG. 2009.

# Interaction Between Genes and Childhood Trauma on the Outcome of Psychiatric Disorders

6

Marta de Castro-Catala, Sergi Papiol, Neus Barrantes-Vidal, and Araceli Rosa

M. de Castro-Catala
Secció de Zoologia i Antropologia Biològica, Departament de Biologia Evolutiva, Ecologia i Ciències Ambientals, Facultat de Biologia, Universitat de Barcelona (UB), Barcelona, Spain

Institut de Biomedicina de la Universitat de Barcelona (IBUB), Barcelona, Spain
e-mail: martadcastrocatala@gmail.com

S. Papiol
Department of Psychiatry and Psychotherapy, University Hospital, Ludwig Maximilian University of Munich, Munich, Germany

Institute of Psychiatric Phenomics and Genomics, University Hospital, Ludwig Maximilian University of Munich, Munich, Germany

Centre for Biomedical Research Network on Mental Health (CIBERSAM), Instituto de Salud Carlos III, Madrid, Spain
e-mail: Sergi.Papiol@med.uni-muenchen.de

N. Barrantes-Vidal
Department of Clinical and Health Psychology, Universitat Autònoma de Barcelona, Barcelona, Spain

Department of Mental Health, Fundació Sanitària Sant Pere Claver, Barcelona, Spain

Centre for Biomedical Research Network on Mental Health (CIBERSAM), Instituto de Salud Carlos III, Madrid, Spain
e-mail: Neus.Barrantes@uab.cat

A. Rosa (✉)
Secció de Zoologia i Antropologia Biològica, Departament de Biologia Evolutiva, Ecologia i Ciències Ambientals, Facultat de Biologia, Universitat de Barcelona (UB), Barcelona, Spain

Institut de Biomedicina de la Universitat de Barcelona (IBUB), Barcelona, Spain

Centre for Biomedical Research Network on Mental Health (CIBERSAM), Instituto de Salud Carlos III, Madrid, Spain
e-mail: araceli.rosa@ub.edu


G. Spalletta et al. (eds.), *Childhood Trauma in Mental Disorders*,
https://doi.org/10.1007/978-3-030-49414-8_6

## 6.1 The Complex Aetiology of Psychiatric Disorders

The study of human diseases has shown that nearly all conditions and diseases have an underlying genetic component. Some disorders obey the standard Mendelian recessive, dominant or X-linked pattern of inheritance, being determined by mutations in a single gene (e.g. cystic fibrosis). However, many others, that are also clearly heritable, are caused by a combination of multiple genetic, environmental and lifestyle factors, most of which have not yet been identified. That seems the case of most psychiatric disorders, complex diseases with a polygenic architecture (i.e. with involvement of multiple genes of minor effect) that could be influenced by other multiple factors (e.g. personal, environmental or lifestyle).

Family and twin studies have become an important tool to disentangle the role of genes and environment in mental disorders, showing evidence of a strong genetic component for many of them. *Heritability* is the parameter that measures the percentage of the phenotype explained by the genetic component. This measure ranges from zero to one, with lower values indicating that almost all the trait variability among people is due to environmental factors with very little influence of genetic factors, and higher values indicating that almost all the variability comes from genetic differences with little contribution from environmental factors. In the case of disorders such as schizophrenia, heritability estimates are between 0.6 and 0.8, and for bipolar disorder around 0.6. The co-morbidity between the disorders was estimated around 63% due to additive genetic effects common to both disorders. Shared environmental effects seem to be small but significant for both disorders [1]. Additionally, the lifetime risk of developing these disorders is related to the degree of biological relatedness with an affected person, showing greater risks associated with higher levels of genes shared in an exponential way that clearly fits with a model of multiple genes of minor effect (Fig. 6.1).

Over the last decades, psychiatric genetics has tried to find the genes responsible for major mental disorders using *candidate-gene research strategies*. These studies focused on genes coding for proteins involved in pathways believed to be disrupted in the physiopathology of the disorder, or genes with a regulatory role on DNA expression (functional candidate genes), and genes located in genomic positions previously associated with the disease (positional candidate genes).

There is a huge number of studies published to date exploring the possible genes involved in mental disorders. They have suggested that some allelic variations in multiple genes coding for proteins involved in neurobiological pathways (e.g. in dopaminergic, serotoninergic or glutamatergic neurotransmitter systems or pathways related with neurodevelopment) may confer lower efficiency to these systems. This reduced efficiency would lead to specific pathophysiological disturbances that may impact neural circuits and cause the development of different phenotypes related with psychosis.

In the last years, advances in our understanding of genetic variation and in the technology to measure it, have enabled large-scale studies analysing up to several million polymorphisms across the human genome in thousands of samples, such as *genome-wide association studies* (GWAS). These studies have successfully identified many genetic variants (mainly single nucleotide polymorphisms, *SNPs*)

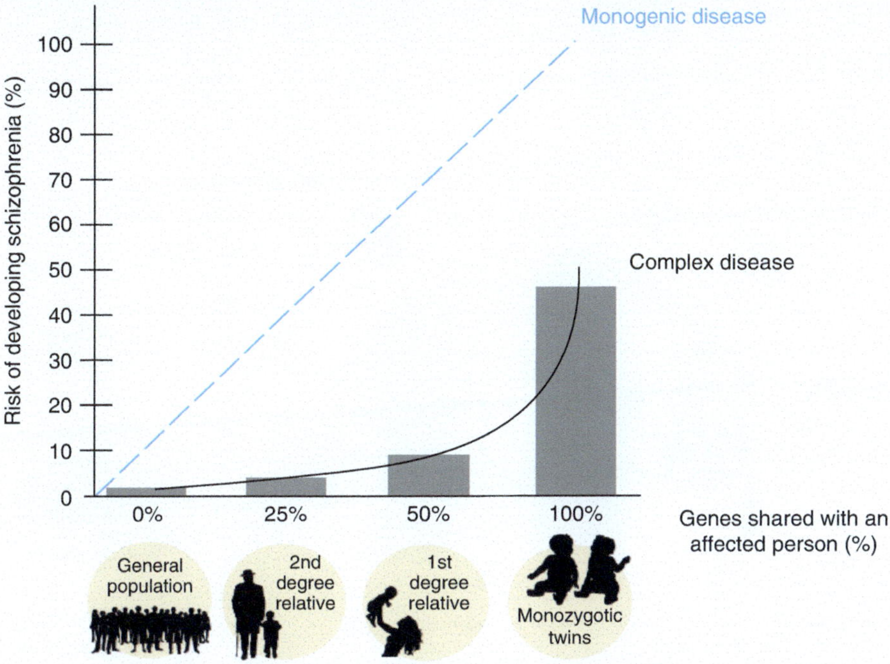


**Fig. 6.1** Lifetime risk for developing a mental disorder depends on the percentage of genes shared with an affected individual. In the case of schizophrenia, whereas the lifetime risk for the illness is about 1% in the general population, this risk increases exponentially depending on the percentage of genes shared with an affected person. In a monogenic or Mendelian disease, we would expect a lineal increase (i.e. blue line). However, according to family data, the distribution of risk for this complex mental disorder clearly fits with a model of multiple genes of minor effect or complex disease (black line). Moreover, the concordances described in monozygotic twins (risk of about 46%) suggest the involvement of environmental influences

contributing to the susceptibility for different psychiatric disorders. However, despite all these efforts, each of the genetic variants discovered to date explains only a small fraction of the overall heritability previously described. One of the conclusions derived from GWAS studies is that common variants *en masse* (SNP-$h^2$) explain a substantial proportion of heritability. However, an important fraction of heritability cannot be explained, the so-called missing heritability. Some researchers suggest that epistasis (i.e. interactions gene–gene) and **gene–environment interplay** may account for at least a part of this missing heritability and should be considered [2].

## 6.2 Gene–Environment Dialogue: Role of Genotype in the Development of Mental Disorders in Individuals Exposed to Childhood Adversity

That genetic and environmental factors are interrelated and jointly modulate the expression of multiple phenotypes is a completely expected and plausible biological mechanism, i.e. genotypes do not exist separately, and their expression must depend

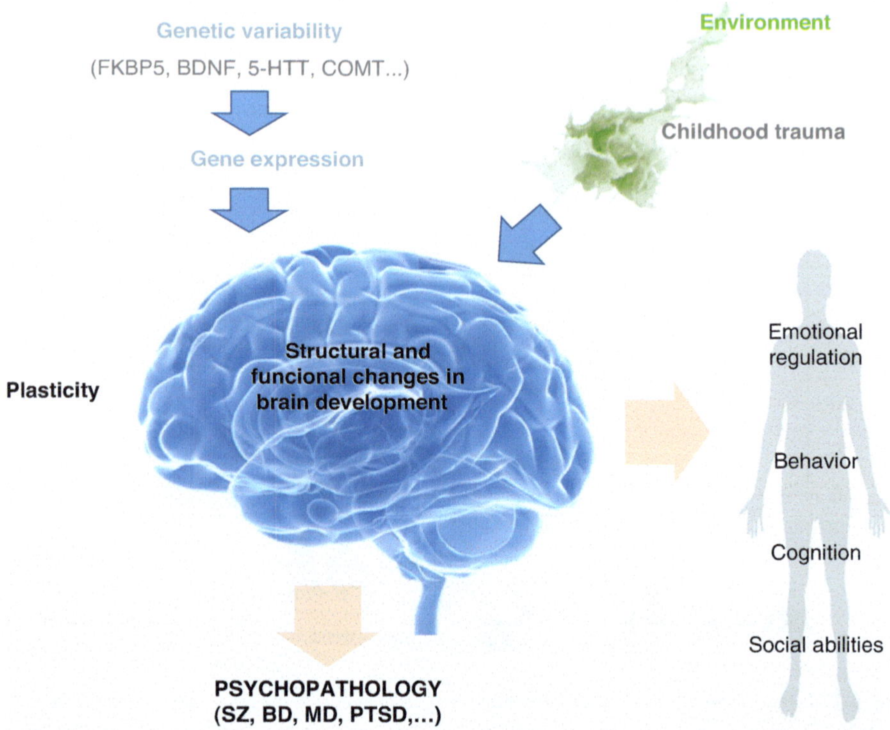


**Fig. 6.2** Genetic liability and environmental factors contribute to structural and functional changes of our brain that will determine our emotional regulation, cognitive functions, social abilities and shape our behaviour, and also can be responsible for psychiatric disorders. This is possible because the plasticity of our brain and its capacity of modifying itself according to external stimuli, and of adjusting its response to the stimulus received. *SZ* schizophrenia, *BD* bipolar disorder, *MD* major depression, *PTSD* post-traumatic stress disorder

to some extent on environmental context (from the cell environment to the social environment of a person) (Fig. 6.2). Ecogenetics is a discipline that studies genetic traits in relation to the response to environmental influences. It demonstrates that genes modify both the exposure and sensitivity to environment, and that environment impacts on gene expression and function [3, 4].

The most used strategies to study the role of genetic factors in the development of mental disorders in individuals that suffer early adversity are the study of: (1) gene–environment correlations (rGE) and (2) gene–environment interactions (GxE) (Fig. 6.3).

### 6.2.1 Gene–Environment Correlation

The first approach (rGE) is concerned with the possibility that the occurrence of the environmental exposure is not entirely random and depends, at least partially, on genetic factors (i.e. genes and environments are correlated) (Fig. 6.3a). A genetic variant, or the genetic background of the individual, may be associated with an

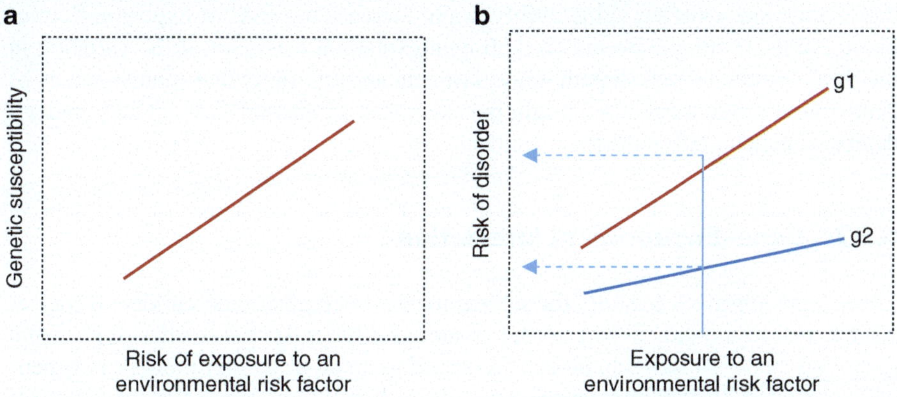


**Fig. 6.3** (**a**) *Gene–environment correlation* (rGE): the genetic control of exposure to the environment. The greater the genetic susceptibility, the higher the probability to be exposed to an environmental risk factor. (**b**) *Gene–environment interaction*: the genetic control of sensitivity to the environment. The effect of an environmental risk factor for a disease is modulated by the genotype. Then, with the same exposition to an environmental factor, one individual with genotype 2 (g2) would have higher risk than one carrying the genotype 1 (g1) (Adapted from van Os and Marcelis 1998)

increase of the likelihood that a person would be exposed to an environmental risk factor, which in turn, increases the likelihood to develop a particular trait or disease.

Three specific ways by which genes may exert an effect on the environment have been delineated: (1) **Passive rGE**: refers to the fact that among biologically related relatives (i.e. non-adoptive families), parents provide not only their children's genotypes but also their rearing environment. Therefore, the child's genotype and the home environment are correlated; (2) **Evocative rGE**: refers to the idea that individuals' genotypes influence the responses they receive from others. For example, a child having a cheerful disposition might be more likely to receive positive attention from others than a child who has an antisocial behaviour that evoke unpleasant responses from others; (3) **Active rGE** refers to the fact that an individual actively selects or avoids certain environments, and these processes are influenced by an individual's genotype.

Although it is not very usual to explore these mechanisms, especially regarding childhood trauma, some studies have tried to analyse them. The published studies have used two genetically informative designs, twin studies and twin-parent studies. For example, using a birth cohort of 1,116 twins and their families who participated in the Environmental Risk Longitudinal Study (E-RISK), Jaffee and colleagues analysed the association between maltreatment in children with or without antisocial behaviour [5]. Their findings minimized the possibility that children who were victims of maltreatment provoked the abusive incidents because of their difficult behaviour. Thus, the authors concluded that the risk factors causing child abuse were more likely to be found within the family environment or the adult abuser.

According to this and more recent studies, while we cannot rule out that gene–environment correlation plays a role, at least part of the association of childhood trauma with psychopathology points to a possible causal explanation. That means that while a pre-existing vulnerability might increase the risk for exposure to traumatic events, traumatic events on their own explain at least part of the variation in the development of psychopathology. The important point is that many sources of behavioural influence that we might consider "environmental" are actually under a degree of genetic influence [6].

### 6.2.2 Gene–Environment Interaction

The second approach to study the interaction between genes and childhood trauma (GxE) tries to explain if part of the observed differential risk to develop mental pathology between individuals early exposed to an adverse environment is genetically mediated. If such interaction exists, the response (e.g. depressive, psychotic or psychotic-like outcome) of an individual to this environmental risk factor will be moderated by genetic variants carried by the individual. Thus, two persons exposed to the same environmental risk factor will develop a different phenotype depending on their genetic background (Fig. 6.3b).

The interplay between genes and environment can be modelled on either *additive* or *multiplicative* models. For example, an additive model between two risk factors E and G would assume that the risk of disease being jointly exposed to E and G is the sum (addition) of the risk of being exposed to E only or exposed to G only (Fig. 6.4a). In multiplicative models, two types of interactions represented in Fig. 6.4b and c are possible.

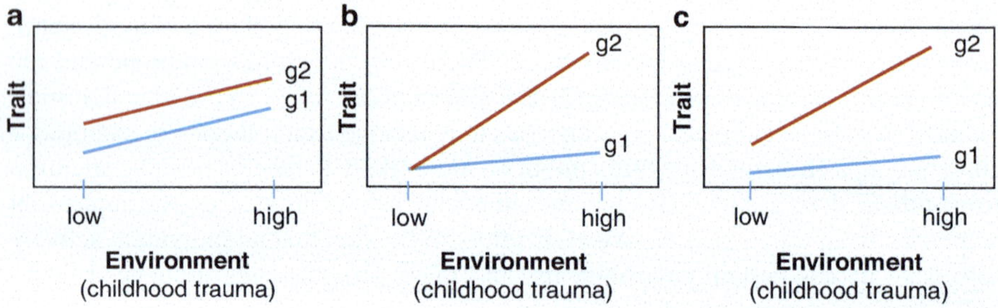


**Fig. 6.4** Three qualitative patterns of gene–environment interplay. The *y*-axis represents a trait value (e.g. depressive symptoms); the *x*-axis represents the environmental condition (e.g. to have experienced childhood trauma), and the red and blue lines depict two different genotypes with varying levels of liability to depression. (**a**) An additive model. The lines are parallel increasing from low- to high-risk environments, the increase in the level of Y is the same across the two genotypes. Genes and environment act independently of one another. (**b**) Represents a "fan-shaped" interaction, where the impact of genes is dependent on the environment, and vice versa. The key characteristic of this interaction is that, in benign environments, the difference in the level of the outcome variable (i.e. depression) as a function of the level of genetic liability is quite modest. That is, genes are not doing that much in a protective environment. However, with increasingly severe environmental exposures, the difference between genotypes increases. (**c**) The effect of an environmental factor on the expression of the trait is modulated by the genotype, with the same exposition to an environmental factor, one individual with the genotype 2 (g2) will have higher risk than one carrying the genotype 1 (g1)

In the first one, the influence of genotype is greater in one environmental context than in another (Fig. 6.4a). In the second, there is no association of genotype with the outcome in the absence of exposure to particular environmental conditions (Fig. 6.4b). In other situations, the effect of genotype is differential in the absence of environmental exposure. When the exposition to severe environmental factors takes place, the difference between genotypes increases in a synergic way (Fig. 6.4c).

Several theoretical models have been proposed to describe GxE. The most accepted and used framework is *the diathesis-stress model or vulnerability-stress model* [7]. This model posits that genetic vulnerability predisposes an individual to the development of a psychiatric disorder and, when a level of stressors exceeds this vulnerability threshold, symptoms emerge (Fig. 6.5a). In this sense, individuals carrying genetic risk variants will be more vulnerable to the effects of adverse environmental factors (e.g. childhood trauma) and thus, more prone to develop psychopathology than those

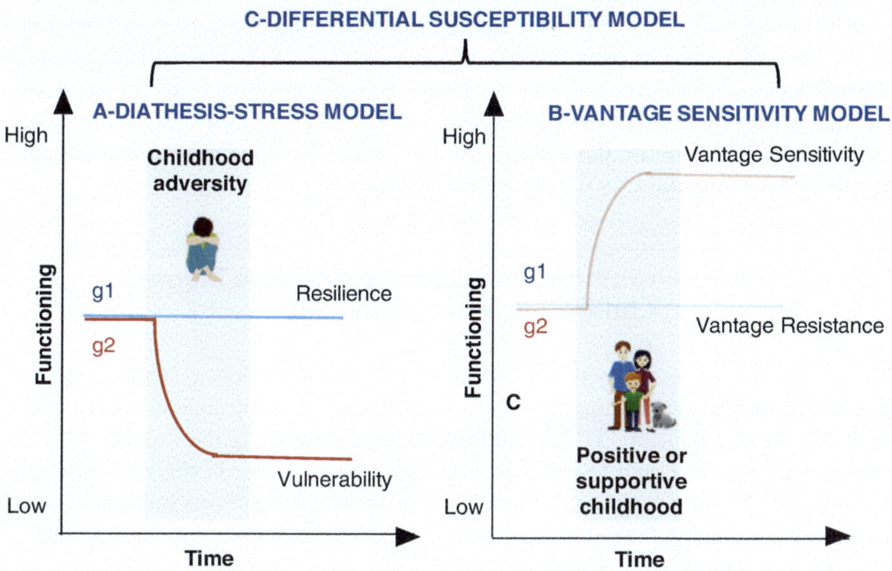


**Fig. 6.5** Different theoretical proposed models to describe gene–environment interactions (environmental sensitivity genetically mediated). (**a**) *The diathesis-stress model* (used in the vast majority of G × E studies) assumes that having a genetic sensitivity (genotype 2) amplifies the probability that exposure to childhood adversity (e.g. abuse, neglect or negative parenting) will result in emotional and behavioural problems, compared to subjects with low genetic sensibility (genotype 1). Thus, the perspective presumes that some individuals are by nature more vulnerable than others, as they possess dysfunctional "risk alleles" in the case of early negative exposures. (**b**) *The Vantage sensitivity model* considers that individuals with high genetic sensitivity (genotype 2) have increased response to positive environmental exposures, compared to less sensitive individuals (genotype 1); (**c**) *The differential susceptibility model*, integrating the two previous models, assumes that persons with plasticity alleles may be more readily shaped by environmental rewards and punishments than other genotypes. Plasticity alleles show significantly poorer adjustment than other genotypes when the environment is adverse, but significantly better adjustment than other genotypes when the environment is supportive (Adapted from Assary et al. 2018 [9])

with the non-risk variants. In the absence of such adversities, these inherent vulnerabilities are not sufficient to lead to psychopathology and subjects recover from the adversity, becoming resilient (Fig. 6.5a). In the *Vantage Sensitivity Model* of GxE, the effects of positive or supportive environments are accentuated for some genotypes (i.e. g2).

The diathesis-stress model and the vantage sensitivity model of GxE have recently been challenged by the *differential-susceptibility model* [8], which suggests that individuals differ in their general susceptibility to both negative and positive environmental influences. According to this model, the subjects may carry "plastic variants" and not "risk" and "non-risk" variants (Fig. 6.5c). In this sense, subjects carrying the plastic variants will be more permeable to the environment (for worse and for better). Thus, the same genetic variants involved in increasing the negative effects of adverse experiences could also be involved in enhancing the likelihood of benefiting from positive ones (Fig. 6.5b).

Although recently it has been increasingly pointed out that individuals may differ in their susceptibility to the environment across a range of exposures (from positive to negative), these approaches have been scarcely considered, and the positive environmental experiences are less explored than the negative ones. In this case, although we will focus on childhood adversity, which is the main topic of this book, it is important to underlie the need of incorporating the full range of positive and negative environmental experiences on GxE studies.

## 6.3 Gene–Environment Interactions in Severe Mental Disorders: Childhood Trauma and GxE

Studies on GxE have mostly relied on candidate-gene approaches. In these studies, researchers choose a particular gene of interest based on its biological function and its involvement on specific biological mechanisms underlying the disorder or phenotype being studied, and test whether the association between this variation and the disorder differs across environments. This approach has been implemented in childhood trauma GxE studies in mental disorders, typically relying on single nucleotide polymorphisms (SNPs).

That childhood trauma underlies the development of severe mental disorders has been largely supported, as explained along several chapters of this book. However, the exact mechanisms involved are not well understood. In this regard, the **neurodevelopmental hypothesis** proposes that mental disorders are related to adverse conditions (genetic background and environmental factors) that lead to abnormal brain development. Neurodevelopmental disturbances occurred in perinatal periods may lead to dysfunction of neuronal circuits and vulnerability to stress during vulnerable brain periods (Fig. 6.6). Subsequent stressors at a later point in time such as childhood trauma may then trigger the disorder [10]. **Stress sensitization** is another plausible hypothesis suggesting that early experiences increase the risk for mental disorders through the dysregulation of the hypothalamic-pituitary-adrenal (HPA) axis (Fig. 6.7). Exposure to stress causes the activation of the HPA axis, which

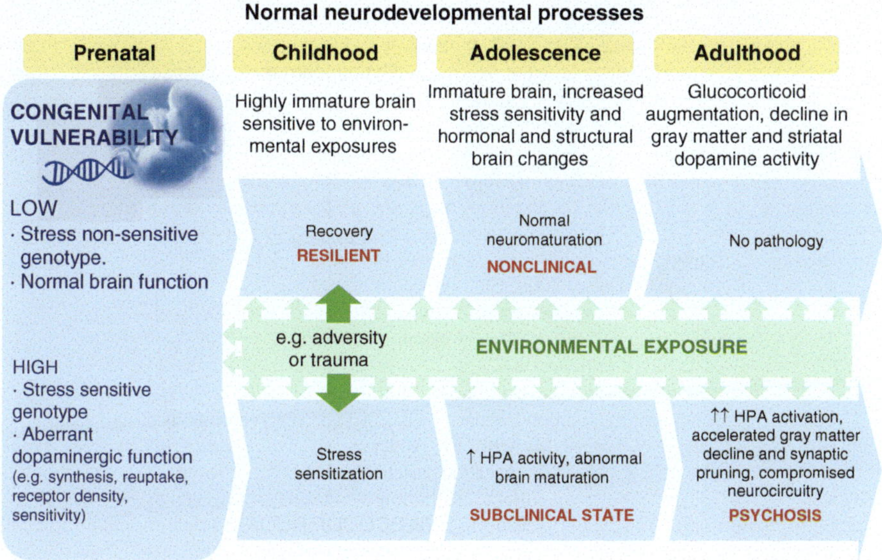


**Fig. 6.6** Gene–environment interaction, stress and neurodevelopmental trajectories. The figure shows a tentative model of neural mechanisms in a subtype of psychosis characterized by greater stress sensitivity. Our brain faces several modifications during the neurodevelopment. Congenital genetic vulnerability may determine our brain structure and function. Genetic vulnerability (low or high) predisposes the individual to the development of a psychiatric disorder. The presence of adverse environmental exposures, such as early adversities or childhood trauma, may lead to abnormal neurodevelopment (i.e. changes in brain and neuronal activity). These alterations may cause latent vulnerabilities (subclinical state) that finally may lead to the development of clinical symptomatology (e.g. psychosis) (Adapted from Holtzman et al. 2013 [16]). *HPA* hypothalamic-pituitary-adrenal axis

triggers several pathways involved in the regulation of gene expression for metabolism, immune function, cognition and brain development, thereby preparing the body to respond to such stress. However, severe and/or prolonged stress exposure, especially in very sensitive periods, can alter this normal response [11] and may precipitate a cascade of events with neurobiological consequences, including aberrant neural circuit changes such as an abnormal increase in dopamine signalling or neurotrophic factors [12, 13] (Fig. 6.6). In this sense, alterations in HPA axis reactivity and levels of related molecules have been found in patients with depression, anxiety and psychosis [14, 15]. In Fig. 6.7 we report a theoretical model of neuronal mechanism in a subtype of psychosis characterized by greater stress sensitivity proposed by Holtzman and colleagues that could be expanded to other psychiatric disorders [16].

According to this, polymorphic variants in genes involved in neurodevelopment or in the modulation of the HPA axis-driven stress response, are potential moderators of the impact of childhood adversity on mental health (Table 6.1). Considering the high co-morbidity of psychiatric disorders [16] and their shared genetic aetiology, it is not surprising that many of these candidate genes have been examined and associated with multiple disorders.

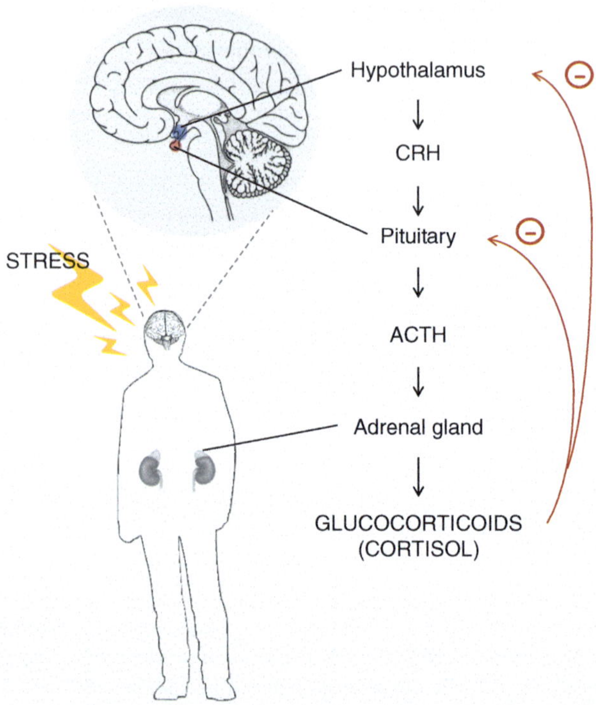


**Fig. 6.7** The hypothalamus-pituitary adrenal (HPA) axis. The adaptive response to perceived stress is mediated by the HPA axis. In the presence of a stressor, corticotropin-releasing hormone (CRH) is secreted from the hypothalamus, which acts on the pituitary gland and results in the release of adrenocorticoid hormone (ACTH). This stimulates the production and release of cortisol from the adrenal cortex, which binds to its receptor, the glucocorticoid receptor (GR), which once activated, exerts a wide range of effects orchestrating the systemic stress response. Besides this, cortisol also inhibits the synthesis and release of CRH and ACTH in the hypothalamus and the pituitary gland, enabling a negative feedback regulation critical for the reduction of HPA axis activation and the restoration of homeostasis once the threat has subsided

### 6.3.1 Psychosis

A number of studies have attempted to identify genetic variants interacting with childhood adversity that increase the risk for psychosis. Some genes classically related to psychosis that were suspected to play a role in the GxE in psychosis vulnerability have been studied. One such gene is the catechol-o-methyltransferase gene (*COMT*). It has been shown that variability in this gene expression modulated the severity of positive symptoms in schizophrenia in patients who experienced physical abuse [17]. Also, it was observed that severity of negative symptoms in those patients exposed to emotional neglect differed between genotypes.

Another gene classically associated with psychosis is the brain-derived neurotrophic factor (*BDNF*). This gene contains a polymorphism, *Val66Met*, which causes a modification in the protein suspected to compromise protein secretion and

**Table 6.1** Main candidate genes in GxE studies analysing interactions between maltreatment and genetic vulnerability during early developmental phases

| Gene | Function |
|---|---|
| BDNF | Supporting the survival of existing neurons and stimulating neurogenesis |
| COMT | Involved in degradation of catecholamines (dopamine, epinephrine and norepinephrine) |
| CRHR1 | Role in the stress response/modulation promoted by the HPA axis |
| FKBP5 | Involved in the regulation of neuroendocrine stress mechanisms since it is a co-chaperone interfering with glucocorticoid receptor activity |
| GABRA2 | Forms GABA receptors |
| MAOA | Involved in the degradation of neurotransmitters like dopamine norepinephrine and serotonin |
| NR3C1 and 2 | Fundamental for HPA stress reactivity |
| SLC6A4 | Function of transcribing proteins which regulate the reuptake of serotonin at brain synapses |

*BDNF* Brain-derived neurotrophic factor, *COMT* Catechol-o-methyl transferase, *CRHR1* Corticotropin-releasing hormone receptor 1, *FKBP5* FK506 binding protein 5, GABA Gamma-aminobutyric acid, *GABRA2* Gamma-aminobutyric acid receptor subunit alpha-2, *HPA* Hypothalamic-pituitary-adrenal axis, *MAOA* Monoamine oxidase A, *NR3C1* Glucocorticoid receptor, *SLC6A4* Serotonin transporter

normal neurotransmission. Studies in clinical samples showed that subjects exposed to childhood trauma differed in terms of schizophrenia risk depending on the polymorphism-carried variant [18]. Similarly, the involvement of *Val66Met* was detected at onset of first psychotic episode in individuals exposed to early trauma [19]. It is interesting to note that this gene was also studied in relation to psychosis proneness in non-clinical samples, while no effect of this GxE interaction on psychotic-like experiences was found [20, 21]. Other studies have detected this GxE effect when exploring sub-phenotypes of psychosis such as cognition and brain abnormalities [22], although, interestingly, this sample included patients with both schizophrenia spectrum disorders and bipolar disorder.

The FK506 binding protein 5 (*FKBP5*) is a more recently emerged gene that has also been explored in GxE studies in psychosis showing interesting results. Variability in this gene expression interacted with childhood trauma and had an effect on positive psychotic symptoms [23]. Moreover, this GxE effect was also detected in non-clinical subjects in two independent studies analysing psychotic-like experiences [23, 24].

### 6.3.2 Depressive Disorders

The first GxE study in depression was published 15 years ago in a pioneering study [25] analysing a functional polymorphism in the serotonin transporter gene (*SLC6A4*), the *5-HTTLPR*. The *SLC6A4* gene plays a key role in regulating serotoninergic neurotransmission. The *5-HTTLPR* polymorphism has two frequent alleles (short and long). The short variant reduces the amount of serotonin transporter produced. In this study, they observed that this short variant in the *5-HTTLPR* was moderating the influence of stressful life events on depression [25] and subsequent studies supported these findings showing similar results when childhood maltreatment was specifically considered [26–28]. However, a more recent meta-analysis could not find this interaction for depression [29].

The *BDNF Val66Met* has also been studied in GxE in depression; however, these studies mostly have analysed depressive symptoms in healthy people. Likewise in psychiatric disorders, this polymorphism seemed to moderate the effect of childhood trauma on depressive symptoms [28] and interestingly, brain structural and arousal changes were detected as possible underpinnings of this GxE interaction [30].

Similarly to what observed for *BDNF*, the genetic variability within the *FKBP5* gene has also been associated with depression in a context of childhood trauma environment, thus supporting the hypothesis of common underpinnings of different psychiatric disorders [31, 32].

Another interesting gene is the corticotropin-releasing hormone receptor 1 (*CRHR1*). The corticotropin-releasing hormone is a key component of the HPA axis, acting in a widespread circuitry and integrating endocrine, autonomic and behavioural responses to stress. Genetic variability within the *CRHR1* has been demonstrated to predict adult depression in interaction with childhood trauma in several independent studies [33, 34].

The glucocorticoid receptor (NR3C1) also plays an important role in the HPA axis. Variability within the *NR3C1* gene has been explored in GxE studies which observed differences on the development of depressive symptoms depending on the subject's genetic variants, and the context of childhood trauma [35].

### 6.3.3 Anxiety Disorders

In the case of anxiety disorders, some preliminary GxE research has been conducted, although some studies have explored depressive and anxiety symptoms together. This is the case of the study by Gatt and colleagues analysing *BDNF* and mentioned in the previous section [30]. Nevertheless, there are also studies only focused on anxiety disorders that have detected the effect of *BDNF* variants in a context of childhood trauma environments [36]. Similarly, the interaction effect between *FKBP5* variations and childhood trauma was observed in a mixed phenotype of depression and anxiety symptoms [37], and also in patients showing only anxiety symptoms [38].

### 6.3.4 Substance-Dependence Disorder

There is limited research exploring the interaction effect between genes and childhood trauma on substance-dependence disorders. Mineralocorticoid and glucocorticoid receptor genes (*NR3C2* and *NR3C1*, respectively) were proved to influence the susceptibility to crack/cocaine addiction and response to detoxification treatment in interaction with childhood trauma [39].

Another gene explored in this regard is the gamma-aminobutyric acid receptor subunit alpha-2 (*GABRA2*). This gene encodes for a subunit that is part of the GABA-α receptors, which are activated by the major inhibitory neurotransmitter in the human brain, the GABA. Variability within *GABRA2* interacted with childhood trauma to influence addiction vulnerability [40].

Variability within the monoamine oxidase gene (*MAOA*) was found to interact with childhood trauma and predicted alcoholism. This protein is important for the breakdown of monoamines, which, among other functions, serve to inactivate monoamine neurotransmitters (i.e. dopamine or serotonin). The *FKBP5* gene was also found to interact with early life trauma in predicting heavy drinking in college students [41].

There are other multiple studies analysing substances use; however, they have explored such disorders as another environmental variable interacting with genetic factors affecting the psychosis phenotype, or other diseases, but not as the outcome of the interaction (e.g. [42–44]).

### 6.3.5 Other Psychiatric Disorders

To date, scarce gene–environment interaction research has been conducted in relation to other disorders such as bipolar, post-traumatic stress, personality or obsessive-compulsive disorders, and the genes associated are mostly the same as the ones associated with the disorders mentioned in the previous sections.

In the case of bipolar disorder, studies have shown that *BDNF* is also involved in the moderation of the effect of childhood trauma on the development of such disorder [45]. Genetic variations within the *SLC6A4* gene have also been explored in one study in relation to bipolar disorder, observing an interaction effect between *5-HTTLPR* variants and childhood trauma on the age at onset [46].

An interaction effect between *5-HTTLPR* genotype and childhood trauma was also observed in post-traumatic stress disorder (PSTD) [47]. Genetic variants within *FKBP5* were observed to interact with childhood trauma predicting adult PSTD symptoms [48, 49].

Personality disorders have also been explored in GxE studies, which have shown the involvement of *MAOA* genetic variants as modulators of the effect of childhood abuse on borderline personality disorder endophenotypes [50], antisocial personality disorder [51] and a broad range of personality pathologies [52]. The *FKBP5*–childhood trauma interaction has also been reported for borderline personality disorder [53].

Finally, regarding obsessive-compulsive disorder, the *BDNF Val66Met* was observed to significantly increase the risk for this disorder in interaction with childhood trauma [54].

Beside the important contribution of studies on specific candidate genes and early adverse environment on the development of all the mentioned disorders, it is important to acknowledge several limitations.

Firstly, the candidate gene approach requires a biological hypothesis. Thus, it is necessary to have previous knowledge of specific pathophysiological disturbances in molecular pathways, which may in turn, impact the neural circuits associated to the pathophysiology of the disorder. However, our knowledge on the biological mechanisms underlying psychiatric disorders remains rather limited.

Second, only a handful of the large number of reported statistically significant interactions has been replicated, and some studies include small sample sizes and inappropriate (or lack of) adjustment for multiple testing.

Additionally, most of the GxE studies have been conducted from a diathesis-stress perspective, which needs to be revisited according to the differential susceptibility model, suggesting that many of the common genetic variants could reflect susceptibility to both positive and negative environmental inputs.

Finally, the simplistic conceptualization of most studies considering one single genetic risk and one unique environmental factor is another important limitation. As we previously commented, most psychiatric disorders are influenced by many thousands of gene variants of small effect likely working together to shape the risk (i.e. polygenic) and also non-genetic factors. For this reason, GxE studies are transitioning towards polygenic and genome-wide approaches in larger better-powered samples.

## 6.4 Polygenic Approaches in GxE Studies

Given the limitations of the candidate gene approach, GxE research in mental disorders has begun to employ polygenic approaches that interrogate the entire genome for variants moderating the effects of the environment on psychiatric disorders.

Polygenic risk scores (PRS) provide a novel opportunity to test the diathesis-stress model since PRS can be conceptualized as an indicator of the diathesis and will likely constitute a much accurate instrument compared to a single risk gene. PRS estimation uses *Genome-Wide Association Study* (GWAS) results to predict the genetic risk burden of each individual in an independent sample. PRS are estimated as the sum of risk alleles weighted by their respective estimated effect sizes (Fig. 6.8) [55, 56].

Using this approach, two previous studies have examined the interaction between the PRS for major depressive disorder (MDD) and childhood trauma in the prediction of MDD [57, 58]. Although both studies did find a significant interaction between the PRS and childhood maltreatment, the nature of this interaction differed markedly for each of the studies. For example, Peyrot and colleagues demonstrated in the NESDA study (Netherlands Study of Depression and Anxiety) that the effects of childhood trauma on MDD were greater for those with a higher PRS for MDD. However, Mullins and colleague found in the RADIANT-UK sample a stronger impact of PRS on MDD risk in those unexposed to CT. These opposing findings, both of which were significant, are not well understood, and it remains unclear whether they reflect actual differences between cultures, differences between recruitment of participants into cohorts, or chance.

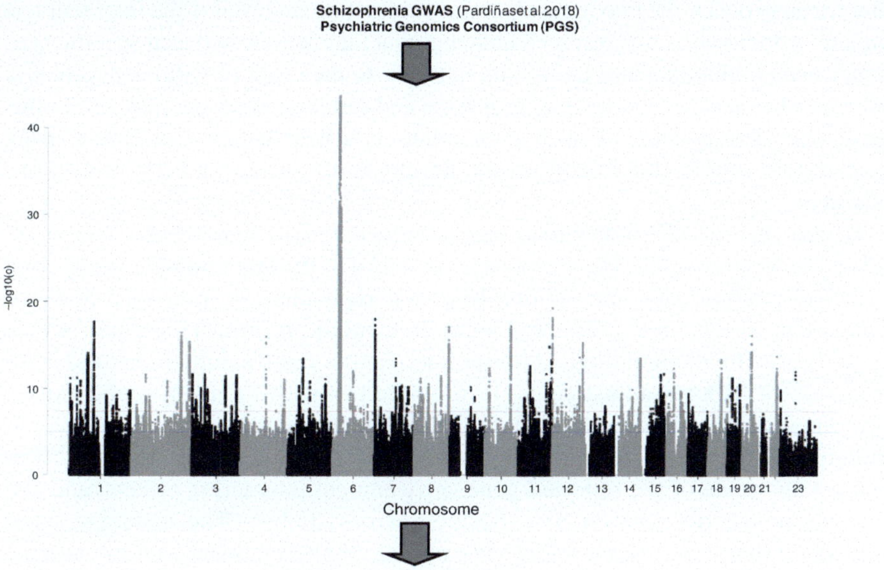


Polygenic Risk Score (PRS)

In an independent sample, we can calculate the individual PRS: the **number of risk variants** carried by the individual (0, 1, or 2) was multiplied by the logarithm of the odds ratio for that particular variant.

snpid hg18chr bp a1 a2 or se pval info ngt CEUaf

| | | | | | OR | | P-value |
|---|---|---|---|---|---|---|---|
| rs3131972 | 1 | 742584 | A | G | 1.0257 | 0.0835 | 0.761033 |
| rs3131969 | 1 | 744045 | A | G | 1.0221 | 0.0801 | 0.784919 |
| rs3131967 | 1 | 744197 | T | C | 1.0227 | 0.0858 | 0.79352 |

Individual PRS (genetic risk load) for each subject, for each P value threshold

**Fig. 6.8** Example of calculation of schizophrenia PRS from the genome-wide association study (GWAS) on schizophrenia. The schizophrenia GWAS conducts a systematic examination in a discovery sample to test whether genotypic frequencies for variants across the genome differ between individuals affected by schizophrenia and controls. The PRS for an individual encompasses the additive effect of multiple common SNPs across the genome. It is computed based on the most recent summary statistics on the GWAS from the Psychiatric Genomic Consortium at different *P*-value thresholds and calculated multiplying the imputation probability for the risk allele by odds ratio for such genetic variant in the discovery sample. The resulting values are summed up in an additive fashion obtaining an individual estimate of the SZ genetic burden in each individual

## 6.5 Conclusions and Future Directions

In this chapter, we have presented several examples of the growing interest in the field of gene–childhood trauma interaction contributing to complex psychiatric phenotypes. These studies, including GxE models, increase the power to detect new genetic or environmental factors that influence the trait considering the potential interaction between the two, a result that would not be achieved if the interaction is ignored. Additionally, this research indicates that interactions between specific genotypes and childhood abuse contribute not only to the onset of various psychiatric disorders, but also to their course, prognosis and refractoriness to treatment. In this sense, the identification of such interactions can help uncover disease-causing mechanisms, which could result in the development of new or better preventive measures.

To date, most GxE studies have been conducted using hypothesis-driven candidate–gene approaches. However, only a handful of the large number of reported statistically significant interactions has been replicated, despite well-powered replication efforts for reproducing some influential preliminary reports, which were affected by methodological limitations such as small sample sizes and inappropriate (or lack of) adjustments for multiple testing. Moreover, replication in the context of gene–environment interaction effects faces additional challenges, including differences in exposure measurement protocols across studies, differences in the scale of the reported gene–environment interaction effects, and differences in the distribution of exposures across studies. The candidate gene interaction literature can therefore only provide limited guidance on the number and size of gene–environment interaction effects expected to truly exist in human populations, although it does suggest that large and pervasive interaction effects are unlikely. For this reason, GxE research in psychiatric disorders is beginning to shift from candidate gene to genome-wide approaches. These studies will need larger samples and at the same time, a more objective and accurate assessment of childhood trauma.

Additionally, given that several mental disorders have been associated to different environmental factors (e.g. being socially disadvantaged in childhood, urbanicity, cannabis use, season of birth, childhood adversity as factors linked to psychosis), future studies should incorporate multiple environmental risk factors rather than specific ones (GxExE).

On the other hand, as we have seen along this book, there is substantial etiological overlap between different disorders, given that the same environmental and genetic factors have been found to contribute to different mental disorders. In this regard, childhood abuse has been linked to schizophrenia, bipolar disorder, major depression and anxiety. In the same way, many of these candidate genes (e.g. *BDNF*) have been examined and associated with multiple disorders. In this regard, some authors have suggested that the use of traits or outcome measures, rather than the diagnosis, could be useful in GxE studies. These endophenotypes or intermediate phenotypes tend to be more proximal to the biological determinants, and their study allows the determination of the aetiological pathways underlying them, avoiding the complexity of a diagnosis-based phenotype.

A better understanding of how genes and environment work together in the aetiology of psychiatric disorders will not only help us to understand the origins of psychopathology, but may also identify novel pathways for intervention and treatment.

## References


1. Lichtenstein P, Yip BH, Björk C, Pawitan Y, Cannon TD, Sullivan PF, Hultman CM. Common genetic determinants of schizophrenia and bipolar disorder in Swedish families: a population-based study. Lancet. 2009;373:234–9.
2. Woo HJ, Yu C, Kumar K, Reifman J. Large-scale interaction effects reveal missing heritability in schizophrenia, bipolar disorder and posttraumatic stress disorder. Transl Psychiatry. 2017;7:e1089.
3. Kendler KS, Eaves LJ. Models for the joint effect of genotype and environment on liability to psychiatric illness. Am J Psychiatry. 1986;143:279–89.
4. Van Os J, Rutten BPF, Poulton R. Gene-environment interactions in schizophrenia: review of epidemiological findings and future directions. Schizophr Bull. 2008;34:1066–82.
5. Jaffee SR, Caspi A, Moffitt TE, Taylor A. Physical maltreatment victim to antisocial child: evidence of an environmentally mediated process. J Abnorm Psychol. 2004;113:44–55.
6. Lecei A, Decoster J, De Hert M, et al. Evidence that the association of childhood trauma with psychosis and related psychopathology is not explained by gene-environment correlation: a monozygotic twin differences approach. Schizophr Res. 2019;205:58–62.
7. Zubin J, Spring B. Vulnerability--a new view of schizophrenia. J Abnorm Psychol. 1977;86:103–26.
8. Belsky J, Pluess M. Beyond diathesis stress: differential susceptibility to environmental influences. Psychol Bull. 2009;135:885–908.
9. Assary E, Vincent JP, Keers R, Pluess M. Gene-environment interaction and psychiatric disorders: review and future directions. Semin Cell Dev Biol. 2018;77:133–43.
10. Schmitt A, Malchow B, Hasan A, Falkai P. The impact of environmental factors in severe psychiatric disorders. Front Neurosci. 2014;8:19.
11. De Bellis MD, Zisk A. The biological effects of childhood trauma. Child Adolesc Psychiatr Clin N Am. 2014;23:185–222.
12. Carbone DL, Handa RJ. Sex and stress hormone influences on the expression and activity of brain-derived neurotrophic factor. Neuroscience. 2013;239:295–303.
13. Van Winkel R, Stefanis NC, Myin-Germeys I. Psychosocial stress and psychosis. A review of the neurobiological mechanisms and the evidence for gene-stress interaction. Schizophr Bull. 2008;34:1095–105.
14. Elzinga BM, Spinhoven P, Berretty E, de Jong P, Roelofs K. The role of childhood abuse in HPA-axis reactivity in social anxiety disorder: a pilot study. Biol Psychol. 2010;83:1–6.
15. Myin-Germeys I, van Os J. Stress-reactivity in psychosis: evidence for an affective pathway to psychosis. Clin Psychol Rev. 2007;27:409–24.
16. Holtzman CW, Trotman HD, Goulding SM, Ryan AT, MacDonald AN, Shapiro DI, Brasfield JL, Walker EF. Stress and neurodevelopmental processes in the emergence of psychosis. Neuroscience. 2013;249:172–91.
17. Green MJ, Chia TY, Cairns MJ, Wu J, Tooney PA, Scott RJ, Carr VJ. Catechol-O-methyltransferase (COMT) genotype moderates the effects of childhood trauma on cognition and symptoms in schizophrenia. J Psychiatr Res. 2014;49:43–50.
18. Bi X, Lv X, Ai X, Sun M, Cui K, Yang L, Wang L, Yin A, Liu L. Childhood trauma interacted with BDNF Val66Met influence schizophrenic symptoms. Medicine (Baltimore). 2018;97:e0160.
19. Theleritis C, Fisher HL, Shäfer I, et al. Brain derived neurotropic factor (BDNF) is associated with childhood abuse but not cognitive domains in first episode psychosis. Schizophr Res. 2014;159:56–61.
20. Alemany S, Arias B, Aguilera M, Villa H, Moya J, Ibáñez MI, Vossen H, Gastó C, Ortet G, Fañanás L. Childhood abuse, the BDNF-Val66Met polymorphism and adult psychotic-like experiences. Br J Psychiatry. 2011;199:38–42.

21. de Castro-Catala M, van Nierop M, Barrantes-Vidal N, et al. Childhood trauma, BDNF Val66Met and subclinical psychotic experiences. Attempt at replication in two independent samples. J Psychiatr Res. 2016. https://doi.org/10.1016/j.jpsychires.2016.08.014.
22. Aas M, Haukvik UK, Djurovic S, et al. BDNF val66met modulates the association between childhood trauma , cognitive and brain abnormalities in psychoses. Prog Neuro-Psychopharmacol Biol Psychiatry. 2013;46:181–8.
23. Collip D, Myin-Germeys I, Wichers M, et al. FKBP5 as a possible moderator of the psychosis-inducing effects of childhood trauma. Br J Psychiatry. 2013;202:261–8.
24. de Castro-Catala M, Peña E, Kwapil TR, Papiol S, Sheinbaum T, Cristóbal-Narváez P, Ballespí S, Barrantes-Vidal N, Rosa A. Interaction between FKBP5 gene and childhood trauma on psychosis, depression and anxiety symptoms in a non-clinical sample. Psychoneuroendocrinology. 2017;85:200–9.
25. Caspi A. Influence of life stress on depression: moderation by a polymorphism in the 5-HTT gene. Science (80). 2003;301:386–9.
26. Uher R, Caspi A, Houts R, Sugden K, Williams B, Poulton R, Moffitt TE. Serotonin transporter gene moderates childhood maltreatment's effects on persistent but not single-episode depression: replications and implications for resolving inconsistent results. J Affect Disord. 2011;135:56–65.
27. Brown GW, Ban M, Craig TKJ, Harris TO, Herbert J, Uher R. Serotonin transporter length polymorphism, childhood maltreatment, and chronic depression: a specific gene-environment interaction. Depress Anxiety. 2013;30:5–13.
28. Aguilera M, Arias B, Wichers M, et al. Early adversity and 5-HTT/BDNF genes: new evidence of gene-environment interactions on depressive symptoms in a general population. Psychol Med. 2009;39:1425–32.
29. Culverhouse RC, Saccone NL, Horton AC, et al. Collaborative meta-analysis finds no evidence of a strong interaction between stress and 5-HTTLPR genotype contributing to the development of depression. Mol Psychiatry. 2018;23:133–42.
30. Gatt JM, Nemeroff CB, Dobson-Stone C, Paul RH, Bryant RA, Schofield PR, Gordon E, Kemp AH, Williams LM. Interactions between BDNF Val66Met polymorphism and early life stress predict brain and arousal pathways to syndromal depression and anxiety. Mol Psychiatry. 2009;14:681–95.
31. Appel K, Schwahn C, Mahler J, et al. Moderation of adult depression by a polymorphism in the FKBP5 gene and childhood physical abuse in the general population. Neuropsychopharmacology. 2011;36:1982–91.
32. Zimmermann P, Brückl T, Nocon A, et al. Interaction of FKBP5 gene variants and adverse life events in predicting depression onset: results from a 10-year prospective community study. Am J Psychiatry. 2011;168:1107–16.
33. Bradley RG, Binder EB, Epstein MP, et al. Influence of child abuse on adult depression. Arch Gen Psychiatry. 2008;65:190.
34. Polanczyk G, Caspi A, Williams B, Price TS, Danese A, Sugden K, Uher R, Poulton R, Moffitt TE. Protective effect of CRHR1 gene variants on the development of adult depression following childhood maltreatment. Arch Gen Psychiatry. 2009;66:978.
35. Bet PM, Penninx BWJH, Bochdanovits Z, Uitterlinden AG, Beekman ATF, van Schoor NM, Deeg DJH, Hoogendijk WJG. Glucocorticoid receptor gene polymorphisms and childhood adversity are associated with depression: new evidence for a gene-environment interaction. Am J Med Genet Part B Neuropsychiatr Genet. 2009;150B:660–9.
36. Chen Z-Y, Jing D, Bath KG, et al. Genetic variant BDNF (Val66Met) polymorphism alters anxiety-related behavior. Science. 2006;314(5796):140–3. https://doi.org/10.1126/science.1129663.
37. Scheuer S, Ising M, Uhr M, Otto Y, von Klitzing K, Klein AM. FKBP5 polymorphisms moderate the influence of adverse life events on the risk of anxiety and depressive disorders in preschool children. J Psychiatr Res. 2016;72:30–6.

38. Isaksson J, Comasco E, Aslund C, Rehn M, Tuvblad C, Andershed H, Nilsson KW. Associations between the FKBP5 haplotype, exposure to violence and anxiety in females. Psychoneuroendocrinology. 2016;72:196–204.
39. Rovaris DL, Mota NR, Bertuzzi GP, Aroche AP, Callegari-Jacques SM, Guimarães LSP, Pezzi JC, Viola TW, Bau CHD, Grassi-Oliveira R. Corticosteroid receptor genes and childhood neglect influence susceptibility to crack/cocaine addiction and response to detoxification treatment. J Psychiatr Res. 2015;68:83–90.
40. Enoch M-A, Hodgkinson CA, Yuan Q, Shen P-H, Goldman D, Roy A. The influence of GABRA2, childhood trauma, and their interaction on alcohol, heroin, and cocaine dependence. Biol Psychiatry. 2010;67:20–7.
41. Lieberman R, Armeli S, Scott DM, Kranzler HR, Tennen H, Covault J. *FKBP5* genotype interacts with early life trauma to predict heavy drinking in college students. Am J Med Genet Part B Neuropsychiatr Genet. 2016;171:879–87.
42. Vinkers CH, Van Gastel WA, Schubart CD, Van Eijk KR, Luykx JJ, Van Winkel R, Joëls M, Ophoff RA, Boks MPM. The effect of childhood maltreatment and cannabis use on adult psychotic symptoms is modified by the COMT Val158Met polymorphism. Schizophr Res. 2013;150:303–11.
43. Soler J, Arias B, Moya J, Ibáñez MI, Ortet G, Fañanás L, Fatjó-Vilas M. The interaction between the ZNF804A gene and cannabis use on the risk of psychosis in a non-clinical sample. Prog Neuro-Psychopharmacology Biol Psychiatry. 2019;89:174–80.
44. Alemany S, Arias B, Fatjó-Vilas M, Villa H, Moya J, Ibáñez MI, Ortet G, Gastó C, Fañanás L. Psychosis-inducing effects of cannabis are related to both childhood abuse and COMT genotypes. Acta Psychiatr Scand. 2014;129:54–62.
45. Miller S, Hallmayer J, Wang PW, Hill SJ, Johnson SL, Ketter TA. Brain-derived neurotrophic factor val66met genotype and early life stress effects upon bipolar course. J Psychiatr Res. 2013;47:252–8.
46. Etain B, Lajnef M, Henrion A, et al. Interaction between SLC6A4 promoter variants and childhood trauma on the age at onset of bipolar disorders. Sci Rep. 2015;5:16301.
47. Xie P, Kranzler HR, Poling J, Stein MB, Anton RF, Brady K, Weiss RD, Farrer L, Gelernter J. Interactive effect of stressful life events and the serotonin transporter 5-HTTLPR genotype on posttraumatic stress disorder diagnosis in 2 independent populations. Arch Gen Psychiatry. 2009;66:1201.
48. Binder EB, Bradley RG, Liu W, et al. Association of FKBP5 polymorphisms and childhood abuse with risk of posttraumatic stress disorder symptoms in adults. JAMA. 2008;299:1291–305.
49. Watkins LE, Han S, Harpaz-Rotem I, Mota NP, Southwick SM, Krystal JH, Gelernter J, Pietrzak RH. FKBP5 polymorphisms, childhood abuse, and PTSD symptoms: results from the National Health and resilience in veterans study. Psychoneuroendocrinology. 2016;69:98–105.
50. Kolla NJ, Meyer J, Sanches M, Charbonneau J. Monoamine oxidase-a genetic variants and childhood abuse predict impulsiveness in borderline personality disorder. Clin Psychopharmacol Neurosci. 2017;15:343–51.
51. Ducci F, Enoch M-A, Hodgkinson C, Xu K, Catena M, Robin RW, Goldman D. Interaction between a functional MAOA locus and childhood sexual abuse predicts alcoholism and antisocial personality disorder in adult women. Mol Psychiatry. 2008;13:334–47.
52. Byrd AL, Manuck SB, Hawes SW, Vebares TJ, Nimgaonkar V, Chowdari KV, Hipwell AE, Keenan K, Stepp SD. The interaction between monoamine oxidase a ( *MAOA* ) and childhood maltreatment as a predictor of personality pathology in females: emotional reactivity as a potential mediating mechanism. Dev Psychopathol. 2019;31:361–77.
53. Amad A, Ramoz N, Peyre H, Thomas P, Gorwood P. FKBP5 gene variants and borderline personality disorder. J Affect Disord. 2019;248:26–8.
54. Hemmings SMJ, Lochner C, van der Merwe L, Cath DC, Seedat S, Stein DJ. BDNF Val66Met modifies the risk of childhood trauma on obsessive-compulsive disorder. J Psychiatr Res. 2013;47:1857–63.

55. Wray NR, Lee SH, Mehta D, Vinkhuyzen AAE, Dudbridge F, Middeldorp CM. Research review: polygenic methods and their application to psychiatric traits. J Child Psychol Psychiatry. 2014;55:1068–87.
56. Maier RM, Visscher PM, Robinson MR, Wray NR. Embracing polygenicity: a review of methods and tools for psychiatric genetics research. Psychol Med. 2018;48:1055–67.
57. Mullins N, Power RA, Fisher HL, et al. Polygenic interactions with environmental adversity in the aetiology of major depressive disorder. Psychol Med. 2016;46:759–70.
58. Peyrot WJ, Milaneschi Y, Abdellaoui A, Sullivan PF, Hottenga JJ, Boomsma DI, Penninx BWJH. Effect of polygenic risk scores on depression in childhood trauma. Br J Psychiatry. 2014;205:113–9.

# Childhood Trauma, Attachment Patterns, and Psychopathology: An Evolutionary Analysis

7

Alfonso Troisi

## 7.1 Introduction

The Dunedin Longitudinal Study is a long-running cohort study of 1037 people born between April 1, 1972 and March 31, 1973 in Dunedin, New Zealand. Participants were assessed at ages 3, 5, 7, 9, 11, 13, 15, 18, 21, 26, 32, and 38 years and followed from birth to midlife with 95% retention. At the most recent follow-up assessment when the cohort reached 38 years of age, a segment comprising 22% of the cohort accounted for 36% of the cohort's injury insurance claims; 40% of excess obese kilograms; 54% of cigarettes smoked; 57% of hospital nights; 66% of welfare benefits; 77% of fatherless child-rearing; 78% of prescription fills; and 81% of criminal convictions [1]. The researchers found they could have predicted which adults were likely to incur such costs as early as age 3 based on assessment of four childhood risk factors that are thought to augur poor adult outcomes: growing up in a socioeconomically deprived family, exposure to maltreatment, low IQ, and poor self-control.

The methodological sophistication of the longitudinal study by Caspi and coworkers is probably unrivaled. Yet, in the last two decades, hundreds of retrospective and observational studies have converged on the same conclusion: children who experience severe chronic stressors are vulnerable to a plethora of medical and psychiatric problems across the life span. Early adverse experiences include a variety of stressful conditions that range from living in a family of low socioeconomic status, to neglect or abuse by caregivers in its various forms (i.e., physical, emotional, sexual). Children maltreated by their parents not only develop psychiatric disorders at higher than expected rates, but also show increased prevalence of metabolic syndrome, coronary heart disease, some cancers, and autoimmune conditions as they age [2].

A. Troisi (✉)
Department of Systems Medicine, University of Rome Tor Vergata, Rome, Italy
e-mail: alfonso.troisi@uniroma2.it



G. Spalletta et al. (eds.), *Childhood Trauma in Mental Disorders*,
https://doi.org/10.1007/978-3-030-49414-8_7

In this chapter, the relationship between childhood trauma and psychopathology will be analyzed from the vantage point of two interrelated theoretical perspectives: *attachment theory* and *evolutionary theory*. The focus on attachment is advantageous for several reasons. Recent studies have challenged the widely held assumption that certain stressful early experiences are linked to specific mental health outcomes [3]. Childhood trauma is associated with a variety of different psychiatric symptoms and syndromes. The most likely explanation for such non-specific effects is that childhood trauma impacts on basic psychobiological mechanisms that underlie various diagnostic entities [4]. Individuals with insecure attachment present one or more of these altered mechanisms including hypervigilance to threat, deficits in emotion recognition, and insensitivity to reward. This makes dysfunctional attachment a strong candidate for the development of a transdiagnostic model of psychopathology linked to early adverse experiences [5]. Patients with the same diagnoses show major clinical and neurobiological differences depending on the presence or absence of stressful experiences during childhood. For example, maltreated individuals with depressive, anxiety, and substance use disorders have an earlier age at onset, greater symptom severity, more comorbidity, a greater risk for suicide, and poorer treatment response than non-maltreated individuals with the same diagnoses. Imaging findings associated with these disorders, such as reduced hippocampal volume and amygdala hyperreactivity, are more consistently observed in maltreated individuals and may represent a maltreatment-related risk factor [6]. Moreover, in patients with a diagnosis of major depression, blood levels of deacetylase sirtuin-1 (a major inhibitor of oxidative stress) are correlated with the severity of depressive symptoms only in the subgroup with a history of traumatic childhood [7].

Linking childhood trauma to dysfunctional attachment patterns rather than to psychiatric diagnoses, is in line with the new approach to clinical observation and mental health research codified in the Research Domain Criteria (RDoC, a research framework for new approaches to investigating mental disorders) [8]. Attachment is included in the RDoC restricted list of psychobiological domains/constructs that are relevant to psychopathology because experimental and clinical data have demonstrated that attachment research can move in two directions: upward from physiological measures to clinically relevant behaviors and social interactions, and downward to the genetic and molecular/cellular processes that underlie the structure and function of the mediating brain circuits. This is possible because attachment is an evolved behavioral system that functionally integrates a variety of components ranging from genes to social relationships. In this chapter, the adoption of the evolutionary perspective will allow me to move across different levels of analysis when I will describe the findings of studies investigating the impact of childhood trauma on mental health.

The evolutionary perspective is not only non-reductionist (i.e., aimed at integrating biological and psychosocial data) but also focused on phylogenetic history and adaptive function [9]. This will emerge clearly from the theoretical models outlined throughout the chapter. In order to achieve a comprehensive understanding of the relationship between childhood trauma and mental health, we need a theoretical framework explaining the adaptive functioning of the human mind, the biological and psychological mechanisms sensitive to early adverse experiences, and the variables that modulate individual differences in stress sensitivity. Evolutionary biology offers such a theoretical framework.

## 7.2 Evolutionary Explanations for Early Environmental Sensitivity

Evolutionary explanations for early environmental sensitivity address the question of why natural selection has produced a trait that can sometimes lead to detrimental health outcomes. The question is an ultimate question, and should not be confused with the proximate question regarding mechanisms that translate early experiences into those physiological, psychological, and behavioral changes compromising adult health. The investigation of proximate mechanisms is complementary to the study of the evolutionary origins of early environmental sensitivity, and it becomes much more informative when conducted within the framework of evolutionary theory. Mechanisms are the "*how*" component of developmental–behavioral science but, in biology, any "how" presupposes a "*why*" [10]. Mechanisms are better understood if their relationships with adaptive and maladaptive outcomes are unveiled (see the section on "Brain mechanisms").

Evolutionary reasoning is based on the comparison of alternative traits. Theoretically, we could imagine the existence of human beings who are completely insensitive to short- and long-term effects of early experiences. Do these individuals exist? If not, why some variable degree of early environmental sensitivity characterizes all human beings? Is it the inevitable result of the limited power of natural selection in buffering environmental insults or is the explanation more complex? Does the negative correlation between early adversity and adult health reflect some kind of adaptation? This part of the chapter is an attempt to answer these questions with particular reference to the relationship between early stressful experiences and psychological and behavioral sequelae in adulthood.

## 7.3 Developmental Plasticity

In evolutionary biology, early environmental sensitivity has been conceptualized in a new way by coining the term "*developmental plasticity*." Developmental plasticity is defined as the capacity of genetically similar individuals to produce substantially different phenotypes depending upon environmental conditions during early life [11]. The concept of developmental plasticity differs in two important ways from traditional models of early environmental sensitivity.

Traditional models are based on the concept of vulnerability and explain the effects of early adversity from the diathesis–stress perspective. According to traditional models, individuals vary in their capacity to buffer the impact of stressful environments, with some being more resilient than others. Individual differences in vulnerability and resilience interact with the severity and duration of early stress in producing short- and long-term effects that impact on adult health. From such a perspective, there is no room for adaptive explanations of early environmental sensitivity. By contrast, both types of evolutionary models (i.e., predictive models and constrains models; see below) posit that developmental plasticity is adaptive and has evolved through natural selection. Developmental plasticity provides the means by which an organism can respond to environmental circumstances without undergoing a change in its genome. It is more efficient for the individual to capitalize on widely available signals from the environment as a mean for guiding development, rather than exclusively relying on genetic signaling alone [12].

Another important difference that distinguishes models based on developmental plasticity from those based on the diathesis–stress perspective is the focus on both negative and positive early experiences. The shift from the concept of vulnerability to the concept of plasticity implies that individuals more sensitive to environmental influences are more affected than others by both negative and positive contextual conditions. Those allegedly "vulnerable" individuals, who are most adversely affected by different stressors, may thus be the very same individuals who reap the most benefit from environmental support and enrichment, including the absence of adversity [13]. The term "differential susceptibility" describes individual differences in the capacity to calibrate developmental pathways in response to both negative and positive early experiences [14]. The term "vantage sensitivity" is used to indicate the specific capacity to benefit from early environmental support and enrichment [15, 16] (Fig. 7.1).

The terms and concepts outlined in this paragraph are explained and discussed below by focusing on epidemiological and clinical studies that have assessed the relevance of developmental plasticity for human physical and mental health.

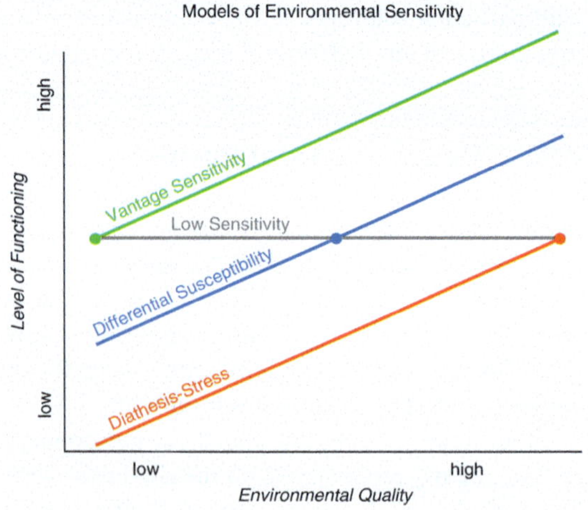


**Fig. 7.1** Graphic illustration of three different models of sensitivity to early experience. See the text for explanation. (Reproduced with permission from reference [16]. Courtesy of Alexia Jolicoeur-Martineau and Michael Pluess)

### 7.3.1 Predictive Models

The basic hypothesis of *predictive models* is that cues received in early life influence the development of a phenotype that will be adapted to the environmental conditions of later life. The validity of predictive models relies on the assumption that, over evolutionary history, early environmental conditions reliably predict the characteristics of the adult environment. Only if such a condition is satisfied, individuals can evolve the ability to modify their phenotype during development, as to maximize adaptation to their adult environment. When early cues fail to correctly predict the adult environment, the resulting mismatch between early-life conditions and adult ambience causes poor later life health [17].

The *predictive adaptive response* (PAR) model was originally developed to explain why inadequate early nutrition is associated with an increased risk for adult metabolic and cardiovascular diseases in contemporary human populations. The explanation was that developmental plasticity leads to physiological (e.g., insulin resistance) and behavioral (e.g., preference for high-fat foods) modifications that maximize adaptation only if adult nutritional conditions closely match those experienced in childhood. In those human societies, where economic circumstances and nutrition are rapidly improving, there is a mismatch between prediction (food shortage) and subsequent reality (food abundance), which leads to later health problems [18].

In the field of mental health, the PAR model has been successfully applied to explain the development of different styles of adult attachment [19]. The attachment theory is one of the most influential model proposed to explain the relationship between early experience and adult personality and social behavior [20]. According to attachment theory, infants develop expectations about their caregivers' availability and responsiveness based on the quality of parental care they receive. These expectations then serve as the basis for the development of mental representations of the self and the other ("internal working models" in the terminology of attachment theory) that influence later psychosocial functioning. Infants with emotionally available caregivers develop a model of the self as loved and valued, and a model of the other as loving. When infants instead have experiences that lead them to expect caregivers to be rejecting or undependable, they develop a model of the self as unloved or rejected, and a model of the other as unloving or rejecting. Parental caregiving is also an indirect marker of the level of harshness and/or unpredictability in the environment [21]. Generally, parenting quality is reduced when the family environment is unstable (e.g., frequent marital conflict or inconsistent financial resources) or under conditions of chronic social stress (e.g., poverty or lower quality neighborhoods).

Adult attachment styles (secure, anxious, avoidant) largely reflect early caregiving experiences. Sensitive and responsive parenting promotes secure attachment, whereas unreliable or unresponsive parenting promotes insecure attachment orientations, which take one of two prototypical forms: anxious attachment (a hyperactivating strategy involving unrestrained attempts at achieving emotional closeness to significant others) and avoidant attachment (a deactivating strategy involving a desire for emotional independence and rejection of emotional closeness and support).

The traditional perspective of clinical psychology views secure attachment as the healthy pattern and the insecure patterns as reflecting some kind of emotional dysfunction. In contrast with such a normative approach, the PAR model postulates that, at least in the ancestral environment, all three patterns of attachment were equally adaptive in terms of promoting reproductive fitness in the ecological niches that gave rise to them. According to this hypothesis, attachment patterns represent evolved psychological mechanisms capable of internalizing early experiences and producing adult phenotypes more likely to increase fitness in specific environments [22].

The PAR model builds on a major midlevel evolutionary theory, the *life history theory* (LHT). According to LHT, allocating resources and energy to important life tasks, such as growth and reproduction, involves trade-offs [23]. In many long-lived species, including humans, when to reproduce, and how many resources to invest in each offspring are two basic "decisions" that have a major impact on reproductive fitness. The "choice" between a parenting effort strategy (characterized by delayed maturation, discriminative sexual behavior, low fertility, and high parental investment) and a mating effort strategy (early maturation, promiscuous sexual behavior, high fertility, and low parental investment) depends on the prevailing ecological conditions. Under adverse environmental conditions where the flow of resources is chronically low or unpredictable, it can be adaptive to "choose" a quantity-oriented strategy (alias a "fast" life history strategy), by channeling more effort through rapid physical development, earlier mating, and multiple, short-term relationships. Delayed maturation and reproduction under conditions of environmental risk and uncertainty may dearly cost individuals since they are more likely to die before reproducing. In contrast, environments in which resources are plentiful and relationship bonds are more reciprocal and enduring should foster a quality-oriented strategy (alias a "slow" life history strategy) consisting in slower physical development, later mating, and long-term pair bonds structured around greater parental investment. In such environments, reproductive fitness should be enhanced by deferring reproduction until prospective parents have acquired the skills and resources needed to maximize the quality of life for each offspring.

According to the PAR model, in the ancestral environment, the principal evolutionary function of attachment patterns was to translate information about the availability and predictability of resources (including parental care and attention) into behavioral strategies for promoting reproductive fitness. Consistently with these predictions, a number of studies have found that secure individuals are likely to engage in enduring romantic relationships and have both the desire and capacity to invest heavily in parental care. In contrast, avoidant individuals frequently report involvement in one-night stands and sex outside established relationships, and they tend to be cold, remote, and unsupportive parents [24]. Even more convincing is the evidence linking early experience, patterns of attachment, somatic growth, and reproductive development e.g., see [25]. If less responsive parental care received during childhood should foster the development of a mating effort strategy, then individuals with an insecure pattern of attachment should show accelerated somatic growth and early pubertal maturation. In line with these predictions, studies of pubertal timing in girls have found that stressful family environments are associated with early menarche, whereas an affectionate and responsive parental style is associated with delayed puberty [26].

Del Giudice [22] further elaborated the LHT model of adult attachment styles by taking into consideration sex differences predicted by parental investment theory (Fig. 7.2). In the 1970s, the American sociobiologist Robert Trivers developed a theoretical model that proved to be very powerful in predicting sexual strategies in animal species [27]. The model is based on the idea that there is a relationship between the relative amount of time and energy that males and females invest in offspring (i.e., parental investment) and their sexual behavior. If parental investment is asymmetrical (i.e., one sex invests a lot and the other very little), then males' and females' sexual strategies, as well as their morphology and physiology, will differ (i.e., the species is sexually dimorphic). Compared to individuals of the more investing sex, those of the sex that invests less tend to be larger in size and have higher metabolic rates and juvenile mortality, later sexual maturity, and shorter lifespans. They compete for mating opportunities, either by winning fights with other individuals of their sex or by presenting displays preferred by individuals of the other sex. They also tend to be sexually promiscuous, preferring quantity of mates over quality. In other words, they adopt the sexual strategy that will allow them to mate with the highest possible number of partners, rather than mating with a few partners who were carefully selected for their reproductive quality. In mammals, including humans, female parental investment is generally much greater than that of males.

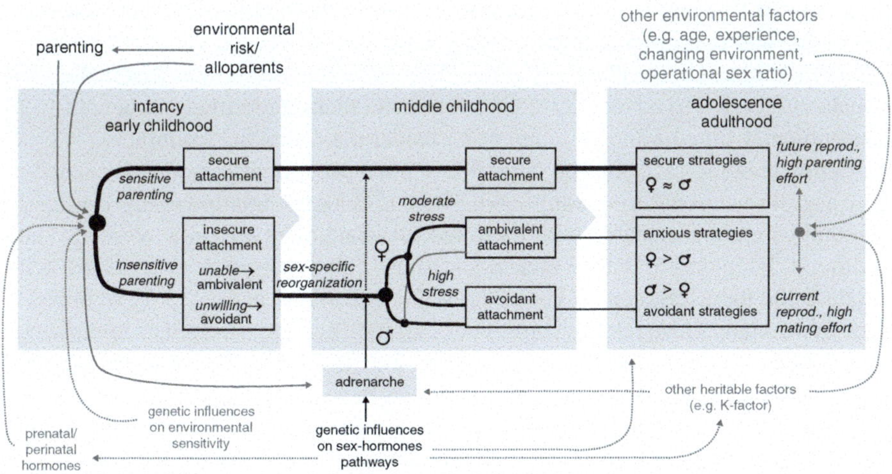


**Fig. 7.2** A causal model of early experience, attachment patterns, and reproductive strategies based on life history theory and sexual selection theory. (Reproduced with permission from reference [22]. Courtesy of Marco Del Giudice). The letter "K" in K-factors refers to slow and high-investment strategies, which are labeled "K-strategies"

The costs of pregnancy and lactation are inevitably at the expense of females, and this affects their sexual strategies. Female mammals tend to be discriminative in mate choice, thus minimizing the risk of wasting a substantial portion of their reproductive budgets by avoiding copulations that would not translate into optimal genetic replication. According to Del Giudice [22], secure attachment is associated with a slow life history strategy in both sexes. Yet, because of asymmetry in parental investment, insecure attachment is likely to take different forms in males and females. Insecure males should readily adopt avoidant strategies, which are most likely to maximize their fitness in a threatening environment. Insecure females, on the other hand, should preferentially adopt care-eliciting strategies (i.e., anxious attachment) that would keep them in close contact with kin and other potential "helpers" in the social group.

Empirical data have only partially confirmed all the predictions of the PAR model of attachment styles e.g., see [28]. However, considering that almost half of the general population can be classified as insecurely attached [29], the complex relationship between early experiences, attachment styles, and adult sociosexual behavior appears to be better explained by adaptive plasticity than by the diathesis–stress model.

### 7.3.2 Differential Susceptibility

*Differential susceptibility* (i.e., the existence of individual differences in response to both negative and positive early experiences) is a form of developmental plasticity that has practical implications for the implementation of preventive programs aimed at modifying early environments to promote mental health. A first major practical implication is that preventive programs should aim at both minimizing adverse early experience and maximizing positive early experiences because vulnerability genes may in fact function like plasticity genes. The same genetic variants that make individuals vulnerable to adversities could also make them more likely to benefit from environmental support and enrichment. Consider the following examples.

A clinical study [30] showed how differential susceptibility modulates sensitivity to social rejection during adulthood. There is evidence that lower levels of early maternal care are associated with more fearful attachment late in adolescence and adulthood. People with fearful attachment show high levels of rejection sensitivity and describe themselves as follows: "I am uncomfortable getting close to others. I want emotionally close relationships, but I find it difficult to trust others completely, or to depend on them. I worry that I will be hurt if I allow myself to become too close to others." [31]. Molecular genetic studies have shown that rejection sensitivity also depends on genetic differences in opioid neurotransmission, with the A118G polymorphism of the *OPRM1* gene playing a major role in modulating individual reactions to social exclusion [32]. The authors of the clinical study used the Parental Bonding Inventory to measure early maternal care and the Relationship Questionnaire (RQ) to measure fearful attachment in a sample of psychiatric patients. The findings of the study fitted well with the differential susceptibility model. The pattern emerging from the RQ data was a crossover interaction between genotype and maternal caregiving. Participants expressing the minor 118G allele had similar scores on fearful attachment, regardless of the quality of maternal care. By contrast, early experience made a major difference for participants carrying the A/A genotype (i.e., the plastic genetic variant). Those who recalled higher levels of maternal care reported the lowest levels of fearful attachment, whereas those who recalled lower levels of maternal care scored highest on fearful attachment.

Another practical implication of differential susceptibility is that the effects of preventive interventions will not necessarily be homogeneous across participants. Identification of plastic genotypes expressing as differential individual sensitivity to environmental conditions is a prerequisite for any successful program of selective prevention. For individuals who are more sensitive to the environment, promoting a rich array of niches and assisting them in finding valued, rewarding places may be a more effective intervention strategy. Inspired by such a perspective, a randomized controlled trial of foster care compared to institutional rearing focused on the genetic contribution to indiscriminate child behavior [33]. Children enrolled in the Bucharest Early Intervention Project were assessed comprehensively before the age of 30 months and subsequently randomized to either care as usual, or high-quality foster care. Indiscriminate social behavior was assessed at four time points (baseline, 30 months, 42 months, and 54 months of age), using caregiver report with the Disturbances of Attachment Interview. Children who were carriers of both the *BDNF* Met allele and the s/s *5HTTPLR* genotype, two plastic genotypes [34], had the greatest decline in indiscriminate behavior by 54 months if they were placed in foster care. However, if they remained in the institutional care they had the highest level of indiscriminate symptoms at 54 months. Children who had neither sensitive genotype had little changes in symptoms.

### 7.3.3 Constraints Models

*Developmental constraints models* posit that, in the face of early adversity, natural selection favors developmental strategies that promote immediate survival, even if other aspects of development are impaired [11]. It is worth emphasizing that constraint models are models of an adaptive process. Avoiding immediate death or impairment is more adaptive than being unable to alter any developmental processes in the face of environmental stressors, even if altered developmental trajectories may generate negative consequences later in life.

Balbernie [35] has applied the constraints model to reinterpret, from an evolutionary perspective, the attachment pattern that DSM-5 describes as disinhibited social attachment disorder [36]. Children with this diagnosis show reduced or absent reticence in approaching and interacting with unfamiliar adults. They seem to lack any wariness of the stranger, a core feature of the attachment system. The etiology of disinhibited attachment is strictly associated with grossly abnormal rearing experiences such as prolonged institutionalization in orphanages or repeated changes of primary caregivers. It has been estimated that in Russian orphanages, a child may experience 50–100 different caregivers in the first 2 years of life [37], and a Canadian study of children removed from Romanian orphanages after the age of 8 months found that 61% would approach strangers and, of these, 52% would willingly go home with them [38].

According to Balbernie [35], disinhibited attachment evolved as a contingent strategy to allow immediate survival following loss of dedicated caregiving, or lack of opportunities to build specific intimate relationships with dedicated caregivers. Disinhibited attachment would reflect an extremization of the innate human capacity to direct attachment behavior to a plurality of caregivers: "For most of our history, if a child's parents came to an untimely end then, if no adult came forward to take responsibility for the child, their best chance of surviving to reach reproductive age would lie in making use of whomsoever came to hand as a temporary source of aid" (pp. 270–271).

The fact that disinhibited attachment is a trade-off between immediate survival and later dysfunctional outcomes is suggested by longitudinal observations of children who fulfill the criteria for the clinical diagnosis of the disorder. According to DSM-5, in preschool children, verbal and social intrusiveness appear most prominent, often accompanied by inappropriate attention-seeking behavior and inauthentic expression of emotion. Even after placement in normative caregiving environments, some children show persistent signs of the disorder, at least through adolescence. "*A means of survival has become a way of life, maladaptive when inflexible*" [35, p. 277].

## 7.4 Brain Mechanisms: Damage or Adaptation?

The psychological (cognitive and emotional) and behavioral profiles that characterize adult people who have experienced stressful environments early in life are associated with specific changes at the level of neuroanatomy and neurophysiology [39, 40, 41]. When discussing the enduring neurobiological effects of child abuse and neglect, Teicher and coworkers [42] have raised a key question regarding the meaning of maltreatment-associated brain differences. The prevailing view posits that the brain is damaged by excessive exposure to stress, and psychopathology is a direct consequence. An alternative view is that the brain, and the way it processes information, is selectively modified by early stressors "*to facilitate survival and reproduction in what seems, so far, to be a threatening and malevolent world*" (p. 653). The authors make it clear that the two explanations are not mutually exclusive. Some experiences might be so severe to damage the brain, whereas others can be construed as evolved adaptations.

Teicher and coworkers [42] list possible examples of neuroplastic adaptive changes caused by child maltreatment, including alterations in auditory cortex and arcuate fasciculus in children experiencing verbal abuse, in visual cortex and visual–limbic pathway in subjects visually witnessing domestic violence, and thinning of the genital representation area in the somatosensory cortex of sexually abused females.

Chester and coworkers [43] have developed a theoretical framework that combines findings from attachment research and clinical neuroscience and merges them through the application of the concept of *adaptive developmental plasticity*. Their core hypothesis is that different types of early stressful environments calibrate the brain social pain network (part of the neural system dedicated to the detection of physical pain) in different ways as to optimize reactions to social rejection during adolescence and adulthood. Individuals with an avoidant attachment style have often experienced chronic social rejection, stemming from cold caregivers who rejected their needs for comfort and acceptance. By contrast, individuals with an anxious attachment style have often experienced an unpredictable pattern of parental care, with caregivers showering affection upon them 1 min and ignoring them the next. If such early life experiences are reliable indicators of what the conditions of adulthood will be like (as assumed by the PAR model), the social pain network requires different calibration in people with avoidant or anxious attachment styles. When social resources are consistently unavailable, chronic and intense responses to social rejection would be associated with high costs (e.g., stress-related suppression of immune system). Therefore, in avoidant individuals, the optimal calibration of the social pain network should consist in a blunted response to social rejection. Anxiously attached individuals should instead display an increased capacity of detecting and responding to social signals of acceptance and rejection.

The neural correlates of the dual social pain response (i.e., diminished in avoidant individuals and increased in anxiously attached individuals) have been demonstrated by brain imaging studies [44]. Based on these findings, the optimal calibration of the brain social pain network involves structural and functional modifications of the dorsal anterior cingulate cortex (dACC) and the anterior insula. The dACC and the insula are important components of the social aversion module, which also includes the amygdala, hippocampus, and anterior temporal lobe [45]. The social aversion module works in a dynamic balance with the social approach module, which includes the ventral tegmental area, hypothalamus, striatum, and the ventral medial orbitofrontal cortex. Human social behavior basically depends on a dynamic "push–pull" alternation between these two opposing neural circuits. Consistently with the optimal calibration hypothesis, high attachment avoidance was found to negatively correlate with responses in reward-related brain areas (ventral striatum and orbitofrontal cortex) in social settings [46]. Avoidant individuals can distance themselves from others and maintain a high level of self-sufficiency because they are less sensitive to both the pain of social rejection (via hypofunctionality of dACC and anterior insula) and the pleasure of interpersonal closeness (via reduced activation of ventral striatum and orbitofrontal cortex).

Since recent studies have expanded the adaptive calibration model to include stress physiology [47, 48, 49], there is growing evidence that many of the physiological and neural modifications caused by child maltreatment reflect short- and long-term adaptive responses, rather than dysfunctional abnormalities related to damaged mechanisms.

## 7.5 Adaptation, Well-Being, and Health

As shown in the Introduction, there is an etiological link between childhood trauma and psychopathology. Such a causal relationship is mediated in part, by insecure attachment. Then, how is it possible that evolutionary studies often use the terms "adaption" or "adaptive" to describe what happens to the unfortunate victims of this chain of tragic events (i.e., childhood trauma → insecure attachment → adult mental disorders)? In common language, "adaptive" is an optimistic word that seems inappropriate for labeling the condition of people who report high levels of subjective suffering, experience severe difficulties in establishing rewarding interpersonal relationships, and are considered by mental health professionals as worth of therapeutic interventions. To explain the apparent paradox, we need to clarify the concept of adaptation in evolutionary biology and its relationships with the concepts of health and subjective well-being.

In evolutionary biology, an adaptation is defined as a trait (anatomical, physiological, psychological, or behavioral) that increases the average inclusive fitness of individuals that express it, relative to individuals who do not express the trait. Adaptations are products of natural selection and occur over generations rather than within an individual's lifetime. The ultimate function of an adaptation is gene propagation via maximization of survival and reproduction. The technical definition of evolutionary adaptation does not imply any explicit reference to health or subjective well-being. Therefore, the expression of an adaptation may, or may not, coincide with health and/or subjective well-being (Fig. 7.3).

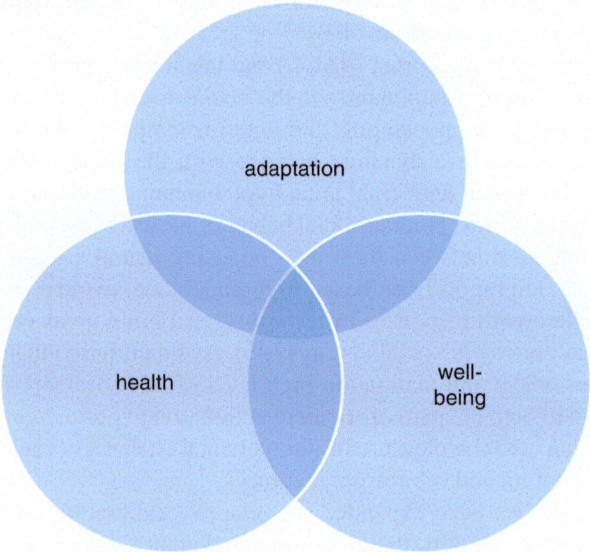


**Fig. 7.3** The complex relationships between biological adaptation, subjective well-being, and health. In most cases, health as defined in medicine and psychiatry corresponds to a full overlapping of the three conditions. Yet, there are cases of partial overlap. Some conditions may be biologically maladaptive even if they are associated with subjective well-being and considered healthy according to prevalent cultural standards (e.g., adoption of unrelated children). Some conditions are labeled as unhealthy even if they are likely to be biological adaptations associated with subjective well-being (e.g., primary psychopathy). Some conditions reflect biological adaptations designed to preserve health even if they jeopardize subjective well-being (e.g., defensive symptoms such as vomit or reactive depression). Some short- and long-term consequences of childhood trauma are likely to belong to the last category

Subjective well-being generally reflects conditions that are biologically adaptive because emotions have evolved to provide information about costs and benefits of past, present, and future behavior. Pleasurable emotions are elicited by situations that are (or were in the evolutionary past) statistically associated with a positive benefit–cost ratio [50]. Conversely, negative emotions are generally associated with maladaptive situations. Yet, negative emotions may be adaptive. They may work as a motivational drive to activate coping mechanisms that restore biological adaptation. A depressive state reactive to unfair treatment at work may prompt the decision to quit a frustrating job. Mental distress can work as a useful defense in the same way such that disturbing physical symptoms counteract disease processes (e.g., vomiting to eliminate toxins or coughing to clear foreign matter from the respiratory tract). In addition, manifestations of distress may function as social signals to others to elicit their assistance in achieving biological goals [51]. Analyzing the evolutionary meaning of emotional dysfunctions associated with insecure attachment, Ein-Dor and Hirschberger [52] have suggested that adaptive explanations should be searched for at the group level. According to their *Social Defense Theory*, each of the three major styles of attachment had unique adaptive advantages that promoted survival in ancestral environments, and the combination of different attachment styles in a group overcomes individual limitations disposition (i.e., the sentinel abilities of anxious members, the rapid response to threats without much deliberation by avoidant members, and the leadership and social-oriented abilities of secure members). Accordingly, a group comprising all three styles of attachment patterns benefited from the combined abilities of each disposition and counterweighed the shortcomings of each individual disposition.

Another aspect to be considered is that evolutionary models of developmental plasticity refer to psychological and behavioral adaptations that evolved in environments vastly different from our modern world. We cannot exclude that the aversive emotions experienced by people with insecure attachment styles are in part caused by the mismatch between the social organization of ancestral human groups and current expectations about what is the appropriate behavior in close relationships. Regardless of the validity of the adaptationist hypotheses reported above, the point we should keep in mind is that subjective well-being is irrelevant when reasoning about evolutionary adaptation. An allele that increases inclusive fitness will increase in prevalence over the generations, even at the cost of individual and societal happiness [53].

The relationship between adaptation and health is even more complex because the concept of health, and especially the concept of mental health, is largely defined by cultural criteria [54]. Theoretically, it is possible to define the distinction between health and disease in purely biological terms. Adaptive failure is a criterion of morbidity which is objective and immune to the perils of cultural relativism. Recently, I have proposed an evolutionary definition of mental disorder based on adaptive failure [55]. The definition includes five criteria. Criterion A defines mental disorder as a maladaptive syndrome, which is a psychological or behavioral syndrome negatively impacting the individual's inclusive fitness. Criteria B and C specify that: [1] individuals who have a mental disorder make choices that penalize their inclusive fitness, relative to all feasible alternative strategies in a specific environment; and [2] at the individual level, the inability to achieve short-term biological goals is a valid proxy indicator of mental disorder when estimates of inclusive fitness cannot be made. Criteria D and E clarify that neither the demonstration of a pathophysiology underlying the syndrome, nor the demonstration of subjective suffering associated with the syndrome are necessary or sufficient for a diagnosis of mental disorder. I refer the reader to my original paper for a methodological discussion of the scientific validity of each criterion [55].

The problem with the definition I have proposed is that it cannot be used in clinical practice because, unlike other biological sciences, medicine and psychiatry are strongly influenced by social values and public expectations. The inflexible application of the criterion of adaptive failure would produce unwanted effects on the classification of some behavioral syndromes. Behaviors that we value should be considered as mental disorders (e.g., adoption of unrelated children), and conditions that we dislike and want to change would turn out to be sophisticated adaptations. A good example is psychopathy. Traditionally, psychopathy is classified as a psychiatric disorder with clear affective, cognitive, behavioral, and neurological correlates [56]. Understandably, others do not like psychopaths because their interpersonal behavior is systematically based on deception, manipulation, callousness, and coerciveness. For most people, it is even difficult to accept the idea that psychopaths are psychiatric patients rather than wicked individuals. Imagine what would be the social reaction to the statement that psychopaths are healthy persons who express a sophisticated set of evolutionary adaptations. Yet, there is preliminary evidence that psychopathy has evolved as a frequency-dependent, alternative strategy that increased reproductive success in ancestral environments through persistent social exploitation. In social environments in which the majority of group members adopt a strategy of cooperation, a small number of individuals may be able to maintain a socially parasitic strategy. The strategy can bring high fitness benefits when rare, but becomes less rewarding at higher frequencies because of anti-cheater vigilance in the population and because of the increased probability that a cheater will encounter another cheater [57].

The discussion above explains why adaptive interpretations of short- and long-term effects of childhood trauma are necessarily independent from clinical assessments that focus on subjective suffering and diagnostic criteria. However, as explained in the next section, this does not mean that evolutionary models are irrelevant for clinical practice.

## 7.6 Clinical and Public Health Implications

When analyzed from an evolutionary perspective, many of the short- and long-term outcomes of childhood trauma can be reinterpreted as adaptive strategies. In evolutionary studies of human behavior, "adaptive" is not synonymous of "good." There are adaptive phenotypes that we want to change for ethical or medical reasons. This is certainly the case for the outcomes of childhood trauma, regardless of their possible adaptive significance. Mental health professionals adopting the evolutionary approach are fully aware that their task is to be healers of the distressed, not watchdogs of biological adaptation. Since childhood trauma is often associated with much individual and interpersonal suffering, its prevention and treatment are absolute priorities for public health worldwide [1].

In a recent review of how childhood trauma becomes "biologically embedded" in altered physiology across body systems, Berens and coworkers [58] highlighted a series of recommendations for implementing clinical and public health interventions. Some of their recommendations can be enriched and partly revised considering findings of evolutionary studies. First, they recommend that screening for early life adversity should become a routine part care for children and adult. Evolutionary research suggests to integrate retrospective assessment of childhood trauma (e.g., by the Adverse Childhood Experiences Questionnaire) with assessment of attachment patterns. According to the *transdiagnostic model of psychopathology*, insecure attachment mediates a variety of health problems linked to early adverse experiences. The combination of childhood trauma and insecure attachment is likely to be a more severe risk factor than a history of childhood trauma per se. Second, they recommend that investigators must develop new intervention strategies based on genotype-dependent response variation. Evolutionary studies of differential susceptibility suggest not to investigate individual differences based on the reductive dichotomy vulnerability/resilience. Vulnerability genes may in fact be plastic genes. The same genetic variants that make individuals vulnerable to adversities could also make them more likely to benefit from environmental support and enrichment. Third, they recommend to consider the moderation effect of socially embedded gender roles in predicting the pathogenic impact of childhood trauma. Evolutionary studies suggest that sex- and gender-related sensitivity to early stress is also moderated by the interaction between life history strategies and sexual selection. For example, those psychiatric disorders that reflect "fast" life history strategies are more likely to be associated with avoidant attachment in males and anxious attachment in females.

In conclusion, an evolutionary understanding of the physical and mental effects of childhood trauma, as summarized in this chapter, may have important implications for clinical practice and public health.

## References

1. Caspi A, Houts RM, Belsky DW, Harrington H, Hogan S, Ramrakha S, Poulton R, Moffitt TE. Childhood forecasting of a small segment of the population with large economic burden. Nat Hum Behav. 2016;1. https://doi.org/10.1038/s41562-016-0005.
2. Nusslock R, Miller GE. Early-life adversity and physical and emotional health across the lifespan: a Neuroimmune network hypothesis. Biol Psychiatry. 2016;80(1):23–32. https://doi.org/10.1016/j.biopsych.2015.05.017.
3. Vachon DD, Krueger RF, Rogosch FA, Cicchetti D. Assessment of the harmful psychiatric and behavioral effects of different forms of child maltreatment. JAMA Psychiat. 2015;72(11):1135–42. https://doi.org/10.1001/jamapsychiatry.2015.1792.

4. Jaffee SR. Child maltreatment and risk for psychopathology in childhood and adulthood. Annu Rev Clin Psychol. 2017;13:525–51. https://doi.org/10.1146/annurev-clinpsy-032816-045005.
5. Ein-Dor T, Viglin D, Doron G. Extending the Transdiagnostic model of attachment and psychopathology. Front Psychol. 2016;7:484. https://doi.org/10.3389/fpsyg.2016.00484.
6. Teicher MH, Samson JA. Childhood maltreatment and psychopathology: a case for ecophenotypic variants as clinically and neurobiologically distinct subtypes. Am J Psychiatry. 2013;170(10):1114–33. https://doi.org/10.1176/appi.ajp.2013.12070957.
7. Lo Iacono L, Visco-Comandini F, Valzania A, Viscomi MT, Coviello M, Giampà A, Roscini L, Bisicchia E, Siracusano A, Troisi A, Puglisi-Allegra S, Carola V. Adversity in childhood and depression: linked through SIRT1. Transl Psychiatry. 2015;5:e629. https://doi.org/10.1038/tp.2015.125.
8. Insel T, Cuthbert B, Garvey M, Heinssen R, Pine DS, Quinn K, Sanislow C, Wang P. Research domain criteria (RDoC): toward a new classification framework for research on mental disorders. Am J Psychiatry. 2010;167(7):748–51. https://doi.org/10.1176/appi.ajp.2010.09091379.
9. Troisi A. The painted mind: behavioral science reflected in great paintings. New York: Oxford University Press; 2017.
10. Mayr E. The philosophical foundations of Darwinism. Proc Am Philos Soc. 2001;145(4):488–95.
11. Lea AJ, Tung J, Archie EA, Alberts SC. Developmental plasticity: bridging research in evolution and human health. Evol Med Public Health. 2018;2017(1):162–75. https://doi.org/10.1093/emph/eox019.
12. Tost H, Champagne FA, Meyer-Lindenberg A. Environmental influence in the brain, human welfare and mental health. Nat Neurosci. 2015;18(10):1421–31. https://doi.org/10.1038/nn.4108.
13. Leighton C, Botto A, Silva JR, Jiménez JP, Luyten P. Vulnerability or sensitivity to the environment? Methodological issues, trends, and recommendations in gene-environment interactions research in human behavior. Front Psych. 2017;8:106. https://doi.org/10.3389/fpsyt.2017.00106.
14. Belsky J. The differential susceptibility hypothesis: sensitivity to the environment for better and for worse. JAMA Pediatr. 2016;170(4):321–2. https://doi.org/10.1001/jamapediatrics.2015.4263.
15. de Villiers B, Lionetti F, Pluess M. Vantage sensitivity: a framework for individual differences in response to psychological intervention. Soc Psychiatry Psychiatr Epidemiol. 2018;53(6):545–54. https://doi.org/10.1007/s00127-017-1471-0.
16. Jolicoeur-Martineau A, Wazana A. Distinguishing differential susceptibility, diathesis-stress and vantage sensitivity: beyond the single gene and Environment model. PsyArXiv. 2018; https://doi.org/10.31234/osf.io/27uw8.
17. Hanson MA, Gluckman PD. Early developmental conditioning of later health and disease: physiology or pathophysiology? Physiol Rev. 2014;94(4):1027–76. https://doi.org/10.1152/physrev.00029.2013.
18. Bateson P, Gluckman P, Hanson M. The biology of developmental plasticity and the predictive adaptive response hypothesis. J Physiol. 2014;592(11):2357–68. https://doi.org/10.1113/jphysiol.2014.271460.
19. Szepsenwol O, Simpson JA. Attachment within life history theory: an evolutionary perspective on individual differences in attachment. Curr Opin Psychol. 2018;25:65–70. https://doi.org/10.1016/j.copsyc.2018.03.005.
20. Mikulincer M, Shaver PR. Attachment in adulthood: structure, dynamics, and change. 2. New York: Guilford Press; 2016.
21. Chisholm JS. The evolutionary ecology of attachment organization. Hum Nat. 1996;7(1):1–37. https://doi.org/10.1007/BF02733488.
22. Del Giudice M. Sex, attachment, and the development of reproductive strategies. Behav Brain Sci. 2009;32(1):1–21.; discussion 21-67. https://doi.org/10.1017/S0140525X09000016.
23. Chua KJ, Lukaszewski AW, Grant DM, Sng O. Human life history strategies. Evol Psychol. 2017;15(1):1474704916677342. https://doi.org/10.1177/1474704916677342.

24. Rholes WS, Simpson JA, Friedman M. Avoidant attachment and the experience of parenting. Personal Soc Psychol Bull. 2006;32(3):275–85.
25. Belsky J, Houts RM, Fearon RM. Infant attachment security and the timing of puberty: testing an evolutionary hypothesis. Psychol Sci. 2010;21(9):1195–201. https://doi.org/10.1177/0956797610379867.
26. Sung S, Simpson JA, Griskevicius V, Kuo SI, Schlomer GL, Belsky J. Secure infant-mother attachment buffers the effect of early-life stress on age of menarche. Psychol Sci. 2016;27(5):667–74. https://doi.org/10.1177/0956797616631958.
27. Trivers RL. Parental investment and sexual selection. In: Campbell B, editor. Sexual selection and the descent of man, 1871–1971. Chicago, IL: Aldine; 1972. p. 136–79.
28. Dunkel CS, Lukaszewski AW, Chua K. The relationships between sex, life history strategy, and adult romantic attachment style. Personal Individ Differ. 2016;98:176–8., ISSN 0191-8869. https://doi.org/10.1016/j.paid.2016.04.040.
29. Ein-Dor T, Mikulincer M, Doron G, Shaver PR. The attachment paradox: how can so many of us (the insecure ones) have no adaptive advantages? Perspect Psychol Sci. 2010;5(2):123–41. https://doi.org/10.1177/1745691610362349.
30. Troisi A, Frazzetto G, Carola V, Di Lorenzo G, Coviello M, Siracusano A, Gross C. Variation in the μ-opioid receptor gene (OPRM1) moderates the influence of early maternal care on fearful attachment. Soc Cogn Affect Neurosci. 2012;7(5):542–7. https://doi.org/10.1093/scan/nsr037.
31. Bartholomew K, Horowitz LM. Attachment styles among young adults: a test of a four-category model. J Pers Soc Psychol. 1991;61(2):226–44.
32. Way BM, Taylor SE, Eisenberger NI. Variation in the mu-opioid receptor gene (OPRM1) is associated with dispositional and neural sensitivity to social rejection. Proc Natl Acad Sci U S A. 2009;106(35):15079–84. https://doi.org/10.1073/pnas.0812612106.
33. Drury SS, Gleason MM, Theall KP, Smyke AT, Nelson CA, Fox NA, Zeanah CH. Genetic sensitivity to the caregiving context: the influence of 5httlpr and BDNF val66met on indiscriminate social behavior. Physiol Behav. 2012;106(5):728–35. https://doi.org/10.1016/j.physbeh.2011.11.014.
34. Mesquita AR, Belsky J, Li Z, Baptista J, Carvalho-Correia E, Maciel P, Soares I. Institutionalization and indiscriminate social behavior: differential-susceptibility versus diathesis-stress models for the 5-HTTLPR and BDNF genotypes. Physiol Behav. 2015;152(Pt A):85–91. https://doi.org/10.1016/j.physbeh.2015.09.015.
35. Balbernie R. Reactive attachment disorder as an evolutionary adaptation. Attach Hum Dev. 2010;12(3):265–81. https://doi.org/10.1080/14616734.2010.482223.
36. American Psychiatric Association. Diagnostic and statistical manual of mental disorders. 5th ed. Washington, DC: DSM-5, American Psychiatric Association; 2013.
37. Groark CJ, Muhamedrahimov RJ, Palmov OI, Nikiforova NV, McCall RB. Improvements in early care in Russian orphanages and their relationship to observed behaviors. Infant Ment Health J. 2005;26(2):96–109. https://doi.org/10.1002/imhj.20041.
38. Chisholm K, Carter M, Ames E, Morison S. Attachment security and indiscriminate friendly behavior in children adopted from Romanian orphanages. Dev Psychopathol. 1995;7:1283–94.
39. Bick J, Nelson CA. Early adverse experiences and the developing brain. Neuropsychopharmacology. 2016;41(1):177–96. https://doi.org/10.1038/npp.2015.252.
40. Nemeroff CB. Paradise lost: the neurobiological and clinical consequences of child abuse and neglect. Neuron. 2016;89(5):892–909. https://doi.org/10.1016/j.neuron.2016.01.019. Review
41. Teicher MH, Samson JA. Annual research review: enduring neurobiological effects of childhood abuse and neglect. J Child Psychol Psychiatry. 2016;57(3):241–66. https://doi.org/10.1111/jcpp.12507.
42. Teicher MH, Samson JA, Anderson CM, Ohashi K. The effects of childhood maltreatment on brain structure, function and connectivity. Nat Rev Neurosci. 2016;17(10):652–66. https://doi.org/10.1038/nrn.2016.111.

43. Chester DS, Pond RS Jr, Richman SB, Dewall CN. The optimal calibration hypothesis: how life history modulates the brain's social pain network. Front Evol Neurosci. 2012;4:10. https://doi.org/10.3389/fnevo.2012.00010.
44. DeWall CN, Masten CL, Powell C, Combs D, Schurtz DR, Eisenberger NI. Do neural responses to rejection depend on attachment style? An fMRI study. Soc Cogn Affect Neurosci. 2012;7(2):184–92. https://doi.org/10.1093/scan/nsq107.
45. Vrticka P. Interpersonal closeness and social reward processing. J Neurosci. 2012;32(37):12649–50. https://doi.org/10.1523/JNEUROSCI.3157-12.2012.
46. Strathearn L, Fonagy P, Amico J, Montague PR. Adult attachment predicts maternal brain and oxytocin response to infant cues. Neuropsychopharmacology. 2009;34(13):2655–66. https://doi.org/10.1038/npp.2009.103.
47. Ellis BJ, Del Giudice M. Beyond allostatic load: rethinking the role of stress in regulating human development. Dev Psychopathol. 2014;26(1):1–20. https://doi.org/10.1017/S0954579413000849.
48. Peckins MK, Susman EJ, Negriff S, Noll J, Trickett PK. Cortisol profiles: a test for adaptive calibration of the stress response system in maltreated and nonmaltreated youth. Dev Psychopathol. 2015;27(4 Pt 2):1461–70. https://doi.org/10.1017/S0954579415000875.
49. Ellis BJ, Oldehinkel AJ, Nederhof E. The adaptive calibration model of stress responsivity: an empirical test in the tracking Adolescents' individual lives survey study. Dev Psychopathol. 2017;29(3):1001–21. https://doi.org/10.1017/S0954579416000985.
50. Troisi A. Psychoactive drug use: expand the scope of outcome assessment. Behav Brain Sci. 2011;34(6):324–5. https://doi.org/10.1017/S0140525X11000793.
51. Troisi A. Psychotraumatology: what researchers and clinicians can learn from an evolutionary perspective. Semin Cell Dev Biol. 2018;77:153–60. https://doi.org/10.1016/j.semcdb.2017.09.001.
52. Ein-Dor T, Hirschberger G. On sentinels and rapid responders: the adaptive functions of emotion dysregulation. In: Lench HC, editor. The function of emotions: when and why emotions help us. Berlin: Springer; 2018. p. 25–43. https://doi.org/10.1007/978-3-319-77619-4.
53. Nesse RM. Natural selection and the elusiveness of happiness. Philos Trans R Soc Lond Ser B Biol Sci. 2004;359(1449):1333–47.
54. Clark J. Medicalization of global health 2: the medicalization of global mental health. Glob Health Action. 2014;7:24000. https://doi.org/10.3402/gha.v7.24000.
55. Troisi A. The evolutionary diagnosis of mental disorder. Wiley Interdiscip Rev Cogn Sci. 2015;6(3):323–31. https://doi.org/10.1002/wcs.1339.
56. Hare RD, Neumann CS. Psychopathy as a clinical and empirical construct. Annu Rev Clin Psychol. 2008;4:217–46. https://doi.org/10.1146/annurev.clinpsy.3.022806.091452.
57. Glenn AL, Kurzban R, Raine A. Evolutionary theory and psychopathy. Aggress Violent Behav. 2011;16(5):371–80.
58. Berens AE, Jensen SKG, Nelson CA 3rd. Biological embedding of childhood adversity: from physiological mechanisms to clinical implications. BMC Med. 2017;15(1):135. https://doi.org/10.1186/s12916-017-0895-4.

# Part III

# Neuropsychiatric Disorders

# Childhood Trauma in Bipolar Disorders

# 8

Delfina Janiri, Gianfranco Spalletta, and Gabriele Sani

## 8.1 Introduction

Emil Kraepelin described manic depressive illness as a periodic and recurrent disorder, emphasizing its biological cyclicity [1]. Nevertheless, he recognized the importance of environmental stressors in individual variations of the clinical expression of bipolar disorder (BD). He proposed a genetically determined irritability of affectivity so that the psychosis itself emerged from certain predisposing "*basic states*", modulated by environmental stressors. In other words, we can say that the endogenous cyclicity does not exclude reactivity to life experiences.

Life experiences and the clinical expression of BD are indissolubly linked. The life of patients with BD is profoundly influenced by their illness. The outcome of BD strongly impacts the patients' quality of life and has often a bearing on their

D. Janiri (✉)
Department of Psychiatry and Neurology, Sapienza University of Rome, Rome, Italy

Fondazione Policlinico Universitario "Agostino Gemelli" IRCCS, Rome, Italy

Centro Lucio Bini, Rome, Italy
e-mail: delfina.janiri@gmail.com

G. Spalletta
Laboratory of Neuropsychiatry, IRCCS Fondazione Santa Lucia, Rome, Italy

Menninger Department of Psychiatry and Behavioral Sciences, Baylor College of Medicine, Houston, TX, USA
e-mail: g.spalletta@hsantalucia.it

G. Sani
Fondazione Policlinico Universitario "Agostino Gemelli" IRCCS, Rome, Italy

Centro Lucio Bini, Rome, Italy

Department of Neuroscience, Section of Psychiatry, Università Cattolica del Sacro Cuore, Rome, Italy
e-mail: Gabriele.sani@uniroma1.it



G. Spalletta et al. (eds.), *Childhood Trauma in Mental Disorders*,
https://doi.org/10.1007/978-3-030-49414-8_8

personal history. In the same way, BD is highly susceptible to life events [2]. In the early nineteenth century, Minkowski highlighted that in manic-depressive illness, "*synchronism*" with life experiences was excessive and inappropriate. He claimed that patients with manic-depressive illness always retain their contact with the surrounding world and can be overwhelmed by their own sensitivity to these life experiences. Today, recent studies conceptualized this observation as a hypersensitivity to emotional stimuli [3].

Among life experiences, previous studies focused on stressful and traumatic events as potential risk factors for BD [2]. Early trauma seems to be particularly relevant in the genesis and the progression of the illness [4]. Several studies found increased prevalence of childhood trauma in BD and pointed out its effect on the clinical outcome of the disease [4, 5]. Further findings linked it to specific psychopathological dimensions or cognitive profiles [6, 7]. Recently, research groups tried to identify neurobiological patterns underpinning the relationship between early adverse events and BD diagnosis [8, 9].

In this chapter, we review all relevant evidence investigating the relationship between childhood trauma and BD. Based on a systematic literature search, we show how childhood trauma can increase the risk for BD. We also provide a detailed description of the relationship between BD and different subtypes of childhood trauma. We discuss specific clinical, psychopathological and neurobiological implications of this relationship. In concluding remarks, we attempt to identify future directions for further investigations.

## 8.2 Childhood Trauma in Bipolar Disorders

Among environmental stressors, childhood trauma has emerged as one of the most important factors associated with BD. Childhood trauma could be conceptualized as abuse or neglect. Childhood abuse includes any act, carried out or omitted, that resulted in actual or potential harm to a child. Neglect refers to the failure of a parent or any other person responsible for the child, to provide material necessities for the child's survival or to provide attention, love and support required for the child's emotional development. About half of previous studies in BD differentiated between childhood abuse (physical, sexual, or emotional) and neglect (emotional or physical) [5]. Among these studies, the childhood trauma questionnaire (CTQ) is the instrument most frequently used to assess early adverse events. The CTQ is a retrospective, self-report questionnaire assessing five different types of childhood trauma: sexual abuse, physical abuse, emotional abuse, emotional neglect and physical neglect [10] (Table 8.1).

Childhood abuse and neglect have been reported by 51% of patients with BD. Specifically, emotional abuse has been reported by 37% of patients, of which 24% reported physical abuse, 24% emotional neglect, 21% sexual abuse and 12% physical neglect. In addition, one-third of those patients presented with a combination of different types of trauma [11]. A recent meta-analysis on 19 eligible studies

**Table 8.1** Childhood traumatic experiences assessed by the CTQ (Bernstein et al. 1994)

| Type of childhood abuse | Description |
|---|---|
| Sexual abuse | Sexual contact or conduct between a child younger than 18 years of age and an adult or older person |
| Physical abuse | Bodily assaults on a child by an adult or older person which posed a risk of, or resulted in, injury |
| Emotional abuse | Verbal assaults on a child's sense of worth or well-being or any humiliating or demeaning behaviour directed towards a child by an adult or older person |
| Physical neglect | Failure of caretakers to provide for a child's basic physical needs, including food, shelter, clothing, safety, and healthcare |
| Emotional neglect | Failure of caretakers to meet children's basic emotional and psychological needs, including love, belonging, nurturance, and emotional support |

demonstrated that childhood adversity was 2.63 times (95% CI 2.00–3.47) more likely to have occurred in patients with BD than in healthy controls (HCs) [4]. Emotional abuse was four times more likely to have occurred in BD than in HCs, an effect seemingly larger than for other types of childhood adversity (OR = 4.04, 95% CI 3.12–5.22). Authors did not find differences in rates of childhood adversity between the two BD main subtypes: BD type I (BD-I) and BD type II (BD-II).

Recently, we specifically addressed the question of whether the occurrence of childhood trauma was differently distributed in BD-I and BD-II. We found that all patients with BD reported more severe traumatic childhood experiences than HCs. Both BD-I and BD-II patients differed significantly from HCs for emotional abuse. However, patients with BD-I differed significantly from HCs for sexual abuse, and patients with BD-II differed from HCs for emotional neglect [12]. Results indicated that the assessment of childhood trauma can unveil differences between BD subtypes and pointed out the importance of separately considering BD subtypes in the evaluation of childhood traumatic experiences.

## 8.3 Childhood Trauma and Clinical Outcomes of Bipolar Disorders

Childhood trauma is a key determinant in the expression and clinical course of BD. There is a body of evidence relating the history of childhood maltreatment to a severe illness progression. In a review of 18 studies performed by Daruy-Filho and colleagues [5], childhood abuse and neglect have been highlighted as specific risk factors associated with worse clinical outcome of BD. In a recent review, Aas and colleagues [13] confirmed this observation.

The association between childhood trauma and earlier age of onset appears consistent across studies. Both Daruy-Filho [5] and Aas [13] underscored this result. The association is consistent across different types of childhood traumatic experience.

The number of studies finding a link between earlier onset of BD and physical abuse [14, 15] or emotional abuse/neglect [16, 17] appears to be nearly equal, whereas there is more evidence indicating sexual abuse as a risk factor for developing early-onset BD [14, 15, 17, 18].

Childhood trauma also influences the recurrence of the illness. It has been associated with an increased number of mood episodes [17, 11] and with a rapid cycle pattern. Patients tend to present a rapid-cycling pattern with an odds ratio (OR) of 1.96 if they reported physical abuse [14], and with an OR of 2.04 if they reported sexual abuse [17]. This increased risk appears to be particularly important in BD, as it is related to the unfavourable response to pharmacological treatment associated with rapid cycling [19].

Early adverse events have also been related to increased severity of mood symptoms. Garno and colleagues [11] found an increased manic/depressive symptom severity in patients who experienced childhood trauma [11]. However, authors did not distinguish the nature of childhood abuse in their analyses. Conversely, Leverich and colleagues [11] specified that increased severity of mania was associated with childhood physical and sexual abuse [14]. Several authors also found that patients reporting childhood trauma presented episodes marked by higher frequency of psychosis [20–22]. In an attempt to explain the relationship between psychosis and sexual abuse, some authors have suggested that affective symptoms, mainly depression and anxiety, may primarily mediate this association [22]. In agreement to this, a recent study found a specific association between childhood abuse and auditory hallucinations, which was strongest for sexual abuse and mood congruent hallucinated voices [23].

Replicated findings have been obtained for the association between substance use and childhood physical/sexual abuse [24, 25] or emotional abuse [26]. A recent study proposed the existence of additive effects between early trauma and cannabis abuse on the clinical expression of BD [26].

The very strong association with suicidal behaviours can be considered the most important clinical outcome of childhood trauma in BD. Daruy-Filho et al. [5] and Aas et al. [13] converged in showing the link between suicidality and early adverse events as a very stable result across studies. Suicide attempts were specifically linked to emotional and sexual abuse [12, 17, 27], in particular to both emotional and sexual abuse in BD-I, and only to emotional abuse in BD-II [12]. A recent study by our group pointed out emotional abuse as a direct predictor, among several other variables, of suicide attempts in BD [27]. The study speculated that the high risk of suicide attempts found in patients reporting emotional abuse could be related to inadequate emotion regulation mechanisms. This in line with studies showing that emotional dysregulation is linked to emotional abuse in BD [28] and is implicated in the neurobiology of suicide risk [29]. Our study [27] specifically found an association between childhood emotional abuse and suicidal attempts, but not ideation. It is therefore possible that childhood trauma can mediate the transition from ideation to action in patients with BD. Further longitudinal studies are needed to clarify this point.

## 8.4 Childhood Trauma and Psychopathological Dimensions of Bipolar Disorders

Childhood trauma might not only impact the clinical course of BD, but also influence specific psychopathological dimensions of the illness. Qualitative psychopathological dimensions represent key features for diagnostically framing BD, and can in turn impact the clinical complexity/severity of the illness.

In the early nineteenth century, Janet identified a close relationship between dissociative symptoms and traumatic experiences [30]. According to the fifth edition of the Diagnostic and Statistical Manual of Mental Disorders (DSM-5) [31], dissociative symptoms are experienced as follows: (a) unbidden intrusions into awareness and behaviour, accompanying losses of continuity in subjective experience, or (b) an inability to access information or to control mental functions that normally are readily amenable to access or control. Several studies indicated pathological dissociation as closely linked to mood symptoms. A recent study investigated childhood trauma in patients with BD using the CTQ, and dissociation using the Dissociative Experience Scale (DES) [32]. Interestingly, authors found that patients presenting with dissociative experiences recalled a greater amount of childhood trauma and reported an earlier age at onset.

Impulsivity and aggression have been described as core features of BD. A path analysis in a large sample (n = 484) of patients with BD demonstrated significant associations between childhood emotional abuse and impulsive dimensions [33]. Garno and colleagues found a specific relationship between aggressive behaviours and childhood trauma [34]. Intriguingly, in the same study they identified diagnosis of comorbid borderline personality disorder as a predictor of trait aggression in BD. These findings bear on unresolved controversies about the differential nosology of borderline personality disorder and BD. Impulsive and aggressive behaviours are typically present in the clinical picture of borderline personality disorder [31], and previous studies demonstrated a strong association between borderline personality disorder and history of abuse in childhood [35]. It is therefore possible that early adverse events might influence mechanisms underpinning both BD and borderline personality disorder [36]. Accordingly, it may be reasonable to include borderline personality disorder within a broader bipolar spectrum.

Affective lability is a further psychopathological feature strongly linked with childhood trauma. It has been conceptualized as frequent and intense fluctuations in affect in response to both pleasant and unpleasant events. As previously stated, already in the early nineteenth century, Minkowski highlighted that in manic-depressive illness "synchronism" with life experiences was excessive and inappropriate [37]. Aas and colleagues demonstrated that exposure to childhood trauma is associated with higher scores in affective lability, with the strongest association being for emotional abuse [28]. Moreover, the authors found that affective lability in BD mediated the relationship between childhood trauma experiences and several clinical variables, including suicide attempts and mixed episodes [6]. The existence of an emotional dysregulation in BD has been currently established, and specifically related to emotional abuse

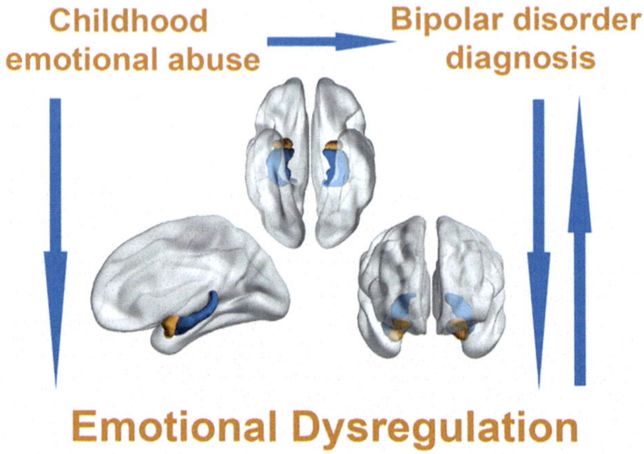


**Fig. 8.1** Childhood emotional abuse in bipolar disorders: the effect on the limbic system

(Fig. 8.1), which is indicated as the most important childhood trauma subtype in BD [4]. Supporting these findings, a recent study showed alterations in brain response to an emotional face-processing task in patients with psychosis (a common symptom in BD) who experienced early trauma [38]. Furthermore, in a study on the effect of childhood trauma in BD on deep grey matter structures, we found that early adverse events modulated volumes of the amygdala and hippocampus, which are central key areas in emotional processing and emotional regulation (Fig. 8.1) [8].

## 8.5 Childhood Trauma and Cognitive Profile in Bipolar Disorders

It is well established that BD is characterized by a state-independent cognitive impairment [39]. Recently, researchers have started to investigate the association between childhood trauma, BD, and altered cognitive development.

Savitz and colleagues [40] conducted the first study assessing early adverse events and neuropsychological tasks in a large sample of patients with BD-I and BD-II, unaffected relatives, and HCs. They found that the BD-I group performed worse than unaffected relatives on tests of verbal recall memory, and highlighted a specific association between childhood trauma and poor cognitive performance. A subsequent study confirmed the association between verbal memory impairment and childhood trauma in BD-I [7]. The study assessed cognitive functions and early adverse events through the CTQ in patients recently recovered from a first manic episode. Childhood trauma was associated with poorer performance in verbal memory, auditory attention and working memory in patients, while a different pattern was observed in HCs.

A further study aimed at investigating executive functions in patients with BD and assessed potential interactions between diagnosis and childhood trauma on a Go/No–Go task. Results showed a significant main effect of early trauma on inhibitory control accuracy, as the trauma group exhibited significantly poorer accuracy on inhibition trials compared to the normative group [41].

In a recent study using a clustering technique, Jiménez and colleagues [42] were able to identify three groups of patients with BD on the basis of their cognitive performance. The study was based on the assumption that cognitive impairment in BD seems to change according to different profiles of severity. Therefore, the aim of the study was to determine which variables contributed to the neurocognitive clustering membership. Results indicated that CTQ total scores and the estimated intelligence quotient significantly contributed to differentiate the three neurocognitive groups.

## 8.6 Neurobiological Consequences of Childhood Trauma in Bipolar Disorders

Previous studies have generated a large body of knowledge regarding the neurobiological correlates of BD. However, few efforts have been directed towards developing biologically informed constructs of childhood trauma in BD.

Information about molecular mechanisms mediating the consequences of childhood trauma in BD mostly derives from studies focused on gene–environment interactions.

Exposure to childhood trauma has been found to interact with various susceptibility genes, and to increase the risk of BD. Met carriers of the brain-derived neurotrophic factor (BDNF) val66met who had been exposed to childhood sexual abuse presented an earlier age at ilness onset [43]. Accordingly, recalling a history of childhood trauma and being a Met carrier of the BDNF val66met was associated with significantly reduced BDNF mRNA levels [44]. A recent review also found that Met allele negatively modulated the association between childhood trauma and memory performance, verbal ability and verbal fluency [45].

Other studies showed significant interactions between genetic variants and childhood trauma on age of onset; two studies in particular, focused on the short variant of the serotonin (5-hydroxytryptamine; 5-HT) transporter-linked polymorphic region (5-HTTLPR) [46, 47], and one study focused on the variant of the toll-like receptor, a protein that play a key role in the innate immune system (TLR2).

Interactions between childhood trauma and candidate genes in BD influenced severity of the illness. A specific impact was found on suicidal risk. Benedetti and colleagues [46] showed a significant effect of 5-HTTLPR on the relationship between stress, depression and suicide. The same group also demonstrated that early stress in combination with specific variants of the CLOCK gene, which modulates changes of circadian behaviours in healthy humans, influenced suicidal risk in bipolar depression [48]. According to this, a recent study found that emotional abuse was associated with poor sleep quality in BD [49]. We can therefore speculate that a history of childhood abuse in BD might underlie changes in circadian behaviour and in turn, impact on sleep alterations, which are primary symptoms in all phases of the illness.

In addition to genetic markers, inflammation might have a role in mediating the consequences of childhood trauma in BD. Several components of the immune pathway become altered *per se* in BD, and they can be further modulated by early adverse events. To our knowledge, no previous studies addressed the issue of the relationship between childhood trauma and inflammatory response in BD. However, Aas and colleagues found an association between childhood trauma, elevated C-reactive protein, and body mass index (BMI) in a combined sample comprising adults with schizophrenia and BD [13]. Consistently, a recent study found a relationship in BD between childhood sexual abuse and BMI [50].

Neuroimaging studies had a transformative influence on the field of the neurobiological correlates of psychiatric disorders. They have firmly established BD as a brain disorder involving multiple, spatially distributed structural and functional brain abnormalities [51, 52]. In the last few years, neuroimaging studies have also provided evidence linking childhood traumatic experiences in BD to brain dysfunction (Table 8.2).

**Table 8.2** Neuroimaging studies on childhood trauma in bipolar disorders

| First author | Year | Sample | Comparison group | Measure to assess childhood trauma | Neuroimaging technique | Major findings |
|---|---|---|---|---|---|---|
| Bücker | 2014 | 53 adults with DSM-IV BD diagnosis recently recovered from their first manic episode, with ($n = 23$) and without ($n = 30$) CT | 16 healthy controls without CT | CTQ | Automatic segmentation (FreeSurfer) Corpus callosum considered as ROI | The total corpus callosum volume was found to be smaller in BD patients with CT compared to BD patients without trauma but not to healthy controls |
| Duarte | 2016 | 39 euthymic DSM-IV BD-I patients with and with ($n = 20$) and without ($n = 19$) CT | 20 healthy controls without CT | CTQ | VBM | BD-I patients had significant negative correlations between CTQ total score and GMV in the right dorsolateral prefrontal cortex (PFC) and the right thalamus; between physical abuse and GMV in the right dorsolateral PFC; between physical neglect and GMV in the thalamus bilaterally; and between emotional neglect and GMV in the right thalamus. |

**Table 8.2** (continued)

| First author | Year | Sample | Comparison group | Measure to assess childhood trauma | Neuroimaging technique | Major findings |
|---|---|---|---|---|---|---|
| Aas | 2017 | 101 patients with a DSM-IV schizophrenia spectrum or bipolar spectrum diagnosis. | – | CTQ | Task-based functional magnetic resonance imaging (response to emotional stimuli) | Higher levels of total childhood trauma were associated with stronger differentiation in brain responses to negative compared with positive faces in clusters comprising the right angular gyrus, supramarginal gyrus, middle temporal gyrus and the lateral occipital cortex |
| Janiri | 2017 | 105 patients with DSM-IV BD-I and BD-II diagnosis with and without childhood trauma. | 113 healthy controls | CTQ | Automatic segmentation (FreeSurfer) | Patients with BP showed a global reduction of deep grey matter volumes compared to HCs. However, childhood trauma modulated the impact of diagnosis specifically on the amygdala and hippocampus. Childhood trauma was associated with bilateral decreased volumes in HCs and increased volumes in patients with BP. |
| Stevelink | 2018 | 251 patients with DSM-IV BD-I diagnosis | 163 healthy controls | CTQ | DTI | BD patients with childhood abuse had lower FA in widespread regions of the brain relative to patients without childhood abuse, no differences were found between healthy individuals with and without abuse. |

*BD* bipolar disorder, *BD-I* bipolar disorder type I, *BD-II* bipolar disorder type II, *CT* childhood trauma, *CTQ* childhood trauma questionnaire, *DSM-IV* diagnostic and statistical manual of mental disorders fourth edition, *GMV* grey matter volume, *VBM* voxel-brain morphometry, *ROI* region of interest, *DTI* diffusion tensor imaging, *FA* fractional anisotropy

Structural volumetric changes have been found in patients with BD who experienced early adverse events. Duarte and colleagues [53], using a voxel-based morphometry approach, found that BD-I patients showed a negative correlation between CTQ total score and grey matter volumes in the right dorsolateral prefrontal cortex (DLPFC) and the right thalamus. Specifically, they found that physical abuse and physical neglect were driving this association. Bücker and colleagues [54] found reduced volume of the corpus callosum in patients with BD with childhood trauma compared to those without trauma. A recent study by our group found that amygdala and hippocampus volumes were differently affected by childhood trauma in patients with BD and HCs [8]. Specifically, childhood trauma was associated with bilateral decreased volumes in HCs and increased volumes in patients with BD. These results emerged in the context of a global reduction of deep grey matter volumes in patients with BD. We speculated that increased amygdala volumes were related to an emotional over-reactivity and increased hippocampal volumes to an over-representation of the traumatic experience. This observation is in line with the above-mentioned hypersensitivity to emotional stimuli and emotional dysregulation found in patients with BD (Fig. 8.1).

Emotional dysregulation could also explain the results of a recent study that used functional magnetic resonance imaging (fMRI) to test an emotion recognition paradigm in a sample of patients with BD and schizophrenia [38]. Authors found that higher levels of childhood trauma were associated with a stronger differentiation in brain response to negative compared to positive faces.

Some authors investigated the influence in BD of childhood trauma on structural brain connectivity. Stevelink and colleagues [55] found a global decrease in white matter integrity in patients with BD who experienced early adverse events compared to those who did not report childhood trauma. Integrity of white matter microstructure was assessed using Diffusion Tensor Imaging (DTI) and quantified using fractional anisotropy (FA) as a metric of white matter integrity. A resting-state fMRI (rs-fMRI) study found specific alterations in the limbic network [9] and a correlation between childhood trauma and decreased ventromedial Prefrontal Cortex-hippocampus and prefronto-limbic functional connectivity (FC). These results confirmed the importance of the hippocampus as a potential key area in modulating early traumatic events in BD (Fig. 8.1).

## 8.7 Conclusions

In this chapter, we reviewed the major findings to date about the relationship between childhood trauma and BD. To summarize, childhood trauma is 2.63 times more likely to have occurred in patients with BD compared with HCs. Among the different subtypes of childhood trauma, emotional abuse emerged as the most important. Several findings demonstrated an association between negative clinical outcomes in BD and early adverse events. According to the highest level of evidence, the strongest association seems to be with early onset of the disease and increased suicidal risk. Previous studies demonstrated a specific link between childhood trauma, cognitive, and psychopathological variables in

BD. In this context, the psychopathological variable standing out above the others seems to be affective lability/interpersonal sensitivity. This could be explained by a causal link between childhood trauma and emotional dysregulation, inducing affective lability/interpersonal sensitivity (Fig. 8.1). Nevertheless, the direction of this link should be properly investigated. As a final note, neurobiological findings, in particular from neuroimaging studies, confirmed the involvement of brain regions specifically related to emotion regulation, such as the limbic structures.

Some comments on future directions arise from this review. Above all, since it significantly alters the disease trajectory, childhood trauma should be systematically assessed in patients with BD using validated clinical interviews or questionnaires. Since different instruments possess different psychometric properties, it is advisable to converge to a homogenous assessment of childhood trauma. This to facilitate comparison and reproducibility across studies, and ultimately, meta-analysability. Consortiums of research groups should carefully address this point in the future.

This review has also uncovered the importance of considering different types of childhood trauma in BD. Many trauma measures focused on only one or two types of maltreatment, typically sexual or physical abuse, while ignoring other forms of victimization, such as emotional maltreatment. In contrast, emotional abuse emerged as the most important type of childhood trauma recalled by patients with BD [4]. In this context, a distinction should be made between childhood abuse and neglect. Childhood neglect refers to the failure of caregivers to meet a child's basic physical or emotional needs, and has been related to increased severity in the clinical course of BD. Childhood neglect can be conceptualized as an indirect form of victimization and should be assessed by specific instruments such as the CTQ. About half of the reviewed studies assessed childhood trauma subtypes, and among these, most were focused on the relationship between early adverse events and clinical variables. Only few neurobiological studies specifically considered the distinct subtypes of childhood trauma. This was often due to the small study sample size that did not allow to stratify patients on the basis of specific subtypes of early trauma. Conversely, we strongly recommend to investigate all subtypes of childhood trauma in patients with BD. Different types of measures can present a different sensitivity to different types of trauma, which could be involved in biologically meaningful BD phenotypes.

Another relevant research gap that needs to be addressed is the differential impact of childhood trauma on different BD subtypes. Childhood trauma occurrence is differently distributed between BD-I and BD-II [12]. Nevertheless, most previous studies did not consider the clinical distinctions between BD subtypes and instead aggregated all patients into a single group. This is a potentially important shortcoming because diagnoses of BD-I and BD-II are associated with different types of childhood trauma [12], which can in turn impact on clinical, psychopathological and neurobiological variables.

In addition to these observations, the data available on the relationship between childhood trauma and BD suggest two directions for further studies.

First, current clinical data urge us to consider the possibility of a specific treatment for patients who reported early trauma. A large body of studies indicated that patients with BD who experienced childhood trauma are part of a specific clinical population. Compared to patients who were never exposed to childhood abuse, patients with at least two childhood trauma abuses were more at risk of not responding to lithium (OR = 4.91 95% CI) [56]. However, there are currently no guidelines on the clinical management of these patients. Researchers should focus on more personalized care plans for those individuals with BD exposed to childhood trauma.

Second, all the studies reviewed in this chapter assessed childhood trauma retrospectively. The reliability of the recall could be consequently influenced by uncontrolled bias. Furthermore, cross-sectional studies are poorly informative about causal mechanisms underpinning the effect of early adverse events on BD. Future longitudinal studies could help to explore the effect of childhood trauma on neurobiological patterns, in order to understand the mechanisms that render neural networks vulnerable to BD. These investigations could be used to model the effect of treatments, thereby assisting in the selection of personalized interventions based on their predicted effect.

**Bipolar Disorder and Childhood Trauma: Drawing from the Life of Virginia Woolf**

"All extremes of feeling are allied with madness" (V. Woolf, Orlando)

Adeline Virginia Woolf was born into an affluent household in South Kensington, London, as the seventh child in a family of eight. Her mother Julia was a celebrated English woman, noted for her beauty as a Pre-Raphaelite model and philanthropist. She suffered from depression and has been described as "haunted, worn down and beautiful". Her father Leslie was an agnostic writer and biographer. He was reported to have unpredictable mood swings, although never very severe, most likely cyclothymic.

Virginia's half sibling, Gerald, 16 years older than Virginia, sexually abused her when she was 6 years old. He molested her when her family was vacationing in Cornwall.

"There was a slab outside the dining room for standing dishes upon. Once, when I was very small, Gerald lifted me to this, and as I sat there he began to explore my body. I can remember the feel of his hand going under my clothes; going firmly and steadily lower and lower. I remember how I hoped that he would stop; how I stiffened and wriggled as his hand approaches my private parts. But it did not stop. I remember resenting, disliking it—what is the word for so dumb and mixed a feeling?" (Woolf, Moments of being).

Virginia experienced her first manic episode at age 13, after her mother's death from influenza. She became painfully excitable and nervous and then intolerably depressed, displaying both manic and depressive symptoms. Virginia's father died of a bowel cancer, which caused Virginia to suffer a second episode of illness at the age of 22. Reports state that she entered a manic state and required three nurses to keep her under control. She was psychotic for several months and she tried to jump out of a window in the family's home in London. Her next suicide attempt was serious and happened when she was 31 years old. She attempted suicide by taking 100 g of barbital. She would have died if it were not for the intervention of her husband, Leonard.

During the later part of her life, Virginia went through several severe depressive and manic episodes. Throughout the last years of her life, symptoms of mood episodes occurred increasingly more frequent and became more severe. Because of the variety of her symptoms and their fluctuating manifestation over the years, Virginia was probably diagnosed with an erratic personality disorder. This diagnosis is in clear contradiction with Woolf's disorder, which included mania. Various drug treatments were administered to Woolf, but she was never treated with lithium, because lithium's effects had yet to be discovered and the drug was not made available during her lifetime.

Virginia Woolf died at the age of 59. One day she filled her coat pockets with heavy stones and headed to the River Ouse to never return. She had been severely depressed during the last 2 weeks of her life. Before she committed suicide, she wrote to her husband Leonard:

"I want to tell you that you have given me complete happiness. No one could have done more than you have done. Please believe that. It is this madness. All I want to say is that until this disease came on we were perfectly happy. It was all due to you. No one could have been so good as you have been, from the very first day till now. Everyone knows that."

(V. Woolf, Letter).

References

Woolf V. Moments of being: a collection of autobiographical writing. 2nd ed. Orlando: Harcourt Brace; 1985

Woolf V. Letter to Ethel Smyth [22 June 1930], letter to Vanessa Bell [23 March 1941] & letter to Leonard Woolf [28 March 1941]. In: Trautmann Banks J, ed. Congenial spirits: the selected letters of Virginia Woolf. London: The Hogarth Press, 1989.

Jamison KR. Touched with Fire: Manic-Depressive Illness and the Artistic Temperament. New York: Free Press Paperbacks; 1993.

Boeira MV, Berni G de Á, Passos IC, Kauer-Sant'Anna M, Kapczinski F. (2017). Virginia Woolf, neuroprogression, and bipolar disorder. Rev Bras Psiquiatr. 39, 69–71.

Koutsantoni K. (2012). Manic depression in literature: the case of Virginia Woolf. Medical Human. 38 (1), 7–14.

## References

1. Kraepelin E. Manic depressive insanity and paranoia. Edinburgh: E. and S. Livingstone; 1921.
2. Lex C, Bäzner E, Meyer TD. Does stress play a significant role in bipolar disorder? A meta-analysis. J Affect Disord. 2017;208(October 2016):298–308. https://doi.org/10.1016/j.jad.2016.08.057.
3. M'bailara K, Demotes-Mainard J, Swendsen J, Mathieu F, Leboyer M, Henry C. Emotional hyper-reactivity in normothymic bipolar patients. Bipolar Disord. 2009;11(1):63–9.
4. Palmier-Claus JE, Berry K, Bucci S, Mansell W, Varese F. Relationship between childhood adversity and bipolar affective disorder: systematic review and meta-analysis. Br J Psychiatry. 2016;209:454–9. http://bjp.rcpsych.org/cgi/doi/10.1192/bjp.bp.115.179655
5. Daruy-Filho L, Brietzke E, Lafer B. Grassi-Oliveira R. Childhood maltreatment and clinical outcomes of bipolar disorder. Acta Psychiatr Scand. 2011;124:427–34. http://onlinelibrary.wiley.com/doi/10.1111/j.1600-0447.2011.01756.x/full
6. Aas M, Henry C, Bellivier F, Lajnef M, Gard S, Kahn JP, et al. Affective lability mediates the association between childhood trauma and suicide attempts, mixed episodes and co-morbid anxiety disorders in bipolar disorders. Psychol Med. 2017;47(5):902–12.
7. Bücker J, Kozicky J, Torres IJ, Kauer-Sant'Anna M, Silveira LE, Bond DJ, et al. The impact of childhood trauma on cognitive functioning in patients recently recovered from a first manic episode: data from the systematic treatment optimization program for early mania (STOP-EM). J Affect Disord. 2013;148(2–3):424–30.
8. Janiri D, Sani G, Rossi P De, Piras F, Iorio M, Banaj N, et al. Amygdala and hippocampus volumes are differently affected by childhood trauma in patients with bipolar disorders and healthy controls. Bipolar Disord. 2017;19:353–62.
9. Souza-Queiroz J, Boisgontier J, Etain B, Poupon C, Duclap D, D'Albis MA, et al. Childhood trauma and the limbic network: a multimodal MRI study in patients with bipolar disorder and controls. J Affect Disord. 2016;200:159–64.
10. Bernstein DP, Fink L, Handelsman L, Foote J, Lovejoy M, Wenzel K, et al. Initial reliability and validity of a new retrospective measure of child abuse and neglect. Am J Psychiatry. 1994;151(8):1132–6. https://doi.org/10.1176/ajp.151.8.1132.
11. Garno JL, Goldberg JF, Ramirez PM, Ritzler BA. Impact of childhood abuse on the clinical course of bipolar disorder. Br J Psychiatry. 2005;186:121–5. http://bjp.rcpsych.org/content/186/2/121.short
12. Janiri D, Sani G, Danese E, Simonetti A, Ambrosi E, Angeletti G, Erbuto D, Caltagirone C, Girardi P, Spalletta G. Childhood traumatic experiences of patients with bipolar disorder type I and type II. J Affect Disord. 2015;175:92–7.
13. Aas M, Henry C, Andreassen OA, Bellivier F, Melle I, Etain B. The role of childhood trauma in bipolar disorders. Int J Bipolar Disord. 2016;4(1):2. http://www.journalbipolardisorders.com/content/4/1/2
14. Leverich GS, McElroy SL, Suppes T, Keck PE, Denicoff KD, Nolen WA, et al. Early physical and sexual abuse associated with an adverse course of bipolar illness. Biol Psychiatry. 2002;51:288–97.
15. Post RM, Altshuler LL, Kupka R, Mcelroy SL, Frye MA, Rowe M, et al. Verbal abuse, like physical and sexual abuse, in childhood is associated with an earlier onset and more difficult course of bipolar disorder. Bipolar Disord. 2015;17(3):323–30.
16. Li X, Bin LJT, Zhu XZ, Zhang L, Tang YL, Wang CY. Childhood trauma associates with clinical features of bipolar disorder in a sample of Chinese patients. J Affect Disord. 2014;168:58–63.
17. Etain B, Aas M, Andreassen OA, Lorentzen S, Dieset I, Gard S, et al. Childhood trauma is associated with severe clinical characteristics of bipolar disorders. J Clin Psychiatry. 2013;74:991–8. http://www.ncbi.nlm.nih.gov/pubmed/24229750
18. Dienes KA, Hammen C, Henry RM, Cohen AN, Daley SE. The stress sensitization hypothesis: understanding the course of bipolar disorder. J Affect Disord. 2006;95(1–3):43–9.

19. Koukopoulos A, Sani G, Koukopoulos AE, Minnai GP, Girardi P, Pani L, et al. Duration and stability of the rapid-cycling course: a long-term personal follow-up of 109 patients. J Affect Disord. 2003;73:75–85.
20. Janssen I, Krabbendam L, Bak M, Hanssen M, Vollebergh W, De Graaf R, et al. Childhood abuse as a risk factor for psychotic experiences. Acta Psychiatr Scand. 2004;109:38–45.
21. Romero S, Birmaher B, Axelson D, Goldstein T, Goldstein BI, Gill MK, et al. Prevalence and correlates of physical and sexual abuse in children and adolescents with bipolar disorder. J Affect Disord. 2009;112:144–50.
22. Bebbington P, Jonas S, Kuipers E, King M, Cooper C, Brugha T, et al. Childhood sexual abuse and psychosis: data from a cross-sectional national psychiatric survey in England. Br J Psychiatry. 2011;199(1):29–37. http://www.ncbi.nlm.nih.gov/pubmed/21508437
23. Upthegrove R, Chard C, Jones L, Gordon-Smith K, Forty L, Jones I, et al. Adverse childhood events and psychosis in bipolar affective disorder. Br J Psychiatry. 2015;206(3):191–7.
24. Goldstein BI, Strober MA, Birmaher B, Axelson DA, Esposito-Smythers C, Goldstein TR, et al. Substance use disorders among adolescents with bipolar spectrum disorders. Bipolar Disord. 2008;10(4):469–78.
25. Brown GR, McBride L, Bauer MS, Williford WO, Cooperative-Studies-Program. Impact of childhood abuse on the course of bipolar disorder: a replication study in U.S. veterans. J Affect Disord. 2005;89:57–67.
26. Aas M, Etain B, Bellivier F, Henry C, Lagerberg T, Ringen A, et al. Additive effects of childhood abuse and cannabis abuse on clinical expressions of bipolar disorders. Psychol Med. 2014;44(8):1653–62.
27. Janiri D, De Rossi P, Kotzalidis GD, Girardi P, Koukopoulos AE, Reginaldi D, et al. Psychopathological characteristics and adverse childhood events are differentially associated with suicidal ideation and suicidal acts in mood disorders. Eur Psychiatry. 2018;53:31–6. http://linkinghub.elsevier.com/retrieve/pii/S0924933818301044
28. Aas M, Aminoff SR, Vik Lagerberg T, Etain B, Agartz I, Andreassen OA, et al. Affective lability in patients with bipolar disorders is associated with high levels of childhood trauma. Psychiatry Res. 2014;218(1–2):252–5. http://www.ncbi.nlm.nih.gov/pubmed/24803185
29. Neacsiu AD, Fang CM, Rodriguez M, Rosenthal MZ. Suicidal behavior and problems with emotion regulation. Suicide and life-threatening behavior. 2017;
30. Janet P. L'Amnésie et la dissociation des souvenirs par l'émotion. J Psycologie. 1904;1:417–53.
31. American Psychiatric Association. DSM 5. Am J Psychiatr. 2013:991. http://ajp.psychiatryonline.org/article.aspx?articleID=158714%5Cnhttp://scholar.google.com/scholar?hl=en&btnG=Search&q=intitle:DSM-5#0
32. Hariri AG, Gulec MY, Orengul FFC, Sumbul EA, Elbay RY, Gulec H. Dissociation in bipolar disorder: relationships between clinical variables and childhood trauma. J Affect Disord. 2015;184:104–10. https://doi.org/10.1016/j.jad.2015.05.023.
33. Raust A, Sportiche S, Geoffroy PA, Aouizerate B, Desage A, Olie E, et al. Childhood trauma, dimensions of psychopathology and the clinical expression of bipolar disorders: a pathway analysis. J Psychiatr Res. 2017;95:37–45.
34. Garno JL, Gunawardane N, Goldberg JF. Predictors of trait aggression in bipolar disorder. Bipolar Disord. 2008;10(2):285–92.
35. Herman JL, Perry JC, van der Kolk B. a. Childhood trauma in borderline personality disorder. Am J Psychiatry. 1989;146(4):490–5.
36. Agius M, Lee J, Gardner J, Wotherspoon D. Bipolar II disorder and borderline personality disorder - co-morbidity or spectrum? In: Psychiatria Danubina; 2012. p. 197–201.
37. Minkoswski E. La schizophrénie: Psychopathologie des schizoïdes et des schizophrènes. Paris: Payot; 1927.
38. Aas M, Kauppi K, Brandt CL, Tesli M, Kaufmann T, Steen NE, et al. Childhood trauma is associated with increased brain responses to emotionally negative as compared with positive faces in patients with psychotic disorders. Psychol Med. 2017;47(4):669–79.
39. Torres IJ, Boudreau VG, Yatham LN. Neuropsychological functioning in euthymic bipolar disorder: a meta-analysis. Acta Psychiatr Scand. 2007;116(SUPPL. 434):17–26.

40. Savitz JB, van der Merwe L, Stein DJ, Solms M, Ramesar RS. Neuropsychological task performance in bipolar spectrum illness: genetics, alcohol abuse, medication and childhood trauma. Bipolar Disord. 2008;10(4):479–94.
41. Marshall DF, Passarotti AM, Ryan KA, Kamali M, Saunders EFH, Pester B, et al. Deficient inhibitory control as an outcome of childhood trauma. Psychiatry Res. 2016;235:7–12. https://doi.org/10.1016/j.psychres.2015.12.013.
42. Jiménez E, Solé B, Arias B, Mitjans M, Varo C, Reinares M, et al. Impact of childhood trauma on cognitive profile in bipolar disorder. Bipolar Disord. 2017;19(5):363–74.
43. Miller S, Hallmayer J, Wang PW, Hill SJ, Johnson SL, Ketter TA. Brain-derived neurotrophic factor val66met genotype and early life stress effects upon bipolar course. J Psychiatr Res. 2013;47(2):252–8. https://doi.org/10.1016/j.jpsychires.2012.10.015.
44. Aas M, Haukvik UK, Djurovic S, Tesli M, Athanasiu L, Bjella T, et al. Interplay between childhood trauma and BDNF val66met variants on blood BDNF mRNA levels and on hippocampus subfields volumes in schizophrenia spectrum and bipolar disorders. J Psychiatr Res. 2014;59:14–21.
45. Mandolini GM, Lazzaretti M, Pigoni A, Delvecchio G, Soares JC, Brambilla P. The impact of BDNF Val66Met polymorphism on cognition in bipolar disorder: a review. J Affect Disord. 2018;243(July 2018):552–8. https://doi.org/10.1016/j.jad.2018.07.054.
46. Benedetti F, Riccaboni R, Poletti S, Radaelli D, Locatelli C, Lorenzi C, et al. The serotonin transporter genotype modulates the relationship between early stress and adult suicidality in bipolar disorder. Bipolar Disord. 2014;16(8):857–66.
47. Etain B, Lajnef M, Henrion A, Dargél AA, Stertz L, Kapczinski F, et al. Interaction between SLC6A4 promoter variants and childhood trauma on the age at onset of bipolar disorders. Sci Rep. 2015;5:1–9. https://doi.org/10.1038/srep16301.
48. Benedetti F, Riccaboni R, Dallaspezia S, Locatelli C, Smeraldi E, Colombo C. Effects of CLOCK gene variants and early stress on hopelessness and suicide in bipolar depression. Chronobiol Int. 2015;32(8):1156–61.
49. Aubert E, Jaussent I, Olié E, Ducasse D, Azorin JM, Bellivier F, et al. Effect of early trauma on the sleep quality of euthymic bipolar patients. J Affect Disord. 2016;206:261–7. https://doi.org/10.1016/j.jad.2016.07.045.
50. Leclerc E, Mansur RB, Grassi-Oliveira R, Cordeiro Q, Kapczinski F, McIntyre RS, et al. The differential association between history of childhood sexual abuse and body mass index in early and late stages of bipolar disorder. J Affect Disord. 2017;227:214–8. https://doi.org/10.1016/j.jad.2017.10.031.
51. Hibar DP, Westlye LT, van Erp TGM, Rasmussen J, Leonardo CD, Faskowitz J, et al. Subcortical volumetric abnormalities in bipolar disorder. Mol Psychiatry. 2016:1–7. http://www.ncbi.nlm.nih.gov/pubmed/26857596
52. Doucet GE, Bassett DS, Yao N, Glahn DC, Frangou S. The role of intrinsic brain functional connectivity in vulnerability and resilience to bipolar disorder. Am J Psychiatry. 2017;174(12):1214–22.
53. Duarte DGG, Neves MDCL, Albuquerque MR, De Souza-Duran FL, Busatto G, Corrêa H. Gray matter brain volumes in childhood-maltreated patients with bipolar disorder type I: a voxel-based morphometric study. J Affect Disord. 2016;197:74–80. https://doi.org/10.1016/j.jad.2016.02.068.
54. Bücker J, Muralidharan K, Torres IJ, Su W, Kozicky J, Silveira LE, et al. Childhood maltreatment and corpus callosum volume in recently diagnosed patients with bipolar I disorder: data from the systematic treatment optimization program for early mania (STOP-EM). J Psychiatr Res. 2014;48(1):65–72. http://www.ncbi.nlm.nih.gov/pubmed/24183241
55. Stevelink R, Abramovic L, Verkooijen S, Begemann MJH, Sommer IEC, Boks MP, et al. Childhood abuse and white matter integrity in bipolar disorder patients and healthy controls. Eur Neuropsychopharmacol. 2018;28(7):807–17. https://doi.org/10.1016/j.euroneuro.2018.05.003.
56. Etain B, Lajnef M, Brichant-Petitjean C, Geoffroy PA, Henry C, Gard S, et al. Childhood trauma and mixed episodes are associated with poor response to lithium in bipolar disorders. Acta Psychiatr Scand. 2016:1–9. http://www.ncbi.nlm.nih.gov/pubmed/27987204

# Childhood Trauma in Depressive Disorders

# 9

Monica Aas and Bruno Etain

## 9.1 Childhood Trauma Increases the Risk of MDD

Childhood trauma is an important risk factor for developing a wide range of severe mental disorders, including major depressive disorders (MDDs) [1]. According to the US Department Human Services [2], in 2017 alone, approximately 683,000 children were victims of maltreatment. Most of these children were victims of neglect (75.3%), with physical abuse (17.2%) and sexual abuse (8.4%) being the next most frequent types of child maltreatment. A large national representative sample of more than 9,000 adults showed that severe childhood adversity accounts for nearly 32% of psychiatric disorders [3], with up to 44% for early onset disorders [3]. Major depressive disorder (MDD) is the leading cause of disability worldwide, and a major contributor to the overall global burden of mental diseases.

For example, it is estimated that up to 50% of the suicides per year worldwide occur within a depressive episode and that patients with MDD are almost 20-fold more likely to die by suicide than the general population [4]. Depressive episodes

M. Aas (✉)
NORMENT, KG Jebsen Centre for Psychosis Research, Institute of Clinical Medicine, University of Oslo, Oslo, Norway

Division of Mental Health and Addiction, Oslo University Hospital, Oslo, Norway
e-mail: monica.aas@gmail.com; monica.aas@medisin.uio.no

B. Etain
Université Paris Diderot, Paris, France

INSERM U1144, Faculté de Pharmacie de Paris, Université Paris Descartes, Université Paris Diderot, Université Sorbonne Paris Cité, Paris, France

Département de Psychiatrie et de Médecine Addictologique, AP-HP, GH Saint-Louis - Lariboisière - F. Widal, Paris, France

ENBREC, European Network of Bipolar Research Expert Centres (ENBREC), Paris, France
e-mail: Bruno.Etain@inserm.fr


G. Spalletta et al. (eds.), *Childhood Trauma in Mental Disorders*,
https://doi.org/10.1007/978-3-030-49414-8_9

have a high risk of reoccurrence with a 50% relapse rate after the first, 70% after the second, and 90% after the third depressive episode. Studies show that genetic factors account for around 40% of variance risk for MDD [5, 6], while large-scale epidemiological studies indicate that exposure to early adversity (i.e., childhood abuse, maltreatment, or other early stressors) accounts for a 54% risk of current depression [7]. A recent study demonstrated that childhood trauma is associated with an increased odds ratio (OR) of 13.7 for developing MDD [8]. Similarly, in a sample of adults with a history of childhood trauma, 62% met *DSM-IV* criteria for lifetime MDD compared to a 28% with lifetime posttraumatic stress disorder (PTSD) [9], indicating a strong link between childhood trauma and MDD. As such, MDD is the most common adult psychopathology following childhood trauma [7]. A 10–25% reduction in maltreatment could potentially prevent 31.4–80.3 million depression and anxiety cases worldwide [10].

## 9.2 Childhood Trauma and the Clinical Expression of MDD

Patients with both MDD and childhood trauma have a more severe illness over time, including increased risk for suicide attempts, earlier age at onset, more depressive episodes, and poor response to antidepressant treatment [1, 11]. As discussed by Nemeroff [1], there is considerable evidence that childhood trauma is associated with a more severe course of depression including chronicity [12], specific features of depression (hypersomnia or sleep abnormalities, interpersonal rejection sensitivity, increased appetite, leaden limb paralysis) [13], and poorer outcome to treatment with psychopharmacology and/or psychotherapy [14]. Indeed, childhood trauma was recently linked to a more chronic subtype of MDD using a large and representative sample in the Netherlands Study of Depression and Anxiety (NESDA). Wiersma and colleagues [12] studied chronicity of MDD and childhood trauma in a large sample of more than 1,000 individuals collected over a 3-year period in the mid-2000. Participants had a current DSM-IV-TR diagnosis of MDD and were recruited from the community, primary care settings, and specialized mental health-care facilities. Chronicity of depression was defined as patients being depressed for 24 months or more in the past 4 years. In this study, chronicity of depression was significantly associated with a higher prevalence of childhood trauma, whereas not linked to any specific types of childhood traumatic events. The strongest association was found for those with the highest score on a cumulative index summarizing frequency of childhood trauma (OR = 3.26). Even after controlling for comorbid anxiety disorders, severity of depressive symptoms, and age at onset of depression, the childhood trauma index was associated with chronicity of depression (OR = 2.06). Hence, these results suggest that multiple childhood traumas might be independent determinants of chronicity in depression. Higher chronicity in MDD with childhood trauma is also supported by a recent meta-analysis showing that depressed patients who were abused during childhood had a poor response to a combination of structured psychological treatment and pharmacotherapy with antidepressants [15]. A large meta-analysis by Nanni and colleagues [14] including 23,544 participants identified childhood trauma as a risk factor for mood reoccurrences in MDD, where recurrence was defined in terms of number of depressive episodes while persistence was defined in terms of duration of a current depressive episode. Treatment outcome was defined in terms of either a response (a 50% reduction in depression severity ratings from baseline), or remission (a decrease in depression severity below a predefined clinical significance level). This meta-analysis of 16 epidemiological studies of more than 20,000 participants concluded that childhood trauma was associated with an elevated risk of developing recurrent and persistent depressive episodes (OR = 2.27, 95% confidence interval [CI] = 1.80–2.87). A meta-analysis of 10 clinical trials (3,098 participants) revealed that childhood maltreatment was associated with lack of response or remission during treatment for MDD (OR = 1.43, 95% CI = 1.11–1.83) [1]. Meta-regression analyses confirmed that the results were not influenced by publication bias, choice of outcome measure, prevalence or incidence in different samples, study quality, age, or lifetime prevalence of depression. The fact that childhood trauma (especially emotional abuse and neglect) is associated with poorer outcome in MDD was recently confirmed by Paterniti and colleagues [16] who assessed 238 individuals suffering from a current major depressive episode. Fifty percent of these participated in a follow-up study over a 2-year period. Among the latter, 45.4% did not recover or remit during the follow-up period. The median time to remission or recovery was 28.9 months and the median time to the first recurrence was 25.7 months. Childhood history of physical neglect predicted a slower time to remission or recovery.

Patients with early onset MDD are more likely to report childhood trauma experiences than patients with a later onset. Indeed, a large national survey of circa 9,000 individuals demonstrated that trauma experiences were observed in 45% of all childhood-onset disorders versus 26% to 32% of later-onset disorders [3]. Interestingly, a recent study by Comijs and colleagues [8] suggested that a history of childhood trauma is not only a risk factor for early onset depression, but also a predictor of late life onset of MDD (after age 60). This sample was collected from 378 depressed and 132 nondepressed individuals, aged from 60 to 93 years. All participants were recruited as part of the Netherlands Study of Depression in Older persons (NESDO). Measures of childhood abuse included psychological, physical and sexual abuse, and emotional neglect. Fifty-three percent of the depressed older adults reported childhood abuse, compared to 16% of the nondepressed older adults. Using a logistic regression adjusted for age, sex, and level of education, depression was strongly associated with physical abuse (OR = 13.71; 95%CI = 3.25–57.91), and sexual abuse (OR = 5.35; 95%CI = 2.36–12.14). Childhood abuse was associated with early onset (OR = 13.73, 95%CI = 7.31–25.80), middle age onset (OR = 5.36, 95%CI = 2.90–9.90), and late onset depression (OR = 4.74, 95%CI = 2.51–8.95). Hence, this study suggests that childhood trauma might have a deleterious influence on the risk for MDD whatever the age at onset, even if the effect appears more pronounced in early onset cases.

### 9.2.1 Childhood Trauma and Subtypes of MDD

Patients with MDD reporting childhood trauma are more likely to present with atypical symptoms of depression including hypersomnia, interpersonal rejection sensitivity, increased appetite, and leaden limb paralysis [13]. Trauma has also been linked to melancholic depression, thus there seem to be some discrepancies in the literature. The study by Withers and colleagues [13] investigated the association between a history of childhood trauma and DSM-IV atypical features of depression in almost 300 patients. Atypical features of depression were more commonly observed in patients reporting childhood trauma. Lifetime trauma both before and after a diagnosis of MDD was reported significantly more often by depressed patients with atypical features, than by those without ($p < 0.001$). Patients with atypical features also reported significantly more traumatic experiences (recent stressors) prior to depression onset. When gender, age at onset, or duration of depression were used as covariates, depressive subtype with atypical features was a significant predictor of reported trauma prior to depression onset. This could indicate that patients with both MDD and a history of childhood trauma have different clinical features of MDD as compared to those without childhood traumatic events. Moreover, Withers and colleagues found that traumatic events independently from time (traumatic events in childhood or adulthood) were associated with atypical features of depression. Another study suggests a preferential association between dysthymia and childhood trauma experiences, rather than separate episodes of depression [17]. Lizardi and colleagues suggested that early adversity is linked to dysthymia rather than to single depressive episodes with remission in between episodes. The study by Lizardi and colleagues [17] included three groups: 97 outpatients with dysthymia, 45 patients with episodic major depression, and 45 healthy controls. Patients with episodic depression and patients with dysthymia reported more physical and sexual abuse, and poorer relationships with both parents compared to healthy controls. Furthermore, patients with dysthymia reported poorer parenting than those with episodic major depression. The results could not be accounted for by mood state effects, comorbidity with borderline, and antisocial personality disorders. This study thus supports poor family relationships as a factor for continuous depressive symptoms over time rather than for a more episodic nature of the disorder. The study by Horwitz and colleagues [18] followed up a group of 600 children over 20 years and found that abused and neglected women reported more symptoms of dysthymia, antisocial personality disorder, and alcohol misuse than healthy individuals, again demonstrating links between childhood trauma and dysthymia. Insecure attachment has also been more strongly linked to dysthymia than to major depression.

### 9.2.2 Childhood Trauma, MDD, and Illness Comorbidity

Patients with MDD with a history of childhood trauma are more likely to present with a dual diagnosis. A prospective cohort study of over 500 children with documented abuse (physical or sexual) or neglect were compared to a matched sample of non-abused and non-neglected children. The children who reported childhood trauma exhibited higher rates of comorbid disorders including PTSD and substance/alcohol abuse [19]. These findings are consistent with those from Putnam and colleagues [20], who found that multiple childhood traumatic events resulted in complex adult psychopathology, as defined by higher rates of comorbidity and a greater number of symptoms. This is also supported by some of our own studies [21] showing that patients with a mood disorder are more likely to also suffer from cannabis and alcohol misuse, compared to patients without childhood traumatic experiences.

Although a relationship is observed between increased risk of substance abuse in individuals experiencing childhood trauma, the literature does not support a clear mediating effect between childhood trauma, substance misuse, and the development of psychopathology. One putative interpretation of the co-occurrence of MDD and substance use (i.e., cannabis use) in individuals with high levels of childhood trauma is the possible opposite effects of the two disorders on the Hypothalamic–Pituitary–Adrenal (HPA) axis, thus potentially related to some sort of "self-medication" aiming to normalize the HPA axis.

As already mentioned, patients with MDD who report a history of childhood trauma are also at increased risk for suicide attempts, and are also more likely to have a comorbid anxiety disorder, and other diagnostic comorbidities. For example, the study by Vitriol and colleagues [22] demonstrated in a sample of almost 400 patients with MDD that childhood trauma was associated with dual diagnosis. Specifically, having a dual social anxiety disorder was associated with having witnessed domestic violence during childhood (OR = 2.2, CI 1.2–3.8), childhood physical abuse (OR = 2.7, CI 1.6–4.4), and sexual abuse by a nonrelative (OR = 2.7, CI 1.3–4.2). A dual PTSD diagnosis was associated with physical injury as a consequence of physical abuse (OR = 1.9, CI 1.1–3.6), sexual abuse by a relative (OR = 3.2, IC 1.8–5.9), and sexual abuse by a nonrelative (OR = 2.2, CI 1.2–4.1). A dual antisocial personality disorder was associated with traumatic separation from a caregiver (OR = 3.2, CI 1.2–8.5), physical abuse (OR = 2.8, CI 1.1–6.9), and sexual abuse by a nonrelative (OR = 4.8, CI 1.2–11.5). A dual panic disorder was associated with sexual abuse by a relative (OR = 1.9, CI 1.1–3.1), and a dual generalized anxiety disorder was associated with sexual abuse by a nonrelative (OR = 1.9, CI 1.1–3.3).

### 9.2.3 Gender Differences and Trauma Subtypes

Females are overrepresented in clinical samples with MDD compared to males (prevalence rates up to twice). Adolescence is a vulnerable time for developing MDD in females, and the higher incidence in females manifests itself during adolescence [23]. Type of depression also differs between genders, with females more often reporting depression with anxiety, with sleep and appetite changes, and loss of energy. In boys, anhedonia and greater variation in mood and energy are more often reported [24]. A recent study also shows that poorer function in females with MDD is mediated by depressive symptoms, while males' poorer function in MDD was mediated by poorer sleep and anxiety [25].

Maltreatment increases risk for depression in both males and females, though some studies suggest greater risk for depression in abused females than in males [26]. Macmillan and colleagues studied traumatic experiences in childhood in more than 7,000 individuals aged 15–64. In females, the association between MDD and a history of sexual abuse in childhood was higher (OR = 3.9) as compared to males (OR = 1.9), with similar findings for physical abuse in females (OR = 3.2) as compared to males (OR = 1.5). The review paper by Norman and colleagues [27] did not find differences in reports of abuse and neglect between females and males; however, their review did not include sexual abuse in childhood. Furthermore, physical abuse, emotional abuse, and neglect were linked to depression, with the strongest ORs for physical abuse.

Only few studies have examined the specific effects of various types of childhood abuse/neglect on MDD. Intriguingly, two recent meta-analyses suggest that emotional abuse may have an even stronger link to depression than sexual abuse [28, 29].

There is emerging data that the psychological and biological consequences of deprivation/neglect differ substantially from those of threat/abuse. This is relevant to disentangle trauma subtypes, as many studies included only a trauma total score, independently from type of trauma, and did not differentiate abuse from neglect, which may have different effects on the developing brain. It should be noted that separating types if trauma is challanging, as individuals who have experienced one type of abuse are also at greater risk of having experienced other types of abuse and neglect. For example, although research shows a strong link between sexual abuse and depression, those with sexual abuse are also more likely to have experienced other types of abuse, for example emotional abuse.

### 9.2.4 Are Multiple Childhood Trauma Events Linked to More Severe Illness Severity in MDD?

The literature supports a dose effect of more types of childhood trauma and of trauma severity on the clinical features of depression. For example, the study by Negele and colleagues [30] evaluated 349 chronically depressed patients using the Childhood Trauma Questionnaire to capture early traumatic experiences. Seventy-six percent of the chronically depressed patients reported clinically significant histories of childhood trauma. In addition, a large proportion of patients (37%) reported multiple childhood traumas. The group who reported multiple types of childhood trauma had also the most severe depressive symptoms. Multiple regression analysis suggested that childhood emotional abuse and sexual abuse were significantly associated with a higher symptom severity in chronically depressed adults. Yet, expanding the regression model for multiple exposures revealed that multiplicity was the only remaining significant predictor for symptom severity in chronically depressed patients. Hence, being exposed to multiple types of childhood trauma may be more consistently linked to depressive symptoms in adulthood than any specific type of trauma the individual might have experienced. In the large Adverse Childhood

Experiences (ACE) Study, consisting of more than 9,000 individuals, a cumulative effect of chronic exposure to multiple childhood adversities was the strongest risk factor for disease burden, including depression [31].

### 9.2.5 Does the Time of Trauma Matter?

Timing of the childhood abuse/neglect is important (for a full overview, see Andersen and Teicher [7]). As discussed in the latter study, the developing brain undergoes a period of overproduction and pruning of synapses and signaling mechanisms between childhood, adolescence, and young adulthood. Windows of vulnerability potentially occur during periods of very rapid development, and synaptic pruning during adolescence might unmask underlying predispositions. It has been suggested that depression in adolescence may be a result of overproduction or overpruning in the prefrontal cortex, amygdala, and hypothalamus. Stress-related hormones such as the glucocorticoids can adversely affect the brain in two major ways. Firstly, abnormal secretion of these hormones can alter trajectories of development, predisposing to the emergence of psychopathology. Abundant preclinical evidence shows that increased glucocorticoid secretion reduces neurogenesis and synaptogenesis, especially in the hippocampus. Secondly, glucocorticoids can directly affect brain function, leading to depressivity [32]. The study by Wei and colleagues studied mice overexpressing glucocorticoid receptor (GR) in forebrain compared to wild mice. The mice that overexpressed GR displayed a significant increase in depressant-like behaviors relative to the wild type and were are also supersensitive to antidepressants. Thus, mice overexpressing GR in the forebrain have a consistently wider than normal range of reactivity in both positive and negative emotionality tests. This phenotype was associated, in specific brain regions, with increased expression of genes relevant to emotionality: corticotropin-releasing hormone, serotonin, norepinephrine, and dopamine transporters, and 5-hydroxytryptamine (1A) receptor. Thus, GR overexpression in the forebrain may cause higher depressivity secondary to a unique pattern of molecular regulation.

Mood is regulated by cortical and limbic regions. As these pathways mature during adolescence, they are affected by exposure to gonadal and adrenal hormones. Episodes of melancholic depression are often associated with increased secretion of glucocorticoids, which might potentially suppress hippocampal neurogenesis. Selective serotonin reuptake inhibitors (SSRIs) increase hippocampal neurogenesis in animal models of depression in parallel with symptoms improvement [33, 34]. These drugs may also block the loss of hippocampal volume associated with depression duration.

Imaging findings associated with MDD, such as reduced hippocampal volume and amygdala hyper-reactivity, are more consistently observed in maltreated individuals, and may represent a marker of the exposure to maltreatment [23]. Maltreatment-associated increased risk of depression has been suggested to be linked to abuse at specific time-points. Andersen et al. studied 26 women with repeated childhood sexual abuse and 17 women without childhood sexual abuse, all participants were between 18 and 22 years of age. Authors found sensitive period effects on hippocampal and amygdala volume, frontal cortex and corpus callosum gray matter volume. Hippocampal volume was reduced in association with childhood sexual abuse at ages 3–5 years and ages 11–13 years. Corpus callosum was reduced with childhood sexual abuse at ages 9–10 years. Depressive symptoms were specifically associated with abuse experiences in the age range 3–6 years. Andersen and colleagues suggest that childhood maltreatment in this time window initiates a cascade of consequences leading to hippocampal morphological abnormalities and vulnerability to depression. Reduced hippocampal volume is one of the most consistent brain finding in depression. Animal studies support that early life stress results in a 34–36% reduction in synaptic density in the hippocampus. Preclinical works by Anacker and colleagues highlight the importance of hippocampal neurogenesis to regulate the hypothalamus-pituitary-adrenal (HPA) axis, which may play a crucial role in the development and in the resolution of depressive symptoms. A recent study shows that increased neurogenesis in the dentate gyrus is linked to resilience to stress, while reduced neurogenesis is linked to increased stress response, depression, and cognitive inflexibility [33].

Another time period sensitive to maltreatment and depressive symptoms is teen age years [7]. Andersen et al. show that exposure to a stressful event between 14 and 16 years is associated with developmental changes in the prefrontal cortex (PFC), including synaptic loss [35]. Synaptic loss was also observed in an adolescent model of social stress in rats, and depressive symptoms manifested immediately [36]. Depressive episodes in adolescents often occur within a year of exposure to one or more significantly stressful life events and might be the result of stress-induced alterations in prefrontal development (see [7]). An important task of the PFC is to regulate the amygdala as already mentioned. By weakening the connection between the prefrontal cortex and the amygdala, the prefrontal cortex may not be able to adequately correct for impulses from the amygdala, which could predispose an individual to depression.

### 9.2.6 Childhood Trauma, MDD, and Physical Health

Patients with MDD have shorter life expectancy than the general population. A recent large register study from Denmark consisting of more than 100,000 people confirms that patients with MDD have on average a 14-year reduction in life expectancy in males and a 10-year shorter life expectancies in females [37]. Yet, not much is known about the mechanisms behind this reduced life expectancy. Intriguingly, the large ACE study of more than 9,000 individuals identified that subjects who had experienced four or more types of childhood exposure had a 4- to 12-fold increased health risks for alcoholism, drug abuse, and a 1.4- to 1.6-fold increased risk for physical inactivity and severe obesity [31]. The more types of childhood adversities the individual reported, the more somatic comorbidities were likely to be present including ischemic heart disease, cancer, chronic lung disease, skeletal fractures, and liver disease. The ACE study confirms that childhood adverse experiences are strongly related to having multiple health risk factors in adulthood [31], which needs to be studied further.

## 9.3 Response to Antidepressants and Psychosocial Interventions

Both psychotherapy and psychopharmacology are effective in treating MDD. However, as many as approximately 30% of patients do not remit from MDD, even after several treatment trials (for an excellent review see [4]). Interestingly, as shown earlier in this chapter, patients with MDD who report childhood trauma are less likely to remit and more likely to have more episodes. Thus, patients with childhood trauma may represent a subgroup with poorer response to treatment. The biological or psychological mechanism linking trauma to diminished response to treatment is still poorly understood.

Studies exploring treatment resistance in MDD could also integrate childhood trauma as a potential predictor of nonresponse. Indeed, childhood trauma has been associated with a poorer response in resistant depression [38, 39]. Randomized controlled trials on the efficacy of interventions should assess childhood trauma at baseline in order to study how this factor is relevant in selecting those who might or might not respond to treatment. For example, in the recent large study ($N > 1.000$) by Williamsen et al., patients with MDD who reported childhood abuse or neglect were 1.6 times less likely over 8 weeks of treatment with one of three randomly assigned antidepressants, to benefit from a typical first-line antidepressant than individuals without trauma. Especially reports of early childhood trauma (occurred at age 7 or earlier) of physical, sexual, or emotional abuse were associated with a worse response to the drugs. By the end of the treatment period, only 18% of depressed participants with a history of abuse from age 7 or earlier had at least a 50% reduction in depression symptoms and only 16% achieved remission. This contrasted with MDD without childhood trauma where 82% had a 50% reduction in symptoms over the time of treatment, and 84% achieved remission [11].

Guidelines indicate that for people with moderate or severe depression, the most effective treatment is a combination of antidepressant medication and psychological intervention, for example, cognitive behavioral therapy (CBT). CBT is based on the assumption that maladaptive believes contribute to the onset and maintenance of MDD. According to Beck's model, a change in these maladaptive representations can lead to changes in emotional regulation and dysfunctional behaviors. Although it is beyond the scope of this chapter to review CBT, a trauma-focused cognitive behavioral therapy (TF-CBT) was developed to address the multiple negative impacts of stressful or traumatic life events for children and adolescence and for their parents/primary caregivers (see the recent paper by de Arellano and colleagues [40] for more details). In short, TF-CBT provides psychoeducation and assists clients in developing coping mechanisms for when they are confronted with any abuse-related memories and feelings. Consequently, TF-CBT aims to reduce the anxiety that underlies PTSD as well as depression.

## 9.4 Biological Mediators of the Links Between Childhood Trauma and MDD

It is beyond the scope of this chapter to review all the relevant biological systems that might play a role in mediating the impact of childhood trauma on the risk of developing MDD or a more severe form of the disorder. Some of the biological systems (neuroplasticity, inflammation, circadian system, or HPA axis) could be much more central as first-line mechanisms linking childhood trauma to MDD susceptibility. Second, independently of psychiatric diagnoses, childhood trauma induces long-term modifications in inflammation processes [41–43]. Patients with MDD show abnormalities in the levels of C-reactive protein (CRP) [44], and of various cytokines such as interleukin (IL)-2 receptor, tumor necrosis factor-α, soluble tumor necrosis factor receptor type 1, IL-6 and IL-4 [45]. A clear majority of studies indicate that patients with MDD have also reduced Brain-Derived Neurotrophic Factor (BDNF) in serum and in plasma and HPA axis abnormalities. The underlying mechanisms are still largely unknown. The role of long-term changes in biomarkers following childhood trauma experiences is warranted. How trauma may influence brain function and structures will also be discussed below.

### 9.4.1 HPA Axis

Patients with depression often show changes in the HPA axis functioning. Hyperactivity of the HPA axis is one of the most commonly observed biological abnormalities in depression. It has been suggested that childhood traumatic events lead to persistent neurobiological abnormalities predisposing patients to depression [46]. One of these neurobiological abnormalities is represented by changes in the HPA axis. Numerous studies have shown that childhood traumatic events are associated with changes in the HPA axis, and both an increase [47] and reduction in cortisol secretion (biological "stress hormone") [48] have been reported. These inconsistent findings may be a result of diurnal rhythm differences (patterns of activity or behavior that follow day–night cycles), cortisol measured as a reactivity to a behavioral stressor as opposed to baseline cortisol levels, and differences in the studied sample, for example, healthy individuals with childhood trauma compared to patients with a severe mental illness with childhood trauma. Cigarette smoking and alcohol use have also been linked to a blunted stress reactivity in individuals reporting childhood trauma [49], and exposure to smoking and alcohol use may vary between studies.

### 9.4.2 BDNF

BDNF is essential for the human brain. BDNF promotes neuronal growth and differentiation during brain development, and increase synaptic plasticity and maintenance of neurons in adult life [50]. It crosses the blood–brain barrier (BBB), and peripheral levels are highly correlated with levels in the cerebrospinal fluid ($r = 0.8$) [51]. Over the last decade, a large number of studies have found low levels of BDNF in neurodegenerative and neuropsychiatric disorders [50]. Low levels are observed in MDD [52, 53]. To this end, mechanisms driving reductions of BDNF levels in MDD are unclear. Later clinical illness course, characterized by more episodes [54], and/or the current mood episode [53] have been suggested to influence BDNF levels. BDNF has also been suggested as a marker of illness stage. Bus et al. [53] recently showed that patients with a treatment-resistant depression were more likely to have lower BDNF levels over a 2-year period, while patients with higher BDNF were more likely to be in remission at follow-up [53]. Interactions between high exposure to childhood trauma experiences and reduced BDNF levels have been proposed. BDNF protects the brain from the toxic effect of glucocorticoids (stress hormone) and support neurogenesis [55]. Although we and others have previously shown that childhood trauma experiences reduce BDNF levels [56, 57], it is possible that reduced BDNF levels prior to childhood trauma increase the adverse effect of childhood trauma experiences on brain development, accompanied by more severe neurodevelopmental changes following trauma. This is supported by a recent study showing emotional reappraisal linked to psychopathology, and smaller hippocampal volume in depression following childhood traumatic events in Met carriers only of the BDNF val66met [58, 59], a specific gene variant linked to reduction of BDNF in plasma. It should be mentioned that one study found higher peripheral BDNF following trauma events [60] in MDD, potentially indicating an increase in BDNF levels as a possible attempt to neutralize the negative effects of childhood trauma on the brain.

### 9.4.3 MRI Changes

Inadequate input (defined as "neglect") as well as abuse have been associated with reduced gray and white matter volumes. Specific disruptions in connectivity between the amygdala and the prefrontal cortex (PFC) have been demonstrated. This is interesting, as the prefrontal cortex can regulate inputs from the amygdala, and if this control function is weakened, children may be predisposed to later depression. Indeed, having a strong functional connectivity between amygdala and prefrontal cortex has been linked to resilience after stressful life events. The effects of deprivation on subcortical structures are mixed, however. For example, neuro-imaging studies of children raised in institutions have found inconsistent effects on amygdala and hippocampal size [61–63] relative to controls. The amygdala has become an increasingly important component of theories of depression, largely based on imaging studies of mood regulation. The amygdala is over-responsive to fearful stimuli in depression [64], without being properly regulated by the prefrontal cortex. Patients with MDD fail to show the evidence of cognitive adjustments (increased activity in bilateral dorsolateral prefrontal cortex on post-error trials), which is observed in healthy control groups. This could lead to an excessive and persistent degree of negative affectivity, which is believed to be a vulnerability factor for depression.

### 9.4.4 Childhood Trauma and the Reward System

Recent studies indicate that a history of childhood trauma influences the reward system, which again, may be at stake in MDD. Impaired reward learning has emerged as a key characteristic of MDD and may be a vulnerability marker in individuals with remitted MDD [65]. This is supported by studies showing diminished reward responsiveness in MDD with acute depressive symptoms [66] and with higher severity of distress/misery symptom [67]. Blunted reward sensitivity is also reported in high-risk individuals for depression [68], and observed in remitted MDD currently in an euthymic phase [65]. Neurobiological pathways through which a history of childhood trauma may contribute to reduced motivation and increased negative mood have been suggested, with a recent study showing a blunted reward responsiveness in women with childhood sexual abuse, after correcting for current symptom levels [69]. Likewise, another study showed a link between cumulative childhood trauma and blunted reward responses [70].

Studies using event-related potentials (ERPs) during a probabilistic reward task in conjunction with functional magnetic resonance imaging (fMRI) [71, 72] demonstrate variations in reward positivity (RewP) amplitude to activation within the "brain reward pathway," particularly the ventral striatum, anterior cingulate cortex (ACC), and midfrontal cortex. A pharmacologic manipulation hypothesized to reduce phasic striatal dopaminergic responses has also been found to affect RewP and the underlying ACC activation [73], suggesting that the RewP may provide an index of phasic reward signaling that originates in the striatum and projects to the ACC. A recent study revealed reduced activity in the ACC in patients with remitted MDD during a reward processing task, supporting reduced reward responsiveness in MDD in euthymic phases [65]. Dysregulation of the reward system, including blunted ACC activation in remitted MDD, could be a long-term neurobiological abnormality predisposing individuals with childhood trauma to depression.

### 9.4.5 Childhood Trauma and Sleep Abnormalities

Patients with MDD also show major sleep disturbances [74, 75] (insomnia, hypersomnia, delayed sleep phase, etc.). Indeed, in the general population, childhood adversity is a risk factor for adult sleep disorders [76, 77]. Since the circadian system modulates the biological responses to stressful environmental factors, this hypothesis requires more attention. To date, only one study has explored such a hypothesis among patients with anxiety and depressive disorders and found that a high stress load in childhood was associated with alterations in several sleep parameters assessed with actigraphy [78]. Schafer and Bader [78] gave 48 psychiatric outpatients a self-report questionnaire assessing current depression, current anxiety symptoms, and stress load to investigate such constructs in childhood (before the age of 13 years), adolescence (between the age of 13 and 18 years), and adulthood (between the age of 19 and current age). Biological measures using actigraphy over a week period were used to assess sleep quality. Participants also kept a diary about sleep patterns. High stress load in childhood, but not in adolescence, was associated with shortened total sleep time, prolonged sleep onset latency, decreased sleep efficiency, and an increased number of body movements in sleep, even after accounting for the effects of later occurring stress and psychopathological symptoms such as depression and anxiety scores. These findings indicate that stress in childhood may have long-term effects on sleep patterns. Results are consistent with findings from previous studies indicating an association between childhood adversities and higher levels of nocturnal activity. They also suggest that high stress load during childhood might be a vulnerability factor for sleep continuity problems in adulthood. To date, few studies have investigated how sleep problems in depression following childhood trauma experiences are associated with cognitive functioning (such as reduced memory and attention) and reduced gray matter volume as observed in depression. For example, a large study by Wan et al. [79] on more than 2,000 individuals followed up for 4 years showed that those who developed long sleep latency (>30 min), long sleep duration (≥7.95 hours), and late midsleep time (after 3:00 am) at baseline were related to the risk of cognitive decline at 4-year follow-up. Numerous studies have been published showing a link between sleep deprivation and cognitive functioning (see, for example, the review article by Alhola et al. [80]), demonstrating that acute and chronic sleep problems are linked to attention and working memory deficits. Interestingly, subjective sleep problems are also linked to gray matter reduction [81]. The study by Chao and colleagues [81] investigated subjective sleep problems in 144 war veterans and found that veterans with depression scored higher on subjective sleep problems assessed with the Pittsburgh Sleep Quality Index (PSQI). In addition, those who reported traumatic events were more likely to report problems with sleep, and those who reported sleep problems were more likely to have reduced total cortical and frontal gray matter volumes compared to those who did not report sleep problems. Some of the cognitive deficits and brain alterations observed in MDD may thus be related not only to abnormalities in the HPA axis, but also to abnormalities in sleep patterns following trauma experiences, although this hypothesis needs further clarifications.

### 9.4.6 Gene–Environment Interactions in MDD

Genome-wide association studies (GWAS) in MDD have typically failed to identify the specific genetic variants involved [82]. Gene–environment interactions (GxEs) whereby a person inherits sensitivity to environmental factors, are believed to have an important role in MDD [83]. In view of the many studies demonstrating altered HPA axis activity in MDD and the prominent role of childhood trauma in risk for MDD, it is not surprising that many research groups have sought to determine whether polymorphisms in candidate genes in the HPA axis and other biological systems that modulate stress responses influence the risk for MDD by interacting with childhood abuse and neglect. Diathesis-stress theories of MDD predict that individuals' sensitivity to stressful events depends on their genetic makeup. This mechanism is called "*gene–environment interaction*" in which the phenotypic response to the environment is conditioned by the genotype of the individuals. Although several potential candidate genes such as the serotonin transporter (*SLC6A4*), CRH receptor 1 (*CRHR1*), and the gene encoding peptidyl-prolyl *cis-trans* isomerase (*FKBP5*) have been identified in the interaction with childhood trauma, replication studies have been inconsistent. Otte and colleagues suggested that differences in the timings and the type of adverse environmental circumstances have hampered replication studies of single candidate genes [4]. Regarding the interaction between SLC6A4 and childhood trauma in MDD, Culverhouse et al. suggested no major interaction [84].

### 9.4.7 Polygenic Risk Score for MDD and Childhood Trauma

More than two decades has passed after the ground-breaking study by Caspi and colleagues showing that individuals who carried at least one short variant of the serotonin transporter gene (5-HTTLPR) were more likely to become depressed after experiencing negative life events compared to the participants who were homozygotic long-long carriers of the serotonin transporter gene [85]. Today most recent studies are investigating combinations of polygenic risk scores for MDD (as the current knowledge is that MDD is a polygenic illness) and stressful life events. MDD is likely to be highly polygenic, arising from the combined effect of many risk variants, each with small effect sizes [82, 86]. Polygenic risk scoring can be used to test the effect of multiple genetic variants simultaneously. Polygenic risk scores are calculated by using subsets of single nucleotide polymorphisms (SNPs) from genome-wide association studies (GWAS), which are selected according to their *p* value and weighted by their effect size, to calculate a polygenic risk score (PRS) for each individual in an independent validation sample. The PRS can then be tested for its ability to differentiate between cases and controls in the validation dataset [87, 88]. These findings have led to the hypothesis that GxEs in a highly polygenic trait such as MDD may involve multiple genetic variants rather than one specific locus, in interaction with environmental risk factors. Indeed, interactions between polygenic scores for MDD and childhood trauma were found to increase risk for depression in the NESDA sample, accounting for 0.6% of variance in MDD status [89]. The latter study included a sample of more than 1,500 individuals with an MDD diagnosis and 340 healthy individuals (controls). More than 900 of the patients had a chronic MDD. Authors found that the polygenic risk scores and childhood trauma independently affected MDD risk, and the effect of polygenic risk scores on depression is increased in the presence of childhood trauma. Hence, the study showed that individuals with both high polygenic risk scores and exposure to childhood trauma were particularly at risk for developing MDD.

A second study that investigated potential interactions using polygenic risk score for MDD and information on childhood trauma was published by Mullin and colleagues in 2016 [90]. PRS significantly predicted depression, explaining 1.1% of variance in phenotype. Significant interaction between PRS for MDD and childhood trauma was found ($p = 0.002$), but showed an inverse association with MDD status, opposite to the one found in previous studies [89]. Authors concluded that cases who experienced more severe CT tended to have a lower PRS than other cases or controls. Thus, such observation would suggest that childhood trauma is a risk factor for MDD, especially in those with low polygenic risk. Hence, individual with multiple childhood experienced trauma may be at risk for developing MDD even if their genetic makeup is low.

It should be noted that a study by Peyrot and colleagues [91], the largest and most recent study on polygenic risk for MDD and childhood trauma, did not find evidence for any interaction. This study suggested that the genetic heterogeneity of MDD is not attributable to genome-wide moderation of genetic effects by childhood trauma. It tested interactions between polygenic risk for MDD and childhood trauma in almost 6,000 individuals from nine cohorts contributing to the Psychiatric Genetic Consortium that had available data on childhood trauma. As expected, patients with MDD were more likely to report a history of childhood trauma compared to healthy individuals. The polygenic risk score explained from 1.18% to 1.71% of variation in MDD risk. No evidence for interaction between PRS and childhood trauma was observed.

### 9.4.8 Epigenetic Molecular Mechanisms

The biological substrate of the deleterious effects of childhood trauma in individuals may also involve epigenetic modifications [23]. One adaptive mechanism to modulate stress response is supposed to act through subtle modifications of gene expression, primarily through epigenetic mechanisms such as methylation and modifications of amino acids in a histone protein. Both pre- [92] and perinatal [93] stress have long-term effects on the HPA axis [94] in rodent models and humans. The altered HPA axis in adult humans exposed to childhood trauma could be related to epigenetic changes of stress regulatory genes, such as the glucocorticoid receptor (GR) gene, also called the "NR3C1 gene" or the FKBP5 gene [95] (a glucocorticoid receptor co-chaperone). It has been suggested that epigenetic modifications in response to early environmental conditions are key mechanisms to explain the effects of childhood adversity on the increasing risk of psychopathology in adulthood. This remains to be thoroughly explored in MDD [96].

One intriguing factor that might explain reduced physical health in MDD patients who report childhood trauma experiences is the observation of shorter telomere length following childhood trauma experiences. Telomeres are DNA–protein structures at the tails of chromosomes that shorten with increasing age in most human tissues [97, 98]. Human telomeres are composed of tandem repeats of the TTAGGG sequence and average between 6 and 12 kilobases (kb) in length [97, 99]. When telomeres become critically short, the risk of cellular apoptosis "cell death" is increased, and proliferation is stopped, which eventually compromises tissue renewal capacity and function [100]. Telomere length (TL) may, therefore, represent a "*molecular clock*" that contributes to aging and altered physical health.

Studies report shorter age-adjusted TL in depression [101]. A growing body of evidence has linked chronic stress and childhood traumatic events to accelerated shortening of TL [102–104] suggesting a link between accelerated aging in maltreated patients. Reduced TL might be a mediator between childhood trauma and poorer physical health in MDD, which needs further explorations.

## 9.5 Clinical Recommendations in Daily Practice

First, a routine assessment of childhood trauma in both the early phases and established cases of MDD should be systematically undertaken due to the heightened risk of developing a more severe illness over time. It has previously been reported that abuse is often inadequately assessed in psychiatric clinics [105], an aspect that needs consideration in the overall assessment process. The assessment of childhood trauma is particularly required for subgroups of patients with early onset MDD, comorbidity with suicide attempts or substance misuse, high level of mood recurrences or greater mood instability (as a categorical pattern, i.e., rapid cycling or as a dimensional one, i.e., affective lability), and lack of response to antidepressants. In routine practice, the assessment should be undertaken using clinical interviews and will also benefit from the systematic use of questionnaires. It is beyond the scope of this chapter to discuss in depth the respective psychometric properties of interviews and self-reports. Since numerous self-reports and interviews exist with some differences in psychometric qualities, none of them can be recommended more than another [106]. Nevertheless, the use of the CTQ may be relevant as it is widely used in clinical research in MDD, and explores different subtypes of trauma and not only physical or sexual abuse [30].

Second, identifying childhood trauma could be the focus of early intervention strategies and approaches to potentially prevent exposed individuals from developing an unalterable chronic illness over time; however, this is only speculation. As described by Thompson and colleagues [107], examples of treatment targets (working directly on the dissociative experiences in response to trauma) could be psychological techniques, coping strategies, body awareness/mindfulness techniques, and stress management. Because a history of childhood trauma is also associated with reduced functioning in childhood, reaching these individuals before illness onset could potentially reduce the severity and (hypothetically) development of the illness. Existing data supports several psychological interventions in effectively preventing or treating the negative consequences of childhood trauma in general (eye moment desensitization and reprocessing [EMDR], trauma-focused Cognitive behavioral therapy [t-CBT] for sexually abused children) [108–110]. Landin-Romero et al. [110] and Mueser et al. [111] assessed using CBT a sample of 108 patients with PTSD and either major mood disorder (85%), schizoaffective disorder, or schizophrenia (15%), of whom 25% also had a borderline personality disorder. Interestingly, CBT patients improved significantly more than patients in the treatment as usual group at a blinded posttreatment and the 3- and 6-month follow-up. As childhood trauma is associated with affective dysregulation [112, 113] and impulsivity [114], psychosocial therapies should target not only traumatic experiences per se but also cognitive defects or the emotional dysregulation linked to traumatic experiences. In this context, therapies that target emotional regulation or cognitive functioning might help counterbalancing the negative effects of trauma. To date, this is an under-researched field, which deserves more attention. This will also open discussions about how those patients with MDD exposed to trauma would respond in a similar manner to "classical" or more targeted psychosocial therapies.

Intriguingly, Eye Movement Desensitization (EMDR) therapy showed good results in treating individuals with a history of trauma. The recent meta-analysis by Chen and colleagues [115] suggests that EMDR reduces symptoms in PTSD symptoms, depression, and/or anxiety both posttreatment and at follow-up compared with all other alternative therapies (cognitive behavioral therapy, individual/group therapy, and fluoxetine) and control treatment (pill placebo, active listening, and treatment as usual). EMDR was originally validated only for trauma-related disorders such as PTSD. A recent review suggests that EMDR is effective in treating trauma in patients cutting across diagnostic groups of MDD, bipolar disorder, and schizophrenia [116]. Recent studies have been positive showing that EMDR reduces depressive symptoms in MDD [117]; however, the studies are until now scarce. EMDR may be a promising treatment for MDD patients with childhood trauma experiences who do not respond to treatment as usual or in combination with traditional treatment.

We have already mentioned that patients with depression tend to have abnormalities in the HPA axis, with studies showing elevated HPA activity. As a potential future perspective, biological systems mentioned above may interact and converge to a high physical burden and reduced life expectancy in patients with a severe mental disorder [118]. Indeed, a meta-analysis by Norman et al. [27] explored the long-term consequences of childhood abuse and neglect, and found associations not only with psychiatric, but also with physical disorders. Through long-lasting consequences on the alterations of immune-inflammatory markers, sleep parameters, and HPA axis, childhood trauma might be associated with poor health conditions that should be addressed in treatment, in addition to regular psychiatric treatment.

## 9.6 Conclusion

This chapter discussed the role of childhood trauma in MDD and contributed to a new understanding of the negative consequences of early life stress. The aim was to frame childhood trauma in a biological context of susceptibility and to discuss novel long-term pathophysiological consequences in MDD. Childhood traumatic events are risk factors for developing MDD, in addition to a more severe clinical presentation over time. Patients with MDD who have been exposed to childhood maltreatment have an earlier age at onset, greater symptom severity, more comorbidities, a greater risk for suicide, and poorer treatment response than non-maltreated individuals with the same diagnosis [23]. Childhood trauma leads to alterations of affect regulation, impulse control, and cognitive functioning that might decrease the ability to cope with later stressors. Childhood trauma interacts with several genes belonging to different biological pathways (HPA axis, serotonergic transmission, neuroplasticity, immunity, calcium signaling, and circadian rhythms) to increase the risk for MDD or a more severe clinical expression of the disorder. Epigenetic factors may also be involved in the neurobiological consequences of childhood trauma in MDD. Biological sequelae such as chronic inflammation sleep disturbances or telomere shortening are potential mediators of the negative effects of childhood trauma in MDD, in particular, regarding physical health. The main clinical implication is to systematically assess childhood trauma in patients with MDD, or at least in those with a severe, instable, or resistant course. The challenge for the next years will be to fill the gap between clinical and scientific research and routine practice since recommendations for managing this specific population are lacking. Little is known on which psychotherapies should be provided or which targets therapists should focus on, as well as on how childhood trauma could explain the resistance to antidepressant and mood stabilizers. Future research should focus on providing evidence-based trauma interventions in MDD. Treating childhood trauma experiences early could potentially reduce MDD cases, thus improving quality of life in patients and reducing society costs.

## References


1. Nemeroff CB. Paradise lost: the neurobiological and clinical consequences of child abuse and neglect. Neuron. 2016;89(5):892–909.
2. Children's Bureau/ACYF/ACF/HHS. Department of Health & Human Services; 2017.
3. Green JG, McLaughlin KA, Berglund PA, Gruber MJ, Sampson NA, Zaslavsky AM, et al. Childhood adversities and adult psychiatric disorders in the national comorbidity survey replication I: associations with first onset of DSM-IV disorders. Arch Gen Psychiatry. 2010;67(2):113–23.
4. Otte C, Gold SM, Penninx BW, Pariante CM, Etkin A, Fava M, et al. Major depressive disorder. Nat Rev Dis Primers. 2016;2:16065.
5. Glowinski AL, Madden PA, Bucholz KK, Lynskey MT, Heath AC. Genetic epidemiology of self-reported lifetime DSM-IV major depressive disorder in a population-based twin sample of female adolescents. J Child Psychol Psychiatry. 2003;44(7):988–96.
6. Sullivan PF, Neale MC, Kendler KS. Genetic epidemiology of major depression: review and meta-analysis. Am J Psychiatry. 2000;157(10):1552–62.
7. Andersen SL, Teicher MH. Stress, sensitive periods and maturational events in adolescent depression. Trends Neurosci. 2008;31(4):183–91.
8. Comijs HC, van Exel E, van der Mast RC, Paauw A, Oude Voshaar R, Stek ML. Childhood abuse in late-life depression. J Affect Disord. 2013;147(1–3):241–6.
9. Teicher MH, Samson JA, Polcari A, Andersen SL. Length of time between onset of childhood sexual abuse and emergence of depression in a young adult sample: a retrospective clinical report. J Clin Psychiatry. 2009;70(5):684–91.
10. Li M, D'Arcy C, Meng X. Maltreatment in childhood substantially increases the risk of adult depression and anxiety in prospective cohort studies: systematic review, meta-analysis, and proportional attributable fractions. Psychol Med. 2016;46(4):717–30.
11. Williams LM, Debattista C, Duchemin AM, Schatzberg AF, Nemeroff CB. Childhood trauma predicts antidepressant response in adults with major depression: data from the randomized international study to predict optimized treatment for depression. Transl Psychiatry. 2016;6:e799.

12. Wiersma JE, Hovens JG, van Oppen P, Giltay EJ, van Schaik DJ, Beekman AT, et al. The importance of childhood trauma and childhood life events for chronicity of depression in adults. J Clin Psychiatry. 2009;70(7):983–9.
13. Withers AC, Tarasoff JM, Stewart JW. Is depression with atypical features associated with trauma history? J Clin Psychiatry. 2013;74(5):500–6.
14. Nanni V, Uher R, Danese A. Childhood maltreatment predicts unfavorable course of illness and treatment outcome in depression: a meta-analysis. Am J Psychiatry. 2012;169(2):141–51.
15. Read J, Bentall RP. Negative childhood experiences and mental health: theoretical, clinical and primary prevention implications. Brit J Psych. 2012;200(2):89–91.
16. Paterniti S, Sterner I, Caldwell C, Bisserbe JC. Childhood neglect predicts the course of major depression in a tertiary care sample: a follow-up study. BMC Psychiatry. 2017;17(1):113.
17. Lizardi H, Klein DN, Ouimette PC, Riso LP, Anderson RL, Donaldson SK. Reports of the childhood home environment in early-onset dysthymia and episodic major depression. J Abnorm Psychol. 1995;104(1):132–9.
18. Horwitz AV, Widom CS, McLaughlin J, White HR. The impact of childhood abuse and neglect on adult mental health: a prospective study. J Health Soc Behav. 2001;42(2):184–201.
19. Widom CS, DuMont K, Czaja SJ. A prospective investigation of major depressive disorder and comorbidity in abused and neglected children grown up. Arch Gen Psychiatry. 2007;64(1):49–56.
20. Putnam KT, Harris WW, Putnam FW. Synergistic childhood adversities and complex adult psychopathology. J Trauma Stress. 2013;26(4):435–42.
21. Aas M, Etain B, Bellivier F, Henry C, Lagerberg T, Ringen A, et al. Additive effects of childhood abuse and cannabis abuse on clinical expressions of bipolar disorders. Psychol Med. 2013:1–10.
22. Vitriol V, Cancino A, Leiva-Bianchi M, Serrano C, Ballesteros S, Asenjo A, et al. Childhood trauma and psychiatric comorbidities in patients with depressive disorder in primary care in Chile. J Trauma Dissociation. 2017;18(2):189–205.
23. Teicher MH, Samson JA. Childhood maltreatment and psychopathology: a case for ecophenotypic variants as clinically and neurobiologically distinct subtypes. Am J Psychiatry. 2013;170(10):1114–33.
24. Bennett DS, Ambrosini PJ, Kudes D, Metz C, Rabinovich H. Gender differences in adolescent depression: do symptoms differ for boys and girls? J Affect Disord. 2005;89(1–3):35–44.
25. Carmona NE, Subramaniapillai M, Mansur RB, Cha DS, Lee Y, Fus D, et al. Sex differences in the mediators of functional disability in major depressive disorder. J Psychiatr Res. 2018;96:108–14.
26. MacMillan HL, Fleming JE, Streiner DL, Lin E, Boyle MH, Jamieson E, et al. Childhood abuse and lifetime psychopathology in a community sample. Am J Psychiatry. 2001;158(11):1878–83.
27. Norman RE, Byambaa M, De R, Butchart A, Scott J, Vos T. The long-term health consequences of child physical abuse, emotional abuse, and neglect: a systematic review and meta-analysis. PLoS Med. 2012;9(11):e1001349.
28. Infurna MR, Reichl C, Parzer P, Schimmenti A, Bifulco A, Kaess M. Associations between depression and specific childhood experiences of abuse and neglect: a meta-analysis. J Affect Disord. 2016;190:47–55.
29. Nelson J, Klumparendt A, Doebler P, Ehring T. Childhood maltreatment and characteristics of adult depression: meta-analysis. Br J Psychiatry. 2017;210(2):96–104.
30. Negele A, Kaufhold J, Kallenbach L, Leuzinger-Bohleber M. Childhood trauma and its relation to chronic depression in adulthood. Depress Res Treat. 2015;2015:650804.
31. Felitti VJ, Anda RF, Nordenberg D, Williamson DF, Spitz AM, Edwards V, et al. Relationship of childhood abuse and household dysfunction to many of the leading causes of death in adults. The adverse childhood experiences (ACE) study. Am J Prev Med. 1998;14(4):245–58.
32. Wei Q, Lu XY, Liu L, Schafer G, Shieh KR, Burke S, et al. Glucocorticoid receptor overexpression in forebrain: a mouse model of increased emotional lability. Proc Natl Acad Sci U S A. 2004;101(32):11851–6.

33. Anacker C, Hen R. Adult hippocampal neurogenesis and cognitive flexibility - linking memory and mood. Nat Rev Neurosci 2017.
34. Anacker C, Zunszain PA, Cattaneo A, Carvalho LA, Garabedian MJ, Thuret S, et al. Antidepressants increase human hippocampal neurogenesis by activating the glucocorticoid receptor. Mol Psychiatry. 2011;16(7):738–50.
35. Andersen SL, Tomada A, Vincow ES, Valente E, Polcari A, Teicher MH. Preliminary evidence for sensitive periods in the effect of childhood sexual abuse on regional brain development. J Neuropsychiatry Clin Neurosci. 2008;20(3):292–301.
36. Leussis MP, Lawson K, Stone K, Andersen SL. The enduring effects of an adolescent social stressor on synaptic density, part II: Poststress reversal of synaptic loss in the cortex by adinazolam and MK-801. Synapse. 2008;62(3):185–92.
37. Laursen TM, Musliner KL, Benros ME, Vestergaard M, Munk-Olsen T. Mortality and life expectancy in persons with severe unipolar depression. J Affect Disord. 2016;193:203–7.
38. Douglas KM, Porter RJ. The effect of childhood trauma on pharmacological treatment response in depressed inpatients. Psychiatry Res. 2012;200(2–3):1058–61.
39. Shamseddeen W, Asarnow JR, Clarke G, Vitiello B, Wagner KD, Birmaher B, et al. Impact of physical and sexual abuse on treatment response in the treatment of resistant depression in adolescent study (TORDIA). J Am Acad Child Adolesc Psychiatry. 2011;50(3):293–301.
40. de Arellano MA, Lyman DR, Jobe-Shields L, George P, Dougherty RH, Daniels AS, et al. Trauma-focused cognitive-behavioral therapy for children and adolescents: assessing the evidence. Psychiatr Serv. 2014;65(5):591–602.
41. Baumeister D, Akhtar R, Ciufolini S, Pariante CM, Mondelli V. Childhood trauma and adulthood inflammation: a meta-analysis of peripheral C-reactive protein, interleukin-6 and tumour necrosis factor-alpha. Mol Psychiatry. 2016;21(5):642–9.
42. Coelho R, Viola TW, Walss-Bass C, Brietzke E, Grassi-Oliveira R. Childhood maltreatment and inflammatory markers: a systematic review. Acta Psychiatr Scand. 2014;129(3):180–92.
43. Tursich M, Neufeld RW, Frewen PA, Harricharan S, Kibler JL, Rhind SG, et al. Association of trauma exposure with proinflammatory activity: a transdiagnostic meta-analysis. Transl Psychiatry. 2014;4:e413.
44. Dargel AA, Godin O, Kapczinski F, Kupfer DJ, Leboyer M. C-reactive protein alterations in bipolar disorder: a meta-analysis. J Clin Psychiatry. 2015;76(2):142–50.
45. Munkholm K, Brauner JV, Kessing LV, Vinberg M. Cytokines in bipolar disorder vs. healthy control subjects: a systematic review and meta-analysis. J Psychiatr Res. 2013;47(9):1119–33.
46. Carmine M, Pariante, Lightman SL. The HPA axis in major depression: classical theories and new developments. Trends Neurosci. 2008;31(9):464–8.
47. Heim C, Newport DJ, Mletzko T, Miller AH, Nemeroff CB. The link between childhood trauma and depression: Insights from HPA axis studies in humans. Psychoneuroendocrinology. 2008;33(6):693–710.
48. Carroll D, Ginty AT, Whittaker AC, Lovallo WR, de Rooij SR. The behavioural, cognitive, and neural corollaries of blunted cardiovascular and cortisol reactions to acute psychological stress. Neurosci Biobehav Rev. 2017;77:74–86.
49. Carr CP, Martins CM, Stingel AM, Lemgruber VB, Juruena MF. The role of early life stress in adult psychiatric disorders: a systematic review according to childhood trauma subtypes. J Nerv Ment Dis. 2013;201(12):1007–20.
50. Nuernberg GL, Aguiar B, Bristot G, Fleck MP, Rocha NS. Brain-derived neurotrophic factor increase during treatment in severe mental illness inpatients. Transl Psychiatry. 2016;6(12):e985.
51. Fernandes BS, Molendijk ML, Kohler CA, Soares JC, Leite CM, Machado-Vieira R, et al. Peripheral brain-derived neurotrophic factor (BDNF) as a biomarker in bipolar disorder: a meta-analysis of 52 studies. BMC Med. 2015;13:289.
52. Bus BA, Molendijk ML, Penninx BJ, Buitelaar JK, Kenis G, Prickaerts J, et al. Determinants of serum brain-derived neurotrophic factor. Psychoneuroendocrinology. 2011;36(2):228–39.

53. Bus BA, Molendijk ML, Tendolkar I, Penninx BW, Prickaerts J, Elzinga BM, et al. Chronic depression is associated with a pronounced decrease in serum brain-derived neurotrophic factor over time. Mol Psychiatry. 2014.
54. Kauer-Sant'Anna M, Kapczinski F, Andreazza AC, Bond DJ, Lam RW, Young LT, et al. Brain-derived neurotrophic factor and inflammatory markers in patients with early- vs. late-stage bipolar disorder. Int J Neuropsychopharmacol. 2009;12(4):447–58.
55. Taliaz D, Stall N, Dar DE, Zangen A. Knockdown of brain-derived neurotrophic factor in specific brain sites precipitates behaviors associated with depression and reduces neurogenesis. Mol Psychiatry. 2010;15(1):80–92.
56. Mondelli V, Cattaneo A, Belvederi MM, Di FM, Handley R, Hepgul N, et al. Stress and inflammation reduce brain-derived neurotrophic factor expression in first-episode psychosis: a pathway to smaller hippocampal volume. J Clin Psychiatry. 2011;72(12):1677–84.
57. Aas M, Haukvik UK, Djurovic S, Tesli M, Athanasiu L, Bjella T, et al. Interplay between childhood trauma and BDNF val66met variants on blood BDNF mRNA levels and on hippocampus subfields volumes in schizophrenia spectrum and bipolar disorders. J Psychiatr Res 2014.
58. Bilc MI, Vulturar R, Chis A, Buciuman M, Nutu D, Bunea I, et al. Childhood trauma and emotion regulation: the moderator role of BDNF Val66Met. Neurosci Lett. 2018;685:7–11.
59. Carballedo A, Morris D, Zill P, Fahey C, Reinhold E, Meisenzahl E, et al. Brain-derived neurotrophic factor Val66Met polymorphism and early life adversity affect hippocampal volume. Am J Med Genet B Neuropsychiatr Genet. 2013;162b(2):183–90.
60. Jeon HJ, Kang ES, Lee EH, Jeong EG, Jeon JR, Mischoulon D, et al. Childhood trauma and platelet brain-derived neurotrophic factor (BDNF) after a three month follow-up in patients with major depressive disorder. J Psychiatr Res. 2012;46(7):966–72.
61. Mehta MA, Golembo NI, Nosarti C, Colvert E, Mota A, Williams SC, et al. Amygdala, hippocampal and corpus callosum size following severe early institutional deprivation: the English and Romanian adoptees study pilot. J Child Psychol Psychiatry. 2009;50(8):943–51.
62. Sheridan MA, McLaughlin KA. Dimensions of early experience and neural development: deprivation and threat. Trends Cogn Sci. 2014;18(11):580–5.
63. Tottenham N, Hare TA, Quinn BT, McCarry TW, Nurse M, Gilhooly T, et al. Prolonged institutional rearing is associated with atypically large amygdala volume and difficulties in emotion regulation. Dev Sci. 2010;13(1):46–61.
64. Roberson-Nay R, McClure EB, Monk CS, Nelson EE, Guyer AE, Fromm SJ, et al. Increased amygdala activity during successful memory encoding in adolescent major depressive disorder: an FMRI study. Biol Psychiatry. 2006;60(9):966–73.
65. Whitton AE, Kakani P, Foti D, Van't Veer A, Haile A, Crowley DJ, et al. Blunted neural responses to reward in remitted major depression: a high-density event-related potential study. Biol Psychiatry Cogn Neurosci Neuroimag. 2016;1(1):87–95.
66. Pizzagalli DA, Iosifescu D, Hallett LA, Ratner KG, Fava M. Reduced hedonic capacity in major depressive disorder: evidence from a probabilistic reward task. J Psychiatr Res. 2008;43(1):76–87.
67. Burkhouse KL, Gorka SM, Afshar K, Phan KL. Neural reactivity to reward and internalizing symptom dimensions. J Affect Disord. 2017;217:73–9.
68. Bress JN, Foti D, Kotov R, Klein DN, Hajcak G. Blunted neural response to rewards prospectively predicts depression in adolescent girls. Psychophysiology. 2013;50(1):74–81.
69. Pechtel P, Pizzagalli DA. Disrupted reinforcement learning and maladaptive behavior in women with a history of childhood sexual abuse: a high-density event-related potential study. JAMA Psychiat. 2013;70(5):499–507.
70. Hanson JL, Albert D, Iselin AM, Carre JM, Dodge KA, Hariri AR. Cumulative stress in childhood is associated with blunted reward-related brain activity in adulthood. Soc Cogn Affect Neurosci. 2016;11(3):405–12.
71. Carlson JM, Foti D, Mujica-Parodi LR, Harmon-Jones E, Hajcak G. Ventral striatal and medial prefrontal BOLD activation is correlated with reward-related electrocortical activity: a combined ERP and fMRI study. NeuroImage. 2011;57(4):1608–16.

72. Becker MP, Nitsch AM, Miltner WH, Straube T. A single-trial estimation of the feedback-related negativity and its relation to BOLD responses in a time-estimation task. J Neurosci. 2014;34(8):3005–12.
73. Santesso DL, Evins AE, Frank MJ, Schetter EC, Bogdan R, Pizzagalli DA. Single dose of a dopamine agonist impairs reinforcement learning in humans: evidence from event-related potentials and computational modeling of striatal-cortical function. Hum Brain Mapp. 2009;30(7):1963–76.
74. Geoffroy PA, Scott J, Boudebesse C, Lajnef M, Henry C, Leboyer M, et al. Sleep in patients with remitted bipolar disorders: a meta-analysis of actigraphy studies. Acta Psychiatr Scand. 2015;131(2):89–99.
75. Ng TH, Chung KF, Ho FY, Yeung WF, Yung KP, Lam TH. Sleep-wake disturbance in interepisode bipolar disorder and high-risk individuals: a systematic review and meta-analysis. Sleep Med Rev. 2015;20:46–58.
76. Baiden P, Fallon B, den Dunnen W, Boateng GO. The enduring effects of early-childhood adversities and troubled sleep among Canadian adults: a population-based study. Sleep Med. 2015;16(6):760–7.
77. Koskenvuo K, Hublin C, Partinen M, Paunio T, Koskenvuo M. Childhood adversities and quality of sleep in adulthood: a population-based study of 26,000 Finns. Sleep Med. 2010;11(1):17–22.
78. Schafer V, Bader K. Relationship between early-life stress load and sleep in psychiatric outpatients: a sleep diary and actigraphy study. Stress Health. 2013;29(3):177–89.
79. Suh SW, Han JW, Lee JR, Byun S, Kwon SJ, Oh SH, et al. Sleep and cognitive decline: a prospective nondemented elderly cohort study. Ann Neurol. 2018;83(3):472–82.
80. Alhola P, Polo-Kantola P. Sleep deprivation: impact on cognitive performance. Neuropsychiatr Dis Treat. 2007;3(5):553–67.
81. Chao LL, Mohlenhoff BS, Weiner MW, Neylan TC. Associations between subjective sleep quality and brain volume in gulf war veterans. Sleep. 2014;37(3):445–52.
82. Ripke S, Wray NR, Lewis CM, Hamilton SP, Weissman MM, Breen G, et al. A mega-analysis of genome-wide association studies for major depressive disorder. Mol Psychiatry. 2013;18(4):497–511.
83. Uher R. Gene-environment interactions in severe mental illness. Front Psych. 2014;5:48.
84. Culverhouse RC, Saccone NL, Horton AC, Ma Y, Anstey KJ, Banaschewski T, et al. Collaborative meta-analysis finds no evidence of a strong interaction between stress and 5-HTTLPR genotype contributing to the development of depression. Mol Psychiatry. 2018;23(1):133–42.
85. Caspi A, Sugden K, Moffitt TE, Taylor A, Craig IW, Harrington H, et al. Influence of life stress on depression: moderation by a polymorphism in the 5-HTT gene. Science. 2003;301(5631):386–9.
86. Wray NR, Pergadia ML, Blackwood DH, Penninx BW, Gordon SD, Nyholt DR, et al. Genome-wide association study of major depressive disorder: new results, meta-analysis, and lessons learned. Mol Psychiatry. 2012;17(1):36–48.
87. Purcell SM, Wray NR, Stone JL, Visscher PM, O'Donovan MC, Sullivan PF, et al. Common polygenic variation contributes to risk of schizophrenia and bipolar disorder. Nature. 2009;460(7256):748–52.
88. Dudbridge F. Power and predictive accuracy of polygenic risk scores. PLoS Genet. 2013;9(3):e1003348.
89. Peyrot WJ, Milaneschi Y, Abdellaoui A, Sullivan PF, Hottenga JJ, Boomsma DI, et al. Effect of polygenic risk scores on depression in childhood trauma. Br J Psychiatry. 2014;205(2):113–9.
90. Mullins N, Power RA, Fisher HL, et al. Polygenic interactions with environmental adversity in the aetiology of major depressive disorder. Psychol Med. 2016;46:759–70.
91. Peyrot WJ, Van der Auwera S, Milaneschi Y, Dolan CV, Madden PAF, Sullivan PF, et al. Does childhood trauma moderate polygenic risk for depression? A Meta-analysis of 5765 subjects from the psychiatric genomics consortium. Biol Psychiatry 2017.

92. Seckl JR, Meaney MJ. Glucocorticoid programming. Ann N Y Acad Sci. 2004;1032:63–84.
93. Fish EW, Shahrokh D, Bagot R, Caldji C, Bredy T, Szyf M, et al. Epigenetic programming of stress responses through variations in maternal care. Ann N Y Acad Sci. 2004;1036:167–80.
94. Herbert J, Goodyer IM, Grossman AB, Hastings MH, de Kloet ER, Lightman SL, et al. Do corticosteroids damage the brain? J Neuroendocrinol. 2006;18(6):393–411.
95. Klengel T, Mehta D, Anacker C, Rex-Haffner M, Pruessner JC, Pariante CM, et al. Allele-specific FKBP5 DNA demethylation mediates gene-childhood trauma interactions. Nat Neurosci. 2013;16(1):33–41.
96. Nemeroff CB, Binder E. The preeminent role of childhood abuse and neglect in vulnerability to major psychiatric disorders: toward elucidating the underlying neurobiological mechanisms. J Am Acad Child Adolesc Psychiatry. 2014;53(4):395–7.
97. Aubert G, Lansdorp PM. Telomeres and aging. Physiol Rev. 2008;88(2):557–79.
98. Blackburn EH. Switching and signaling at the telomere. Cell. 2001;106(6):661–73.
99. Yamaguchi H, Calado RT, Ly H, Kajigaya S, Baerlocher GM, Chanock SJ, et al. Mutations in TERT, the gene for telomerase reverse transcriptase, in aplastic anemia. N Engl J Med. 2005;352(14):1413–24.
100. Blasco MA. Telomere length, stem cells and aging. Nat Chem Biol. 2007;3(10):640–9.
101. Vance MC, Bui E, Hoeppner SS, Kovachy B, Prescott J, Mischoulon D, et al. Prospective association between major depressive disorder and leukocyte telomere length over two years. Psychoneuroendocrinology. 2018;90:157–64.
102. Epel ES, Blackburn EH, Lin J, Dhabhar FS, Adler NE, Morrow JD, et al. Accelerated telomere shortening in response to life stress. Proc Natl Acad Sci U S A. 2004; 101(49):17312–5.
103. Price LH, Kao HT, Burgers DE, Carpenter LL, Tyrka AR. Telomeres and early-life stress: an overview. Biol Psychiatry. 2013;73(1):15–23.
104. Shalev I, Entringer S, Wadhwa PD, Wolkowitz OM, Puterman E, Lin J, et al. Stress and telomere biology: a lifespan perspective. Psychoneuroendocrinology. 2013;38(9):1835–42.
105. Read J, Fraser A. Staff response to abuse histories of psychiatric inpatients. Aust N Z J Psychiatry. 1998;32(2):206–13.
106. Roy CA, Perry JC. Instruments for the assessment of childhood trauma in adults. J Nerv Ment Dis. 2004;192(5):343–51.
107. Thompson AD, Nelson B, Yuen HP, Lin A, Amminger GP, McGorry PD, et al. Sexual trauma increases the risk of developing psychosis in an ultra high-risk "prodromal" population. Schizophr Bull. 2014;40(3):697–706.
108. Ehring T, Welboren R, Morina N, Wicherts JM, Freitag J, Emmelkamp PM. Meta-analysis of psychological treatments for posttraumatic stress disorder in adult survivors of childhood abuse. Clin Psychol Rev. 2014;34(8):645–57.
109. Macdonald G, Higgins JP, Ramchandani P, Valentine JC, Bronger LP, Klein P, et al. Cognitive-behavioural interventions for children who have been sexually abused. Cochrane Database Syst Rev. 2012;5:CD001930.
110. Landin-Romero R, Novo P, Vicens V, McKenna PJ, Santed A, Pomarol-Clotet E, et al. EMDR therapy modulates the default mode network in a subsyndromal, traumatized bipolar patient. Neuropsychobiology. 2013;67(3):181–4.
111. Mueser KT, Rosenberg SD, Xie H, Jankowski MK, Bolton EE, Lu W, et al. A randomized controlled trial of cognitive-behavioral treatment for posttraumatic stress disorder in severe mental illness. J Consult Clin Psychol. 2008;76(2):259–71.
112. Aas M, Aminoff SR, Vik Lagerberg T, Etain B, Agartz I, Andreassen OA, et al. Affective lability in patients with bipolar disorders is associated with high levels of childhood trauma. Psychiatry Res. 2014;218(1–2):252–5.
113. Aas M, Henry C, Bellivier F, Lajnef M, Gard S, Kahn JP, et al. Affective lability mediates the association between childhood trauma and suicide attempts, mixed episodes and co-morbid anxiety disorders in bipolar disorders. Psychol Med. 2017;47(5):902–12.

114. Etain B, Mathieu F, Liquet S, Raust A, Cochet B, Richard JR, et al. Clinical features associated with trait-impulsiveness in euthymic bipolar disorder patients. J Affect Disord 2012.
115. Chen R, Gillespie A, Zhao Y, Xi Y, Ren Y, McLean L. The efficacy of eye movement desensitization and reprocessing in children and adults who have experienced complex childhood trauma: a systematic review of randomized controlled trials. Front Psychol. 2018;9:534.
116. Valiente-Gomez A, Moreno-Alcazar A, Treen D, Cedron C, Colom F, Perez V, et al. EMDR beyond PTSD: a systematic literature review. Front Psychol. 2017;8:1668.
117. Ostacoli L, Carletto S, Cavallo M, Baldomir-Gago P, Di Lorenzo G, Fernandez I, et al. Comparison of eye movement desensitization reprocessing and cognitive behavioral therapy as adjunctive treatments for recurrent depression: the European depression EMDR network (EDEN) randomized controlled trial. Front Psychol. 2018;9:74.
118. Kessing LV, Vradi E, McIntyre RS, Andersen PK. Causes of decreased life expectancy over the life span in bipolar disorder. J Affect Disord. 2015;180:142–7.

# Childhood Trauma in Psychoses 10

Ruud van Winkel and Aleksandra Lecei

## 10.1 Introduction

"I was convinced that we were on the verge of a world war. The public defense sirens were tested every first Monday of the month. To me, those sirens meant that the cruise missiles had already been launched. We had just minutes to do what had to be done: find each other and wait to die. Sometimes I could actually feel the radiation taking effect. I felt nauseous and believed that my hair was falling out. The rest of the world pretended to carry on as normal, but I could see that everyone was afraid. They knew that we were all about to suffer a slow, painful death, but nobody knew how to prepare for it. [...] The strange thing is that time did not exist for me in that situation. [...] It is a kind of vacuum, no-man's-land. Not in real life, but also not dead. [...] Now I no longer see my psychoses as isolated psychopathology. [...] My psychoses are my way of reacting to my life history. They are my response to the unpredictable abuse I had to face as a child. I hit my father back when I finally was angry enough, after years of submission. My father left the house after threatening to kill himself, after which the whole family turned against me. [...] In the years to follow I lost all my strength and exchanged it for guilt, fear, and incomprehensible psychotic experiences. I became the problem that had to be solved. I don't think that abuse itself is a strong cause for psychosis. [...] I think that the threat and the betrayal that come with it feed psychosis. The betrayal of the family that says, "you must have asked for it, " instead of stand-

Author "Aleksandra Lecei" contributed equally to this manuscript.

R. van Winkel (✉)
Dept. of Neurosciences, Research Group Psychiatry, Center for Clinical Psychiatry, KU Leuven, Leuven, Belgium

KU Leuven, University Psychiatric Center KU Leuven, Leuven, Belgium

Dept. of Psychiatry and Psychology, School for Mental Health and Neuroscience, Maastricht University, Maastricht, The Netherlands
e-mail: ruud.vanwinkel@upckuleuven.be

A. Lecei
Dept. of Neurosciences, Research Group Psychiatry, Center for Clinical Psychiatry, KU Leuven, Leuven, Belgium
e-mail: aleksandra.lecei@kuleuven.be



G. Spalletta et al. (eds.), *Childhood Trauma in Mental Disorders*,
https://doi.org/10.1007/978-3-030-49414-8_10

ing up for you. That excuses the offender and accuses the victim. And forces the child to accept the reality of the adults. That forces the child to say that the air is green, while she sees clearly it is not green but blue. That is a distortion of reality that is very hard to deal with when you're a child. You are forced to betray yourself [1]."

This piece is a first person account of a woman with psychosis and a history of childhood trauma. Psychosis is a mental health disturbance that can cause people to perceive and interpret the world differently from those around them. Symptoms can fall into one or more of two categories: positive and negative. Positive symptoms include hallucinations (i.e., seeing, hearing, and feeling things that are not there), delusions (i.e., false beliefs), and disorganized speech and/or behavior. Negative symptoms relate to functions that are diminished or absent in patients with psychosis, such as loss of motivation, flat affect, and apathy. If more than two symptoms are present during at least 6 months and cause functional dysfunction, the psychotic disorder is referred to as schizophrenia (Fig. 10.1).

Evidence accrued in recent years that psychotic experiences are common in the general population with a lifetime prevalence of 7.2% [2]. Psychotic experiences can be fleeting hallucinations, suspiciousness, paranoia, and magical thinking. These are associated with a greater risk for later development of psychotic disorder (or other disorders) and have been reported in young children and adolescents, too [3]. It is, then, important to look into early childhood experiences that may be on the etiological pathway to psychosis. A possible causal role of childhood trauma has been proposed decades ago, not just because a significant proportion of people with psychosis reports a history of adversity [4]. Childhood trauma refers to a broad range of adverse experiences usually before the age of 16, including life events (e.g., death of a parent, environmental catastrophes) or single and/or repeated maltreatment (e.g., sexual abuse, physical abuse, physical neglect, emotional abuse, emotional neglect, peer bullying, i.e., maltreatment with the intention to harm). In this chapter, we will focus on the latter and discuss the different levels through which childhood trauma (which in the literature is sometimes also referred to as "adversity" or "maltreatment") might exert its influence on the development of psychosis.

**Fig. 10.1** Positive and negative symptoms of schizophrenia

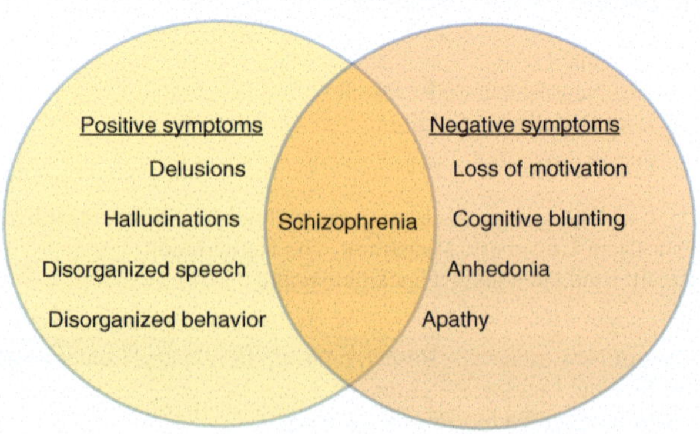

## 10.2 The Epidemiological Evidence for a Link Between Childhood Trauma and Psychosis

Evidence is amassing that consistently points in the same direction: childhood adversities, specifically with the intention to harm, are associated with an increased risk for psychosis. In a comprehensive meta-analysis, Varese and colleagues identified 36 studies with a combined sample size of 80,000, and concluded that childhood adversities were associated with a two- to fourfold increased risk for later psychosis [5]. The overall association was 2.78 increased odds, irrespective of study design. More specifically, zooming in on different forms of adversity, the odds rations reported were 2.4 for bullying, 3.4 for emotional abuse, 2.9 for neglect, 3.0 for physical abuse, and 2.4 for sexual abuse. Only parental death was non-significantly associated with psychosis (odds 1.7) [5]. There is, furthermore, some evidence for a dose–response relationship [6–10]. Research suggests that for each additional indicator of maltreatment, there would be a modest linear increase in risk for psychosis [11].

Since the review by Varese and colleagues, additional evidence has been published suggesting that a history of childhood trauma is not just common among those with a psychotic disorder, but also among those on the psychosis spectrum, including children and adolescents [7, 12–26]. In a prospective study using a large twin cohort, researchers assessed childhood trauma and psychotic experiences at age 5, 7, 10, and 12 years. They found that maltreatment by an adult, as well as peer victimization with intention to harm were associated with a 3.3 increased risk for psychotic experiences [27]. This is further underlined in a study conducted by Kelleher and colleagues who found that after cessation of trauma in adolescence, psychotic experiences may also stop [7]. Thus, when studying the association between childhood trauma and psychosis, it may be important to move away from categorical diagnoses and consider the whole psychosis spectrum.

## 10.3 The Question of Causality

It is generally assumed that childhood trauma is a causal risk factor in bringing on later psychosis. However, before such a claim can be substantiated, a number of issues need to be addressed. These include, among others, the question of robustness, temporality and genetic confounding when studies are summarized in a review [7, 28]. As evident from the meta-analysis by Varese and colleagues [5], there seems to be a well-established robust and strong association between childhood trauma and psychosis across different samples along the psychosis continuum. The same meta-analysis also summarized findings that point toward a dose–response relationship [5]. However, these studies were unable to determine the directionality of the association between childhood trauma and psychosis. While the assumption may be that childhood trauma increases the risk for psychosis, other explanations may also need to be considered [28, 29] (Fig. 10.2).

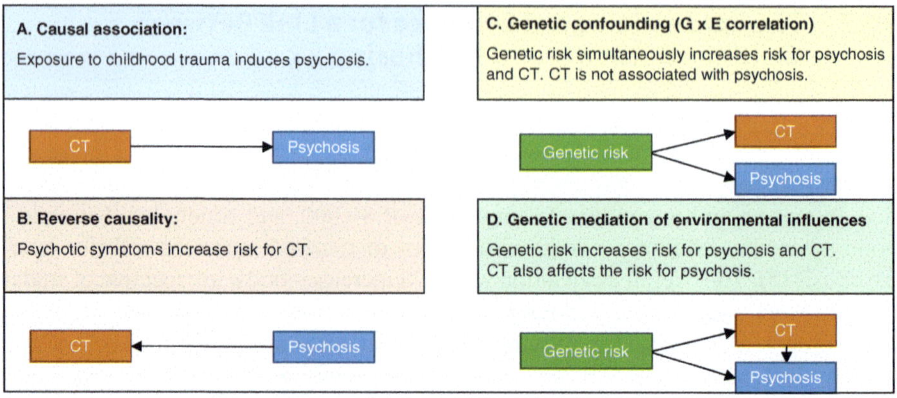


**Fig. 10.2** Etiological models explaining the association between childhood trauma (CT) and psychosis [28]

A possible alternative could be the hypothesis of "*reverse causation*." This would be the case if early symptoms of psychosis were to lead to an increased risk of exposure to childhood adversity. Psychotic symptoms would then have to be caused by other factors, and experiences of childhood adversity would only follow later in a person's development [28]. To test this hypothesis, temporality needs to be determined, but obviously, this is very difficult in the context of childhood trauma since prospective studies are ethically difficult. Therefore, only a few studies to date have been able to successfully adopt this approach, and even fewer have assessed whether psychotic experiences were present before the exposure to childhood traumatic events [28]. In one study, researchers found that individuals who were sexually abused before the age of 16 had a twofold increased risk for developing any diagnosis of psychotic disorder, and a 2.6-fold increased risk for schizophrenia, compared to age- and gender-matched non-abused individuals [30]. Similarly, another prospective study found that those maltreated before the age of 12 had a 3.16-fold increased odds to report psychotic symptoms later in life [27]. Neither one of these studies, however, was designed to establish temporality of onset of psychotic symptoms. Recently, two studies have been published that prospectively examined both exposure to childhood trauma and psychotic symptoms. The findings exemplify that, whereas childhood traumatic experiences increased the odds for later psychotic symptoms, psychotic experiences also predicted a greater risk for exposure to traumatic events [7, 31]. The observation of a bidirectional relation calls to consider another, though not necessarily mutually exclusive, explanation.

Before causality, or any relationship can be established, the possibility that both childhood trauma and psychotic symptoms are caused by a third factor needs to be ruled out. It is conceivable that the exposure to early life adversity is not random, but may develop out of a preexisting vulnerability, such as genetic vulnerability. The hypothesis of a "*gene-environment correlation*" postulates that a genetic predisposition would lead to both an increased likelihood of exposure to traumatic events, as well as to the later expression of psychotic symptoms. In this case, individuals with an increased genetic risk for psychosis would be more likely to be exposed to adversity due to inborn traits associated with psychosis, such as impaired social functioning or cognitive abilities [32–34]. In this case, the association between childhood trauma and psychosis may seem statistically plausible, when in fact this association could be explained by underlying genetic vulnerability. In other words, the relation between childhood trauma and psychosis would be "genetically confounded" [19, 28, 29].

There are different ways of controlling for "genetic confounding." An extensively applied approach is to use family psychiatric history as a proxy for genetic risk. Studies that adopted this approach did not find evidence that family liability confounded the association between childhood trauma and psychosis. Neither did they find evidence that combined family liability and childhood trauma increased the odds for later psychosis beyond the effect of each individually [8, 35], (for a review see [36]). Another method is to use direct measures of genetic vulnerability. A number of studies looked at specific candidate genes (i.e., genes that are thought to be either implicated in psychosis itself, or in possible mechanisms underlying psychosis onset) [37–40]. Again, others computed a polygenic risk score for schizophrenia. A polygenic risk score is a way to combine genome-wide genetic liability into one variable by summing up and weighting all risk alleles carried by one individual [41]. This approach could not confirm (passive) gene–environment correlation either [36]. However, neither of these concepts is currently able to capture all relevant genetic risks [19, 28, 29]. To overcome this limitation, the authors of three recently published studies used a monozygotic twin-differences approach. The idea behind this approach is that if two genetically identical (monozygotic) twins differ in environmental exposure, and that exposure is associated with psychotic symptoms, the association cannot be attributed to underlying genetic risk alone as monozygotic twins share 100% of their genes [19, 29, 31]. In agreement with the other approaches, these studies found that increased levels of exposure to childhood trauma were associated with more symptoms of psychosis, thus showing that the association cannot be solely attributed to genetic confounding.

A last consideration concerns the fact that not all individuals with a history of childhood trauma continue to experience psychotic symptoms. At the same time, whereas a great number of those with a psychotic disorder report having been victimized, not all persons with psychosis have. It is therefore crucial to consider that childhood trauma is neither sufficient nor necessary for the onset of psychosis [8, 28, 42]. Nevertheless, while other genetic and nongenetic factors might play a role in the etiology of psychosis, the current literature suggests that at least part of the association between childhood traumatic experiences and psychosis is genuine, and in agreement with a possible causal explanation [29].

## 10.4 Differential Sensitivity to Trauma: The Role of Genetic and Epigenetic Variation

While multiple studies have shown that the association between childhood trauma and psychosis persists even after controlling for genetic risk, it is currently unclear how genetic vulnerability plays a role in one's individual risk to develop psychosis following exposure, and which genes are likely to be involved [28, 43, 44]. The current literature, however limited, points to genes that are not specific to psychosis, but associated with mechanisms such as neuroplasticity and stress regulation. Again, the genetic contribution to the association between childhood trauma and psychosis can be either directly or indirectly examined.

Studies that adopted an indirect approach used family history or a twin-design and focused on heritable traits that are thought to be associated with psychosis. Based on data from a twin sample from the general population, the authors reported that worse cognitive speed, as well as a lifetime diagnosis of depression, moderated the effects of childhood trauma on the development of psychosis [45]. However, other studies, using family psychiatric history by either looking at parental psychiatric history, or at co-twin's symptoms, found no evidence of an interaction between indirect genetic liability and childhood trauma on the emergence of psychotic symptoms [27, 31, 35].

A number of studies have used a direct approach by looking at genetic variation to examine the interaction between childhood trauma and psychosis. Most of these studies examined candidate genes that are involved in regulating neurotransmitters (e.g., serotonin transporter gene SLC6A4/5-HTT/, specifically a functional polymorphism in the promoter region 5-HTTLPR) [46], neuroplasticity and cell survival following stress (brain-derived neurotrophic factor; BDNF) [40, 47], and the stress response system (FKBP5) [37]. For example, there is some evidence that a short-(s-) allele polymorphism in the promoter region of 5-HTTLPR is associated with several psychiatric disorders, possibly through its effects on the stress response system [48, 49]. One study found an interaction between 5-HTTLPR variations and physical neglect and abuse on cognitive functioning in patients with psychosis [46], in line with previous work suggesting that an increased stress response may be associated with cognitive impairments [50]. Another study looked at the moderating effects of the BDNF-Val66Met polymorphism between childhood trauma on positive and negative symptoms of psychosis [51]. Valine (Val) and methionine (Met) are amino acids that are used in the biosynthesis of proteins. The substitution of valine to methionine, by changes in a single base pair of the BDNF gene, may lead to differences in the activity of BDNF. The val/met polymorphism has been linked to changes in intracellular trafficking (i.e., distribution and release of macromolecules throughout and outside of a cell), as well as changes in activity-dependent secretion of BDNF [52]. Met carriers with a history of abuse reported more positive symptoms of psychosis compared to those with a Val/Val genotype [51]. However, others failed to replicate this interaction [40, 47]. A different gene of interest is FKBP5, which is thought to be involved in modulating the feedback loop determining glucocorticoid receptor sensitivity. Collip and colleagues [37] found an interaction between two FKBP5 single nucleotide polymorphisms and childhood trauma on psychotic symptoms and increased stress sensitivity in a twin sample from the general population. In agreement with these findings, a study of two independent samples from the general population found this polymorphism to be related to increased early stress sensitivity and the expression of positive psychotic symptoms [53]. Another replication study found an interaction between childhood trauma and the same polymorphism on schizotypy, psychotic experiences, as well as symptoms of anxiety and depression, in the general population [47]. Green and colleagues

[38], albeit looking at a different polymorphism, found a variation in combination with maltreatment to affect cognition, in those with schizophrenia and healthy controls. For a more extensive overview of relevant gene–environment interaction studies, we also refer to the chapter by Araceli Rosa and colleagues in this handbook.

While the mentioned studies have looked at variations in DNA sequence, it is also possible that childhood trauma exerts its influence through epigenetics [44, 54]. Epigenetics refers to changes in gene expression rather than modifications of the genetic code itself [54]. While this process is not yet fully understood, vigorous scientific attention has been directed to DNA methylation as a mechanism through which early life stress may become "embedded" in the genome [55]. An animal study found that the level of postnatal maternal care in rodents was associated with alterations in the methylation of the glucocorticoid NR3C1 receptor [56]. A human postmortem study [57] reported increased methylation of the NR3C1 promoter and changes in mRNA (i.e., messenger RNA that communicates information from DNA to the ribosome about what amino acid sequence should be read, thereby specifying gene expression) in those with a childhood trauma history, compared to suicide victims and healthy controls. However, a second study failed to replicate these findings [56]. A number of in vivo studies have reported altered DNA methylation as a consequence of childhood trauma (for a review see [42]). To our knowledge, there is only one study that examined methylation in first-episode psychosis patients with a history of childhood trauma within a considerable sample ($n_{\text{first-episode-psychosis}} = 48$; $n_{\text{control}} = 48$) [58]. Researchers found lower DNA methylation of LINE-1 (class I transposable elements in the DNA) sequences in those patients. However, in a recently published article on the E-Risk dataset (n = 1.658) (Environmental Risk longitudinal study), authors found very limited evidence for an association between childhood adversity and changes in epigenetic variation (in peripheral blood) [59]. They also found that methylation associations for adversities overlapped with those of other health behaviors such as tobacco smoking. Another study found similar results in that DNA methylation in whole blood was associated with tobacco smoking, and also BMI and additive genetic effects [60]. This highlights the methodological difficulties of disentangling biological effects of environmental influences from other health-related behaviors associated with a trauma history [59].

The evidence for genetic moderation of the association between childhood trauma and psychosis, thus, is mixed. Nevertheless, these studies exemplify the numerous genetic mechanisms through which childhood trauma could potentially influence the development of psychosis, and other related mental health disorders [54, 59, 61].

## 10.5 Biological Mechanisms Underlying the Link Between Trauma and Psychosis

The *diathesis-stress model* of mental illness postulates that major life stressors, such as childhood trauma, can be a serious neurodevelopmental insult on the developing brain and contribute to the onset of psychiatric disorders [62]. The idea that abnormalities in the limbic system could be related to the effects of childhood abuse on brain development dates back to van der Kolk and Greenberg in 1987 [63]. Since then, a number of studies have found structural, functional, and molecular changes in brain structures and persistent associations with childhood trauma, e.g., [64–66]. Below, we will discuss implicated alterations of the brain (more specifically hippocampus, amygdala, and findings from studies on cortical thickness), the stress response system and the immune system. Intact stress response and immune systems are of crucial importance for mental and physical health, behavioral adaptation, and brain development; the dysregulation of these systems may therefore have long-lasting effects on the developing brain and one's mental and physical health [67].

### 10.5.1 The Brain

#### 10.5.1.1 Hippocampus

The hippocampus may be the most likely place to reflect the effects of childhood trauma [68]. This brain area is densely packed with glucocorticoid receptors, which makes it highly susceptible to early life stress [69]. Structural imaging studies found reduced bilateral hippocampal volume in those with a history of childhood trauma, those with a psychotic disorder, and in individuals with both [67, 68, 70, 71]. Interestingly, volume reductions were also found in nonpsychotic siblings, individuals at risk for schizophrenia, as well as in people with psychotic bipolar disorder (for a review see [72]). Studies found that childhood trauma was associated with reduced hippocampal volume in patients with first-episode psychosis, which may be associated with worse cognitive performance compared to patients with a diagnosis of schizophrenia, but no history of adversity [39, 73, 74]. There is some evidence that the hippocampus is involved in regulating the hypothalamic-pituitary-adrenal (HPA) axis stress response and that smaller hippocampal volume in those with a first-psychotic episode may partially be explained by stress-related processes [39]. In animal models, decreased hippocampal volumes, specifically in dentate gyrus (DG) and cornu ammonis 3 and 1 (CA3;CA1), were found in mice exposed to maternal deprivation (i.e., traumatic event) [75]. This fits well with previous findings from human postmortem studies that suggest that the DG and CA3 subfields appear most affected by molecular and cellular changes when comparing tissue of previously healthy individuals with persons with schizophrenia [76]. Additionally, on a cellular level, it was shown that the hippocampus contains "place cells" and "grid cells" in the entorhinal cortex. These cells are responsible for the spatiotemporal representation of places, routes, and associated experiences [77]. As spatiotemporal disorientation may be a symptom of psychosis in some patients, it may not be farfetched that childhood trauma could exert its effects on the development of psychotic symptoms by affecting hippocampal development [78].

#### 10.5.1.2 Amygdala

The amygdala is a key limbic structure for prioritizing and encoding emotionally salient information, and detecting and responding to facial expressions and potential threats [79, 80]. Given the altered emotional responses observed in patients with schizophrenia, an affective pathway to psychosis has been previously proposed [81–83]. This also makes sense based on the finding that, like the hippocampus, the amygdala is rich of glucocorticoid receptors, which would make it especially vulnerable to early stress, including childhood trauma. Most studies have examined emotional reactivity following childhood trauma on a behavioral level, for example, by examining reactivity to daily life stress [82, 83]. Other studies have examined structural and functional abnormalities of the amygdala in a variety of mental disorders, including psychosis [84–86]. Structural studies report contradicting results. Some found decreased amygdala volumes in those with a first-episode psychosis and a trauma history. Others found increased volumes and argue that this may be due to the stimulating effects of stress on pyramidal cells in this brain structure [87, 88]. The same discrepant findings were found on a functional level, in that some show either patterns of hypo- or hyper-activation during emotional tasks in patients with psychosis [89]. Therefore, Cancel and colleagues [90] suggested to examine functional connectivity between the amygdala and other related brain areas. They found that in patients with schizophrenia and a history of childhood trauma—specifically sexual abuse and physical neglect—connectivity between amygdala and posterior cingulate/precuneus was decreased [90]. These results are similar to previous studies that used a face recognition task and found decreased connectivity in response to fearful faces [91].

#### 10.5.1.3 Findings from Studies on Cortical Thickness

There is some evidence that gray matter density is correlated with childhood trauma in patients with psychosis. Sheffield and colleagues [92] found that overall gray matter differed between psychotic patients with and without a history of trauma, and healthy participants. More specifically, they found that individuals with a history of sexual abuse and a diagnosis of psychosis differed from healthy controls in medial and inferior frontal, inferior and superior temporal, precentral gyri and inferior parietal lobe grey matter density. Yet, psychosis patients without sexual abuse only differed from healthy controls in cerebellar gray matter volume. Furthermore, within the psychosis patients group, those with and without a history of sexual abuse differed in gray matter volume in bilateral anterior cingulate cortices and left inferior frontal gyrus only [92]. In another study, authors reported anterior cingulate volumetric differences (compared to healthy controls) in patients with schizophrenia and antisocial personality disorder after controlling for childhood abuse [93]. Moreover, at least three studies demonstrated that prefrontal cortical deficits can be observed in victims of childhood trauma without mental health problems, too [94–96]. This fits well with cognitive models that propose that childhood trauma is associated with cognitive impairments in those with a psychotic disorder, suggesting that neurocognitive alterations mediate the development of psychosis following trauma exposure [39, 73, 97]. However, in a large prospective cohort study using the Dunedin dataset (Dunedin Multidisciplinary Health and Development Study, a longitudinal study that started in 1972), Danese and colleagues found that whereas those exposed to victimization had indeed impaired cognitive functions (including general intelligence, executive function, processing speed, memory, perceptual reasoning, and verbal comprehension), these deficits were explained by preexisting cognitive deficits prior to the victimization event(s) [98].

These studies underscore the effect of childhood trauma on brain development and also raise caution when trying to unravel the neurobiological correlates of psychopathology, as some may be specific to the disorder itself, and others may represent alterations associated with exposure to childhood trauma, or may even predate the exposure to trauma [68, 92].

### 10.5.2 The Stress Response System

One of the most commonly proposed idea is that childhood trauma increases the probability for later psychosis through its impact on the stress response system, which is in growing children, still under development. Early traumatic experiences may cause heightened sensitivity of the HPA-axis to subsequent stress, which eventually could increase the risk for psychosis (and related mental disorders) [99, 100]. Similar results were reported for the association between childhood trauma and subclinical psychotic experiences in a general population sample, through a pathway of heightened subjective stress appraisal [101].

At the behavioral level, studies found that individuals with a history of trauma reacted more strongly, with more negative affect, and more paranoia, to minor daily life stressors [82, 102, 103]. In an experimental study, Valmaggia and colleagues used virtual reality and reported stronger paranoid ideation to a neutral social environment in people at ultra-high risk for psychosis, and a history of bullying [104]. In agreement with these results, another virtual reality study indicated heightened social stress reactivity as a possible link between childhood trauma, psychosis liability, and paranoid ideation. Those with a trauma history were found to report more paranoid ideation and subjective distress, compared to individuals without such experiences [105].

In addition, there is a growing body of literature that links exposure to childhood adversities, and psychosis, to hyper-activation and sensitization of the HPA axis [67, 71, 100]. HPA-axis activity is often measured by examining salivary cortisol levels. There is some evidence for both elevated baseline cortisol secretion and blunted cortisol awakening response in patients with psychosis, including those with childhood trauma [106–109]. Others found that the pituitary gland was enlarged in those with a psychotic disorder [110, 111]. This fits in well with the "*traumagenic neurodevelopmental model*" which assumes that HPA-axis dysregulation may mediate the relationship between childhood trauma and the development of psychosis [97, 112]. However, there is also some evidence that stressful events reduce HPA-axis activation [113, 114] and that pituitary gland volume in children at risk for psychosis due to adversity is reduced [115]. Irrespective of these discrepancies, there seems to be a correlation between elevated dopaminergic brain response associated with psychosis and salivary cortisol, although the direction is still to be determined [108, 116]. It could be that cortisol exerts its influence on the expression of symptoms through its effects on dopaminergic pathways ("bottom-up" model). It is also plausible that underlying symptoms and corresponding neurotransmitter activity affect HPA-axis activity ("top-down" model) [117].

Childhood trauma may increase the risk for psychosis by affecting the mesolimbic dopamine system through a mechanism of exaggerated dopamine release to subsequent social stressors later in life [99, 118]. There may be a feedback loop at play: prolonged stress, including adversity, may increase glucocorticoid release, which in turn may increase dopamine secretion. This may start a positive feedback loop through which the increased dopaminergic activity elevates HPA activity and glucocorticoid release. Since it has previously been shown that dopaminergic hyperreactivity plays a role in psychosis, this may be a mechanism through which childhood trauma increases the risk for the subsequent development of psychotic symptoms [67]. Yet, most evidence that dopamine release is elevated following exposure to stress still comes mainly from animal models [119], but studies are beginning to emerge which provide some evidence in humans as well [97, 107, 117].

### 10.5.3 Immune System

There is accumulating evidence that alterations in the innate immune system are associated with psychotic disorders, and that childhood traumatic events are associated with a pro-inflammatory state in adulthood. These alterations include increased chemokines (e.g., CCL-11), acute-phase proteins (e.g., C-reactive protein [CRP]), and cytokines [120]. Some cytokines are pro-inflammatory, such as interleukin IL-6, IL-8, and tumor necrosis factor (TNF-α). Consequently, there are cytokines with a dampening effect as to keep a right balance [121]. A recent meta-analysis reported an association between childhood trauma and inflammatory markers. The biggest effect was found for TNF-α, followed by IL-6 and C-reactive protein [122]. Interestingly, Dennison and colleagues found that increased levels of pro-inflammatory markers were found only in patients with schizophrenia who also had a history of childhood trauma, as compared to patients without [123]. In another study, elevated CRP levels were shown in first-episode psychosis patients with a trauma history, but not in patients without a trauma, or healthy controls [124]. Yet, another study found the same group differences for elevated TNF-α levels [125]. These results have led to speculations that early childhood adversities "*bring about epigenetic changes that lead to a pro-inflammatory phenotype in adulthood*," which has been associated with a range of mental and physical health issues, including psychosis [123]. While this sounds like a plausible interpretation, there is still little direct evidence, to date, supporting this hypothesis.

Interestingly, the association between HPA-axis (dys-)functioning and childhood trauma on the one hand, and the association between childhood trauma and the immune system on the other, has led to the exploration of interactive pathways [126]. It was found that receptors for one or more of the stress hormones are expressed on lymphocytes. Moreover, lymphocytes are also able to synthesize the adrenocorticotropic hormone (ACTH), which is secreted by the pituitary gland and is important in the regulation of the stress response system [127]. While much is still unknown, it may indeed be the case that early life stress is a neurodevelopmental insult on multiple, interacting systems [42, 68].

## 10.6 Psychological Mechanisms Linking Trauma and Psychosis

In addition to alterations at the biological level, it is likely that the exposure to severe traumatic events in childhood impact on one's psychological development, and that psychological factors may co-determine a person's vulnerability, or resilience, to these events. A number of possible psychological mechanisms linking exposure to psychopathological outcomes have been proposed, including dysfunctional cognitive schemas, affective dysregulation, insecure attachment styles, and dissociative mechanisms.

Cognitive models propose that negative beliefs about the self and others, and increased threat anticipation may mediate the association between childhood trauma and psychosis [128–130]. The *social defeat theory* postulates that it is not the experience of adversity itself, but the enduring feeling of defeat and the subordinate position experienced during adverse events that increase the risk for psychosis. Such feelings may stem from childhood trauma [131–133]. Animal studies report that long-term isolation can lead to reductions in whole-brain volume, hippocampus, or medial prefrontal cortex, and that social defeat can reduce neurogenesis [131]. However, Selten and colleagues note that to date, there are no studies that have examined possible brain changes in humans at psychosis onset in relation to proxy measures of social defeat (for a review see [131]). Preliminary support for the social defeat theory was found by a number of behavioral studies that showed that perceptions of defeat and entrapment were associated with positive symptoms in a clinical group of patients with schizophrenia [134–136]. Results from the NEMESIS-2 study (Netherlands Mental Health Survey and Incidence Study) indicate that self-reported feelings of social defeat may act as a mediator between childhood trauma and the expression of psychosis. This was found for individuals with a psychotic disorder, and also for psychotic experiences in the general population [137]. A virtual reality study found that those at ultra-high risk for psychosis had higher levels of social defeat and entrapment, and were more likely to react with paranoid appraisals to the virtual reality environment than their healthy counterparts [138]. However, there is some debate on the conceptualization of social defeat. The theory is built on observations in animals, in which animals are attacked by others and may actually die from this attack, even if they show subordinate behavior. This type of "subordination" is likely to be different compared to the subordinate position caused by social exclusion that humans can experience. Other concepts used in the context of social defeat studies in humans are discrimination, negative social evaluation, social adversity, social fragmentation, or social disadvantage [131], which is why the term "social defeat" in relation to human studies has previously been indicated as a "misnomer" (i.e., an inaccurate name or designation) [8]. Karlsen and colleagues looked at the association between (racial) discrimination and psychosis and found it to be strongest for discrimination involving physical assault. It has been proposed that not defeat itself, but hostility, threat, and violence in the context of social disadvantage and discrimination might explain the high rates of psychosis in migrants and minority groups [139].

Some researchers have also pointed to the possible relevance of attachment. Attachment styles are thought to reflect early cognitive-affective representations of the self and others, as well as strategies for regulating distress [140]. Children are dependent on their caregivers and adults around them to keep them safe. In his *attachment theory*, Bowlby [141] proposed that infants internalize experiences of interaction with their caregivers, and that this representation is carried forward into adulthood. Early trauma can affect a child's attachment, which can consequently influence expectations and believes about the self and others in interpersonal interactions later in life [141, 142]. There is some evidence that an insecure attachment style (see Fig. 10.3) might act as mediator between childhood trauma and psychotic experiences in clinical samples and the general population [143–145]. On the other hand, an insecure avoidant attachment style mediated the association between neglect and paranoia in data from the US National Comorbidity Survey [146]. Overall, it looks like that an anxious and avoidant attachment style might mediate the relationship between childhood trauma and positive symptoms of psychosis [144–149]. Interestingly, this association was not found for negative symptoms [42].

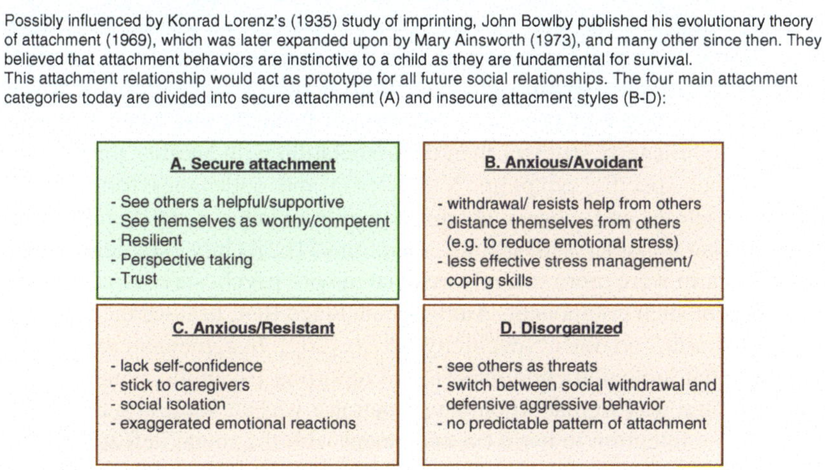


**Fig. 10.3** Attachment styles

Finally, it may be that a dissociative response to childhood trauma paves the way to psychosis. Dissociation can be defined as "*a disruption in the usually integrated functions of consciousness, memory, identity or perception of the environment*" [112, 150]. Research showed that individuals who apply a dissociative coping style following trauma are more likely to have impaired reality testing and may develop subsequent psychotic experiences [151]. In a clinical sample of psychotic patients, it was shown that those with a positive history of trauma had increased dissociative tendencies compared to patients without a trauma history [152, 153]. Moreover, it was found that the risk for psychosis through a history of physical neglect could be explained by increased dissociation in patients with psychosis [154]. Lastly, both in nonclinical and clinical groups, the association between childhood trauma and hallucination proneness was mediated by dissociative tendencies [23, 153].

## 10.7 Specificity of the Association Between Types of Trauma and Distinct Psychotic Symptoms

When moving away from broad diagnostic categories to a symptom-specific level, there is a discussion in the literature whether there are associations between specific forms of childhood trauma and specific symptoms of psychosis (e.g., 154). It has, for example, been proposed that different forms of adversities may exert differential influences upon affective and cognitive processes [6, 23, 155]. However, empirical findings have provided mixed support for this hypothesis. A number of studies found an association between sexual abuse and hallucinations and delusions in that the content may be related to patients' past traumatic experiences [10, 16]. Other studies found emotional abuse and neglect to be most strongly associated with dissociative symptoms [152, 156–158]. Another study, however, found sexual abuse to be most strongly associated with dissociative symptoms [159], while other studies found no evidence for specificity at all [160, 161]. Previous work had some statistical limitations in that they merely relied on the presence or absence of a statistically significant associations between a given form of adversity and a specific symptom of psychosis, but did not formally test whether this association was significantly stronger than the association with another form of adversity, or another specific symptom. Van Nierop and colleagues formally tested specific associations and showed that no form of trauma had a statistically stronger association than the other with specific symptoms [162]. Only experiences with an intention to harm were more strongly associated with psychosis compared to experiences without such component. Authors concluded that, besides the presence of "intention to harm," no other "specificity" exists [162]. In agreement with this finding, a study that looked at children's risk of reporting psychotic symptoms found that the outcome was similar when the perpetrator was an adult or a peer, as long as there was an intention to harm (in agreement with the social defeat hypothesis) [132]. This suggests that the only specific predictor is an element of threat intentionally induced by others [162].

Moreover, it was found that about 45% of the general population's attributable risk for childhood onset psychiatric disorders is accounted for by early maltreatment [163]. As convincingly outlined in this handbook, individuals with a history of early adverse experiences do not only have a higher risk for psychosis, but also show a higher prevalence of a number of disorders, including depression, anxiety, substance abuse, eating disorders, suicidal symptomatology, personality disorder, dissociative disorders, and posttraumatic stress disorder [4, 164–168]. Thus, childhood trauma seems to be characterized by an admixture of affective, anxious, and psychotic symptoms [17, 169, 170]. Investigating symptoms across these different, but related disorders (i.e., internalizing, externalizing, thought disorders) led to suggest that maltreatment was associated with greater "general psychopathology" (i.e., p-factor), but not with any specific endophenotype (i.e., measurable biomarkers that are correlated with a disorder, partly because of a shared genetic disposition) [171].

## 10.8 Treatment Considerations for Patients with a Trauma History

A history of childhood trauma is associated with a greater risk for psychiatric disorders, more comorbidity, worse social functioning, lower remission rates, and less favorable treatment outcomes in general [168, 169, 172–175]. There is some evidence, albeit limited, that greater (supportive) networks act as protective factors, but that social skills in those with trauma and/or psychosis are impaired, which may reduce the likelihood of having such networks [33, 176]. Therefore, social skills, as well as underlying mechanisms related to childhood trauma, such as stress sensitivity and cognitive biases, may be viable targets for psychotherapeutic interventions. Research on possible therapeutic outcomes in clinical groups converges to the necessity of distinguishing between patients with and without a history of childhood trauma [42]. It is not surprising, then, that treatments based on cognitive-behavioral therapy and exposure therapy, and focusing on reducing sensitivity to stress, readjusting cognitive biases, teaching social skills, psychoeducation, stabilization, and development of safe coping skills may be effective for those with a trauma history [177–179].

## 10.9 Conclusion

In this chapter, we have reviewed a body of research on the association between childhood traumatic experiences and psychoses. Evidence suggests that, at least in some people, exposure to childhood trauma with the intention to harm contributes to the onset of psychotic experiences and psychotic disorders. There is, however, also evidence suggesting that underlying traits, such as genetic predisposition, lower cognitive and social skills, may increase the risk for exposure to adversity. Whereas childhood trauma may be neither sufficient nor necessary to explain the onset of psychosis, at least part of that association seems to be causal. Nevertheless, it is important to note that this relationship is not restricted to psychosis only, but holds for a whole myriad of psychiatric disorders. Yet, findings that neurobiological changes can also be observed in individuals with a history of maltreatment, but no psychiatric disorder is perplexing. It may be that there are certain alterations that early life stress brings about, and also that there are compensatory mechanisms in resilient individuals, which enables them to balance out such neurodevelopmental insults [68, 92–96]. To make things more complex, the different pathways from childhood trauma to psychosis discussed in this chapter may act at different and complementary conceptual levels of examination (i.e., neurobiological, psychological), which, when taken together, might converge into an integrated model of psychosis [155, 180, 181].

## References

1. Boevink WA. From being a disorder to dealing with life: an experiential exploration of the association between trauma and psychosis. Schizophr Bull. 2006;32(1):17–9.
2. Linscott RJ, Van Os J. An updated and conservative systematic review and meta-analysis of epidemiological evidence on psychotic experiences in children and adults: on the pathway from proneness to persistence to dimensional expression across mental disorders. Psychol Med. 2013;43(6):1133–49.
3. Fisher HL, Caspi A, Poulton A, Meier MH, Houts R, Harrington H, et al. Specificity of childhood psychotic symptoms for predicting schizophrenia by 38 years of age: a birth cohort study. Psychol Med. 2013;6(8):2077–86.
4. Bendall S, Jackson HJ, Hulbert CA, McGorry PD. Childhood trauma and psychotic disorders: a systematic, critical review of the evidence. Schizophr Bull. 2008;34(3):568–79.
5. Varese F, Smeets F, Drukker M, Lieverse R, Lataster T, Viechtbauer W, et al. Childhood adversities increase the Risk of psychosis: a meta-analysis of patient-control, prospective- and cross-sectional cohort studies. Schizophr Bull. 2012;38(4):661–71.
6. Heins M, Simons C, Lataster T, Pfeifer S, Versmissen D, Lardinois M, et al. Childhood trauma and psychosis: a case-control and case-sibling comparison across different levels of genetic liability, psychopathology, and type of trauma. Am J Psychiatry. 2011;168(12):1286–94.
7. Kelleher I, Keeley H, Corcoran P, Ramsay H, Wasserman C, Carli V, et al. Childhood trauma and psychosis in a prospective cohort study: cause, effect, and directionality. Am J Psychiatry. 2013;170(7):734–41.
8. Morgan C, Gayer-Anderson C. Childhood adversities and psychosis: evidence, challenges, implications. World Psychiatry. 2016;6:93–102.
9. Rössler W, Hengartner MP, Ajdacic-Gross V, Haker H, Angst J. Impact of childhood adversity on the onset and course of subclinical psychosis symptoms—results from a 30-year prospective community study. Schizophr Res. 2014;153(1–3):189–95. https://doi.org/10.1016/j.schres.2014.01.040.
10. Shevlin M, Murphy J, Read J, Mallett J, Adamson G, Houston JE. Childhood adversity and hallucinations: a community-based study using the National Comorbidity Survey Replication. Soc Psychiatry Psychiatr Epidemiol. 2011;46:1203–10.
11. Wicks S, Sc B, Hjern A, Ph D, Gunnell D, Lewis G, et al. Social adversity in childhood and the Risk of developing psychosis: a National Cohort Study. Am J Psychiatry. 2005;(September):1652–7.
12. Addington J, Stowkowy J, Cadenhead KS, Cornblatt BA, Mcglashan TH, Perkins DO, et al. Early traumatic experiences in those at Clinica high Risk for psychosis. Early Interv Psychiatry. 2013;7(3):300–5.

13. Holshausen K, Bowie CR, Harkness KL, Holshausen K, Bowie CR, Harkness KL, et al. The relation of childhood maltreatment to psychotic symptoms in adolescents and young adults with depression the relation of childhood maltreatment to psychotic symptoms in adolescents and young adults with depression. 2016;4416
14. Paksarian D, Eaton WW, Mortensen PB, Merikangas KR, Pedersen CB. A population-based study of the risk of schizophrenia and bipolar disorder associated with parent – child separation during development. 2015;2018:2825–37.
15. Shevlin M, Mcanee G, Bentall RP, Murphy J. Specificity of association between adversities and the occurrence and co-occurrence paranoia and hallucinations : evaluating the stability of childhood risk in an adverse adult environment. Psychosis [internet]. Routledge. 2015;7(3):206–16. https://doi.org/10.1080/17522439.2014.980308.
16. Thompson A, Nelson B, Mcnab C, Simmons M, Leicester S, Mcgorry PD, et al. Psychotic symptoms with sexual content in the " ultra high risk " for psychosis population : frequency and association with sexual trauma. Psychiatry Res. 2010;177(1–2):84–91. https://doi.org/10.1016/j.psychres.2010.02.011.
17. van Nierop M, Viechtbauer W, Gunther N, van Zelst C, de Graaf R, ten Have M, et al. Childhood trauma is associated with a specific admixture of affective, anxiety, and psychosis symptoms cutting across traditional diagnostic boundaries. Psychol Med. 2015;45(06):1277–88.
18. Wolke D, Lereya ST, Fisher HL, Lewis G, Zammit S. Bullying in elementary school and psychotic experiences at 18 years : a longitudinal , population-based cohort study. Psychol Med. 2014;44:2199–211.
19. Alemany S, Goldberg X, van Winkel R, Gastó C, Peralta V, Fañanás L. Childhood adversity and psychosis: examining whether the association is due to genetic confounding using a monozygotic twin differences approach. Eur Psychiatry. 2013;28(4):207–12.
20. Alemany S, Ayesa-Arriola R, Arias B, Fatjó-Vilas M, Ibáñez MI, Ortet G, et al. Childhood abuse in the etiological continuum underlying psychosis from first-episode psychosis to psychotic experiences. Eur Psychiatry. 2015;30(1):38–42.
21. Barrigon ML, DIas FJ, Gurpegui M, Ferrin M, SM D, Moreno-Granados J, et al. Childhood trauma as a risk factor for psychosis : a sib-pair study. J Psychiatr Res. 2015;70:130–6.
22. Bartels-Velthuis AA, Van De Willige G, Jenner JA, Wiersma D, Van Os J. Auditory hallucinations in childhood: associations with adversity and delusional ideation. Psychol Med. 2012;42(3):583–93.
23. Bentall RP, Wickham S, Shevlin M, Varese F. Do specific early-life adversities lead to specific symptoms of psychosis? A study from the 2007 the adult psychiatric morbidity survey. Schizophr Bull. 2012;38(4):734–40.
24. Bratlien U, Øie M, Haug E, Møller P, Andreassen OA, Lien L, et al. Environmental factors during adolescence associated with later development of psychotic disorders - a nested case-control study. Psychiatry Res. 2014;215(3):579–85. https://doi.org/10.1016/j.psychres.2013.12.048.
25. Daalman K, Diederen KMJ, Derks EM, Van Lutterveld R, Kahn RS, Sommer IEC. Childhood trauma and auditory verbal hallucinations. Psychol Med. 2012;42(12):2475–84.
26. DeRosse P, Nitzburg GC, Kompancaril B, Malhotra AK. The relation between childhood maltreatment and psychosis in patients with schizophrenia and non-psychiatric controls Pamela. Schizophr Res. 2014;155:66–71.
27. Arseneault L, Cannon M, Fisher HL, Guilherme D, Polanczyk G, Moffit TE, et al. Childhood trauma and Children's emerging psychotic symptoms: a genetically sensitive longitudinal cohort study Louise. Magn Reson Imaging. 2011;31(3):477–9.
28. Van Winkel R, Van Nierop M, Myin-Germeys I, Van Os J. Childhood trauma as a cause of psychosis: linking genes, psychology, and biology. Can J Psychiatr. 2013;58(1):44–51.
29. Lecei A, Decoster J, Hert M, De Derom C, Jacobs N, Menne-lothmann C, et al. Evidence that the association of childhood trauma with psychosis and related psychopathology is not explained by gene-environment correlation : a monozygotic twin differences approach. Schizophr Res. 2018; https://doi.org/10.1016/j.schres.2018.05.025.

30. Cutajar MC, Mullen PE, Ogloff JRP, Thomas SD, Wells DL, Spataro J. Schizophrenia and other psychotic disorders in a cohort of sexually abused children. Arch Gen Psychiatry. 2010;67(11):1114–9.
31. Schaefer JD, Moffitt TE, Arseneault L, Danese A, Fisher HL, Houts R, et al. Adolescent victimization and early-adult psychopathology: approaching causal inference using a longitudinal twin study to rule out noncausal explanations. Clin Psychol Sci. 2017;6(3):352–71.
32. Danese A, Moffitt TE, Arseneault L, Ben A, Dinardo PB, Gandelman SB, et al. The origins of cognitive deficits in victimized children: implications for neuroscientists and clinicians. Am J Psychiatry. 2018;174(4):349–61.
33. Velthorst E, Ph D, Fett AJ, Ph D, Reichenberg A, Ph D, et al. The 20-year longitudinal trajectories of social functioning in individuals with psychotic disorders. Soc Funct psychotic Disord. 2017;(November):1075–85.
34. Velthorst E, Fett AKJ, Reichenberg A, Perlman G, Van Os J, Bromet EJ, et al. The 20-year longitudinal trajectories of social functioning in individuals with psychotic disorders. Am J Psychiatry. 2017;174(11):1075–85.
35. Trotta A, Di Forti M, Iyegbe C, Green P, Dazzan P, Mondelli V, et al. Familial risk and childhood adversity interplay in the onset of psychosis. Bi Psych Open. 2015:6–13.
36. Trotta A, Iyegbe C, Di Forti M, Sham PC, Campbell DD, Cherny SS, et al. Interplay between schizophrenia polygenic risk score and childhood adversity in first-presentation psychotic disorder: a pilot study. PLoS One. 2016;11(9):1–14.
37. Collip D, Myin-germeys I, Wichers M, Jacobs N, Derom C, Thiery E, et al. FKBP5 as a possible moderator of the psychosis-inducing effects of childhood trauma. 2013;261–268.
38. Green MJ, Chia T, Cairns MJ, Wu J, Tooney PA, Scott RJ, et al. Catechol-O-methyltransferase ( COMT ) genotype moderates the effects of childhood trauma on cognition and symptoms in schizophrenia. J Psychiatr Res. 2014;49:43–50. https://doi.org/10.1016/j.jpsychires.2013.10.018.
39. Mondelli V, Cattaneo A, Murri MB, Papadopoulos AS, Aitchison KJ. Stress and inflammation reduce BDNF expression in first- episode psychosis : a pathway to smaller hippocampal volume. J Clin Psychiatry. 2011;72(12):1677–84.
40. Ramsay H, Kelleher I, Flannery P, Clarke MC, Lynch F, Harley M, et al. Relationship between the COMT-Val158Met and BDNF- Val66Met polymorphisms, childhood trauma and psychotic experiences in an adolescent general population sample. 2013;8(11):1–10.
41. Lewis CM, Vassos E. Prospects for using risk scores in polygenic medicine. Genome Med. 2017:9–11.
42. Misiak B, Krefft M, Bielawski T, Ahmed AM, Sasiadek MM, Frydecka D. Toward a unified theory of childhood trauma and psychosis: a comprehensive review of epidemiological , clinical , neuropsychological and biological findings. Neurosci Biobehav Rev. 2017;75: 393–406.
43. Van Os J, Kenis G, BPF R. The environment and schizophrenia. Nature. 2010; 468(7321):203–12. https://doi.org/10.1038/nature09563.
44. Van Winkel R. Aetiological stratification as a conceptual framework for gene-by-environment interaction research in psychiatry. Epidemiol Psychiatr Sci. 2010;616(2015):6–11.
45. Pfeifer S, Krabbendam L, Germeys IM, Wichers M, Jacobs N, Thiery EW, et al. A cognitive intermediate phenotype study confirming possible gene – early adversity interaction in psychosis outcome: a general population twin study. 2010;2439
46. Aas M, Djurovic S, Athanasiu L, Steen NE, Agartz I, Lorentzen S, et al. Serotonin transporter gene polymorphism, childhood trauma, and cognition in patients with psychotic disorders. Schizophr Bull. 2012;38(1):15–22.
47. De Castro-catala M, Van Nierop M, Barrantes-vidal N, Sheinbaum T, Kwapil TR, Pe E, et al. Childhood trauma, BDNF Val66Met and subclinical psychotic experiences. Attempt at replication in two independent samples. 2016;83:121–9.
48. Caspi A, Hariri AR, Holmes A, Uher R, Moffitt RE. Genetic sensitivity to the environment: the case of the serotonin transporter gene and its implications for studying complex diseases and traits. 2010;167(10):509–27.

49. Mueller A, Brocke B, Fries E, Lesch KP, Kirschbaum C. The role of the serotonin transporter polymorphism for the endocrine stress response in newborns. Psychoneuroendocrinology. 2010;35(2):289–96.
50. Lupien SJ. The effects of stress and stress hormones on human cognition : implications for the field of brain and cognition. 2007;65:209–37.
51. Alemany S, Arias B, Aguilera M, Villa H, Moya J, Ibáñez MI, et al. Childhood abuse, the BDNF-Val66Met polymorphism and adult psychotic-like experiences. Br J Psychiatry. 2011;199(1):38–42.
52. Egan MF, Kojima M, Callicott JH, Goldberg TE, Kolachana BS, Bertolino A, et al. The BDNF val66met polymorphism affects activity-dependent secretion of BDNF and human memory and hippocampal function. Cell. 2003;112:257–69.
53. Alemany S, Moya J, Ibanez MI, Villa H, Mezqita L, Ortet G, et al. Research letter: childhood trauma and the rs1360780 SNP of FKBP5 gene in psychosis: a replication in two general population samples. 2016, 2016;46:221–3.
54. Rutten BPF, Mill J. Epigenetic mediation of environmental influences in major psychotic disorders. Schizophr Bull. 2009;35(6):1045–56.
55. Szyf M, Bick J. DNA methylation: a mechanism for embedding early life experiences in the genome. Child Dev. 2013;84(1):49–57.
56. Weaver ICG, Cervoni N, Champagne FA, Alessio ACD, Sharma S, Seckl JR, et al. Epigenetic programming by maternal behavior. 2004;7(8):847–54.
57. McGowan PO, Sasaki A, D'Alessio AC, Dymov S, Labonte B, Szyf M, et al. Epigenetic regulation of the glucocorticoid receptor in human brain associates with childhood abuse. Nat Neurosci. 2009;12(3):342–8.
58. Misiak B, Szmida E, Karpiński P, Loska O, Sąsiadek MM, Frydecka D. Lower LINE-1 methylation in first-episode schizophrenia patients with the history of childhood trauma. Epigenomics. 2015;7(8):1275–85. https://doi.org/10.2217/epi.15.68.
59. Marzi SJ, Ph D, Sugden K, Ph D, Arseneault L, Ph D, et al. Analysis of DNA methylation in Young people: limited evidence for an association between victimization stress and epigenetic variation in blood. Am J Psychiatry. 2018;175(13):1–13.
60. Hannon E, Knox O, Sugden K, Burrage J, Wong CCY, Belsky W, et al. Characterizing genetic and environmental influences on variable DNA methylation using monozygotic and dizygotic twins. 2018;1–27.
61. Van Winkel R, Esquivel G, Kenis G, Wichers M, Collip D, Peerbooms O, et al. Genome-wide findings in schizophrenia and the role of gene – environment interplay. 2010;16:185–92.
62. Bebbington P. Misery and beyond: the pursuit of disease theories of depression. 1987;13–20.
63. van der Kolk B, Greenberg MS. The psychobiology of the trauma response: hyperarousal, constriction, and addiction to traumatic reexposure. Psychological trauma. Washington, DC: American Psychiatry Press; 1987. p. 63–87.
64. Teicher MH, Samson JA, Polcari A, McGreenery CE. Sticks, stones, and hurtful words: relative effects of various forms of childhood maltreatment. Am J Psychiatry. 2006;163(11): 993–1000.
65. Teicher MH, Parigger A. The 'maltreatment and abuse chronology of exposure' (MACE) scale for the retrospective assessment of abuse and neglect during development. PLoS One. 2015:1–37.
66. Teicher MH, Vitaliano GD. Witnessing violence toward siblings: an understudied but potent form of early adversity. PLoS One. 2011;6(12).
67. Ruby E, Polito S, McMahon K, Gorovitz M, Corcoran C, Malaspina D. Pathways associating childhood trauma to the neurobiology of schizophrenia. Front Psychol Behav Sci. 2014;3(1):1–23.
68. Teicher MH, Samson JA. Annual research review : enduring neurobiological effects of childhood abuse and neglect. J Child Psychol Psychiatry. 2016;3:241–66.
69. Uematsu A, Matsui M, Tanaka C, Takahashi T, Noguchi K, Suzuki M, et al. Developmental trajectories of amygdala and Hippocampus from infancy to early adulthood in healthy individuals. PLoS One. 2012;7(10)

70. Adriano F, Caltagirone C, Spalletta G. Hippocampal volume reduction in first-episode and chronic schizophrenia: a review and meta-analysis. Prog Clin Neurosci. 2012;18(2):180–200.
71. Teicher MH, Andersen SL, Polcari A, Anderson CM, Navalta CP, Kim DM. The neurobiological consequences of early stress and childhood maltreatment. Neurosci Biobehav Rev. 2003;27:33–44.
72. Tamminga CA, Stan AD, Wagner AD. The hippocampal formation in schizophrenia. Am J Psychiatry. 2010;10(167):1178–93.
73. Aas M, Dazzan P, Fisher HL, Morgan C, Morgan K, Reichenberg A, et al. Childhood trauma and cognitive function in first-episode affective and non-affective psychosis. Schizophr Res. 2011;129(1):12–9. https://doi.org/10.1016/j.schres.2011.03.017.
74. Shannon C, Douse K, Mccusker C, Feeney L, Barrett S. The association between childhood trauma and memory functioning in schizophrenia. 2011;37(3):531–7.
75. Scott D, Tamminga CA. Effects of genetic and environmental risk for schizophrenia on hippocampal activity and psychosis-like behavior in mice. Behav Brain Res. 2018;339(October 2017):114–23. https://doi.org/10.1016/j.bbr.2017.10.039.
76. Li W, Ghose S, Gleason K, Begovic A, Perez MS. Bartko J, et al, Synaptic proteins in schizophrenia hippocampus indicate increased neuronal activity in CA3. 2015;172(4):373–82.
77. Moser EI, Kropff E, Moser M. Place cells, grid cells, and the brain' s spatial representation system. 2008.
78. Geuze E, Vermetten E, Bremner JD. MR-based in vivo hippocampal volumetrics: 2. Findings in neuropsychiatric disorders. Mol Psychiatry. 2005;10(2):160–84.
79. Derntl B, Habel U, Windischberger C, Robinson S, Kryspin-Exner I, Gur RC, et al. General and specific responsiveness of the amygdala during explicit emotion recognition in females and males. BMC Neurosci. 2009;10:1–14.
80. Zheng J, Anderson KL, Leal SL, Shestyuk A, Gulsen G, Mnatsakanyan L, et al. Amygdala-hippocampal dynamics during salient information processing. Nat Commun. 2017;8:1–11. https://doi.org/10.1038/ncomms14413.
81. Isvoranu A-M, Van Borkulo CD, Boyette L-L, Wigman JTW, Vinkers CH, Borsboom D, et al. A network approach to psychosis : pathways between childhood trauma and psychotic symptoms. Schizophr Bull. 2017;43(1):187–96.
82. Lardinois M, Lataster T, Mengelers R, Van Os J, Myin-Germeys I. Childhood trauma and increased stress sensitivity in psychosis. Acta Psychiatr Scand. 2011;123(1):28–35.
83. Lataster J, Myin-Germeys I, Lieb R, Wittchen HU, van Os J. Adversity and psychosis: a 10-year prospective study investigating synergism between early and recent adversity in psychosis. Acta Psychiatr Scand. 2012;125(5):388–99.
84. Aas M, Navari S, Gibbs A, Mondelli V, Fisher HL, Morgan C, et al. Is there a link between childhood trauma, cognition, and amygdala and hippocampus volume in first-episode psychosis? Schizophr Res. 2012;137(1–3):73–9. https://doi.org/10.1016/j.schres.2012.01.035.
85. Hoy K, Barrett S, Shannon C, Campbell C, Watson D, Rushe T, et al. Childhood trauma and hippocampal and amygdalar volumes in first-episode psychosis. Schizophr Bull. 2012;38(6):1162–9.
86. Suslow T, Lindner C, Dannlowski U, Walhöfer K, Rödiger M, Maisch B, et al. Automatic amygdala response to facial expression in schizophrenia : initial hyperresponsivity followed by hyporesponsivity. BMC Neurosci. 2013;14(140)
87. Mitra R, Jadhav S, McEwen BS, Vyas A, Chattarji S. Stress duration modulates the spatiotemporal patterns of spine formation in the basolateral amygdala. Proc Natl Acad Sci. 2005;102(26):9371–6. https://doi.org/10.1073/pnas.0504011102.
88. Vyas A, Jadhav S, Chattarji S. Prolonged behavioral stress enhances synaptic connectivity in the basolateral amygdala. 2006;143:387–93.
89. Anticevic A, Van Snellenberg JX, Cohen RE, Repovs G, Dowd EC, Barch DM. Amygdala recruitment in schizophrenia in response to aversive emotional material: a meta-analysis of neuroimaging studies. Schizophr Bull. 2010;38(3):608–21.
90. Cancel A, Comte M, Boutet C, Schneider FC, Rousseau P, Boukezzi S, et al. Childhood trauma and emotional processing circuits in schizophrenia : a functional connectivity study. Schizophr Res. 2017;184:69–72. https://doi.org/10.1016/j.schres.2016.12.003.

91. Mukherjee P, Sabharwal A, Kotov R, Szekely A, Parsey R, Barch DM, et al. Disconnection between amygdala and medial prefrontal cortex in psychotic disorders. Schizophr Bull. 2016;42(4):1056–67.
92. Sheffield JM, Williams LE, Woodwads ND, Heckers S. Reduced gray matter volume in psychotic disorder patients with a history of childhood sexual abuse. Schizophr Res. 2013;143(1):185–91.
93. Kumari V, Uddin S, Premkumar P, Young S, Gudjonsson GH, Raghuvanshi S, et al. Lower anterior cingulate volume in seriously violent men with antisocial personality disorder or schizophrenia and a history of childhood abuse. Aust N Z J Psychiatry. 2014;48(2):153–61.
94. Carballedo A, Lisiecka D, Fagan A, Saleh K, Ferguson Y, Connolly G, et al. Early life adversity is associated with brain changes in subjects at family risk for depression. World J Biol Psychiatry. 2012;13(8):569–78.
95. Edmiston EE, Wang F, Mazure CM, Guiney J, Sinha R, Mayes LC, et al. Corticostriatal-limbic gray matter morphology in adolescents with self-reported exposure to childhood maltreatment. Arch Pediatr Adolesc Med. 2011;165(12):1069–77.
96. Gerritsen L, Tendolkar I, Franke B, Vasquez AA, Kooijman S, Buitelaar J, et al. BDNF Val66Met genotype modulates the effect of childhood adversity on subgenual anterior cingulate cortex volume in healthy subjects. Mol Psychiatry. 2012;17(6):597–603.
97. Read J, Fosse R, Moskowitz A, Perry B. The traumagenic neurodevelopmental model of psychosis revisited. Neuropsychiatry. 2014;4(1):65–79.
98. Danese A, Baldwin JR. Hidden wounds? Inflammatory links between childhood trauma and psychopathology. Annu Rev Psychol. 2017;68:517–44.
99. Collip D, Myin-Germeys I, Van Os J. Does the concept of "sensitization" provide a plausible mechanism for the putative link between the environment and schizophrenia? Schizophr Bull. 2008;34(2):220–5.
100. Heim C, Nemeroff CB. The role of childhood trauma in the neurobiology of mood and anxiety disorders: preclinical and clinical studies. Biol Psychiatry. 2001;49(12):1023–39.
101. Rössler W, Ajdacic-Gross V, Rodgers S, Haker H, Müller M. Childhood trauma as a risk factor for the onset of subclinical psychotic experiences: exploring the mediating effect of stress sensitivity in a cross-sectional epidemiological community study. Schizophr Res. 2016;172(1–3):46–53. https://doi.org/10.1016/j.schres.2016.02.006.
102. Collip D, Oorschot M, Thewissen V, Van Os J, Bentall R, Myin-Germeys I. Social world interactions: how company connects to paranoia. Psychol Med. 2011;41(5):911–21.
103. Reininghaus U, Kempton MJ, Valmaggia L, Craig TKJ, Garety P, Onyejiaka A, et al. Stress sensitivity, aberrant salience, and threat anticipation in early psychosis: an experience sampling study. Schizophr Bull. 2016;42(3):712–22.
104. Valmaggia LR, Day FL, Kroll J, Laing J, Byrne M, Fusar-poli P, et al. Bullying victimisation and paranoid ideation in people at ultra high risk for psychosis. Schizophr Res. 2015;168(1–2):68–73. https://doi.org/10.1016/j.schres.2015.08.029.
105. Veling W, Counotte J, Pot-Kolder R, Van Os J, Van Der Gaag M. Childhood trauma, psychosis liability and social stress reactivity: a virtual reality study. Psychol Med. 2016;46(16):3339–48. http://www.ncbi.nlm.nih.gov/pubmed/27619196
106. Gallagher P, Watson S, Smith MS, Young AH, Ferrier IN. Plasma cortisol-dehydroepiandrosterone (DHEA) ratios in schizophrenia and bipolar disorder. 2007;90:258–65.
107. Mondelli V, Dazzan P, Hepgul N, Di Forti M, Aas M, D'Albenzio A, et al. Abnormal cortisol levels during the day and cortisol awakening response in first-episode psychosis: the role of stress and of antipsychotic treatment. Schizophr Res. 2010;116(2–3):234–42. https://doi.org/10.1016/j.schres.2009.08.013.
108. Pruessner JC, Champagne F, Meaney MJ, Dagher A. Dopamine release in response to a psychological stress in humans and its relationship to early life maternal care: a positron emission tomography study using [11 C] Raclopride. 2004;24(11):2825–31.

109. Ritsner M, Gibel A, Maayan R, Ratner Y, Ram E, Modai I, et al. State and trait related predictors of serum cortisol to DHEA(S) molar ratios and hormone concentrations in schizophrenia patients. Eur Neuropsychopharmacol. 2007;17(4):257–64.
110. Pariante CM, Dazzan P, Danese A, Morgan KD, Brudaglio F, Morgan C, et al. Increased pituitary volume in antipsychotic-free and antipsychotic-treated patients of the Æsop first-onset psychosis study. Neuropsychopharmacology. 2005;30(10):1923–31.
111. Pariante CM. Pituitary volume in psychosis: the first review of the evidence. J Psychopharmacol. 2008;22(2 Suppl.):76–81.
112. Read J, Mental A, Board H, Connolly J, Zealand N. The contribution of early traumatic events to schizophrenia in some patients: a traumagenic neurodevelopmental model. Psychiatry: Interpers Biol Process. 2001;64(4)
113. Mondelli V, Dazzan P, Hepgul N, Di Forti M, Aas M, Albenzio D, et al. Europe PMC funders group abnormal cortisol levels during the day and cortisol awakening response in first-episode psychosis: the role of stress and of antipsychotic treatment. Schizophrenia Res. 2012;116:234–42.
114. Yehunda R. Biology of posttraumatic stress disorder. J Clin Psychiatry. 2001;62(2001):2018.
115. Cullen AE, Day FL, Roberts RE, Pariante CM, Laurens KR. Pituitary gland volume and psychosocial stress among children at elevated risk for schizophrenia. Psychol Med. 2015;45(15):3281–92.
116. Pruessner M, Cullen AE, Aas M, Walker EF. Neuroscience and biobehavioral reviews the neural diathesis-stress model of schizophrenia revisited: an update on recent findings considering illness stage and neurobiological and methodological complexities. Neurosci Biobehav Rev. 2017;73:191–218. https://doi.org/10.1016/j.neubiorev.2016.12.013.
117. Belvederi Murri M, Pariante CM, Dazzan P, Hepgul N, Papadopoulos AS, Zunszain P, et al. Hypothalamic-pituitary-adrenal axis and clinical symptoms in first-episode psychosis. Psychoneuroendocrinology. 2012;37(5):629–44. https://doi.org/10.1016/j.psyneuen.2011.08.013.
118. Van Winkel R, Stefanis NC, Myin-Germeys I. Psychosocial stress and psychosis. A review of the neurobiological mechanisms and the evidence for gene-stress interaction. Schizophr Bull. 2008;34(6):1095–105.
119. Meaney MJ, Szyf M. Environmental programming of stress responses through DNA methylation: life at the interface between a dynamic environment and a fixed genome. Dialogues Clin Neurosci. 2005;7(2):103–23.
120. Strous RD, Shoenfeld Y. Schizophrenia, autoimmunity and immune system dysregulation: a comprehensive model updated and revisited. J Autoimmunity. 2006;27:71–80.
121. Commins SP, Borish L, Steinke JW. Immunologic messenger molecules: cytokines, interferons, and chemokines. J Allergy Clin Immunol. 2010;125(2 Suppl. 2):S53–72. https://doi.org/10.1016/j.jaci.2009.07.008.
122. Baumeister D, Akhtar R, Ciufolini S, Pariante CM, Mondelli V. Childhood trauma and adulthood inflammation: a meta-analysis of peripheral C-reactive protein, interleukin-6 and tumour necrosis factor-α. Mol Psychiatry. 2016;21(5):642–9. https://doi.org/10.1038/mp.2015.67.
123. Dennison U, McKernan D, Cryan J, Dinan T. Schizophrenia patients with a history of childhood trauma have a pro-inflammatory phenotype. Psychol Med. 2012;42(9):1865–71.
124. Hepgul N, Pariante CM, Dipasquale S, Diforti M, Taylor H, Marques TR, et al. Childhood maltreatment is associated with increased body mass index and increased C-reactive protein levels in first-episode psychosis patients. Psychol Med. 2012;42(9):1893–901.
125. Di Nicola M, Cattaneo A, Hepgul N, Di Forti M, Aitchison KJ, Janiri L, et al. Serum and gene expression profile of cytokines in first-episode psychosis. Brain Behav Immun. 2013;31:90–5.
126. Coelho R, Viola TW, Walss-Bass C, Brietzke E, Grassi-Oliveira R. Childhood maltreatment and inflammatory markers: a systematic review. Acta Psychiatr Scand. 2014;129(3):180–92.
127. Webster JI, Tonelli L, Sternberg EM. Neuroendocrine regulation of immunity. Annu Rev Immunol. 2002;20(1):125–63.
128. Gracie A, Freeman D, Green S, Garety PA, Kuipers E, Hardy A, et al. The association between traumatic experience, paranoia and hallucinations: a test of the predictions of psychological models. Acta Psychiatr Scand. 2007;116(4):280–9.

129. Morrison AP. A cognitive behavioural perspective on the relationship between childhood trauma and psychosis. Epidemiol Psychiatr Sci. 2009;18(2009):2018.
130. Smith B, Fowler DG, Freeman D, Bebbington P, Bashforth H, Garety P, et al. Emotion and psychosis : links between depression, self-esteem, negative schematic beliefs and delusions and hallucinations. 2006;86:181–8.
131. Selten J, Booij J, Buwalda B, Meyer-lindenberg A. Biological mechanisms whereby social exclusion may contribute to the etiology of psychosis: a narrative review. 2017;43(2):287–92.
132. Selten J, Van Der Ven E, Rutten BPF, Cantor-graae E. The social defeat hypothesis of schizophrenia : an update. 2013;39(6):1180–6.
133. Selten J, Cantor-Graae E. Social defeat: risk factor for schizophrenia? Brit J Psychiatry. 2005;1–2.
134. Birchwood M, Gilbert P, Gilbert J, Trower P, Meaden A, Hay J. Interpersonal and role related scheme influence the relationship with the dominant "voice" in schizophrenia: a comparison of three models. Psychol Med. 2004;34(8):1571–80.
135. Birchwood M, Meaden A, Trower P, Gilbert P, Plaistow J. The power and omnipotence of voices: subordination and entrapment by voices and significant others. Psychol Med. 2000;30(2):337–44.
136. Taylor PJ, Gooding PA, Wood AM, Johnson J, Pratt D, Tarrier N. Defeat and entrapment in schizophrenia : the relationship with suicidal ideation and positive psychotic symptoms. Psychiatry Res. 2010;178(2):244–8. https://doi.org/10.1016/j.psychres.2009.10.015.
137. van Nierop M, van Os J, Gunther N, van Zelst C, de Graaf R, ten Have M, et al. Does social defeat mediate the association between childhood trauma and psychosis? Evidence from the NEMESIS-2 study. Acta Psychiatr Scand. 2014;129(6):467–76.
138. Valmaggia LR, Day F, Garety P, Freeman D, Antley A, Slater M, et al. Social defeat predicts paranoid appraisals in people at high risk for psychosis. Schizophr Res. 2015;168(1–2):16–22. https://doi.org/10.1016/j.schres.2015.07.050.
139. Karlsen S, Nazroo JY, McKenzie K, Bhui K, Weich S. Racism, psychosis and common mental disorder among ethnic minority groups in England SAFFRON. Psychol Med. 2005;35:1795–803.
140. Bartholomew K, Horowitz LM. Attachment styles among young adults: a test of four-category model. J Pers Soc Psychol. 1991;40:30–40.
141. Bowlby J. Attachment and loss, vol. II. Separation: anxiety and anger. London: Penguin Books; 1973.
142. Howard JA. Attachment and trauma. In: Distress or deliberately defiant? Managing challenging student behaviour due to trauma and disorganised attachment. Toowong, QLD: Australian Academic Press; 2013.
143. Korver-Nieberg N, Berry K, Meijer CJ, De Haan L. Adult attachment and psychotic phenomenology in clinical and non-clinical samples: a systematic review. Psychol Psychother Theory Res Pract. 2014;87(2):127–54.
144. Sheinbaum T, Bifulco A, Ballespi S, Mitjavila M, Kwapil TR, Barrentes-Vidal N. Interview investigation of insecure attachment styles as mediators between poor childhood care and Schizophrenia—spectrum phenomenology. PLoS One. 2015;1–12.
145. Sheinbaum T, Kwapil TR, Barrantes-vidal N. Fearful attachment mediates the association of childhood trauma with schizotypy and psychotic-like experiences. Psychiatry Res. 2014;220(1–2):691–3. https://doi.org/10.1016/j.psychres.2014.07.030.
146. Sitko K, Bentall RP, Shevlin M, Sullivan NO, Sellwood W. Associations between speci fi c psychotic symptoms and speci fi c childhood adversities are mediated by attachment styles : an analysis of the National Comorbidity Survey. 2014;217:202–9.
147. Van Dam DS, Velthorst E, Meijer CJ, De Haan L, Risk FG. Childhood maltreatment , adult attachment and psychotic symptomatology : a study in patients , siblings and controls. Soc Psychiatry Psychiatr Epidemiol. 2014:1759–67.
148. Goodall K, Rush R, Grünwald L, Darling S, Tiliopoulos N. Attachment as a partial mediator of the relationship between emotional abuse and schizotypy. Psychiatry Res. 2015;230(2):531–6. https://doi.org/10.1016/j.psychres.2015.09.050.

149. Pearce J, Simpson J, Berry K, Bucci S, Moskowitz A, Varese F. Attachment and dissociation as mediators of the link between childhood trauma and psychotic experiences. Clin Psychol Psychother. 2017;24(6):1304–12.
150. Anketell C, Dorahy MJ, Shannon M, Elder R, Hamilton G, Corry M, et al. An exploratory analysis of voice hearing in chronic PTSD: potential associated mechanisms. J Trauma Dissociation. 2010;11(1):93–107.
151. Kilcommons AM, Morrison AP. Relationships between trauma and psychosis: an exploration of cognitive and dissociative factors. Acta Psychiatr Scand. 2005;112(5):351–9.
152. Holowka DW, King S, Saheb D, Pukall M, Brunet A. Childhood abuse and dissociative symptoms in adult schizophrenia. Schizophr Res. 2003;60(1):87–90.
153. Perona-Garcelán S, García-Montes JM, Francisco J, López-Jiménez AM, Ruiz-Veguilla M, Ductor-Recuerda MJ, et al. Relationship between childhood trauma , mindfulness , and dissociation in subjects with and without hallucination proneness relationship between childhood trauma , mindfulness , and dissociation in subjects with. J Trauma Dissociation. 2014;9732
154. Evans GJ, Reid G, Preston P, Palmier-Claus J, Sellwood W. Trauma and psychosis : the mediating role of self-concept clarity and dissociation. Psychiatry Res. 2015;228:626–32.
155. Bentall RP, De Sousa P, Varese F, Wickham S, Sitko K, Haarmans M, et al. From adversity to psychosis: pathways and mechanisms from specific adversities to specific symptoms. Soc Psychiatry Psychiatr Epidemiol. 2014;49(7):1011–22.
156. Schäfer I, Harfst T, Aderhold V, Briken P, Lehmann M, Moritz S, et al. Childhood trauma and dissociation in female patients. J Nerv Ment Dis. 2006;194(2):135–8.
157. Schäfer I, Fisher HL, Aderhold V, Huber B, Hoffmann-Langer L, Golks D, et al. Dissociative symptoms in patients with schizophrenia: relationships with childhood trauma and psychotic symptoms. Compr Psychiatry. 2012;53(4):364–71. https://doi.org/10.1016/j.comppsych.2011.05.010.
158. Vogel M, Spitzer C, Kuwert P, Möller B, Freyberger HJ, Jörgen H. Association of Childhood Neglect with adult dissociation in schizophrenic inpatients. 2009;124–130.
159. Sar V, Taycan O, Bolat N, Oezmen M, Duran A, Ozturk E, et al. Childhood trauma and dissociation in schizophrenia. Psychopathology. 2010;43:33–40.
160. Longden E, Sampson M, Read J. Childhood adversity and psychosis: generalised or specific effects? Epidemiol Psychiatr Sci. 2016;25(4):349–59.
161. Van Nierop M, Lataster T, Smeets F, Gunther N, Van Zelst C, De Graaf R, et al. Psychopathological mechanisms linking childhood traumatic experiences to risk of psychotic symptoms: analysis of a large, representative population-based sample. Schizophr Bull. 2014;40(Suppl. 2):123–30.
162. Van Nierop M, Lataster T, Smeets F, Gunther N, Van Zelst C, De Graaf R, et al. Psychopathological mechanisms linking childhood traumatic experiences to risk of psychotic symptoms: analysis of a large, representative population-based sample. Schizophr Bull. 2014;40(Suppl. 2):123–30.
163. Green JG, McLaughlin KA, Berglund PA, Gruber MJ, Sampson NA, Zaslavsky AM, et al. Childhood adversities and adult psychopathology in the National Comorbidity Survey Replication (NCS-R) I: associations with first onset of DSM-IV disorders. Arch Genet Psychiatry. 2010;67(2):1–21.
164. Ball JS, Links PS. Borderline personality disorder and childhood trauma: evidence for a causal relationship. 2009.
165. Dean K, Stevens H, Mortensen PB, Murray RM, Walsh E, Pedersen CB. Full Spectrum of psychiatric outcomes among offspring with parental history of Mental disorder, August 2010. Dean et al. 2010;67(8):822–9.
166. Mortensen PB, Pedersen MG, Pedersen CB. Psychiatric family history and schizophrenia risk in Denmark: which mental disorders are relevant? 2018;(2010):201–210.
167. Norman RE, Byambaa M, De R, Butchart A, Scott J, Vos T. The long-term health consequences of child physical abuse, emotional abuse, and neglect: a systematic review and meta-analysis. PLoS Med. 2012;9(11)

168. Teicher MH, Samson JA. Childhood maltreatment and psychopathology: a case for eco-phenotypic variants as clinically and neurobiologically distinct subtypes. Am J Psychiatry. 2013;170(10):1114–33.
169. Van Nierop M, Bak M, De Graaf R, Ten Have M, Van Dorsselaer S, Van Winkel R, et al. The functional and clinical relevance of childhood trauma-related admixture of affective, anxious and psychosis symptoms. 2016;91–101.
170. Wigman JTW, Van Winkel R, Ormel J, Verhulst FC, van Os J, Vollebergh WAM. Early trauma and familial risk in the development of the extended psychosis phenotype in adolescence. 2012;266–273.
171. Caspi A, Houts RM, Belsky DW, Goldman-mellor SJ. The p factor: one general psychopathology factor in the. Clin Psychol Sci. 2015;2(2):119–37.
172. Álvarez M-J, Roura P, Osés A, Foguet Q, Solà J, Arrufat F-X. Prevalence and clinical impact of childhood trauma in patients with severe Mental disorders. J Nerv Ment Dis. 2011;199(3):156–61.
173. Conus P, Cotton S, Bg S, Berk M, Daglas R. Pretreatment and outcome correlates of past sexual and physical trauma in 118 bipolar I disorder patients with a first episode of psychotic mania. Bipolar Disord. 2010;1:244–52.
174. Gil A, Gama CS, de Jesus DR, Lobato MI, Zimmer M, Belmonte-de-Abreu P. The association of child abuse and neglect with adult disability in schizophrenia and the prominent role of physical neglect. Child Abus Negl. 2009;33(9):618–24.
175. Hodgins S, Lincoln T, Mak T. Experiences of victimisation and depression are associated with community functioning among men with schizophrenia. Soc Psychiatry Psychiatr Epidemiol. 2009;44(6):448–57.
176. Gayer-Anderson C, Fisher HL, Fearon P, Hutchinson G, Morgan K, Dazzan P, et al. Gender differences in the association between childhood physical and sexual abuse, social support and psychosis. Soc Psychiatry Psychiatr Epidemiol. 2015;50(10):1489–500. https://doi.org/10.1007/s00127-015-1058-6.
177. Frueh BC, Grubaugh AL, Cusack KJ, Kimble MO, Elhai JD, Knapp RG. Exposure-based cognitive behavioral treatment of PTSD in adults with schizophrenia or schizoaffective disorder: a pilot study. J Anxiety Disord. 2010;23(5):665–75.
178. Morgan C, Fisher H. Environment and schizophrenia: environmental factors in schizophrenia: childhood trauma—a critical review. Schizophr Bull. 2007;33(1):3–10.
179. Trappler B, Newville ÆH. Trauma healing via cognitive behavior therapy in chronically hospitalized patients. 2007;317–325.
180. Garety PA, Bebbington P, Fowler D, Freeman D, Kuipers E. Implications for neurobiological research of cognitive models of psychosis: a theoretical paper. Psychol Med. 2007;37(10):1377–91.
181. Howes OD, Mcdonald C, Cannon M, Arseneault L, Boydell J, Murray RM. Pathways to schizophrenia : the impact of environmental factors. 2004;7(August)

# PTSD During Childhood, Childhood Trauma, Childhood Maltreatment and How They Relate to Adult PTSD

11

Dominik Andreas Moser, Francesca Suardi, and Daniel Scott Schechter

## 11.1 Introduction

There is no form of psychopathology that is more directly linked to traumatic psychological experiences than that of *post-traumatic stress disorder* (PTSD). After all, by its very definition, one cannot be diagnosed with PTSD without having had a traumatic experience, that is an experience perceived as being threatening to life or health of oneself or someone to whom one is attached. The inverse, however, is not true: the majority of people who undergo an experience that is threatening to life or health of the self or an attachment figure will not develop PTSD [1–3]. The prevalence of PTSD is estimated to be about 3% lifetime, with estimates reaching up to 8% or more, in some communities [4]. Types of traumatic experiences that can lead to PTSD vary across a wide spectrum and include natural disasters, vehicle accidents, medical events that are lived as

D. A. Moser (✉)
Institute of Psychology, University of Bern, Bern, Switzerland

University Service of Child and Adolescent Psychiatry, Lausanne University Hospital, Lausanne, Switzerland
e-mail: domamoser@gmail.com

F. Suardi
Association VIRES, Center for Psychotherapy, Research and Prevention of Violence, Geneva, Switzerland

Relais Enfants Parents Romands (REPR), Geneva, Switzerland
e-mail: francesca.suardi@gmail.com

D. S. Schechter
University Service of Child and Adolescent Psychiatry, Lausanne University Hospital, Lausanne, Switzerland

Department of Child and Adolescent Psychiatry, New York University Grossman School of Medicine, New York, NY, USA

Department of Psychiatry, University of Geneva Faculty of Medicine, Geneva, Switzerland
e-mail: Daniel.Schechter@chuv.ch



G. Spalletta et al. (eds.), *Childhood Trauma in Mental Disorders*,
https://doi.org/10.1007/978-3-030-49414-8_11

potentially life-threatening, and also war, terrorism and domestic violent experiences, as well as other attacks to the physical, sexual (and psychological) integrity of the individual. Among these potential causes, events of interpersonal violence appear to put their victims at the highest risk of developing PTSD [5]. Symptoms for PTSD fall into four general categories and need to be persistent for more than a month to satisfy the Diagnostic and Statistical Manual of Mental Disorders-Fifth Edition (DSM 5, [6]) criteria for diagnosis: (1) intrusions of the traumatic event into the current life, such as flashbacks, nightmares, intrusive memories and high stress when faced with reminders of the event; (2) avoidance of anything related to the traumatic event; (3) negative emotionality and cognition, which can take a wide variety of forms including reduced self-esteem and trust, persistent negative emotionality and motivation; (4) hyperarousal and hypervigilance for anything that may be a reminder of the trauma, and also problems with sleep and concentration. Additionally, dissociative symptoms can be part of PTSD, such as selective amnesia, and altered experience of reality in response to traumatic reminders. Because of the frequency of these symptoms among individuals suffering from PTSD, a dissociative subtype of PTSD has been identified [6].

Childhood trauma can influence PTSD during adulthood in several ways: (1) PTSD can be directly caused by childhood trauma. This happens primarily in two ways: (1a) as a consequence of childhood trauma, PTSD as a pathology can already be developed during childhood (paediatric PTSD), and persist into adulthood; (1b) patients may live through childhood trauma and experience PTSD onset later on once they are adults; (2) similarly to other psychopathologies, childhood trauma can indirectly increase the risk of later developing PTSD. This too can take several avenues, which can overlap being intertwined in complex ways. The most commonly discussed ways childhood trauma can increase the risk for later PTSD include: (2a) an increase in the number of cumulative traumatic events over time from birth on, increasing the chances of developing PTSD particularly when affecting the organism during formative critical and sensitive developmental periods; (2b) by altering psychological processes and environmental factors (both through environmental factors chosen by the patient as well as those that are imposed on the patient) (2c) altered psychobiology including neural, hormonal and (epi)genetic changes resulting form the interaction with other changes directly resulting from childhood trauma. It is important to clarify that this chapter will consider childhood trauma and child maltreatment including neglect and abuse in combination, given that their combined effects are sometimes hard to disentangle in the current literature.

## 11.2 PTSD in Adulthood, as a Disorder Developed During Childhood

There is less scientific evidence and knowledge about how PTSD expresses in infants and young children compared to what is known in adolescents and adults, due to the fact that the research focus on PTSD during the earlier life period is recent [7, 8]. To wit, a PTSD subtype for children below 6 years was introduced only in the latest edition of the Diagnostic Statistical Manual of Mental Disorders [6]. Diagnosis of PTSD depends on the ability to ascribe causes of symptoms to traumatic events, which can be difficult in young children with absent or limited ability to verbalize their emotions and experiences. This difficulty may potentially lead to abused children receiving different diagnosis, such as anxiety disorder or disruptive behaviour disorder. Still, there is clear evidence that traumatic events during childhood can cause PTSD, which can be assessed and diagnosed as early as 2 years of age [9, 10]. The course and duration of PTSD beginning during childhood is less studied. For example, compared to other traumatic causes of PTSD, there are few empirical studies looking at symptoms remission rates following childhood-onset trauma [11]. One Australian study found that for childhood PTSD due to sexual abuse, the average duration to remission was 11.4 years, with considerable variability even though patients had not yet reached their thirties [12]. The possibility that untreated PTSD could create a risk for other comorbidities such as mood and anxiety disorder, substance abuse and personality disorders remains also to be empirically investigated more in depth. There is a large variability in the duration, course and comorbidity of PTSD developed during childhood. An example of how childhood-onset PTSD can endure into adulthood was demonstrated in a study that investigated the consequences of institutional abuse in foster care among a population of older adults about 40 years after the abuse [13]. In this study, even though other events may have caused PTSD in that group, a PTSD rate of 35% was found. This rate is notable if we consider that the rate in an age-matched comparison group without childhood trauma is less than 1% [13].

A meta-analysis showed that childhood trauma itself, even in the absence of related psychopathology measurements, is associated with impairments across a number of cognitive and perceptual domains. These domains include intelligence (as measured by IQ testing), language and visuospatial processing, working memory and executive skills. Children who develop PTSD as a consequence of trauma or maltreatment presented even lower performance scores in these cognitive domains, compared to children who experienced trauma or maltreatment, but did not subsequently develop PTSD [14]. A recent study on more than 2.000 children in England and Wales indicated that not only more than 30% of children experienced trauma exposure, but also that 7.8% of them had experienced some form of PTSD by age 18. Given the general tendency for an overlap among psychopathologies, it is unsurprising that youths who had experienced PTSD were more likely to be diagnosed with other comorbid forms of psychopathology, such as depression, suicidal and para-suicidal behaviour, alcohol and other substance abuse, social anxiety and disruptive behaviour disorders involving impulsivity and violent behaviour [15]. Taking into account the large range of behaviours clinically expressed, PTSD can be considered as an alteration/dysregulation of physiology and emotion regulation.

## 11.3 Childhood Trauma as a Risk Factor Due to Accumulation of Traumatic Experiences

A model often applied to the understanding of PTSD is the *dose-response theory*. This theory posits that the severity and likelihood of developing PTSD increases as the number and severity of traumatic experiences increase [16]. Despite the difficulty of precisely defining and quantifying events and their severity in traumatized populations, a number of studies suggest that the number and types of traumatic events is indeed linked to a higher likelihood of developing PTSD [17]. Unsurprisingly, a fairly sizeable population of PTSD patients did not experience only one traumatic event, but was rather exposed to multiple events, all of which can contribute to PTSD symptomatology [18, 19]. Repeated exposure (linked to duration and enduring feeling of helplessness felt by the subject) also affects the expression of PTSD, and when occurring during formative early development can greatly increase the risk for mental and physical diseases [20, 21]. Taking into account this growing evidence the newest edition of the International Classification of Diagnoses (ICD-11) included complex or developmental PTSD [22] as a subtype of the original PTSD category [23]. According to this approach, exposure to childhood trauma, increases the likelihood of vulnerability to psychiatric and physical morbidity above and beyond PTSD. Quite often this vulnerability is also associated to subsequent additional trauma exposure [24]. From a biological perspective, in a vicious cycle, additional traumatic exposure may then easily bring the already primed neurophysiological system to the point of developing or reigniting post-traumatic stress among other forms of psychopathology, including substance use. That being said, it seems likely that the mechanisms through which this increased vulnerability develops are rooted in psychobiological traces that interact with environmental factors. The lenses of child development and attachment help to see more clearly mechanisms of increased risk that are linked to this process as further explored in the following section.

## 11.4 Psychological Processes and Environmental Factors as a Consequence of Childhood Trauma and How They Relate to PTSD

Among potentially traumatic experiences, some are statistically more likely to be pathogenic for the development of PTSD and later re-victimization than others. For example, childhood abuse and maltreatment are associated with increased risk of subsequent re-victimization (i.e. physical or sexual assault) [25]. Rape or otherwise being coerced into unwanted sex as an adult, for example, is roughly twice as likely to happen to victims of child maltreatment and is itself, highly traumatogenic [26]. Particularly, participants who experienced multiple forms of abuse and neglect are at risk for being subjected to interpersonal violence [27], with sexual abuse being especially important in that regard as an unfortunate precedent [13, 28–30]. Victims of childhood sexual abuse, in particular, have been found to have a greater number of, and more severe PTSD symptoms when being revictimized during adulthood than other PTSD patients [31].

Child abuse indirectly increases the risk both for further traumatization and for the development of PTSD during adulthood probably by increasing the individual's allostatic load, defined as a stress-related long-term strain on brain and body [32]. Evidence points to such an increase in allostatic load being due to childhood trauma, which likely sets a psychobiological precedent for subsequent abusive relationships

[33]. Childhood trauma further complicates close relationships, particularly when childhood trauma is linked to behaviors perpetrated by a primary caregiver. Such caregiver related trauma may be setting the stage for future intense attachments, this through the formation of disorganized, insecure attachment and heightened dependence on non-abusing or violent caregivers [34]. The precise causes for this association between childhood adversity and subsequent morbidity are, however, often difficult to determine. In fact, risk factors are often inter-correlated and their respective directions of effect are unclear. For example, it is common to observe lower socio-economic status (SES) among traumatized individuals and yet trauma-associated psychopathology can undermine the subject's ability to achieve satisfactory employment status, thereby lowering the SES [35, 36]. However, directionality of causation can also be reversed, as reduced SES can also increase the risk for revictimization in multiple ways, including living in neighbourhoods with increased crime, and not having sufficient personal living space to avoid conflicts within a family [37].

Interpersonal violence and related PTSD, which can pass from one generation to another within families—albeit through mechanisms as yet to be elucidated-, might be transmitted at least in part, via assortative mating [38]. Assortative mating refers to the tendency of partner selection with a preference for those who hail from similar cultural, socio-economic areas as the subject, also sharing other psychosocial and phenotypical similarities. Similarly, the scientific literature shows a correlative relationship between childhood abuse and later violence exposure, but the direction of causality is, again, difficult to discern. This is true for several types of trauma exposure, as victims of childhood trauma and children who witnessed intimate partner violence find themselves more often in violent relationships, both as perpetrators and victims [39]. Child physical abuse, clearly associated with prosocial deficits and social cognitive impairments, was even suggested to be correlated with increased autistic traits in intimate relationships in one study [40].

The fact that reduced capacities for learning, memory, executive functioning, language and even lower intelligence are related to early childhood adversities and across child development is equally difficult to interpret [41–45]. One might hypothesize that both reduced cognitive and social-emotional capacities may impair situational appraisal and decision-making in such a way that increases the probability for interpersonal violence, thus contributing to a vicious cycle of violence within high-risk populations.

One potential interpretation of altered decision-making in abused subjects is that the image of "normalcy" in family and relationships may be altered by the experience of witnessing interpersonal violence during childhood. In this context, the concept of child–parent attachment [46] is a key notion to consider since it refers to the process of developing a secure base for the infant, which permits protection for survival and exploration of the environment. This would ideally lead to the internalization of a secure attachment model as a way of being in interaction with others in life. When the caregiver(s) cannot provide protection and/or use their affective communication to help regulating the child's emotions in order to sustain his/her development, the child is at risk for psychopathology [47]. Related to this, attachment disorganization represents a risk factor for later development of psychopathology

[48]. Disrupted parental affective communication is a predictor of later dissociation and potentially disorganized attachment [49, 50]. Given that dissociative symptoms tend to be most prominent among adults who experienced child maltreatment [51], this seems of particular importance for the dissociative subtype of PTSD. This is particularly true for victims of sexual and physical abuse. In this context, PTSD patients who have experienced early life trauma need specific interventions, and this seems to be particularly crucial for the dissociative subtype of PTSD.

At the same time, increased risk for dissociative symptoms is not specific to adults. Following sexual abuse, dissociation already strongly mediates how childhood abuse affects severity and type of psychopathological symptom expression in young children [52].

## 11.5 Increased Risk for Adult PTSD Due to Psychobiologic Changes as a Result of Childhood Trauma

Research on PTSD has identified a number of objectively measured and validated indicators or "biomarkers" of normative biological processes or pathogenic processes that are associated with the development of the disorder. A partial list of these markers includes the following domains: brain structure, brain function, neurophysiology, neurochemistry changes and (epi) genetic changes [53, 54]. However, while there are numerous biomarkers that are associated with PTSD, none appear to be specific to PTSD alone [53–55]. Some of these biomarkers may represent consequences of PTSD and/or represent risk factors for the development of PTSD [55–57], thus rendering the possibility of making any causal interpretation highly unlikely, if not impossible.

Although not without some contradictory findings, the majority of available evidence on brain structure points towards childhood trauma and paediatric PTSD—as well as adult PTSD [58]—being linked to reduced hippocampal and amygdala volumes, which probably stem from a process that originates very early during child development with possible precursors during the prenatal period [59]. This non-uniformity in the literature may be due to either a lack of power or to differential effects of abuse and neglect as a function of both age and sex. Alternatively, a delayed effect of childhood abuse on development may explain heterogeneity in findings.

Two brain structures have been the focus of research in both human studies and animal model studies of development under stress, namely the hippocampus and the amygdala. Childhood trauma has a deleterious effect on the hippocampus, and this may be due to physiological changes. Animal models support the hypothesis that excessive glucocorticoid output as a consequence of stress during sensitive early developmental periods is a likely central mechanism for these findings [60, 61]. It has been postulated that similarly, childhood trauma puts individuals, who end up with lower cortisol levels and decreased cortisol response to stressors, at increased risk for developing chronic post-traumatic stress reactions [62]. There may also be sex-dependent effects on how the type of trauma affects the hippocampus and the amygdala. A recent study indicated that hippocampal volumes of young adult men were predicted by neglect, but not by abuse during young childhood; while for women abuse during puberty/adolescence rather than neglect was the primary predictor [63]. The latter requires further study. The suggestion that preserved hippocampal volume may be bolstering resilience and/or recovery to and from PTSD in adults is at the origin of the high interest in how the hippocampal structure is linked to childhood trauma [64, 65]. The subsequent development of PTSD could be a direct consequence of biological reactions to childhood trauma and/or of impairments in cognitive abilities linked to hippocampal function. Impaired hippocampal functioning, for example, likely makes the proper contextual processing and later recall of traumatic events difficult [66–68]. Testing these hypotheses will necessitate further translational research with animal and human models. The hippocampus and amygdala are not the only brain structures that have been implicated in post-traumatic stress responses to childhood trauma. Two meta-analyses indicate that there are reductions of total brain grey matter linked to paediatric PTSD [69, 70].

Brain function and connectivity, in addition to brain structure, has been linked to adult PTSD. This is true for multiple areas of the brain, including the hippocampus, the amygdala, the insula, the anterior cingulate cortex as well as the ventromedial prefrontal cortex [70–72]. The brain functional network most affected in adult PTSD is the frontal-limbic system, which is responsible for control and expression of emotions, with negative emotions such as fear and anger being particularly affected. Changes in functional connectivity networks (i.e. networks of brain regions that usually co-activate at the same time) have also been linked to PTSD, with increased activity in the salience network (i.e. brain regions that react to information novelty) and decreased anti-correlation between the salience network and the (self-referential) default mode network, as well as the central executive network. The central executive network comprises brain regions that support planning, regulating and strategizing behavior, functions that bear upon PTSD symptom expression and impairment [73].

Beyond the observation that childhood trauma affects brain structure and function, several studies have demonstrated that different types of traumatic events occurring at different times during formative brain development can exert very specific effects. Maladaptive threat-processing, inadequate processing and contextualizing of aversive information and internal states are all aspects of PTSD symptomatology that are addressed by different forms of psychotherapy. The experience of child abuse and exposure to interpersonal violence can directly interfere with developmental processes in the brain that otherwise lead to greater resilience to psychopathology including, of course, PTSD. This is particularly the case when traumatic experiences occur early during formative development of brain structures important for learning, memory and socialization.

Apart from hippocampus, amygdala volume appears most affected by childhood abuse at 10–11 years of age [74], while this is true for prefrontal cortical volume between 14 and 16 years [75]. As noted by Teicher and colleagues, the periods where childhood trauma and abuse appear to have the most predictive effect on volume coincide with periods of maturation in these brain regions [76]. The structure of white matter bundles that serve to connect different areas of the brain has been shown to differ among PTSD patients depending on whether or not they specifically experienced child sexual abuse. Adolescents at 16 years of age who had PTSD and a history of sexual abuse showed macro-structural abnormalities in the corpus callosum compared to PTSD patients who not experienced abuse [77]. The corpus callosum is a particularly important region in this context as it connects some of the most important brain regions involved in PTSD (i.e. the anterior cingulate, and the ventromedial prefrontal cortex and key limbic areas such as the amygdala). In sum, childhood trauma likely affects the developmental trajectory of many brain regions that are linked to PTSD diagnosis during adulthood. Yet, specific effects of trauma on the developing brain are differentially dependent on the type and timing of trauma.

Childhood maltreatment has been linked to changes in an additional network fairly overlapping with that linked to adult PTSD. This additional network is marked by altered activation patterns of the insula and activation of the amygdala in response to emotional faces [78] and reduced connectivity between the salience network and the default mode network [79]. While the experience of child abuse without PTSD has been linked to changes in amygdala activation, such a link has not (yet) been shown in samples of children or adolescents with a PTSD diagnosis, possibly due to lack of studies with adequate sample sizes (see [80–83], and also [84]).

Overall, the experience of childhood abuse affects (1) the brain's threat processing ability, as indicated by changes in the amygdala-prefrontal connectivity, (2) the brain's ability to contextualize aversive information, as indicated by changes in hippocampal volume [59] and (3) the ability for adequate processing of internal states and external aversive stimuli, as indicated by altered insula activation and default mode connectivity [76].

Given that PTSD—by definition—necessitates a highly threatening post-birth life-event to have occurred, classical genetic studies are at disadvantage when researching PTSD as genetics can by definition, not be the primary cause of the pathology. However, genetic studies involving PTSD have found numerous genes to be associated with PTSD; but these genes generally strongly overlap with those of other forms of psychopathology, including both major depressive disorder and schizophrenia [85, 86]. PTSD-related epigenetic research—studying how genes are expressed—could be instructive. One of the more interesting studies on childhood abuse and epigenetics [87] compared 32 PTSD patients who had childhood abuse experience, with 29 subjects who had an overall similar level of trauma, but no childhood abuse, while controlling for age, sex, ethnicity and substance abuse. That study indicated that numerous genetic transcripts in peripheral blood were differentially methylated when PTSD patients with a history of childhood abuse and PTSD patients without such a history were compared. These differences in methylation appear to have particularly strong links to gene networks that are responsible for enrichment of the central nervous system, as well as pathways for tolerance induction among PTSD patients with a history of childhood abuse. Cell death and cell growth rates appeared to be increased among PTSD patients without childhood abuse

[87]. One could infer from this finding that some of the changes in brain structure and function that place abused children at heightened risk for later PTSD may be due to epigenetic changes. Further studies along these lines with adequate sample sizes are needed.

## 11.6 Intergenerational Transmission of Childhood Trauma to PTSD and Other Forms of Psychopathology

Transmission of violence from one generation to the next has been a research topic akin to that of PTSD [88]. A recent study from Croatia showed that parents who reported being slapped or hit as children were also more likely to report doing so to their own children [89]. Childhood trauma and neglect, however, beyond its relationship to later adult PTSD, may also be related to the intergenerational transmission of psychopathology to the next generation, possibly even without children directly experiencing trauma themselves [90–92]. A growing number of studies show how maternal and paternal PTSD can adversely affect the caregiving environment with effects on the parent–infant relationship and its implicit regulatory components, among others [93–96].

Beyond parent-to-child violence, multiple studies have addressed intergenerational transmission of intimate partner violence (and related psychopathology) with an increasing amount of research focusing on non-western study samples [25, 97–103]. While social learning from an abusive or violent parent or identifying with a caregiver-aggressor may well be a mechanism of intergenerational transmission, possibly involving the hippocampus among other brain regions [104], this phenomenon is likely multi-determined.

An (epi)genetic effect may well be another mechanism that could contribute to intergenerational transmission. This view is based on studies that have indicated that children of trauma survivors have altered epigenetic regulation of glucocorticoid receptors and moderators [105, 106]. A number of studies on maternal brain activation and epigenetic markers related to glucocorticoids, serotonin and brain-derived neurotropic factor have also indicated that maternal brain function is possibly linked to epigenetic markers. These markers, in turn, were then linked to brain activation patterns of the frontal-limbic circuit, which activation patterns are significantly associated with altered parental behaviour [107–109]. In other words, one could take from these findings that traumatic experiences and associated PTSD very possibly alter peripherally derived epigenetic markers that signal changes in maternal psychobiology.

In a physiological perspective of PTSD, the peripheral stress hormone cortisol was one of the first biomarkers identified as implicated in the transmission of violence and related PTSD across generations. Maternal PTSD has been associated with alterations of several different cortisol measures among their offspring [55, 62], including measurement of the diurnal rhythm curve [110]. Thus, dysregulation of circadian rhythms and cortisol reactivity may well be related to increased risk of developing PTSD following trauma [62]. Additionally, these significant group differences and associations with respect to cortisol measures may be related to altered appraisal of emotions and caregiving behaviour among traumatized mothers [110]. In sum, physiological, brain function and structure, and epigenetic changes are possibly part of a more complex interaction between biological and psychological factors, which can affect maternal behaviour in a way that may unfavourably alter the caregiving environment.

While these complex biological associations require further translational testing using animal models and well-controlled prospective longitudinal human studies, the empirical evidence in both animal and human models shows that the relational experience of the child with the traumatized caregiver plays a crucial role in child's development [48, 111, 112]. Evidence exists on how trauma can impact more directly parental behaviours, particularly regarding mutual regulation of emotion, and the child's social learning from the parent's interpersonal behaviour. Both mother–child interaction and the caregiver's capacity for mentalization are two key processes to take into account when considering risk and protective factors with regard to intergenerational transmission of violence and PTSD [113]. Mentalization, which is considered to reflect the quality of an individual's early attachment(s), is the ability to infer and wonder about mental states and about their link to behaviour in self and others, and contributes to mutual emotion regulation [114, 115]. Parental reflective functioning is a specific operationalization of mentalization and refers to the quantifiable measure of a parent's capacity to recognize mental states in oneself as well as one's child and relate these states to each other's behaviour [116]. In this way parent–child interactions, supported by parental mentalizing capacity, could be considered another psychologically and behaviourally oriented pathway of intergenerational transmission of psychopathology.

Social context and social experiences also play an important role in intergenerational transmission of PTSD. An example of this form of transmission might be elucidated by a study that indicated that higher PTSD and dissociative symptoms among traumatized mothers are associated with greater exposure to violent media among their young children [117]. Furthermore, the fact that victims of childhood trauma are more likely to find themselves in violent relationships as adults [39], can affect the following generation, even if further direct trauma experiences can be avoided. Even in the absence of child abuse, having witnessed interparental violence as children increased the likelihood of those children later having abusive partnerships [118].

Biological, psychological, social and behaviour-oriented pathways are all likely to interact and can be equally effective in transmitting psychopathology to the next generation. An illustration of how these pathways can be intertwined comes from an fMRI study on parental brain functioning in mothers with and without PTSD. Results of this study showed enhanced brain activation in the anterior cingulate cortex when seeing other adults emotionally interacting with each other. The study also demonstrated that this brain activation correlated with the mothers' capacity to be appropriately sensitive when interacting with their toddlers, independently of whether the mothers suffered from PTSD or not [119]. Further research is needed to elucidate how the brain and social context interact in this complex model of PTSD transmission.

Given the multiple pathways of intergenerational transmission of psychopathology, the parental impact needs to be considered for childhood trauma intervention. In this sense it has to be considered if (1) the parent has experienced traumatic events and (2) if a young child experiences a potentially traumatic event. Such events more often necessitate psychological interventions taking into account the caregiver-child relationships in particular situations, both individually and as a family. An important aspect of such interventions involves the understanding of the multiple psychobiological effects of trauma and how they might affect parents and their child, their relationship and relationships in subsequent generations. Intervention during infancy and early childhood that addresses parental psychopathology, individual developmental needs of the child and the parent–child relationship (i.e. through assessment of parental mental representations, reflective functioning and parent–child interactive behaviour) could increase resilience and reduce risk for the development of subsequent post-traumatic psychopathology. The practitioners themselves should also be highly motivated, informed and comfortable enough to screen for a history of trauma and the presence of PTSD and then be able to talk about intervention. In this regard, paediatricians play a crucial role, since they perform both physical, developmental and mental health screening on children, in order to decide whether there is a need for intervention. Generally, it is the paediatrician's or family practitioner's role to address the parents and motivate them to seek help for the child and/or for themselves as parents.

The effectiveness of existing psychological interventions for PTSD has been proved for multiple approaches including cognitive behavioural therapy and prolonged exposure therapy [120, 121], interpersonal psychotherapy [122], for younger children and parents, psychodynamically oriented child–parent psychotherapy [123] and most recently, for both adults and older children, Eye Movement Desensitization and Reprocessing or "EMDR" [124] with adaptation for complex and relational PTSD [125]. Particularly the latter has been shown to be especially effective in PTSD adult and paediatric patients [126]. Clinical intervention for treatment of complex PTSD in adults that pays attention to the physical body and its connection to the psyche during therapeutic process also exist, and elements of this approach are often integrated into treatments that are primarily based on other theoretical orientation [127]. Taking into account the existing evidence of the parent's PTSD impact on the child during development, parent-child interventions are a key process in prevention of the intergenerational transmission of PTSD. Parent–child interventions with a focus on the impact of parental trauma on the parent–child relationship are being developed [128–130].

Furthermore, understanding whether or not parents have themselves experienced trauma during childhood and later developed PTSD should always be a clinical question when evaluating parents and their children. If the parent had difficult childhood experiences involving trauma and/or problematic early attachment(s), then the parent is particularly likely to benefit from interventions supporting the transition to parenthood. Giving help to the parent represents a first step towards the goal of prevention of child symptomatology [131, 132].

## 11.7 Discussion

The consequences of childhood trauma on PTSD can be both short- and long term. On one end of the spectrum, PTSD may follow from a childhood incident that can leave enduring traces well into adulthood. On the other end of the spectrum, PTSD symptoms can last only weeks to months and remit completely. Beyond that, PTSD as a consequence of trauma may contribute to transmission of psychopathology to the next generation. Such variability in disease onset, course, trauma type and severity makes very difficult to properly track the consequences of childhood trauma on PTSD. This difficulty is made even worse by the frequent use of retrospective measures to assess childhood trauma when investigating adult PTSD patients. This is problematic because altered memory function in relation to traumatic events can actually be a symptom of PTSD. Furthermore, recollection of emotional events decades later is notoriously unreliable, and the same event may be encoded very differently by different individuals.

In conclusion, more prospective, longitudinal studies, tracking children into adulthood are needed, but face the challenge of having to account for both present and future courses of psychopathology, related risk factors and biomarkers. To achieve this, and to fully profit from the considerable acquired knowledge as to translate findings into clinical practice, experienced clinicians in collaboration with researchers, would benefit from the development of models that integrate both psychological and biological concepts and their respective complexities and trajectories. At present, longitudinal studies face the challenge of an increased need for guaranteed long-term financial and institutional stability, a big challenge in the current scientific climate in which funding agencies are less frequently willing to assure long-term financing beyond 4 years.

## References

1. Tang B, Deng Q, Glik D, Dong J, Zhang L. A meta-analysis of risk factors for post-traumatic stress disorder (PTSD) in adults and children after earthquakes. Int J Environ Res Public Health. 2017;14(12)
2. Karstoft KI, Galatzer-Levy IR, Statnikov A, Li Z, Shalev AY, members of Jerusalem Trauma O, et al. Bridging a translational gap: using machine learning to improve the prediction of PTSD. BMC Psychiatry. 2015;15:30.
3. Xue C, Ge Y, Tang B, Liu Y, Kang P, Wang M, et al. A meta-analysis of risk factors for combat-related PTSD among military personnel and veterans. PLoS One. 2015;10(3):e0120270.
4. Bisson JI, Cosgrove S, Lewis C, Robert NP. Post-traumatic stress disorder. BMJ. 2015;351:h6161.

5. White J, Pearce J, Morrison S, Dunstan F, Bisson JI, Fone DL. Risk of post-traumatic stress disorder following traumatic events in a community sample. Epidemiol Psychiatr Sci. 2015;24(3):249–57.
6. American Psychiatric Association. Diagnostic and statistical manual of mental disorders. 5th ed. Arlington, VA: American Psychiatric Publishing; 2013.
7. Alisic E, Zalta AK, van Wesel F, Larsen SE, Hafstad GS, Hassanpour K, et al. Rates of post-traumatic stress disorder in trauma-exposed children and adolescents: meta-analysis. Br J Psychiatry. 2014;204:335–40.
8. Scheeringa MS, Zeanah CH, Myers L, Putnam FW. New findings on alternative criteria for PTSD in preschool children. J Am Acad Child Adolesc Psychiatry. 2003;42(5):561–70.
9. Scheeringa MS, Peebles CD, Cook CA, Zeanah CH. Toward establishing procedural, criterion, and discriminant validity for PTSD in early childhood. J Am Acad Child Adolesc Psychiatry. 2001;40(1):52–60.
10. Egger HL, Emde RN. Developmentally sensitive diagnostic criteria for mental health disorders in early childhood: the diagnostic and statistical manual of mental disorders-IV, the research diagnostic criteria-preschool age, and the diagnostic classification of mental health and developmental disorders of infancy and early childhood-revised. Am Psychol. 2011;66(2):95–106.
11. Morina N, Wicherts JM, Lobbrecht J, Priebe S. Remission from post-traumatic stress disorder in adults: a systematic review and meta-analysis of long term outcome studies. Clin Psychol Rev. 2014;34(3):249–55.
12. Gospodarevskaya E. Post-traumatic stress disorder and quality of life in sexually abused Australian children. J Child Sex Abus. 2013;22(3):277–96.
13. Lueger-Schuster B, Knefel M, Gluck TM, Jagsch R, Kantor V, Weindl D. Child abuse and neglect in institutional settings, cumulative lifetime traumatization, and psychopathological long-term correlates in adult survivors: the Vienna institutional abuse study. Child Abuse Negl. 2018;76:488–501.
14. Malarbi S, Abu-Rayya HM, Muscara F, Stargatt R. Neuropsychological functioning of childhood trauma and post-traumatic stress disorder: a meta-analysis. Neurosci Biobehav Rev. 2017;72:68–86.
15. Lewis SJ, Arseneault L, Caspi A, Fisher HL, Matthews T, Moffitt TE, et al. The epidemiology of trauma and post-traumatic stress disorder in a representative cohort of young people in England and Wales. Lancet Psychiatry. 2019;6(3):247–56.
16. Kolassa IT, Ertl V, Eckart C, Kolassa S, Onyut LP, Elbert T. Spontaneous remission from PTSD depends on the number of traumatic event types experienced. Psychol Trauma-Us. 2010;2(3):169–74.
17. Wilker S, Pfeiffer A, Kolassa S, Koslowski D, Elbert T, Kolassa IT. How to quantify exposure to traumatic stress? Reliability and predictive validity of measures for cumulative trauma exposure in a post-conflict population. Eur J Psychotraumatol. 2015;6:28306.
18. Wolf EJ, Miller MW, Kilpatrick D, Resnick HS, Badour CL, Marx BP, et al. ICD-11 complex PTSD in US national and veteran samples: prevalence and structural associations with PTSD. Clin Psychol Sci. 2015;3(2):215–29.
19. Maercker A, Hecker T, Augsburger M, Kliem S. ICD-11 prevalence rates of posttraumatic stress disorder and complex posttraumatic stress disorder in a German Nationwide sample. J Nerv Ment Dis. 2018;206(4):270–6.
20. Dube SR, Fairweather D, Pearson WS, Felitti VJ, Anda RF, Croft JB. Cumulative childhood stress and autoimmune diseases in adults. Psychosom Med. 2009;71(2):243–50.
21. Jonson-Reid M, Kohl PL, Drake B. Child and adult outcomes of chronic child maltreatment. Pediatrics. 2012;129(5):839–45.
22. van der Kolk BA, Roth S, Pelcovitz D, Sunday S, Spinazzola J. Disorders of extreme stress: the empirical foundation of a complex adaptation to trauma. J Trauma Stress. 2005;18(5):389–99.
23. Ford JD. Complex PTSD: research directions for nosology/assessment, treatment, and public health. Eur J Psychotraumatol. 2015;6:27584.

24. Anda RF, Felitti VJ, Bremner JD, Walker JD, Whitfield C, Perry BD, et al. The enduring effects of abuse and related adverse experiences in childhood. A convergence of evidence from neurobiology and epidemiology. Eur Arch Psychiatry Clin Neurosci. 2006;256(3):174–86.
25. Cannon EA, Bonomi AE, Anderson ML, Rivara FP. The intergenerational transmission of witnessing intimate partner violence. Arch Pediatr Adolesc Med. 2009;163(8):706–8.
26. Kilpatrick DG, Ruggiero KJ, Acierno R, Saunders BE, Resnick HS, Best CL. Violence and risk of PTSD, major depression, substance abuse/dependence, and comorbidity: results from the National Survey of adolescents. J Consult Clin Psychol. 2003;71(4):692–700.
27. Widom CS, Czaja SJ, Dutton MA. Childhood victimization and lifetime revictimization. Child Abuse Negl. 2008;32(8):785–96.
28. Ullman SE. Sexual revictimization, PTSD, and problem drinking in sexual assault survivors. Addict Behav. 2016;53:7–10.
29. Ullman SE, Najdowski CJ, Filipas HH. Child sexual abuse, post-traumatic stress disorder, and substance use: predictors of revictimization in adult sexual assault survivors. J Child Sex Abus. 2009;18(4):367–85.
30. Ports KA, Ford DC, Merrick MT. Adverse childhood experiences and sexual victimization in adulthood. Child Abuse Negl. 2016;51:313–22.
31. Ullman SE, Peter-Hagene LC. Longitudinal relationships of social reactions, PTSD, and Revictimization in sexual assault survivors. J Interpers Violence. 2016;31(6):1074–94.
32. McEwen BS. Allostasis and allostatic load: implications for neuropsychopharmacology. Neuropsychopharmacology. 2000;22(2):108–24.
33. Moriceau S, Sullivan RM. Neurobiology of infant attachment. Dev Psychobiol. 2005;47(3):230–42.
34. Moriceau S, Roth TL, Okotoghaide T, Sullivan RM. Corticosterone controls the developmental emergence of fear and amygdala function to predator odors in infant rat pups. Int J Dev Neurosci. 2004;22(5–6):415–22.
35. Brattstrom O, Eriksson M, Larsson E, Oldner A. Socio-economic status and co-morbidity as risk factors for trauma. Eur J Epidemiol. 2015;30(2):151–7.
36. Lowe SR, Galea S, Uddin M, Koenen KC. Trajectories of posttraumatic stress among urban residents. Am J Commun Psychol. 2014;53(1–2):159–72.
37. Graif C, Gladfelter AS, Matthews SA. Urban poverty and neighborhood effects on crime: incorporating spatial and network perspectives. Sociol Compass. 2014;8(9):1140–55.
38. Pollak RA. An intergenerational model of domestic violence. J Popul Econ. 2004;17(2):311–29.
39. Rivera PM, Fincham F. Forgiveness as a mediator of the intergenerational transmission of violence. J Interpers Violence. 2015;30(6):895–910.
40. Roberts AL, Lyall K, Weisskopf MG. Maternal exposure to childhood abuse is associated with mate selection: implications for autism in offspring. J Autism Dev Disord. 2017;47(7):1998–2009.
41. Torrisi R, Arnautovic E, Pointet Perizzolo VC, Vital M, Manini A, Suardi F, et al. Developmental delay in communication among toddlers and its relationship to caregiving behavior among violence-exposed, posttraumatically stressed mothers. Res Dev Disabil. 2018;82:67–78.
42. Martin L, Kidd M, Seedat S. The effects of childhood maltreatment and anxiety proneness on neuropsychological test performance in non-clinical older adolescents. J Affect Disord. 2019;243:133–44.
43. Sheridan MA, Peverill M, Finn AS, McLaughlin KA. Dimensions of childhood adversity have distinct associations with neural systems underlying executive functioning. Dev Psychopathol. 2017;29(5):1777–94.
44. Biedermann SV, Meliss S, Simmons C, Nothling J, Suliman S, Seedat S. Sexual abuse but not posttraumatic stress disorder is associated with neurocognitive deficits in south African traumatized adolescents. Child Abuse Negl. 2018;80:257–67.
45. Perna RB, Kiefner M. Long-term cognitive sequelae: abused children without PTSD. Appl Neuropsychol Child. 2013;2(1):1–5.
46. Bowlby J. Attachment and loss. New York: Basic Books; 1969.

47. Fonagy P, Steele H, Moran G, Steele M, Higgitt A. The capacity for understanding mental states: the reflective self in parent and child and its significance for security of attachment. Infant Ment Health J. 1991;13:200–17.
48. Lyons-Ruth K, Yellin C, Melnick S, Atwood G. Expanding the concept of unresolved mental states: hostile/helpless states of mind on the adult attachment interview are associated with disrupted mother-infant communication and infant disorganization. Dev Psychopathol. 2005;17(1):1–23.
49. Lyons-Ruth K, Dutra L, Schuder MR, Bianchi I. From infant attachment disorganization to adult dissociation: relational adaptations or traumatic experiences? Psychiatr Clin North Am. 2006;29(1):63–86. viii
50. Byun S, Brumariu LE, Lyons-Ruth K. Disorganized attachment in young adulthood as a partial mediator of relations between severity of childhood abuse and dissociation. J Trauma Dissociation. 2016;17(4):460–79.
51. Vonderlin R, Kleindienst N, Alpers GW, Bohus M, Lyssenko L, Schmahl C. Dissociation in victims of childhood abuse or neglect: a meta-analytic review. Psychol Med. 2018;48(15):2467–76.
52. Ensink K, Berthelot N, Begin M, Maheux J, Normandin L. Dissociation mediates the relationship between sexual abuse and child psychological difficulties. Child Abuse Negl. 2017;69:116–24.
53. Bandelow B, Baldwin D, Abelli M, Bolea-Alamanac B, Bourin M, Chamberlain SR, et al. Biological markers for anxiety disorders, OCD and PTSD: a consensus statement. Part II: neurochemistry, neurophysiology and neurocognition. World J Biol Psychiatry. 2017;18(3):162–214.
54. Bandelow B, Baldwin D, Abelli M, Altamura C, Dell'Osso B, Domschke K, et al. Biological markers for anxiety disorders, OCD and PTSD - a consensus statement. Part I: neuroimaging and genetics. World J Biol Psychiatry. 2016;17(5):321–65.
55. Lehrner A, Bierer LM, Passarelli V, Pratchett LC, Flory JD, Bader HN, et al. Maternal PTSD associates with greater glucocorticoid sensitivity in offspring of holocaust survivors. Psychoneuroendocrinology. 2014;40:213–20.
56. Binder EB, Bradley RG, Liu W, Epstein MP, Deveau TC, Mercer KB, et al. Association of FKBP5 polymorphisms and childhood abuse with risk of posttraumatic stress disorder symptoms in adults. JAMA. 2008;299(11):1291–305.
57. Walker SE, Sandi C. Long-term programing of psychopathology-like behaviors in male rats by peripubertal stress depends on individual's glucocorticoid responsiveness to stress. Stress. 2018;21(5):433–42.
58. Nelson MD, Tumpap AM. Posttraumatic stress disorder symptom severity is associated with left hippocampal volume reduction: a meta-analytic study. CNS Spectr. 2017;22(4):363–72.
59. Paquola C, Bennett MR, Lagopoulos J. Understanding heterogeneity in grey matter research of adults with childhood maltreatment-a meta-analysis and review. Neurosci Biobehav Rev. 2016;69:299–312.
60. Champagne DL, Bagot RC, van Hasselt F, Ramakers G, Meaney MJ, de Kloet ER, et al. Maternal care and hippocampal plasticity: evidence for experience-dependent structural plasticity, altered synaptic functioning, and differential responsiveness to glucocorticoids and stress. J Neurosci. 2008;28(23):6037–45.
61. Suderman M, McGowan PO, Sasaki A, Huang TC, Hallett MT, Meaney MJ, et al. Conserved epigenetic sensitivity to early life experience in the rat and human hippocampus. Proc Natl Acad Sci U S A. 2012;109(Suppl 2):17266–72.
62. Yehuda R, Bierer LM. Transgenerational transmission of cortisol and PTSD risk. Prog Brain Res. 2008;167:121–35.
63. Teicher MH, Anderson CM, Ohashi K, Khan A, McGreenery CE, Bolger EA, et al. Differential effects of childhood neglect and abuse during sensitive exposure periods on male and female hippocampus. NeuroImage. 2018;169:443–52.
64. Rubin M, Shvil E, Papini S, Chhetry BT, Helpman L, Markowitz JC, et al. Greater hippocampal volume is associated with PTSD treatment response. Psychiatry Res Neuroimaging. 2016;252:36–9.

65. Apfel BA, Ross J, Hlavin J, Meyerhoff DJ, Metzler TJ, Marmar CR, et al. Hippocampal volume differences in gulf war veterans with current versus lifetime posttraumatic stress disorder symptoms. Biol Psychiatry. 2011;69(6):541–8.
66. Liberzon I, Abelson JL. Context processing and the neurobiology of post-traumatic stress disorder. Neuron. 2016;92(1):14–30.
67. Lambert HK, McLaughlin KA. Impaired Hippocampus-dependent associative learning as a mechanism underlying PTSD: a meta-analysis. Neurosci Biobehav Rev 2019.
68. Lambert HK, Sheridan MA, Sambrook KA, Rosen ML, Askren MK, McLaughlin KA. Hippocampal contribution to context encoding across development is disrupted following early-life adversity. J Neurosci. 2017;37(7):1925–34.
69. Milani AC, Hoffmann EV, Fossaluza V, Jackowski AP, Mello MF. Does pediatric posttraumatic stress disorder alter the brain? Systematic review and meta-analysis of structural and functional magnetic resonance imaging studies. Psychiatry Clin Neurosci. 2017;71(3):154–69.
70. Patel R, Spreng RN, Shin LM, Girard TA. Neurocircuitry models of posttraumatic stress disorder and beyond: a meta-analysis of functional neuroimaging studies. Neurosci Biobehav Rev. 2012;36(9):2130–42.
71. Etkin A, Wager TD. Functional neuroimaging of anxiety: a meta-analysis of emotional processing in PTSD, social anxiety disorder, and specific phobia. Am J Psychiatry. 2007;164(10):1476–88.
72. Chen AC, Etkin A. Hippocampal network connectivity and activation differentiates posttraumatic stress disorder from generalized anxiety disorder. Neuropsychopharmacology. 2013;38(10):1889–98.
73. Akiki TJ, Averill CL, Abdallah CG. A network-based neurobiological model of PTSD: evidence from structural and functional neuroimaging studies. Curr Psychiatry Rep. 2017;19(11):81.
74. Pechtel P, Lyons-Ruth K, Anderson CM, Teicher MH. Sensitive periods of amygdala development: the role of maltreatment in preadolescence. NeuroImage. 2014;97:236–44.
75. Andersen SL, Tomada A, Vincow ES, Valente E, Polcari A, Teicher MH. Preliminary evidence for sensitive periods in the effect of childhood sexual abuse on regional brain development. J Neuropsychiatry Clin Neurosci. 2008;20(3):292–301.
76. Teicher MH, Samson JA, Anderson CM, Ohashi K. The effects of childhood maltreatment on brain structure, function and connectivity. Nat Rev Neurosci. 2016;17(10):652–66.
77. Rinne-Albers MA, van der Werff SJ, van Hoof MJ, van Lang ND, Lamers-Winkelman F, Rombouts SA, et al. Abnormalities of white matter integrity in the corpus callosum of adolescents with PTSD after childhood sexual abuse: a DTI study. Eur Child Adolesc Psychiatry. 2016;25(8):869–78.
78. Hein TC, Monk CS. Research review: neural response to threat in children, adolescents, and adults after child maltreatment—a quantitative meta-analysis. J Child Psychol Psychiatry. 2017;58(3):222–30.
79. Marusak HA, Etkin A, Thomason ME. Disrupted insula-based neural circuit organization and conflict interference in trauma-exposed youth. Neuroimage Clin. 2015;8:516–25.
80. Herringa RJ. Trauma, PTSD, and the developing brain. Curr Psychiatry Rep. 2017;19(10):69.
81. Crozier JC, Wang L, Huettel SA, De Bellis MD. Neural correlates of cognitive and affective processing in maltreated youth with posttraumatic stress symptoms: does gender matter? Dev Psychopathol. 2014;26(2):491–513.
82. Yang P, Wu MT, Hsu CC, Ker JH. Evidence of early neurobiological alternations in adolescents with posttraumatic stress disorder: a functional MRI study. Neurosci Lett. 2004;370(1):13–8.
83. Wolf MS, Badea R, Berezovsky J. Fast nanoscale addressability of nitrogen-vacancy spins via coupling to a dynamic ferromagnetic vortex. Nat Commun. 2016;7:11584.
84. Garrett AS, Carrion V, Kletter H, Karchemskiy A, Weems CF, Reiss A. Brain activation to facial expressions in youth with PTSD symptoms. Depress Anxiety. 2012;29(5):449–59.
85. Duncan LE, Ratanatharathorn A, Aiello AE, Almli LM, Amstadter AB, Ashley-Koch AE, et al. Largest GWAS of PTSD (N=20 070) yields genetic overlap with schizophrenia and sex differences in heritability. Mol Psychiatry. 2018;23(3):666–73.

86. Wang Y, Karstoft KI, Nievergelt CM, Maihofer AX, Stein MB, Ursano RJ, et al. Posttraumatic stress following military deployment: genetic associations and cross-disorder genetic correlations. J Affect Disord. 2019;252:350–7.
87. Mehta D, Klengel T, Conneely KN, Smith AK, Altmann A, Pace TW, et al. Childhood maltreatment is associated with distinct genomic and epigenetic profiles in posttraumatic stress disorder. Proc Natl Acad Sci U S A. 2013;110(20):8302–7.
88. Oliver JE. Intergenerational transmission of child abuse: rates, research, and clinical implications. Am J Psychiatry. 1993;150(9):1315–24.
89. Rikic J, Beljan P, Milosevic M, Miskulin I, Miskulin M, Mujkic A. Transgenerational transmission of violence among parents of preschool children in Croatia. Acta Clin Croat. 2017;56(3):478–86.
90. Goodman ML, Hindman A, Keiser PH, Gitari S, Ackerman Porter K, Raimer BG. Neglect, sexual abuse, and witnessing intimate partner violence during childhood predicts later life violent attitudes against children among Kenyan women: evidence of intergenerational risk transmission from cross-sectional data. J Interpers Violence. 2017;886260516689777
91. Riva Crugnola C, Ierardi E, Bottini M, Verganti C, Albizzati A. Childhood experiences of maltreatment, reflective functioning and attachment in adolescent and young adult mothers: effects on mother-infant interaction and emotion regulation. Child Abuse Negl. 2019;93:277–90.
92. Talmon A, Horovitz M, Shabat N, Haramati OS, Ginzburg K. "neglected moms"—the implications of emotional neglect in childhood for the transition to motherhood. Child Abuse Negl. 2019;88:445–54.
93. Lyons-Ruth K, Block D. The disturbed caregiving system: relations among childhood trauma, maternal caregiving, and infant affect and attachment. Infant Mental Hearth J. 1996;17:257–75.
94. Muller-Nix C, Forcada-Guex M, Pierrehumbert B, Jaunin L, Borghini A, Ansermet F. Prematurity, maternal stress and mother-child interactions. Early Hum Dev. 2004;79(2):145–58.
95. Gewirtz AH, Polusny MA, DeGarmo DS, Khaylis A, Erbes CR. Posttraumatic stress symptoms among National Guard soldiers deployed to Iraq: associations with parenting behaviors and couple adjustment. J Consult Clin Psychol. 2010;78(5):599–610.
96. Schechter DS, Moser DA, Aue T, Gex-Fabry M, Pointet VC, Cordero MI, et al. Maternal PTSD and corresponding neural activity mediate effects of child exposure to violence on child PTSD symptoms. PLoS One. 2017;12(8):e0181066.
97. McFarlane J, Symes L, Binder BK, Maddoux J, Paulson R. Maternal-child dyads of functioning: the intergenerational impact of violence against women on children. Matern Child Health J 2014.
98. Miranda JK, de la Osa N, Granero R, Ezpeleta L. Multiple mediators of the relationships among maternal childhood abuse, intimate partner violence, and offspring psychopathology. J Interpers Violence. 2013;28(14):2941–65.
99. Adams TR, Handley ED, Manly JT, Cicchetti D, Toth SL. Intimate partner violence as a mechanism underlying the intergenerational transmission of maltreatment among economically disadvantaged mothers and their adolescent daughters. Dev Psychopathol. 2018:1–11.
100. Bosquet Enlow M, Egeland B, Carlson E, Blood E, Wright RJ. Mother-infant attachment and the intergenerational transmission of posttraumatic stress disorder. Dev Psychopathol. 2014;26(1):41–65.
101. Islam TM, Tareque MI, Tiedt AD, Hoque N. The intergenerational transmission of intimate partner violence in Bangladesh. Glob Health Action. 2014;7:23591.
102. Fehringer JA, Hindin MJ. Like parent, like child: intergenerational transmission of partner violence in Cebu, the Philippines. J Adolesc Health. 2009;44(4):363–71.
103. Ehrensaft MK, Cohen P, Brown J, Smailes E, Chen H, Johnson JG. Intergenerational transmission of partner violence: a 20-year prospective study. J Consult Clin Psychol. 2003;71(4):741–53.
104. Alberini CM, Ledoux JE. Memory reconsolidation. Curr Biol. 2013;23(17):R746–50.

105. Perroud N, Rutembesa E, Paoloni-Giacobino A, Mutabaruka J, Mutesa L, Stenz L, et al. The Tutsi genocide and transgenerational transmission of maternal stress: epigenetics and biology of the HPA axis. World J Biol Psychiatry. 2014;15(4):334–45.
106. Yehuda R, Daskalakis NP, Bierer LM, Bader HN, Klengel T, Holsboer F, et al. Holocaust exposure induced intergenerational effects on FKBP5 methylation. Biol Psychiatry. 2016;80(5):372–80.
107. Moser DA, Paoloni-Giacobino A, Stenz L, Adouan W, Manini A, Suardi F, et al. BDNF methylation and maternal brain activity in a violence-related sample. PLoS One. 2015;10(12):e0143427.
108. Schechter DS, Moser DA, Paoloni-Giacobino A, Stenz L, Gex-Fabry M, Aue T, et al. Methylation of NR3C1 is related to maternal PTSD, parenting stress and maternal medial prefrontal cortical activity in response to child separation among mothers with histories of violence exposure. Front Psychol. 2015;6:690.
109. Schechter DS, Moser DA, Pointet VC, Aue T, Stenz L, Paoloni-Giacobino A, et al. The association of serotonin receptor 3A methylation with maternal violence exposure, neural activity, and child aggression. Behav Brain Res. 2017;325(Pt B):268–77.
110. Cordero MI, Moser DA, Manini A, Suardi F, Sancho-Rossignol A, Torrisi R, et al. Effects of interpersonal violence-related post-traumatic stress disorder (PTSD) on mother and child diurnal cortisol rhythm and cortisol reactivity to a laboratory stressor involving separation. Horm Behav. 2017;90:15–24.
111. Schechter DS, Moser DA, Reliford A, McCaw JE, Coates SW, Turner JB, et al. Negative and distorted attributions towards child, self, and primary attachment figure among posttraumatically stressed mothers: what changes with clinician assisted videofeedback exposure sessions (CAVES). Child Psychiatry Hum Dev. 2015;46(1):10–20.
112. Berthelot N, Ensink K, Bernazzani O, Normandin L, Luyten P, Fonagy P. Intergenerational transmission of attachment in abused and neglected mothers: the role of trauma-specific reflective functioning. Infant Ment Health J. 2015;36(2):200–12.
113. Suardi F, Moser DA, Sancho Rossignol A, Manini A, Vital M, Merminod G, et al. Maternal reflective functioning, interpersonal violence-related posttraumatic stress disorder, and risk for psychopathology in early childhood. Attach Hum Dev. 2018:1–21.
114. Fonagy P, Steele H, Moran G, Steele H, Higgit A. The capacity for understanding mental states: the reflective self in parent and child and its significance for security of attachment. Infant Ment Health J. 1991;13:200–16.
115. Allen J. Mentalizing as a conceptual bridge fromPsychodynamic to cognitive-behavioral therapies. Eur Psychother. 2008;8(1):103–21.
116. Slade A. Parental reflective functioning: an introduction. Attach Hum Dev. 2005;7(3):269–81.
117. Schechter DS, Gross A, Willheim E, McCaw J, Turner JB, Myers MM, et al. Is maternal PTSD associated with greater exposure of very young children to violent media? J Trauma Stress. 2009;22(6):658–62.
118. Solanke BL. Does exposure to interparental violence increase women's risk of intimate partner violence? Evidence from Nigeria demographic and health survey. BMC Int Health Hum R. 2018;18
119. Moser DA, Aue T, Suardi F, Manini A, Sancho Rossignol A, Cordero MI, et al. The relation of general socio-emotional processing to parenting specific behavior: a study of mothers with and without posttraumatic stress disorder. Front Psychol. 2015;6:1575.
120. Gillies D, Taylor F, Gray C, O'Brien L, D'Abrew N. Psychological therapies for the treatment of post-traumatic stress disorder in children and adolescents (review). Evid Based Child Health. 2013;8(3):1004–116.
121. Foa EB, Dancu CV, Hembree EA, Jaycox LH, Meadows EA, Street GP. A comparison of exposure therapy, stress inoculation training, and their combination for reducing posttraumatic stress disorder in female assault victims. J Consult Clin Psychol. 1999;67(2):194–200.
122. Markowitz JC, Choo TH, Neria Y. Do acute benefits of interpersonal psychotherapy for posttraumatic stress disorder endure? Can J Psychiatr. 2018;63(1):37–43.

123. Lieberman AF, Van Horn P, Ippen CG. Toward evidence-based treatment: child-parent psychotherapy with preschoolers exposed to marital violence. J Am Acad Child Adolesc Psychiatry. 2005;44(12):1241–8.
124. Wilson G, Farrell D, Barron I, Hutchins J, Whybrow D, Kiernan MD. The use of eye-movement desensitization reprocessing (EMDR) therapy in treating post-traumatic stress disorder-a systematic narrative review. Front Psychol. 2018;9:923.
125. Dellucci H. No matter how slow you go, as long as you don't stop: a six gear mechanics for a safe therapy jour-ney through complex trauma. 11th EMDRIA Conferenc; Hamburg, Germany 2010.
126. Chen R, Gillespie A, Zhao Y, Xi Y, Ren Y, McLean L. The efficacy of eye movement desensitization and reprocessing in children and adults who have experienced complex childhood trauma: a systematic review of randomized controlled trials. Front Psychol. 2018;9:534.
127. Ogden P, Pain C, Fisher J. A sensorimotor approach to the treatment of trauma and dissociation. Psychiatr Clin North Am. 2006;29(1):263–79. xi-xii
128. Slade A, Sadler L, De Dios-Kenn C, Webb D, Currier-Ezepchick J, Mayes L. Minding the baby a reflective parenting program. Psychoanal Study Child. 2005;60:74–100.
129. Slade A, Holland ML, Ordway MR, Carlson EA, Jeon S, Close N, et al. Minding the baby(R): enhancing parental reflective functioning and infant attachment in an attachment-based, interdisciplinary home visiting program. Dev Psychopathol. 2019:1–15.
130. Schechter DS, Rusconi SS. Understanding how traumatised mothers process their toddlers' affective communication under stress: towards preventive intervention for families at high risk for intergenerational violence. In: Emde R, Leuzinger-Bohleber M, editors. Early parenting research and prevention of disorder: psychoanalytic research at interdisciplinary frontiers. London: Karnac Books; 2014.
131. Berthelot N, Lemieux R, Lacharite C. Development of a prenatal program for adults with personal histories of childhood abuse or neglect: a Delphi consensus consultation study. Health Promot Chronic Dis Prev Can. 2018;38(11):393–403.
132. Muzik M, Rosenblum KL. Motherhood in the face of trauma. Pathways toward healing and growth. Berlin: Springer; 2018.

# Childhood Trauma and Personality Disorder

# 12

Claire Perry and Royce Lee

## 12.1 Introduction

In the clinic, the relationship between childhood trauma and personality disorder can be a seemingly omnipresent issue. Although the correlation is maintained in the laboratory, recent work has revealed a more complex relationship between childhood trauma and personality disorder, casting some doubt about the impression gleaned from the clinic. This discrepancy is the source of considerable uncertainty and controversy. Fortunately, a large body of research now clarifies the real but complex relationship between childhood trauma and personality disorder. This information is indispensable in the clinic, where such knowledge is critical to an appropriately nuanced understanding of cases of personality disorder with a history of trauma. This chapter reviews this relationship and summarizes the increasingly sophisticated understanding of how childhood trauma impacts development of the person and the brain.

To understand modern theories of childhood trauma and personality disorders, it is useful to place the body of this work in the larger historical context. In the early twentieth century, a foundational psychological theory of the sequelae of childhood trauma was formed by the work of Pierre Janet and Sigmund Freud. This psychodynamic theory posited that trauma leads to the repression of unwanted memories of childhood abuse. Repressing some drives enhances others in a compensatory fashion, resulting in intrusive psychiatric symptoms [1, 2]. Although state-dependent memory likely has a role in trauma-related psychopathology [3], the centrality of repressed memories in the development of psychopathology has largely been disproven [4, 5]. The reverberations of this important debate are still reflected in general attitudes of specialists and laypeople alike. At the end of the twentieth century,

C. Perry · R. Lee (✉)
The Department of Psychiatry and Behavioral Neuroscience, The University of Chicago, Chicago, IL, USA
e-mail: cperry4@yoda.bsd.uchicago.edu; rlee@yoda.bsd.uchicago.edu



G. Spalletta et al. (eds.), *Childhood Trauma in Mental Disorders*,
https://doi.org/10.1007/978-3-030-49414-8_12

this was exemplified in the difference in attitudes toward recovered memories among psychologists, as clinical psychologists were more likely than experimental psychologists to accept the validity of recovered memories [6]. A recent survey of 2326 American adults found that 8% consulted with therapists who discussed repressed childhood abuse and 4% report recovering memories of past abuse, a trend with decreasing rates in those beginning therapy after the 1990s [6]. These results suggest that an effect of the memory wars was a lagged but pervasive change regarding the therapist attitudes toward recovered memories. Also of interest was the finding that the majority of abuse cases involving repressed memories were physical and emotional, not sexual.

In a concurrent late twentieth century development, the basic neurosciences began to uncover a biological link between parent–infant disruptions, trauma, and stress reactivity. This work revealed how maternal–infant interactions lastingly shape the developing nervous system [7, 8]. Most importantly, the systems in which this plasticity was discovered, the neuroendocrine and limbic brain systems, are poised in the body to produce psychological symptoms such as anxiety, hostility, and altered social cognition that characterize personality disorder. Thus, just as the memory wars led to a re-evaluation of the role of trauma in psychopathology, neuroscience ironically discovered a biological link between early life trauma and psychopathology.

It is in this complex context that the contemporary understanding of the relationship between childhood trauma and personality disorder must be considered. The research reviewed here culminates in what we here refer to as the *General Theory* of trauma and personality disorder: childhood trauma has wide-ranging effects on the development of brain systems. These in turn are expressed phenotypically as personality disorder symptoms common to many different personality disorders. We will review the evidence supporting this theory in the first section of the chapter. This will be followed by a discussion of descriptive data regarding the relationship between subtypes of childhood trauma and dimensions of personality psychopathology. Finally, we will discuss counterarguments to the General Theory based on the problem of PTSD and its relationship to personality disorder.

## 12.2 The Association of Childhood Trauma and Personality Disorder—Empirical Studies

Psychodynamic theories of personality development were based heavily on the experience of the developing self with parental figures. Stern's foundational description of borderline personality disorder (BPD) in the *Psychoanalytic Quarterly* [9] included the clinical observation that borderline pathology was associated with a history of maternal neglect: "…mothers inflicted injuries on their children by virtue of a deficiency of spontaneous maternal affection." Refuting the idea that fantasies of abuse, rather than abuse itself, are of importance, Stern goes on to clarify that "Actual cruelty, neglect and brutality by the parents of many years' duration are factors found in these patients." Although concerns about childhood sexual abuse would later dominate studies of borderline personality disorder, it is of historical interest to note that this early report did not mention it.

Object relations theory in psychoanalytic metapsychology was concerned with the development of the social self in attachment relationships. The phenomenon of splitting, in which an individual thinks of people with whom has a relationship as all-good or all-bad, retained in the DSM-5 criteria for BPD, was first described in relation to infantile frustration with maternal attachment [10]. Such concept was later operationalized by Otto Kernberg into a schematic description of Borderline Personality Organization [11], encompassing what we might today describe as paranoid, borderline, schizotypal, antisocial, and narcissistic personality disorders. The role of trauma in the theory of Borderline Personality Organization was qualified as an environmental factor that interacted with a genetically mediated, inborn aggressive drive. The impact of object relations on contemporary theory, nosology, and treatment continues to be felt today. Perhaps most importantly, the theory of Borderline Personality Organization led directly to a milestone in psychoanalytic psychology: the empirical validation of transference focused psychotherapy for borderline personality disorder [12]. The hypothesized interaction of trauma and trait-like aggression was an idea that remarkably anticipated contemporary etiological theories of personality psychopathology that emerged nearly a half-century later.

### 12.2.1 Cross-Sectional Studies

In the late twentieth century, research regarding the relationship between borderline personality disorder and childhood trauma eschewed theory in favor of empirical and statistical approaches to the question. A comprehensive review of pre-2000 studies of the relationship between childhood abuse and borderline personality disorder organized this empirically focused literature into three waves [13]. In the first wave of studies, borderline personality disorder was found to be associated with structural problems in family life, such as parental separation and loss [14]. In a second wave of studies, sexual abuse was found to be overrepresented in borderline personality disorder vs. Axis II controls [15–17]. At this stage, the relationship between sexual abuse and borderline personality disorder appeared to be trauma and disorder specific [15]. A third wave of studies, using multivariate statistical approaches in cohorts with mixed personality disorder syndromes, identified a complex mixture of abuse experiences, including sexual, physical abuse, and emotional neglect which were associated with a range of personality disorders [18, 19].

In the largest study from this "third wave" cross-sectional data from the Collaborative Longitudinal Personality Disorder Study was examined for multivariate statistical associations between retrospective accounts of childhood abuse and personality disorder diagnosis [20]. In the 668 participants, there were 86 with schizotypal, 167 with borderline, 153 with avoidant, and 153 with obsessive-compulsive personality disorder. Ninety-four participants with major depressive disorder (MDD) and no personality disorder were also studied. The BPD group had the highest rate of childhood sexual abuse (CSA) (37.7%) relative to the other diagnostic categories. CSA in BPD patients was also a significant covariate in a regression analysis predicting suicide attempts though it did not predict other suicidal behaviors, suggesting a highly specific interaction between CSA and BPD prognosis [20]. Furthermore, 39.7% of schizotypal and 51% of BPD participants reported lifetime PTSD, compared to only 20.3% of MDD participants. These results suggest an important but complex relationship between trauma and schizotypal and borderline personality disorders.

A second notable study utilizing a similar approach was conducted in 182 adults with personality disorder. Using the Childhood Trauma Questionnaire, the relationship between abuse subtype and personality disorder diagnosis was examined. History of trauma was widely prevalent across diagnoses, with 78% of subjects meeting criteria for childhood trauma, indicating that trauma is a risk factor shared across most personality disorder diagnoses. Paranoid personality disorder was related to sexual, physical, and emotional abuse; antisocial personality disorder was related to sexual and physical abuse, and BPD was related to emotional abuse. No unique relationship was found between BPD and childhood sexual abuse [21].

The third wave of studies seemed to point to the effects of childhood trauma across a variety of different dimensions and experiences. While this wave of studies did outline several specific relationships between abuse experiences and personality disorder subtypes, the overall data trends toward a more general association between childhood trauma and personality disorders.

#### 12.2.1.1 Recent Developments

DSM-5 eliminated the axial system of psychiatric diagnosis, but carried forward the Cluster system and categorical diagnostic approach from DSM-IV. Although dimensional approaches were proposed as replacements to the categorical diagnoses, lack of consensus led to the dimensional approaches remaining in an alternative section, awaiting further validation. Cluster A contains personality disorders with odd or eccentric features. It encompasses paranoid, schizoid, and schizotypal personality disorders. Cluster B contains personality disorders with dramatic, emotional, or erratic features. It includes antisocial, borderline, histrionic, and narcissistic personality disorders. Cluster C contains personality disorders with anxious and fearful features. It includes avoidant, dependent, and obsessive-compulsive personality disorders.

Using a similar approach to previous studies, a study in 231 psychiatric patients tested a series of trauma-personality disorder-specific hypotheses using a multivariate statistical approach [22]. The results confirmed three out of five *a priori* hypotheses: (1) physical abuse was associated with antisocial personality disorder; (2) emotional neglect was associated with Cluster A disorders; (3) emotional abuse was related to Cluster C disorders. No support was found for a unique association of sexual abuse and borderline symptom or for emotional abuse and narcissistic symptoms as was initially hypothesized.

Two multivariate studies found a more general relationship between childhood trauma exposure and personality disorder symptoms. In a study of 70 abused and 35 non-abused adults, only a general relationship was found between childhood trauma (CTQ) and personality disorder symptoms, although elevated levels of paranoid, borderline, avoidant, and dependent personality disorder were seen in the abused group [23]. In a larger study of 409 patients and non-patient volunteers in Belgium and the Netherlands, a similar multivariate approach was taken [24]. History of childhood trauma was taken with the Interview for Traumatic Events in Childhood (ITEC) assessing sexual abuse, physical abuse, emotional abuse, emotional neglect, and physical neglect [25]. This, combined with the large sample size, represents a strength of the study. Perhaps due to increased statistical power, this study did find specific relationships between trauma subtype and personality disorders. Sexual abuse was found to be associated with paranoid, schizoid, borderline, and avoidant personality disorder, with the largest beta coefficient for BPD. Physical abuse was associated with antisocial personality disorder, as was found previously [22]. Emotional abuse was associated with paranoid, schizotypal, borderline, and Cluster C personality disorder. Emotional neglect was associated with histrionic and BPD.

In the largest and most representative study to date, data from 34,653 adult individuals from the National Epidemiologic Survey on Alcohol and Related Conditions (NESARC) was analyzed to examine the relationship between childhood trauma and personality disorders [26]. All types of abuse and neglect were associated with Cluster A personality disorders, with schizotypal, followed by paranoid, having the strongest relationship. All forms of abuse were related to Cluster B disorders. Antisocial, borderline, and narcissistic personality disorders were associated with all forms of abuse, although no relationship was found for histrionic personality disorder. No strong relationship was found between child abuse and Cluster C personality disorders.

In summary, the empirical literature begins with disorder and abuse-specific associations (childhood sexual abuse and BPD) and evolves to a more general association between abuse of various kinds and a wide range of personality disorder symptoms. Before turning to longitudinal studies, we will review disorder-specific findings.

#### 12.2.1.2 Disorder-Specific Empirical Findings

*Schizotypy*: A review of 25 empirical studies found overwhelming support for the association of childhood trauma and schizotypy [27], with odds ratios in traumatized groups between 2.01 and 4.15. Some evidence suggests that the association, even if present, is reduced in strength compared to BPD [20]. Another cross-sectional study found that PTSD mediated the relationship between a history of emotional abuse and symptoms of schizotypal personality disorder [28]. A study taking a dimensional approach to schizotypy in a random sample from the community appears to confirm a link with childhood trauma using 1510 adults who completed telephone interviews. A history of child abuse and acute trauma were both associated with schizotypal personality disorder symptoms. Interestingly, in males there was some evidence that the relationship between childhood maltreatment and schizotypy was moderated by neurodevelopmental disturbance [29]. The moderation by neurodevelopmental disturbance raises the possibility that common genetic factors may account for both childhood maltreatment and schizotypy, a possibility reinforced by findings from behavioral genetic studies (reviewed below).

*Paranoid Personality Disorder*: Both acute and chronic childhood trauma have been associated with paranoid personality disorder. Individual studies have found that paranoid personality disorder is associated with nearly all forms of abuse, including physical, sexual, and emotional abuse as well as emotional neglect [30]. These findings are consistent with the results of the multivariate studies reported above, in which paranoid personality disorder emerges as perhaps the personality disorder most consistently associated with childhood trauma.

For instance, among young adult victims of burn injury, paranoid personality disorder was the most prevalent, with 19.4% prevalence within the cohort found following semi-structured diagnostic interviews [31]. Furthermore, population-wide risk factors, such as socioeconomic stress, have sufficiently strong effects to lead to detection of elevated paranoid personality disorder rates in certain ethnic populations, such as African Americans, in the United States [32]. This could perhaps be related to data from cognitive psychological experiments which found that increased negative beliefs about the self lead to paranoid anxiety [33, 34]. This data provides important clues regarding the psychological pathways leading to paranoid anxiety.

*Antisocial Personality Disorder (ASPD)*: A robust developmental literature has identified negligent child rearing practices and abuse as an important risk factor for childhood and adolescent conduct disorder [35]. In the forensic setting, physical and crime victimization, but not sexual trauma, is associated with up to a fivefold increase in the risk for ASPD. The overlapping but clinically distinct diagnosis of psychopathy was related only to a history of physical trauma [36]. Interestingly, traits of psychopathy may increase the probability of crime victimization, working against the generally negative relationship between psychopathy and fear to equalize the potential for post-traumatic psychopathology [37]. This data points to the importance of bidirectional causal pathways between personality psychopathology and trauma exposure.

*Borderline Personality Disorder*: A 1999 meta-analysis of 21 published studies from 1980 to 1995 on the association of BPD and CSA, including data from 2479 subjects, found a pooled effect size of $r = 0.28$ of CSA on BPD [38]. This relatively small effect size is likely due to the fact that studies sampling nonclinical populations are less likely to find high rates of PTSD and CSA [39]. Smaller studies, recruited largely from clinical samples, tended to have a larger effect size ($r = -0.775$, $p < 0.001$). The authors conclude that CSA is unlikely to have a major etiological role in the development of PTSD, but may be related to features within the BPD construct such as dissociation. Dissociative symptoms appear in both BPD and PTSD criteria. These symptoms represent a discontinuous experience of the self emotionally, relationally, or spatiotemporally. Dissociative symptoms are orthogonal to temperamental traits such as novelty seeking and harm avoidance, and are correlated instead with character traits such as self-transcendence and self-directedness [40]. Dissociation is closely related to trauma exposure [41]. In fact its relationship to trauma is stronger than to BPD [15]. Dissociative symptoms have been proposed to be an important component of BPD [42, 43]. Proneness to dissociative experiences interacts with personality in a nonlinear fashion, increasing the risk for BPD, suggesting that trauma-related dissociative symptoms [40] can play a critical role in the etiology of BPD [44].

### 12.2.2 Longitudinal Studies

There have been three longitudinal studies directly relevant to the issue of childhood trauma and personality disorder. A cohort of maltreated youth identified by the Department of Human Services (DHS), along with a cohort of non-maltreated but similarly low socioeconomic status youth, have been followed in a set of studies. In the first set, the investigators found that 40% of maltreated children showed a non-resilient trajectory when followed from ages 6 to 10. Ego resiliency is a trait-like flexibility in impulse control that is adjusted according to the social context [45]. Early childhood maltreatment, in individuals high in emotional intensity and low in emotional regulation, was found to precede increases in internalizing (anxious and depressive) symptoms [46]. In an analysis of personality mediators of maltreatment history on adolescent psychopathology, maltreatment was found to be associated with either an overcontrolled (compulsive) or undercontrolled (impulsive) phenotype during early childhood. The overcontrolled phenotype was more likely to develop internalizing psychopathology in adolescence, while the undercontrolled phenotype was more likely to develop externalizing psychopathology [47]. This study is a landmark of developmental psychology and provided a glimpse at the interaction of trauma and personality traits.

A 47-year longitudinal study in the Barbados [48] followed 129 children who had experienced moderate to severe malnutrition but normal birthweight in the first year of life, along with 129 matched controls. When followed into adulthood, childhood malnutrition was associated with elevated symptoms levels of paranoid, schizoid, avoidant, and dependent personality disorder. Childhood maltreatment was associated with paranoid, schizoid, schizotypal, and avoidant personality disorder. Those who had experienced both malnutrition and maltreatment had the highest level of personality disorder symptoms, suggesting additive effects of physiological and psychological stress on the development of personality psychopathology.

In the *Children in the Community Study* [49], a community-based sample of 593 families were followed from childhood (6 years) to adulthood (mean age 33 years). Measures of parental behavior during childhood were examined as risk factors for early adult personality disorder. Of the sample, 12% met criteria for at least one personality disorder at age 22. Generally, the number of problematic parental behaviors was correlated with the severity of personality disorder symptoms. After controlling for covariates, low parental care was associated with symptom levels of antisocial, avoidant, borderline, paranoid, schizoid, and schizotypal personality disorder. High parental abuse was associated with symptom levels of borderline, schizotypal, and paranoid personality disorder. Interestingly, maltreatment mediated the relationship between childhood emotional disturbances and adult personality disorders.

## 12.3 Behavioral Genetics

### 12.3.1 Twin Studies

In the quest to understand the etiology of personality disorder, the field of behavioral genetics has used family, twin, and hybrid research designs to tease apart its genetic and environmental components. In the first small twin study of 10 BPD, 15 schizotypal + BPD, and 44 schizotypal probands, no co-twins had BPD, and thus no heritable component of BPD was found [50]. However, 92 monozygotic and 129 dizygotic twin pairs, about half of whom had a clinical diagnosis of personality disorder, were interviewed with the Structured Clinical Interview for DSM-III-R Personality Disorders (SCID-II) in a subsequent study [51]. The study found that heritability was 0.69 for borderline, 0.78 for obsessive-compulsive, 0.61 for schizotypal, 0.57 for dependent, 0.67 for histrionic, and 0.79 for narcissistic personality disorder. Heritability was relatively lower for paranoid (0.28), and schizoid (0.29) personality disorder. Unlike the previous study, no shared environmental effects were detected for any of the personality disorders.

In a sample of 2794 young adult Norwegian twins, dimensional representations of the DSM-IV personality disorders were examined by measuring criterion counts of personality disorder symptoms (actual personality disorder diagnosis was rare at 3%). A broad genetic risk factor for all personality disorders was found (paranoid, histrionic, borderline, narcissistic, dependent, and obsessive-compulsive), followed by a second genetic factor with loadings only on borderline and antisocial, and a third genetic factor with loadings on schizoid and avoidant [52]. Three environmental factors were identified. The first loaded on the Cluster B personality disorders (along with paranoid and obsessive-compulsive), all of the Cluster A personality disorders, and the Cluster C personality disorders (avoidant and dependent). The heritability of individual personality disorders ranged from a lower limit of 0.34 for paranoid to an upper limit of 0.61 for BPD, with no shared environmental effects (for review see [53]). The estimated heritability of antisocial personality disorder from this sample (0.41) was close to that from a meta-analysis of 51 twin and adoption studies of antisocial behavior, which found additive genetic (0.32) and non-additive genetic (0.09) influences on antisocial behavior [54].

Non-additive, or dominant, genetic effects along with additive genetic effects in risk for BPD were examined using Dutch twin registry that included siblings, using a so-called extended twin design. In 5017 twins, 1266 siblings, 3064 parents, and 949 spouses from 4015 families [55], additive (21.3%), non-additive (23.9%), and unique environmental (54.9%) sources of variance were found for BPD symptom count. Once again, models with no shared environment prevailed.

In summary, the predominance of non-shared environmental sources of variance points toward the likely relevance of gene × environment interaction in the development of BPD [53]. Furthermore, the consistent lack of shared environment effects on personality disorder echoes the results of behavioral genetic research on personality traits such as neuroticism, as seen in a very large study of 20,945 Australian and 24,905 US twins [56]. Variance in neuroticism was accounted for by additive genetic (28–36%), non-additive genetic (13–17%), and non-shared environmental (37–43%) influences.

### 12.3.2 Gene × Environment Interaction

The absence of any major effects of shared environment on risk for personality disorder refutes simple models of trauma-related personality psychopathology, while the large role of genetic factors in the etiology of personality disorder supports a biological model of personality psychopathology. As detected by non-shared environmental effects, the environment may have an impact, but rather than directly affecting psychopathology, it affects psychopathology in interaction with genetic factors [57–59]. This suggests that the neurobiological impacts of trauma may be of particular importance. Adding to the complexity, resilience to trauma exposure is determined by genetic factors, a phenomenon known as *gene × environment correlation*. Genes can influence emotional responses to assault exposure [60]. They can also influence the experience of parental care [61]. Personality traits themselves influence the response to exposure to traumatic events [62].

Thus, two important corollaries of the General Theory are, firstly, the mediation of childhood trauma's effect on personality psychopathology is likely through gene × environment effects on the developing brain, and secondly, exposure to trauma may be partly mediated by genetic factors. The field of epigenetics has opened a new vista on how the environment can have long-lasting effects on phenotype and behavior [63]. Preliminary epigenetic findings have been reported regarding the gene of FKBP5, a co-chaperone regulator of glucocorticoid receptors [64], the glucocorticoid receptor gene (NR3C1) [65], and the monoamine oxidase A promotor in BPD [57, 59, 66]. These epigenetic effects may have pronounced neurobiological implications for the psychopathology of personality disorders and are discussed in the following section. Some of the epigenetics effects of maternal care may be reversible in animal models [63], pointing toward potentially novel treatment approaches targeting epigenetic modifications in the future.

### 12.3.3 Contextual Factors of Culture, Race, and Ethnicity

The absence of shared environmental influences on personality disorder raises questions about the nature of the influence of culture and ethnicity on the development of personality disorder. For instance, there are race differences in personality disorder prevalence, with elevated rates in Native Americans, decreased rates in Asians, and elevated rates of some diagnoses in African Americans [67]. Qualitative work found connections between structural violence, interpersonal violence, and the presentation of BPD symptoms in native populations of Australia [68]. The effect of natural experiments, such as random allocation of resources to different sectors of society, has been studied for insight into this question. Unexpected supplemental income is associated with reduction of mental illness in children, an effect mediated fully by level of parental supervision, and strongest for externalizing symptoms of conduct disorder and oppositional disorder [69]. The effects of structural and interpersonal violence have been found to account for race differences in paranoid personality disorder [32]. Environmental factors that interact with the developing brain such as childhood lead exposure have been longitudinally associated with neuroticism, and decreases in agreeableness and conscientiousness [70]. These findings, combined with the results of twin studies, would suggest that cultural and other broadly contextual factors that impact personality disorder may best be understood as the interaction between genetics and the individual's unique experience of the environment.

## 12.4 Neurobiology: At the Crossroads of Gene × Environment Interaction

Biological clues provide the critical link between animal models and the clinic. Here we would propose that this research provides broad support for the General Theory: that childhood trauma has wide diverse effects on the development of brain systems, and these nonspecifically increase overall levels of personality psychopathology. Trauma has epigenetic effects on stress reactivity [71, 72], itself a largely genetically determined process. These epigenetic effects have important implications for personality disorders as stress reactivity is a common denominator of most personality disorder-related constructs, such as BPD-related symptoms of anxiety, aggression, and shame [73].

### 12.4.1 Cortisol

Converging evidence has pointed to the relevance of the hypothalamic pituitary-adrenal (HPA) axis in understanding BPD. Cortisol, the main end product of HPA axis activation in humans plays an important role in regulating both the magnitude and duration of the stress response via feedback inhibition of the HPA axis. Altered basal cortisol levels have been associated with HPA axis dysregulation with important implications for stress reactivity [74]. Despite somewhat inconsistent results, overall evidence suggests that patients with BPD have decreased levels of basal cortisol compared to healthy controls [75]. Notably, patients with PTSD typically also exhibit blunted levels of basal cortisol and increased sensitivity to the dexamethasone suppression test (DST), a measure of sensitivity of the HPA axis to negative feedback. In contrast, patients with MDD show a profile of elevated basal cortisol and blunted sensitivity to the DST (Thomas et al., 2019). As such, the cortisol profile of BPD patients may be closely tied to interacting depressive and PTSD symptoms. Indeed [76], found that basal cortisol levels in BPD patients were positively correlated with MDD symptoms. Also, patients with both BPD comorbid with PTSD and BPD with high prevalence of PTSD symptoms have been shown to have increased sensitivity to the DST compared to BPD patients with low PTSD symptoms [76, 77], supporting this theoretical framework. Additionally, BPD patients reporting more childhood trauma on the Childhood Trauma Questionnaire

(CTQ), a retrospective measure of childhood trauma exposure, showed lower basal cortisol levels, consistent with a profile more similar to PTSD patients, who have generally been found to have reduced peripheral cortisol levels [78]. The influence of these different symptom profiles could account for inconsistencies across cortisol studies in BPD groups. The cortisol profile of BPD seems to be closely linked to the patient's depressive and PTSD symptom profile and reflects the complex interactions of trauma and mood symptoms on stress reactivity and regulation. As cortisol has been shown to interact with the HPA axis in a concentration-dependent mechanism, understanding how MDD and PTSD symptoms interact in BPD patients may be critical for understanding differences in HPA axis dysregulation and stress reactivity across symptom groups.

### 12.4.2 Corticotropin-Releasing Hormone

Corticotropin-releasing hormone (CRH) levels, as measured in cerebrospinal fluid, have been found to be elevated in patients with PTSD. CRH is an important positive driver of the HPA axis, and has independent, direct effects on brain function through its action on brain CRH receptors. The central effects of CRH signaling include increases in emotional responding to stimuli and altered social behavior [79]. History of childhood trauma, specifically, emotional neglect, has been found across three studies to be associated with increased CRH drive in adults with personality disorder [80–82]. Central CRH drive is inversely associated with reactivity of the pituitary and adrenal glands to stimulation with combined dexamethasone/CRH challenge, consistent with peripheral downregulation in the face of enhanced central CRH drive. Indeed, history of childhood trauma exposure was found to be associated with blunted, rather than enhanced cortisol and adrenocorticotropic hormone (ACTH) response to dexamethasone/CRH challenge [82]. Although none of the subjects tested in these studies met criteria for PTSD, the combination of increased central CRH drive and downregulated peripheral responsiveness resembles that seen in PTSD. Indeed, comorbid PTSD has also been found to be associated with blunted ACTH response to exogenous dexamethasone/corticotropin-releasing hormone (CRH) challenge [83]. Furthermore, as expected with chronic altered set point in hypothalamic release of CRH, hypothalamic gray matter volume is increased in BPD and is positively correlated with CTQ score, another implication of the effects of HPA axis abnormalities in BPD [84]. Dysfunction of the HPA axis seems to be a critical mechanism for understanding trauma-induced changes to stress reactivity in personality disorders.

### 12.4.3 NR3C1 and FKBP5

Yet another mechanism of trauma-induced alterations to HPA axis function comes from the glucocorticoid receptor gene (NR3C1) which is important in providing negative feedback to the HPA axis. Childhood physical abuse in BPD is associated with methylation, a form of epigenetic regulation of gene expression, of the NR3C1 promoter. This methylation has been shown to positively correlate with BPD symptom severity as measured by the Revised Diagnostic Interview for Borderlines (DIB-R), self-injurious behavior, and hospitalizations which may relate to its effects on HPA axis dysregulation [85]. Additionally, the intracellular function of the glucocorticoid receptor is affected by the intracellular chaperone protein FKBP5. Studies have linked FKBP5 candidate genes and the BPD phenotype [86]. Exposure to early life stress, such as intimate partner violence, has interactive effects on FKBP5 risk haplotypes. The CATT haplotype, which is associated with enhanced upregulation of FKBP5 and decreased glucocorticoid receptor sensitivity, is associated with increased stress sensitivity in children [87].

Preliminary evidence of gene × environment and epigenetic effects has also been found for the brain-derived neurotrophic factor (BDNF) gene in BPD [88]. BDNF is an important signaling molecule that affects the developing brain on a macrostructural and microstructural level and whose plasma levels have been shown to be diminished in BPD patients [89]. Interestingly, responders to the psychotherapeutic intervention of dialectical behavioral therapy (DBT) showed decreases in methylation, while most BPD subjects showed increasing methylation over time pointing to potential epigenetics benefits of cognitive therapy [88]. Furthermore, the BDNF 196A allele valine to methionine ($Val^{66}Met$) polymorphism has been implicated in decreased hippocampal volume and decreased BDNF levels in response to childhood abuse and exposure to early life stress and may decrease stress resiliency and increase the risk of developing psychiatric disorders later in life [90, 91]. In addition [92], reported decreased impulsive aggression in BPD patients with the $Val^{66}Val$ variant as compared to those with the $Val^{66}Met$ variant suggests a mediating influence of the BDNF 196A allele on aggressive symptoms in BPD. More research is needed to describe the complex relationship between BDNF, gene × environment, and neurobiological interactions in BPD.

## 12.4.4 Trauma-Related Brain Structural Changes

Trauma-related brain structural changes have also been detected in BPD. Meta-analytic studies confirm that amygdala and hippocampal structural volume is decreased in BPD [64]. These findings are important because of the centrality of the hippocampus and amygdala in the stress reactive behaviors at the core of personality disorder symptoms. Importantly, some of these changes in traumatized individuals with BPD may be accounted for by comorbid PTSD [64]. Patients with PTSD have been characterized by reduced hippocampal and left amygdala volumes [93]. When comparing patients with BPD and patients with BPD comorbid with PTSD, the results have been somewhat inconclusive with some groups reporting decreased hippocampal volume in the comorbid group compared to controls [94, 95] and one reporting no effect of comorbid PTSD on hippocampal volume [96]. Additionally, researchers have found increased amygdala deactivation in response to pain in the comorbid group [97]. Like the neuroendocrine studies, structural brain imaging research suggests that comorbidity with PTSD affects the biological signature of trauma.

Outside of the limbic system, frontal-parietal circuits that mediate social cognitive functions, such as theory of mind and dissociative symptoms, are of interest in trauma and personality disorder. Trauma may be related to decreased detection of facial expression in BPD, a form of affective empathy [98]. Trauma may also impact function of the temporoparietal junction during the neural process of perspective taking in cognitive empathy [99]. Preliminary data points to abnormal development of white matter tracts [100] and gray matter volume of parietal circuits in BPD [101]. It is known that these circuits may play a key role in symptoms found in BPD such as dissociation [102]. More research is needed to understand how the environment works with genetic and biological factors to shape the expression of personality disorder symptoms.

### 12.4.5 Personality Disorder-Specific Effects of Trauma

An important question remains regarding personality disorder-specific effects of trauma. This is an area currently under investigation. One example of this may be impulsive aggression, which is associated with disrupted parental care and childhood trauma [103, 104]. Evidence has been found for genetic moderation of environmental effects with the monoamine oxidase A (MAO-A) gene, whereby those with lower level expression of the MAO-A enzyme are more sensitive to the effects of trauma on aggressive behavior [57, 59, 66]. Genetic deficiency of MAO-A has been linked to increased aggression and plays a moderating role in response to childhood trauma especially in regard to antisocial behavior [57]. Furthermore, serotonin signaling in the brain has a key role in computations important for decision-making and impulsivity [105], and plays a role in modulating anger and reactivity to social cues [106, 107]. The serotonin (5-HT) system clearly plays a central role in impulsive aggression in personality disorder [108]. How the environment impacts development of brain 5-HT signaling is an important topic for future research.

### 12.4.6 Paranoia/Mistrust and CRH

Finally, paranoia and mistrust, as found in paranoid personality disorder, appear important to trauma-related personality disorder symptoms. Adults with paranoid personality disorder, perhaps more than any other personality disorder, consistently report high levels of childhood trauma in both cross-sectional [21] and longitudinal studies [49]. They are also the only personality disorder subgroup identified thus far with elevated CSF CRH levels [30]. In humans, intranasally administered exogenous CRH enhances early components of the event-related potential to neutral faces, a socially ambiguous signal [109]. This suggests that the hostile and suspicious social cognitive profile of paranoid personality disorder may be related to cortical arousal.

In total, biological data provides support for the General Theory. Available evidence points toward trauma-related epigenetic changes in the neurobiological mechanisms important for adaptation to stress that likely underlie a wide range of psychopathological syndromes. The lasting nature of these developmentally expressed changes makes them of heightened importance in personality disorder. At the same time, some evidence, such as the relationships between Paranoid Personality Disorder, trauma, and CRH, counters the General Theory, pointing to specific relationships between trauma and personality disorder subtype.

## 12.5 Comorbid PTSD and Complex PTSD: Limitations of the General Theory

Available evidence indicates that PTSD and BPD are frequently comorbid, with approximately a third to a half of individuals with borderline personality disorder meeting criteria for PTSD [20, 110]. The higher estimates are from clinical, as opposed to epidemiological samples. About a quarter of patients with PTSD have borderline personality disorder [110], with the comorbidity predicting more severe trauma, suicide attempt rates, and lower quality of life. The high prevalence of comorbidity between the two conditions likely reflects complex etiological pathways. Gene–environment correlation may play a role as it has been shown that persons with borderline personality disorder are more likely to experience sexual trauma and rape as adults in comparison with other personality disorders, thus putting them at elevated risk for PTSD [20, 111]. Several aspects of the BPD construct may affect how experiencing trauma leads to PTSD in susceptible individuals. These include sensitivity to fear conditioning [112], decreased habituation [113], and altered limbic brain morphology [114]. Adding to the complexity, ICD-11 includes the diagnostic construct complex PTSD. Complex PTSD includes the defining symptoms of PTSD in addition to disturbed self-regulatory capacities. These are grouped into the five domains of (a) emotion regulation difficulties, (b) disturbances in relational capacities, (c) alterations in attention and consciousness (e.g., dissociation), (d) adversely affected belief systems, and (e) somatic distress or disorganization. Complex PTSD has an estimated 1% prevalence in the community [115]. In the clinic, about half of PTSD cases can be conceptualized as complex PTSD [116]. The distinction is important because complex PTSD predicts less improvement with psychotherapy [117]. Staging treatment based on complexity is required to strike a balance between skill-building and trauma-focused approaches [118]. Skill-building is an essential component of dialectical behavioral therapy. Because trauma therapy activates intense affects, skill-building provides a buffer between distress and impulsive acting-out. One study approached this question by detecting symptom profiles in a large sample of treatment seeking women with a history of child abuse. Four latent classes were identified: (1) a PTSD class with high PTSD symptoms but low BPD symptoms, (2) a complex PTSD class had high levels of PTSD and self-organization symptoms, but low levels of relational BPD symptoms, (3) a BPD class with a high percentage of BPD symptoms, self-organization disturbances, and PTSD symptoms, and (4) a resilient class with low symptoms [119]. About 44.6% of the BPD class cases met criteria for complex PTSD, but in fact, individuals in the BPD class were more likely to meet criteria for PTSD (54%) than complex PTSD. Complex PTSD overlapped with BPD in the symptom of "feelings of emptiness" and relationship disturbances, but could be distinguished from it by the specificity of abandonment, instability of relationships, unstable sense of self, impulsiveness, and suicidality to BPD.

Thus, comorbidity between BPD and PTSD is very real and pervasive. It would be premature to fold one diagnosis into the other, as has been proposed, as neither the BPD nor the complex PTSD construct can account for the range of symptoms these two categories encompass. On the other hand, a 50% comorbidity rate between two disorders is high enough to question the distinctness between the two conditions. One must seriously consider the possibility that historical contingencies, rather than nature itself, have determined how the chips have fallen in terms of diagnostic distinctions, an argument made previously about psychiatric nosological boundaries [120]. The nosological implications are difficult to resolve but point to the need for new approaches for a classification that can account for (1) the range of clinical symptoms and implicated neural circuits, (2) longitudinal course, and (3) etiology (genes and environment). Unfortunately, none of the current and proposed nosological systems such as the Research Domain Criteria [121], DSM-5, and ICD-11 are poised to satisfy all three conditions. It is also not clear if any of the three systems would help to simplify patient understanding of each diagnosis. It is our experience that patients find the overlap of these three conditions unnecessarily confusing, given that useful information is often available for each condition separately. An alternative approach may be to modify current constructs to incorporate some clinical information spread across diagnoses, as is done with the clinical staging of cancer, which systemizes severity and prognosis by biological spread of disease.

## 12.6 Childhood Trauma and Personality Disorder: Dimensional Relationships

Given the mixed evidence for and against the General Theory, the remaining question is the degree to which the relationships between childhood trauma and personality disorder exhibit both generalness and specificity. Answering this would require examining personality disorder subtypes in a dimensional manner, given the frequent comorbidity between personality disorder subtypes. To do so, we applied exploratory factor analysis using principal components analysis (PCA) to personality disorder criterion counts on a sample of male and female adults to identify latent personality disorder factors. Each factor was then modeled using linear regression analysis with levels of child abuse subtypes to identify unique relationships between psychopathology and abuse.

In the study, 1675 male and female adults between the ages of 18–82 (mean = 34.82, SD = 11.065) provided written, informed consent to participate in studies of mood and personality approved by the institutional review board. In the entire sample, exploratory factor analysis was performed, using PCA, on criterion counts of DSM-IV personality disorder symptoms. All personality disorder symptom counts had correlations >0.5 for at least one component and so all were included in the model. Factor 1, *Impulsive/Hostile*, loaded onto antisocial, borderline, narcissistic, and paranoid personality disorder. Factor 2, *Schizotypy*, loaded onto schizoid and schizotypal personality disorder. Factor 3, *Anxious*, loaded onto avoidant and dependent personality disorder (Table 12.1).

Table 12.1: from "Dimensional Relationships between PD and Trauma." Factor loadings of personality disorder severity on the three extracted factors. Note the similarity of Factor 1 to Cluster B, Factor 2 to Cluster A, and Factor 3 to Cluster B personality disorders.

In a subsample of 1063 subjects who completed the Childhood Trauma Questionnaire (CTQ), each of the three extracted factors was entered into separate stepwise linear regression. Analyses including age and gender in the first step, and the CTQ subscales (sexual abuse, emotional neglect, physical neglect, physical abuse, emotional abuse) in the second step. The linear regression analysis for the *Impulsive/Hostile* factor (Table 12.2) resulted in a significant model with addition of CTQ subscale scores. The CTQ subscale scores remained significant after correction for multiple measures: sexual abuse, physical abuse, and emotional abuse. The

**Table 12.1** Factor loadings on personality disorder symptom counts

| | Component | | |
|---|---|---|---|
| | 1 | 2 | 3 |
| Number of schizoid criteria met | −0.148 | 0.887 | 0.045 |
| Number of schizotypal criteria met | 0.344 | 0.711 | −0.021 |
| Number of antisocial criteria met | 0.753 | 0.086 | −0.094 |
| Number of borderline criteria met | 0.580 | 0.122 | 0.345 |
| Number of histrionic criteria met | 0.790 | −0.174 | 0.053 |
| Number of narcissistic criteria met | 0.855 | −0.027 | −0.056 |
| Number of avoidant criteria met | −0.123 | 0.159 | 0.862 |
| Number of dependent criteria met | 0.093 | −0.140 | 0.835 |
| Number of obsessive-compulsive criteria met | 0.388 | 0.032 | 0.309 |
| Number of paranoid criteria met | 0.615 | 0.369 | 0.080 |

Extraction method: Principal component analysis.
Rotation method: Oblimin with Kaiser normalization.
(a) Rotation converged in 7 iterations.

**Table 12.2** Factor 1, Impulsive/Hostile

| | *R* | *F* | Sig | *β* | *T* | Sig. |
|---|---|---|---|---|---|---|
| Model 1 | 0.54 | 1.548 | 0.213 | | | |
| Model 2 | 0.444 | 37.1 | <0.001 | | | |
| Age | | | | −0.009 | −0.316 | 0.752 |
| Gender | | | | −0.089 | −3.149 | 0.002 |
| Sexual abuse | | | | 0.088 | 2.727 | 0.007 |
| Emotional neglect | | | | 0.033 | 0.749 | 0.454 |
| Physical neglect | | | | −0.051 | −1.291 | 0.187 |
| Physical abuse | | | | 0.116 | 2.893 | 0.004 |
| Emotional abuse | | | | 0.317 | 6.601 | <0.001 |

**Table 12.3** Factor 2, Schizotypy

| | *R* | *F* | Sig | *β* | *t* | Sig. |
|---|---|---|---|---|---|---|
| Model 1 | 0.094 | 4.772 | 0.009 | | | |
| Model 2 | 0.273 | 12.168 | <0.001 | | | |
| Age | | | | 0.073 | 2.438 | 0.015 |
| Gender | | | | −0.003 | −0.113 | 0.910 |
| Sexual abuse | | | | −0.080 | −2.284 | 0.023 |
| Emotional neglect | | | | 0.053 | 1.136 | 0.256 |
| Physical neglect | | | | 0.028 | 0.660 | 0.510 |
| Physical abuse | | | | 0.015 | 0.344 | 0.731 |
| Emotional abuse | | | | 0.214 | 4.150 | <0.001 |

**Table 12.4** Factor 3, Anxious

| | R | F | Sig | β | T | Sig. |
|---|---|---|---|---|---|---|
| Model 1 | 0.060 | 1.911 | 0.149 | | | |
| Model 2 | 0.386 | 26.427 | <0.001 | | | |
| Age | | | | −0.058 | −2.006 | 0.045 |
| Gender | | | | 0.014 | 0.471 | 0.637 |
| Sexual abuse | | | | −0.091 | 2.714 | 0.007 |
| Emotional neglect | | | | 0.158 | 3.535 | 0.000 |
| Physical neglect | | | | −0.016 | −0.386 | 0.700 |
| Physical abuse | | | | −0.077 | −1.871 | 0.062 |
| Emotional abuse | | | | 0.272 | 5.508 | <0.001 |

linear regression analysis for the *Schizotypy* factor also resulted in a significant model with the addition of CTQ scores (Table 12.3). After correction for multiple measures, emotional abuse was significantly associated with schizotypy. A significant model was found for the *Anxious* factor with the addition of CTQ scores (Table 12.4). Emotional neglect and emotional abuse were both uniquely associated with the anxious factor.

In summary, three underlying factors were identified, explaining approximately 65% of the variance in a personality disorder symptoms severity: (1) *Impulsive/Hostility*, (2) *Schizotypy*, (3) *Anxious*. Somewhat surprisingly, these resembled Clusters A, B, and C as seen in the DSM personality disorder section, with the exception that paranoid and obsessive-compulsive personality disorder (OCPD) symptoms belonged to the Cluster B-like factor. All three factors were predicted by history of emotional abuse. *Impulsive/Hostility* was also predicted by history of sexual abuse and physical abuse. *Schizotypy* was predicted only by emotional abuse. *Anxious* was predicted by sexual abuse, emotional neglect, and emotional abuse. Some evidence was found for the General Theory: emotional abuse appeared to be related to all three factors. However, evidence was also found for at least some specificity in the relationship between abuse subtypes and personality disorder symptoms.

Table 12.2: from "Dimensional Relationships between PD and Trauma." Note significant relationships between emotional abuse and Cluster B-like symptoms, as well as smaller relationships with sexual abuse and physical abuse.

Table 12.3: from "Dimensional Relationships between PD and Trauma." Note the significant relationships between emotional abuse and Cluster A-like symptoms, common to all PD dimensions.

Table 12.4: from "Dimensional Relationships between PD and Trauma." Note the significant relationships between emotional abuse and Cluster A-like symptoms, as well as smaller relationships with emotional neglect and sexual abuse.

## 12.7 Synthesis and Conclusions

The preceding sections have raised the question of the role of childhood trauma in personality disorder. The initial theory was that childhood sexual abuse, and the repression of the memory of its experience, resulted in the delayed intrusion of symptoms. This meta-psychological theory of trauma has been largely disproven, at least concerning personality disorder. A large body of cross-sectional and longitudinal research has confirmed that the childhood trauma is an important risk factor for personality disorder, but on the whole, relationships between trauma subtype and personality disorder subtype do not appear to be specific. This has led to what we have termed the General Theory. The results of behavioral genetics have pointed to fact that the relationship between childhood trauma and personality disorder is bidirectional. Gene × environment correlation accounts for some of this relationship and has important implications for prevention. Non-shared environmental effects and gene × environment interactions point to the importance of neurobiology in accounting for the lasting effects of trauma on adult phenotype. While some biological systems, such as those underlying learning, stress reactivity, and plasticity, are expected to be broadly relevant to psychopathology, others would be expected to have specificity to emotional and social functions mediated by specific neural circuits.

Evidence suggests that the overlap of PTSD with BPD is a priority area for research. Patients with both conditions experience distress at the ambiguity and arbitrariness of psychiatric nosology. Additionally, treatment follows from diagnosis, and current complicated diagnostic schemes can make treatment planning difficult. Thus, we think that clinicians should be cautious in attempting to understand their patients through their diagnoses, rather than understanding them through the stories they tell and their clinical presentation.

Biological and psychological research suggests that the General Theory may have limitations. In a large group of adults, we presented data showing that there may be some degree of specificity between trauma subtype exposure and personality disorder symptoms. Biological research will likely play a key role in understanding these complex relationships, given the probable importance of gene × environment interaction in the development of personality disorder.

## References


1. Freud S. *The psychotherapy of hysteria.* (J. Strachey, Ed.). 1953rd ed. London: Hogarth Press; 1893.
2. Van der Kolk BA, Van der Hart O. Pierre Janet and the breakdown of adaptation in psychological trauma. Am J Psychiatr. 1989;146(12):1530–40. https://doi.org/10.1176/ajp.146.12.1530.
3. Radulovic J, Lee R, Ortony A. State-dependent memory: neurobiological advances and prospects for translation to dissociative amnesia. Front Behav Neurosci. 2018;12:259. https://doi.org/10.3389/fnbeh.2018.00259.
4. Crews F. The memory wars. New York: New York Review of Books; 1995.
5. Loftus EF, Davis D. Recovered memories. Annu Rev Clin Psychol. 2006;2(1):469–98. https://doi.org/10.1146/annurev.clinpsy.2.022305.095315.
6. Patihis L, Pendergrast MH. Reports of recovered memories of abuse in therapy in a large age-representative U.S. National Sample: therapy type and decade comparisons. Clin Psychol Sci. 2018;7(1):3–21. https://doi.org/10.1177/2167702618773315.
7. Levine S. Maternal behavior as a mediator of pup adrenocortical function. Ann N Y Acad Sci. 1994;746:260–75.
8. Meaney MJ, Aitken DH, Bodnoff SR, Iny LJ, Tatarewicz JE, Sapolsky RM. Early postnatal handling alters glucocorticoid receptor concentrations in selected brain regions. In: Behavioral Neuroscience. US: American Psychological Association; 1985. https://doi.org/10.1037/0735-7044.99.4.765.
9. Stern A. Psychoanalytic investigation of and therapy in the border line group of neuroses. Psychoanal Q. 1938;7:467–89. https://doi.org/10.1080/21674086.1938.11925367.
10. Klein M. Notes on some schizoid mechanisms. Int J Psychoanal. 1946;27:99–110.. 10.978.14464/50697
11. Kernberg O. Borderline personality organization. J Am Psychoanal Assoc. 1967;15(3):641–85. https://doi.org/10.1177/000306516701500309.
12. Clarkin JF, Levy KN, Lenzenweger MF, Kernberg OF. Evaluating three treatments for borderline personality disorder: a multiwave study. Am J Psychiatr. 2007;164(6):922–8. https://doi.org/10.1176/appi.ajp.164.6.922.
13. Zanarini MC. Childhood experiences associated with the development of borderline personality disorder. Psychiatr Clin N Am. 2000;23(1):89–101. https://doi.org/10.1016/S0193-953X(05)70145-3.
14. Grinker RRS, Webble B, Drye RC. The borderline syndrome. A behavioral study of Ego-functions. New York: Basic Books; 1968. https://opus4.kobv.de/opus4-Fromm/frontdoor/index/index/docId/27904
15. Herman J, BP, Van Der Kolk J. Childhood trauma in borderline personality disorder. Am J Psychiatr. 1989;146(4):490–5. https://doi.org/10.1176/ajp.146.4.490.
16. Links PS, Steiner M, Offord DR, Eppel A. Characteristics of borderline personality disorder: A Canadian study. Can J Psychiatr. 1988;33(5):336–40. https://doi.org/10.1177/070674378803300504.
17. Zanarini MC, Gunderson JG, Marino MF, Schwartz EO, Frankenburg FR. Childhood experiences of borderline patients. Compr Psychiatry. 1989;30(1):18–25. https://doi.org/10.1016/0010-440X(89)90114-4.
18. Paris J, Zweig-Frank H, Guzder J. Risk factors for borderline personality in male outpatients. J Nerv Ment Dis. 1994;182(7):375–80. https://doi.org/10.1097/00005053-199407000-00002.
19. Zanarini MC, Williams AA, Lewis RE, Bradford Reich R, Vera SC, Marino MF, et al. Reported pathological childhood experiences associated with the development of borderline personality disorder. Am J Psychiatr. 1997;154(8):1101–6. https://doi.org/10.1176/ajp.154.8.1101.
20. Yen S, Shea MT, Sanislow CA, Grilo CM, Skodol AE, Gunderson JG, et al. Borderline personality disorder criteria associated with prospectively observed suicidal behavior. Am J Psychiatr. 2004;161(7):1296–8. https://doi.org/10.1176/appi.ajp.161.7.1296.

21. Bierer LM, Yehuda R, Schmeidler J, Mitropoulou V, New AS, Silverman JM, Siever LJ. Abuse and neglect in childhood: relationship to personality disorder diagnoses. CNS Spectr. 2003;8(10):737–54. https://doi.org/10.1017/S1092852900019118.
22. Cohen LJ, Tanis T, Bhattacharjee R, Nesci C, Halmi W, Galynker I. Are there differential relationships between different types of childhood maltreatment and different types of adult personality pathology? Psychiatry Res. 2014;215(1):192–201. https://doi.org/10.1016/j.psychres.2013.10.036.
23. Tyrka AR, Wyche MC, Kelly MM, Price LH, Carpenter LL. Childhood maltreatment and adult personality disorder symptoms: influence of maltreatment type. Psychiatry Res. 2009;165(3):281–7. https://doi.org/10.1016/j.psychres.2007.10.017.
24. Lobbestael J, Arntz A, Bernstein DP. Disentangling the relationship between different types of childhood maltreatment and personality disorders. J Personal Disord. 2010;24(3):285–95. https://doi.org/10.1521/pedi.2010.24.3.285.
25. Lobbestael J, Arntz A, Harkema-Schouten P, Bernstein D. Development and psychometric evaluation of a new assessment method for childhood maltreatment experiences: the interview for traumatic events in childhood (ITEC). Child Abuse Negl. 2009;33(8):505–17. https://doi.org/10.1016/j.chiabu.2009.03.002.
26. Afifi TO, Mather A, Boman J, Fleisher W, Enns MW, MacMillan H, Sareen J. Childhood adversity and personality disorders: results from a nationally representative population-based study. J Psychiatr Res. 2011;45(6):814–22. https://doi.org/10.1016/j.jpsychires.2010.11.008.
27. Velikonja T, Fisher HL, Mason O, Johnson S. Childhood trauma and schizotypy: a systematic literature review. Psychol Med. 2015;45(5):947–63. https://doi.org/10.1017/S0033291714002086.
28. Powers AD, Thomas KM, Ressler KJ, Bradley B. The differential effects of child abuse and posttraumatic stress disorder on schizotypal personality disorder. Compr Psychiatry. 2011;52(4):438–45. https://doi.org/10.1016/j.comppsych.2010.08.001.
29. Berenbaum H, Thompson RJ, Milanak ME, Boden MT, Bredemeier K. Psychological trauma and schizotypal personality disorder. J Abnorm Psychol. Berenbaum, Howard: Department of Psychology, University of Illinois at Urbana-Champaign, 603 East Daniel Street, Champaign, IL, US, 61820, hberenba@uiuc.edu: American Psychological Association. 2008; https://doi.org/10.1037/0021-843X.117.3.502.
30. Lee RJ. Mistrustful and misunderstood: a review of paranoid personality disorder. Curr Behav Neurosci Rep. 2017;4:151–65.
31. Thomas CR, Russell W, Robert RS, Holzer CE, Blakeney P, Meyer WJ. Personality disorders in young adult survivors of pediatric burn injury. J Personal Disord. 2012;26(2):255–66. https://doi.org/10.1521/pedi.2012.26.2.255.
32. Raza GT, DeMarce JM, Lash SJ, Parker JD. Paranoid personality disorder in the United States: the role of race, illicit drug use, and income. J Ethn Subst Abus. 2014;13(3):247–57. https://doi.org/10.1080/15332640.2013.850463.
33. Gracie A, Freeman D, Green S, Garety PA, Kuipers E, Hardy A, et al. The association between traumatic experience, paranoia and hallucinations: a test of the predictions of psychological models. Acta Psychiatr Scand. 2007;116(4):280–9. https://doi.org/10.1111/j.1600-0447.2007.01011.x.
34. Turkat ID, Banks DS. Paranoid personality and its disorder. J Psychopathol Behav Assess. 1987;9(3):295–304.
35. Farrington DP. Childhood origins of antisocial behavior. Clin Psychol Psychother. 2005;12(3):177–90. https://doi.org/10.1002/cpp.448.
36. Reddy MK, Johnson JE, Gobin RL, Zlotnick C. Lifetime trauma victimization and PTSD in relation to psychopathy and antisocial personality disorder in a sample of incarcerated women and men. Int J Prison Health. 2015;11(2):64–74. https://doi.org/10.1108/IJPH-06-2014-0016.
37. Willemsen J, De Ganck J, Verhaeghe P. Psychopathy, traumatic exposure, and lifetime posttraumatic stress. Int J Offender Ther Comp Criminol. 2011;56(4):505–24. https://doi.org/10.1177/0306624X11407443.

38. Fossati A, Madeddu F, Maffei C. Borderline personality disorder and childhood sexual abuse: a meta-analytic study. J Personal Disord. 1999;13(3):268–80. https://doi.org/10.1521/pedi.1999.13.3.268.
39. Salzman JP, Salzman C, Wolfson AN, Albanese M, Looper J, Ostacher M, et al. Association between borderline personality structure and history of childhood abuse in adult volunteers. Compr Psychiatry. 1993;34(4):254–7. https://doi.org/10.1016/0010-440X(93)90007-Q.
40. Grabe H-J, Spitzer C, Freyberger HJ. Relationship of dissociation to temperament and character in men and women. Am J Psychiatr. 1999;156(11):1811–3. https://doi.org/10.1176/ajp.156.11.1811.
41. Nijenhuis ERS, Spinhoven P, van Dyck R, van der Hart O, Vanderlinden J. Degree of somatoform and psychological dissociation in dissociative disorder is correlated with reported trauma. J Trauma Stress. 1998;11(4):711–30. https://doi.org/10.1023/A:1024493332751.
42. Denys D, de Vries F, Cath D, Figee M, Vulink N, Veltman DJ, et al. Dopaminergic activity in Tourette syndrome and obsessive-compulsive disorder. Eur Neuropsychopharmacol. 2013;23(11):1423–31. https://doi.org/10.1016/j.euroneuro.2013.05.012.
43. van der Kolk BA, Hostetler A, Herron N, Fisler RE. Trauma and the development of borderline personality disorder. Psychiatr Clin. 1994;17(4):715–30. https://doi.org/10.1016/S0193-953X(18)30082-0.
44. Maldonato, N. M., Sperandeo, R., Moretto, E., & Dell'Orco, S. (2018). A Non-linear predictive model of borderline personality disorder based on multilayer perceptron . Front Psychol. https://www.frontiersin.org/article/10.3389/fpsyg.2018.00447.
45. Kim J, Cicchetti D, Rogosch FA, Manly JT. Child maltreatment and trajectories of personality and behavioral functioning: implications for the development of personality disorder. Dev Psychopathol. 2009;21(3):889–912. https://doi.org/10.1017/S0954579409000480.
46. Kim-Spoon J, Cicchetti D, Rogosch FA. A longitudinal study of emotion regulation, emotion lability-negativity, and internalizing symptomatology in maltreated and nonmaltreated children. Child Dev. 2013;84(2):512–27. https://doi.org/10.1111/j.1467-8624.2012.01857.x.
47. Oshri A, Rogosch FA, Cicchetti D. Child maltreatment and mediating influences of childhood personality types on the development of adolescent psychopathology. J Clin Child Adolesc Psychol. 2013;42(3):2013. https://doi.org/10.1080/15374416.2012.715366.
48. Hock RS, Bryce CP, Fischer L, First MB, Fitzmaurice GM, Costa PT, Galler JR. Childhood malnutrition and maltreatment are linked with personality disorder symptoms in adulthood: results from a Barbados lifespan cohort. Psychiatry Res. 2018;269:301–8. https://doi.org/10.1016/j.psychres.2018.05.085.
49. Johnson JG, Cohen P, Chen H, Kasen S, Brook JS. Parenting behaviors associated with risk for offspring personality disorder during adulthood. Arch Gen Psychiatry. 2006;63(5):579–87. https://doi.org/10.1001/archpsyc.63.5.579.
50. Torgersen S, Kringlen E, Cramer V. The prevalence of personality disorders in a community sample. Arch Gen Psychiatry. 2001;58(6):590–6. https://doi.org/10.1001/archpsyc.58.6.590.
51. Torgersen S, Lygren S, Oien PA, Skre I, Onstad S, Edvardsen J, et al. A twin study of personality disorders. Compr Psychiatry. 2000;41(6):416–25. https://doi.org/10.1053/comp.2000.16560.
52. Kendler KS, Aggen SH, Czajkowski N, Røysamb E, Tambs K, Torgersen S, et al. The structure of genetic and environmental risk factors for DSM-IV personality disorders: a multivariate twin study. Arch Gen Psychiatry. 2008;65(12):1438–46. https://doi.org/10.1001/archpsyc.65.12.1438.
53. Torgersen S. The nature (and nurture) of personality disorders. Scand J Psychol. 2009;50(6):624–32. https://doi.org/10.1111/j.1467-9450.2009.00788.x.
54. Rhee SH, Waldman ID. Genetic and environmental influences on antisocial behavior: A meta-analysis of twin and adoption studies. Psychol Bull. Rhee, Soo Hyun: U Colorado, Inst for Behavioral Genetics, Campus Box 447, Boulder, CO, US, 80309, soo.rhee@colorado.edu: American Psychological Association. 2002; https://doi.org/10.1037/0033-2909.128.3.490.

55. Distel MA, Rebollo-Mesa I, Willemsen G, Derom CA, Trull TJ, Martin NG, Boomsma DI. Familial resemblance of borderline personality disorder features: genetic or cultural transmission? PLoS One. 2009;4(4):e5334. https://doi.org/10.1371/journal.pone.0005334.
56. Lake RIE, Eaves LJ, Maes HHM, Heath AC, Martin NG. Further evidence against the environmental transmission of individual differences in neuroticism from a collaborative study of 45,850 twins and relatives on two continents. Behav Genet. 2000;30:223–33. https://doi.org/10.1023/A:1001918408984.
57. Caspi A, McClay J, Moffitt TE, Mill J, Martin J, Craig IW, et al. Role of genotype in the cycle of violence in maltreated children. Science. 2002;297(5582):851 LP–854. https://doi.org/10.1126/science.1072290.
58. Crowe RR. An adoption study of antisocial personality. Arch Gen Psychiatry. 1974;31(6):785–91. https://doi.org/10.1001/archpsyc.1974.01760180027003.
59. Foley DL, Eaves LJ, Wormley B, Silberg JL, Maes HH, Kuhn J, Riley B. Childhood adversity, monoamine oxidase A genotype, and risk for conduct disorder. Arch Gen Psychiatry. 2004;61(7):738–44. https://doi.org/10.1001/archpsyc.61.7.738.
60. Stein MB, Jang KL, Taylor S, Vernon PA, Livesley WJ. Genetic and environmental influences on trauma exposure and posttraumatic stress disorder symptoms: a twin study. Am J Psychiatr. 2002;159(10):1675–81. https://doi.org/10.1176/appi.ajp.159.10.1675.
61. Kendler KS, Baker JH. Genetic influences on measures of the environment: a systematic review. Psychol Med. 2007;37:615–26. https://doi.org/10.1017/S0033291706009524.
62. Koenen KC, Hitsman B, Lyons MJ, Niaura R, McCaffery J, Goldberg J, et al. A twin registry study of the relationship between posttraumatic stress disorder and nicotine dependence in men. Arch Gen Psychiatry. 2005;62(11):1258–65. https://doi.org/10.1001/archpsyc.62.11.1258.
63. Weaver ICG, Champagne FA, Brown SE, Dymov S, Sharma S, Meaney MJ, Szyf M. Reversal of maternal programming of stress responses in adult offspring through methyl supplementation: altering epigenetic marking later in life. J Neurosci. 2005;25(47):11045 LP–11054. https://doi.org/10.1523/JNEUROSCI.3652-05.2005.
64. Cattane N, Rossi R, Lanfredi M, Cattaneo A. Borderline personality disorder and childhood trauma: exploring the affected biological systems and mechanisms. BMC Psychiatry. 2017. https://doi.org/10.1186/s12888-017-1383-2.
65. Perroud N, Paoloni-Giacobino A, Prada P, Olié E, Salzmann A, Nicastro R, et al. Increased methylation of glucocorticoid receptor gene (NR3C1) in adults with a history of childhood maltreatment: A link with the severity and type of trauma. Transl Psychiatry. 2011;1:e59. https://doi.org/10.1038/tp.2011.60.
66. Kim-Cohen J, Caspi A, Taylor A, Williams B, Newcombe R, Craig IW, Moffitt TE. MAOA, maltreatment, and gene–environment interaction predicting children's mental health: new evidence and a meta-analysis. Mol Psychiatry. 2006;11:903. https://doi.org/10.1038/sj.mp.4001851.
67. Trull TJ, Jahng S, Tomko RL, Wood PK, Sher KJ. Revised NESARC personality disorder diagnoses: gender, prevalence, and comorbidity with substance dependence disorders. J Personal Disord. 2010;24(4):412–26. https://doi.org/10.1521/pedi.2010.24.4.412.
68. Fromene R, Guerin B. Talking with Australian indigenous clients with a borderline personality disorder diagnosis: finding the context behind the label. Psychol Rec. 2014;64(3):569–79. https://doi.org/10.1007/s40732-014-0058-3.
69. Costello EJ, Compton SN, Keeler G, Angold A. Relationships between poverty and PsychopathologyA natural experiment. JAMA. 2003;290(15):2023–9. https://doi.org/10.1001/jama.290.15.2023.
70. Poulton R, Caspi A, Moffitt TE, Cannon M, Murray RM, Harrington H. Children's self-reported psychotic symptoms and adult schizophreniform disorder: a 15-year longitudinal study. Arch Gen Psychiatry. 2000;57(11):1053–8. https://doi.org/10.1001/archpsyc.57.11.1053.
71. McGowan PO, Suderman M, Sasaki A, Huang TCT, Hallett M, Meaney MJ, Szyf M. Broad epigenetic signature of maternal Care in the Brain of adult rats. PLoS One. 2011;6(2):e14739. https://doi.org/10.1371/journal.pone.0014739.

72. Meaney MJ, Aitken DH, Bhatnagar S, Sapolsky RM. Postnatal handling attenuates certain neuroendocrine, anatomical, and cognitive dysfunctions associated with aging in female rats. Neurobiol Aging. 1991;12(1):31–8. https://doi.org/10.1016/0197-4580(91)90036-J.
73. Carpenter RW, Tomko RL, Trull TJ, Boomsma DI. Gene-environment studies and borderline personality disorder: a review. Curr Psychiatry Rep. 2013;15(1):336. https://doi.org/10.1007/s11920-012-0336-1.
74. Stephens MAC, Wand G. Stress and the HPA axis: role of glucocorticoids in alcohol dependence. Alcohol Res. 2012;34:468–83.
75. Thomas N, Gurvich C, Kulkarni J. Borderline personality disorder, trauma, and the hypothalamus-pituitary-adrenal axis. Neuropsychiatr Dis Treat. 2019;15:2601–12. Published 2019 Sep 9. https://doi.org/10.2147/NDT.S198804.
76. Wingenfeld K, Wolf OT. Effects of cortisol on cognition in major depressive disorder, posttraumatic stress disorder and borderline personality disorder. Psychoneuroendocrinology. 2015;51:282–95.
77. Lange W, Wulff H, Berea C, et al. Dexamethasone suppression test in borderline personality disorder–effects of posttraumatic stress disorder. Psychoneuroendocrinology. 2005;30(9):919–23.
78. Aleknaviciute J, Tulen JH, Kamperman AM, de Rijke YB, Kooiman CG, Kushner SA. Borderline and cluster C personality disorders manifest distinct physiological responses to psychosocial stress. Psychoneuroendocrinology. 2016;72:131–13.
79. Strome EM, Wheler GHT, Higley JD, Loriaux DL, Suomi SJ, Doudet DJ. Intracerebroventricular corticotropin-releasing factor increases limbic glucose metabolism and has social context-dependent behavioral effects in nonhuman primates. Proc Natl Acad Sci U S A. 2002;99(24):15749–54. https://doi.org/10.1073/pnas.232480899.
80. Lee R, Geracioti TD Jr, Kasckow JW, Coccaro EF. Childhood trauma and personality disorder: positive correlation with adult CSF corticotropin-releasing factor concentrations. Am J Psychiatr. 2005;162(5). https://doi.org/10.1176/appi.ajp.162.5.995.
81. Lee RJ, Gollan J, Kasckow J, Geracioti T, Coccaro EF. CSF corticotropin-releasing factor in personality disorder: relationship with self-reported parental care. Neuropsychopharmacology. 2006;31(10). https://doi.org/10.1038/sj.npp.1301104.
82. Lee RJ, Hempel J, TenHarmsel A, Liu T, Mathé AA, Klock A. The neuroendocrinology of childhood trauma in personality disorder. Psychoneuroendocrinology. 2012b;37(1). https://doi.org/10.1016/j.psyneuen.2011.05.006.
83. Rinne T, de Kloet ER, Wouters L, Goekoop JG, DeRijk RH, van den Brink W. Hyperresponsiveness of hypothalamic-pituitary-adrenal axis to combined dexamethasone/corticotropin-releasing hormone challenge in female borderline personality disorder subjects with a history of sustained childhood abuse. Biol Psychiatry. 2002;52(11):1102–12. https://doi.org/10.1016/S0006-3223(02)01395-1.
84. Kuhlmann A, Bertsch K, Schmidinger I, Thomann PA, Herpertz SC. Morphometric differences in central stress-regulating structures between women with and without borderline personality disorder. J Psych Neurosci. 2013;38(2):129–37. https://doi.org/10.1503/jpn.120039.
85. Martín-Blanco A, Ferrer M, Soler J, Salazar J, Vega D, Andión O, et al. Association between methylation of the glucocorticoid receptor gene, childhood maltreatment, and clinical severity in borderline personality disorder. J Psychiatr Res. 2014;57:34–40. https://doi.org/10.1016/j.jpsychires.2014.06.011.
86. Amad A, Ramoz N, Peyre H, Thomas P, Gorwood P. FKBP5 gene variants and borderline personality disorder. J Affect Disord. 2019;248:26–8. https://doi.org/10.1016/j.jad.2019.01.025.
87. Halldorsdottir T, Kurtoic D, Müller-Myhsok B, Binder EB, Blair C. Neurobiology of self-regulation: longitudinal influence of FKBP5 and intimate partner violence on emotional and cognitive development in childhood. Am J Psychiatr. 2019;appi.ajp.2019.18091018. https://doi.org/10.1176/appi.ajp.2019.18091018.
88. Perroud N, Salzmann A, Prada P, Nicastro R, Hoeppli M-E, Furrer S, et al. Response to psychotherapy in borderline personality disorder and methylation status of the BDNF gene. Transl Psychiatry. 2013;3:e207. https://doi.org/10.1038/tp.2012.140.

89. Koenigsberg HW, Yuan P, Diaz GA, Guerreri S, Corantes C, Mayson SJ, Zamfirescu C, New AS, Goodman M, Manji HK, Siever LJ. Platelet protein kinase C and brain-derived neurotrophic factor levels in borderline personality diosrder patients. Psychiatry Res. 2012;199:92–7.
90. Bueller JA, Aftab M, Sen S, Gomez-Hassan D, Murmeister M, Zubieta JK. BDNF Val66Met allele is associated with reduced hippocampal volume in healthy subjects. Biol Psychiatry. 2006;59:812–5.
91. Elzinga BM, Molendijk ML, Oude Voshaar RC, Bus BA, Prickaerts J, Spinhoven P, Pennix BJ. The impact of childhood abuse and recent stress on serum braindrived neurotrophic factor and the moderating role of BDNF VAl66Met. Psychopharmacology. 2011;214:319–28.
92. Wagner S, Baskaya O, Dahmen N, Lieb K, Tadic A. Modulatory role of the brain-derived neurotrophic factor Val66Met polymorphism on the effects of serious life events on impulsive aggression in borderline personality disorder. Genes Brain Behav. 2010;9:97–102.
93. Karl A, Schaefer M, Malta LS, Dorfel D, Rohleder N, Werner A. A meta-analysis of structural brain abnormalities in PTSD. Neurosci Biobehav Rev. 2006;30:1004–31.
94. Schmahl C, Berne K, Krause A, Kleindienst N, Valerius G, Vermetten E, Bohus M. Hippocampus and amygdala volumes in patients with borderline personality disorder with or without posttraumatic stress disorder. J Psychiatry Neurosci. 2009;34:289–95.
95. Kreisel SH, Labudda K, Kurlandchikov O, Beblo T, Mertens M, Thomas C, Rullkotter N, Wingenfeld K, Mensebach C, Woermann FG, Driessen M. Volume of hippocampal substructures in borderline personality disorder. Psychiatry Res. 2015;231:218–26.
96. Ruocco AC, Amirthavasagam S, Zakzanis KK. Amygdala and hippocammpal volume reductions as candidate endophenotypes for borderline personality disorder: a meta-analysis of magnetic resonance imaging studies. Psychiatry Res. 2012;201:245–52.
97. Kraus A, Esposito F, Seifritz E, Di Salle F, Ruf M, Valerius G, Ludaescher P, Bohus M, Schmahl C. Amygdala deactivation as a neural correlate of pain processing in patients with borderline personality disorder and co-occurent posttraumatic stress disorder. Biol Psychiatry. 2009;65:819–22.
98. Van Heel M, Luyten P, De Meulemeester C, Vanwalleghem D, Vermote R, Lowyck B. Mentalizing based on external features in borderline personality disorder compared with healthy controls: the role of attachment dimensions and childhood trauma. J Personal Disord. 2019:1–15. https://doi.org/10.1521/pedi_2019_33_373.
99. Hudson A, Van Hamme C, Maeyens L, Brass M, Mueller S. Spontaneous mentalizing after early interpersonal trauma: evidence for hypoactivation of the temporoparietal junction. BioRxiv. 2018;487363. https://doi.org/10.1101/487363.
100. Lee R, Arfanakis K, Evia AM, Fanning J, Keedy S, Coccaro EF. White matter integrity reductions in intermittent explosive disorder. Neuropsychopharmacology. 2016;41(11):2697–703. https://doi.org/10.1038/npp.2016.74.
101. Kimmel CL, Alhassoon OM, Wollman SC, Stern MJ, Perez-Figueroa A, Hall MG, et al. Age-related parieto-occipital and other gray matter changes in borderline personality disorder: A meta-analysis of cortical and subcortical structures. Psychiatry Res Neuroimaging. 2016;251:15–25. https://doi.org/10.1016/j.pscychresns.2016.04.005.
102. Krause-Utz A, Frost R, Winter D, Elzinga BM. Dissociation and alterations in brain function and structure: implications for borderline personality disorder. Curr Psychiatry Rep. 2017;19(1):6. https://doi.org/10.1007/s11920-017-0757-y.
103. Chen P, Coccaro EF, Lee R, Jacobson KC. Moderating effects of childhood maltreatment on associations between social information processing and adult aggression. Psychol Med. 2012;42(6). https://doi.org/10.1017/S0033291711002212.
104. Lee R, Meyerhoff J, Coccaro EF. Intermittent explosive disorder and aversive parental care. Psychiatry Res. 2014;220(1–2). https://doi.org/10.1016/j.psychres.2014.05.059.
105. Balasubramani PP, Chakvravarthy VS, Ravindran B, MA. (2015). A network model of basal ganglia for understanding the roles of dopamine and serotonin in reward-punishment-risk based decision making. Front Comput Neurosci.

106. Lee RJ, Gill A, Chen B, McCloskey M, Coccaro EF. Modulation of central serotonin affects emotional information processing in impulsive aggressive personality disorder. J Clin Psychopharmacol. 2012a;32(3) https://doi.org/10.1097/JCP.0b013e31825368b7.
107. McCloskey MS, Ben-Zeev D, Lee R, Berman ME, Coccaro EF. Acute tryptophan depletion and self-injurious behavior in aggressive patients and healthy volunteers. Psychopharmacology. 2009;203(1) https://doi.org/10.1007/s00213-008-1374-6.
108. Berman ME, Tracy JI, Coccaro EF. The serotonin hypothesis of aggression revisited. Clin Psychol Rev. 1997;17(6):651–65. https://doi.org/10.1016/S0272-7358(97)00039-1.
109. Lee R, Nosa LE, Pinto J, Wermeling D, Towle V. Effects of intranasal Corticotropin-releasing factor on EEG/ERP to threatening faces. In: American College of Neuropsychopharmacology. Hollywood: Amercan College of Neuropsychopharmacology; 2013.
110. Pagura J, Stein MB, Bolton JM, Cox BJ, Grant B, Sareen J. Comorbidity of borderline personality disorder and posttraumatic stress disorder in the U.S. population. J Psychiatr Res. 2010;44(16):1190–8. https://doi.org/10.1016/j.jpsychires.2010.04.016.
111. McGowan A, King H, Frankenburg FR, Fitzmaurice G, Zanarini MC. The course of adult experiences of abuse in patients with borderline personality disorder and Axis II comparison subjects: a 10-year follow-up study. J Personal Disord. 2012;26(2):192–202. https://doi.org/10.1521/pedi.2012.26.2.192.
112. Palomares N, Cuesta-Diaz A, Burke J, Fisher A, Kaur S, Perez-Rodriquez M. Fear conditioning in borderline personality disorder. Curr Behav Neurosci Rep. 2016;3:10–8.
113. Denny BT, Fan J, Fels S, Galitzer H, Schiller D, Koenigsberg HW. Sensitization of the neural salience network to repeated emotional stimuli following initial habituation in patients with borderline personality disorder. Am J Psychiatr. 2018;175(7):657–64. https://doi.org/10.1176/appi.ajp.2018.17030367.
114. Wenzel A, Borges KT, Porto CR, Caminha RM, de Oliveira IR. Volumes of the hippocampus and amygdala in patients with borderline personality disorder: a meta-analysis. J Personal Disord. 2009;23(4):333–45. https://doi.org/10.1521/pedi.2009.23.4.333.
115. Ford JD, Stockton P, Kaltman S, Green BL. Disorders of extreme stress (DESNOS) symptoms are associated with type and severity of interpersonal trauma exposure in a sample of healthy young women. J Interpers Violence. 2006;21(11):1399–416. https://doi.org/10.1177/0886260506292992.
116. Sack M, Sachsse U, Overkamp B, Dulz B. Traumafolgestörungen bei Patienten mit Borderline-Persönlichkeitsstörung. Nervenarzt. 2013;84(5):608–14. https://doi.org/10.1007/s00115-012-3489-6.
117. Dorrepaal E, Thomaes K, Hoogendoorn AW, Veltman DJ, Draijer N, van Balkom AJLM. Evidence-based treatment for adult women with child abuse-related complex PTSD: a quantitative review. Eur J Psychotraumatol. 2014;5:23613. https://doi.org/10.3402/ejpt.v5.23613.
118. Cloitre M, Courtois CA, Charuvastra A, Carapezza R, Stolbach BC, Green BL. Treatment of complex PTSD: results of the ISTSS expert clinician survey on best practices. J Trauma Stress. 2011;24(6):615–27. https://doi.org/10.1002/jts.20697.
119. Cloitre M, Garvert DW, Weiss B, Carlson EB, Bryant RA. Distinguishing PTSD, complex PTSD, and borderline personality disorder: a latent class analysis. Eur J Psychotraumatol. 2014;5 https://doi.org/10.3402/ejpt.v5.25097.
120. Kendler KS. An historical framework for psychiatric nosology. Psychol Med. 2009;39(12):1935–41. https://doi.org/10.1017/S0033291709005753.
121. Carcone D, Ruocco AC. Six years of research on the National Institute of Mental Health's research domain criteria (RDoC) initiative: a systematic review. Front Cell Neurosci. 2017;11 https://doi.org/10.3389/fncel.2017.00046.

# 13 Childhood Trauma and Substance Dependence

Hanie Edalati

## 13.1 Introduction

Exposure to childhood maltreatment and trauma, such as neglect and abuse, has been related to substance use disorders (SUDs) across various populations including community and clinical samples [1–11]. Longitudinal and cross-sectional studies have confirmed that childhood maltreatment is a robust risk factor for development of SUDs, regardless of the population studied, type of maltreatment, and measures used [2, 12–15]. A review article of 31 population-based studies has indicated that nearly all included studies confirmed significant positive associations between childhood maltreatment and adolescent substance abuse [16]. In addition, all types of childhood maltreatment have been reported to increase the risk of substance use in adulthood [2]. There is also a high co-occurrence of different types of childhood maltreatment in relation to the risk of substance use [17] such that exposure to multiple types of maltreatment is associated with cumulative risk of substance use [14]. For instance, exposure to combined physical and sexual abuse, compared to no history of maltreatment, is associated with a fivefold increase in using alcohol and a tenfold increase in illicit drug use [7].

Findings from the Adverse Childhood Experiences (ACEs) study on 8613 adults who attended a primary care clinic in California indicated that exposure to ACEs, such as neglect, abuse and household dysfunction, accounts for 64% of the population attributable risk for illicit drug use and addiction with each ACE enhancing the likelihood for early onset of drug use by two to fourfold [14]. In addition, the ACE

H. Edalati (✉)
CHU Sainte-Justine Research Center, Department of Psychiatry, University of Montreal, Montreal, QC, Canada

Centre for Research on Children and Families, School of Social Work, McGill University, Montreal, QC, Canada
e-mail: hanie.edalati@umontreal.ca; hanie.edalati@alumni.ubc.ca



G. Spalletta et al. (eds.), *Childhood Trauma in Mental Disorders*,
https://doi.org/10.1007/978-3-030-49414-8_13

score showed a strong graded association with the onset of drug use, including initiation of illicit drug use during three age categories (i.e. early adolescence ($\leq$14 years), mid-adolescence (15–18 years), and adulthood ($\geq$19 years)), drug addiction, drug use problems, and parenteral drug use. Individuals who reported five or more ACEs showed a seven- to tenfold higher risk of illicit drug use problems, addiction to illicit drugs, and parenteral drug use compared with those without histories of ACE. ACE score also indicated a strong graded association with lifetime drug use for each of the four birth cohorts examined in this study [14].

In the general population, people who have experienced two or more types of childhood maltreatment compared to none are at higher risks of alcohol abuse regardless of demographic status and other variables such as age of drinking initiation, binge drinking and history of alcoholism in parents and grandparents of participants [18]. Childhood maltreatment is a strong risk factor for the earlier initiation of alcohol and illicit drugs during adolescence [3, 5, 19] and for higher rates of illegal substance use in adulthood [20]. Current or former drinkers with histories of childhood maltreatment initiate drinking earlier, and are more likely to report drinking to cope with problems compared to those without ACEs [21]. In substance abusing and dependent patients, childhood maltreatment is associated with younger age of onset of substance misuse, more frequent use of substances, more experiences of blackouts [22], and current psychological distress symptoms [23]. In addition, childhood maltreatment negatively affects the course of SUDs in patients under treatment [24, 25] and those with histories of childhood maltreatment indicate worse treatment outcomes [11]. These include greater severity of substance use, more lifetime treatment histories, higher rates of morbid course of substance abuse in adulthood [11, 25] and earlier dropout from treatment [26]. History of childhood maltreatment in patients with substance-dependent diagnosis is associated with higher rate and severity of comorbid psychiatric disorders, such as post-traumatic stress disorder (PTSD), major depressive disorder (MDD), anxiety disorders, phobia, personality disorders and suicide attempt [10, 24, 27–31].

Although numerous studies have indicated the relationship between childhood maltreatment and SUDs, the nature and characteristics underlying this relationship remain unclear. In particular, longitudinal studies that elucidated the developmental pathways from childhood maltreatment to subsequent SUDs are lacking. Previous studies have recognised some factors as mediators of this relationship. These factors include, although not being limited to, stressful life events and PTSD symptoms [32], impairments in neurological and stress response systems [33–35], symptoms of social phobia [36] and drinking motives [37, 38]. This chapter integrates findings from several cross-sectional and longitudinal studies into three main categories of explanatory factors and models to better explain how exposure to childhood maltreatment and trauma increases individual's vulnerability to SUDs. Some important models that have been suggested in previous studies to explain this path are also explained. In the final section, the clinical implications and directions for future studies are presented.

## 13.2 Mechanisms Linking Childhood Trauma and Substance Use Disorders

A number of models and theories have been proposed to explain the relationship between childhood trauma and SUDs (e.g. [39–41]). This chapter reviews three main categories of these explanatory mechanisms and models: (1) *brain and neurocognitive mechanisms*; (2) *negative reinforcement and self-medication,* and (3) *psychosocial development.* Please see Table 13.1 for a summary of the mechanisms linking childhood trauma and SUDs. *Neurocognitive models* focus on the effect of childhood trauma on the development of brain structure, function and connectivity during critical periods of childhood and adolescence [35, 42] and examine the role of neurocognitive mechanisms in the path from childhood trauma to SUDs [40, 42–44]. *Negative reinforcement and self-medication models* largely focus on the coping effects of substances, and the use of alcohol and drugs to reduce negative affect and enhance positive affect [45, 46]. Such models explain how trauma-inducing effects of childhood maltreatment negatively influence the stress response systems across lifespan [33, 47] and consequently increase vulnerability to PTSD and internalising disorders (e.g. depression and anxiety). According to these models, substances may be used as a way to self-medicate, to cope with post-traumatic distress, and to regulate emotions [8, 29, 32, 48, 49]. *Psychosocial development models* largely focus on the effect of childhood trauma and growing up in an unfavourable environment on the psychosocial development of a child (e.g. relationship difficulties, reduced self-esteem, development of personality risk profiles) and consequent vulnerability to SUDs [50–54]. Each section explains dominant models and explanatory mediators that link the experience of childhood trauma to the risk of SUDs. It is important to note that these models have several overlaps in their concepts and constructs and occasionally have used similar mediating factors to explain the relationship between childhood trauma and the risk of SUDs.

### 13.2.1 Brain and Neurocognitive Models and Mechanisms

Evidence from animal and human studies has indicated that childhood trauma dramatically impacts the normal development of the brain [44]. Exposure to maltreatment in childhood, when the brain has the highest level of plasticity, can lead to permanent changes in multiple brain circuits involved in the processing of environmental stimuli, also impacting the normal regulation of autonomic, behavioural and endocrine responses to stress (For a review, see [55, 56]). In a series of studies, Teicher and colleagues (see [43, 44, 57–59]) have proposed a cascade model to explain the effect of childhood maltreatment on later psychopathology, and have provided evidence for sensitive periods in development in which specific brain regions are highly vulnerable to the adverse effects of maltreatment and early stress. According to this model, exposure to early stress and maltreatment activates systems involved in stress response, including glucocorticoid, vasopressin-oxytocin

**Table 13.1** Summary of the mechanisms linking childhood trauma and substance use disorders

| Main categories of explanatory models and mechanisms | Models of interest | Focus and objective of related research | Potential mechanisms/ mediators |
|---|---|---|---|
| Brain and neurocognitive difficulties | Cascade Model [43, 44, 57–59]<br>Dual-Process Models [40, 51] | Investigation of the effect of childhood trauma on the normal development of brain structure, function and connectivity during critical periods of childhood and adolescence and its impact on SUDs[a] and other psychopathology<br>Study of the role of neurocognitive mechanisms in the path from childhood trauma to SUDs | Alterations in brain structure, function, and connectivity |
| | | | Behavioural inhibition, impulsivity |
| | | | Memory (formation, content, storage and retrieval) |
| | | | Attention |
| | | | Intellectual performance |
| Negative reinforcement and self-medication | Developmental Traumatology Model [33, 132, 154]<br>Negative Reinforcement Model [45]<br>Self-medication Model [46] | Investigation of the trauma-inducing effects of childhood maltreatment on stress response systems across lifespan and consequent vulnerability to PTSD[b] and internalising/externalising disorders as a risk factor for increased SUDs<br>Study of substances use as a way to self-medicate, to cope with post-traumatic distress, and to regulate emotions/affect among maltreated individuals | Difficulties in stress response systems and regulation |
| | | | Emotion/affect regulation motives for substance use (coping, enhancing positive affect, decreasing negative affect) |
| | | | Comorbid psychiatric disorders (PTSD, anxiety and depressive disorders, externalising disorders) |
| Psychosocial development | Attachment and Relationship Difficulty Models [54, 163, 167]<br>Self-dysfunction models [52, 186] | Investigation of the effect of childhood trauma and adverse environment on the psychosocial development of a child and consequent vulnerability to SUDs. | Insecure attachment styles (avoidant and fearful) |
| | | | Parent–child relationship difficulties |
| | | | Low self-esteem |
| | | | Personality risk profiles (impulsivity, sensation seeking, hopelessness, urgency, anxiety sensitivity) |

[a]*SUDs* substance use disorders
[b]*PTSD* post-traumatic stress disorder

and noradrenergic systems, and consequently enhances the stress response. Increases in stress hormones adversely influence neural morphology, neurogenesis, synaptogenesis and myelination, which result in alterations in the structure, function and connectivity of various brain areas. The sensitivity of the brain regions to early stress and consequent alterations depends on various factors including genetics, the density of glucocorticoid receptors, timing, the rate of development and gender. Permanent consequences include reduced development of hippocampus, amygdala and left hemisphere, reduced size of the corpus callosum, diminished left/right hemisphere incorporation, reduced activity of the cerebellar vermis, and enhanced electrical activity of limbic system circuits [58–62]. These brain alterations are associated with neurocognitive dysfunctions, and increase the vulnerability to psychiatric disorders including SUDs [59].

Over the past year, the first part of this model (i.e. associations between childhood maltreatment and brain alterations) has received strong support in the literature; however, the second part connecting brain alterations to psychiatric disorders has not been replicated in all studies (see review articles [35, 42]). The association between brain alterations due to childhood maltreatment and subsequent psychopathology has been challenged by studies indicating that brain alterations observed in maltreated susceptible individuals (i.e. those who develop psychopathology) were also detectable in maltreated individuals who do not develop substance use and psychiatric disorders (i.e. clinically resilient individuals), and might not be directly related to the higher risk of psychopathology (e.g. [63–68]). Some recent studies indicated that the maltreated resilient individuals may have other differences in their brain that enable them to adaptively regulate their response to stress and maintain their mental health and well-being despite the alterations in stress-susceptible areas of the brain [42, 69]. Altogether, the relationship between childhood maltreatment, brain alterations and psychopathology remains complex and unclear and requires further investigations.

#### 13.2.1.1 Neurocognitive Markers

Exposure to childhood maltreatment is associated with a wide range of impairments in various measures of neurocognitive functions including executive function [70, 71], memory [72–75], attention [76] and intellectual performance [73, 77–80]. Similar cognitive impairments associated with childhood maltreatment have been reported as cognitive vulnerability markers for substance misuse and SUDs in longitudinal studies examining patients prior SUDs onset [81, 82]. In this line, Edalati and Krank [40] reviewed the evidence that cognitive dysfunctions mediate, at least in part, the relationship between childhood maltreatment and vulnerability to SUDs and proposed a model that explains how childhood maltreatment increases the risk for SUDs through the development of a cognitive framework of vulnerability (see Sect. 13.2.1.2 of this chapter).

The most consistently reported cognitive dysfunction is related to behavioural inhibition and impulsivity. Exposure to childhood maltreatment is associated with alterations in the normal development of brain regions underlying executive function [83] including the prefrontal cortex, which continue to develop until late adolescence and young adulthood [84]. Exposure to childhood maltreatment negatively influences executive functioning and related abilities, such as problem solving and planning [85], abstract reasoning [76] and inhibitory control [72, 73]. It has been reported that individuals who were exposed to maltreatment are more impulsive [86, 87] and tend to produce a weaker response to reward cues compared to those without such a history [88]. A history of childhood maltreatment is also associated with impairments in inhibitory control during childhood [72, 73], adolescence [89], and adulthood [90]. Longitudinal studies of children with parental substance use have examined the impact of childhood behavioural disinhibition on prospective SUDs [82, 91]. Parental substance use and dependence have been categorised as one type of adverse childhood experiences (ACEs) [92], and has been linked to both cognitive impairments in childhood and elevated rates of SUDs later in life [93]. These studies have indicated that difficulties in behavioural inhibition and higher rates of impulsivity are prevalent before the onset of substance use in children with parental substance use and that such cognitive dysfunctions strongly predict subsequent SUDs and related problems [91]. Similarly, a longitudinal study on the link between child maltreatment and adult substance use ($N$ = 9421) indicated that the relationship between child maltreatment and smoking and marijuana use in young adults were mediated by higher levels of impulsivity during adolescence [94]. Delayed reward discounting and impulsivity also mediated the links between child maltreatment and adult substance use/abuse using a cross-sectional sample of adults [94]. Another study using a sample of 10- to 15-year-old children and adolescents ($N$ = 865) indicated that exposure to victimisation, whether through parental physical abuse or peer bullying, increases cognitive impulsivity, which in turn enhances the risk of early substance misuse. This effect was significant regardless of several important protective and risk factors including family and neighbourhood characteristics [95].

Memory is another cognitive function which has received extensive attention in studies investigating the association of childhood maltreatment and SUDs. Childhood maltreatment has distinctive effects on different memory systems due to their developmental stages and processes (memory formation, storage and retrieval) [58, 89, 96, 97]. The hippocampus is one of the most critical regions in the brain which plays an important role in memory formation, storage, and retrieval [98]. The hippocampus is very sensitive to the destructive impacts of childhood maltreatment [59], and alterations in its structure may lead to increased consolidation of memory traces and continuation of intrusive memories, which are some of the main characteristic features of PTSD [96, 99]. Experience of childhood maltreatment increases the automatic (implicit) self-anxiety and self-depression memory associations assessed using the Implicit Association Task [100]. In particular, emotional maltreatment, including emotional neglect and emotional abuse, has a strong relationship with increased automatic self-anxiety and self-depression memory associations, compared with sexual and physical abuse [100]. In this line, Potthast and colleagues [101] studied the associative memory network in alcohol dependent patients with experiences of emotional abuse and alcohol dependent patients without experiences of emotional abuse, using a priming paradigm. The priming paradigm included maltreatment-related words (i.e. socially and physically threatening words) that preceded alcohol-related words. This study found a specific priming effect for maltreatment-related cues only in alcohol dependent patients with histories of emotional abuse suggesting that maltreatment-related cues automatically activate a specific associative memory network in alcoholics with emotional abuse experiences which relates the experience of childhood trauma to alcohol abuse [101]. A study by Elwyn and Smith [102] examined whether self-reported memory of childhood maltreatment mediates the association between official prospective reports of childhood maltreatment and adult substance use problems using data from longitudinal Rochester Youth Development Study (RYDS). Using Official Child Protective Services reports and adult retrospective recall (self-report) of childhood maltreatment, these authors indicated that memory of maltreatment was a key factor in linking the prospectively measured childhood maltreatment (official records) and alcohol problems and illegal drug use, regardless of other factors, such as parental substance use, adolescent substance use, and adolescent relationship with parent [102].

Other cognitive functions including intellectual performance and attention have also been suggested to explain the path from childhood maltreatment to the risk of SUDs. However, there are limited studies that specifically examined the mediating effect of these cognitive functions in the association between childhood maltreatment and SUDs. In addition, the majority of prior research has focused on the cognitive dysfunction in clinical samples following excessive and long-term substance use, and cognitive dysfunctions that are the indirect consequences of substance use on brain neuronal systems [103]. Therefore, it is not clear whether these cognitive dysfunctions existed prior to the onset of substance abuse or whether they are the consequences of long-term exposure to drugs.

#### 13.2.1.2 Dual-Process Models

A series of neurodevelopmental models of adolescent behaviours suggested that addiction is a neurodevelopmental disorder that emerges during adolescence [104]. The increased risk of substance use and risky behaviours during this stage of development might result from differential developmental rates of certain brain structures and functions [105, 106]. Brain regions involved in reward processing, motivation and emotional reactivity, such as basal ganglia/limbic structures and orbitofrontal cortex (OFC), appear to develop early during adolescence [107], whereas brain structures responsible for the top-down control of cognition, emotion and behaviour (i.e. the dorso-lateral regions of the prefrontal cortex) develop gradually into late adolescence and young adulthood [84]. In line with this differing timetable of neural development, dual-process models have suggested that the early maturity of subcortical circuitry in contrast to the late maturation of frontal cortical circuitry may result in increased reward-seeking behaviours, in the absence of an adequate behavioural control, thus predisposing adolescents to risk-taking behaviours, including substance abuse and misuse [105, 106]. In addition, these neural changes increase brain vulnerability to the direct pharmacological effects of drugs, and place adolescents who initiate substance use at early ages at a particular risk for early heavy or uncontrollable use, sensitising them to the reinforcing properties of substances and increased risk of future addictive behaviours [105].

A review article of the cognitive pathways between childhood maltreatment and SUDs [40] suggested a model to expand the dual-process models to incorporate the role of childhood maltreatment and associated impairments in cognitive functions, as to further explain vulnerability to early initiation and development of SUDs. According to this model, exposure to childhood maltreatment and growing up in an unfavourable environment play an important role in how adolescent vulnerability is translated into actual risk. Earlier in this chapter, it was explained that exposure to childhood maltreatment may lead to alterations in brain regions implicated in executive functions [70, 71], behavioural control [86, 87], and reward processing [88, 108]. Difficulties in behavioural control have been reported as one of the vulnerability markers of alcohol and drug abuse [93, 109]. Childhood maltreatment has also been related to increased maladaptive self-attitudes [110], and automatic self-anxiety and self-depression memory associations [100]. Maltreated individuals may use substances to regulate these negative self-associations and to relieve tension related to memories associated with traumatic experiences. This may act as a potential reinforcement for compulsive substance misuse (see [40]). Research shows that individuals with histories of childhood maltreatment expect more positive effects, such as relief of negative emotions, from drinking alcohol and using drugs [37], as compared to those with no childhood maltreatment history. The reinforcing effects of substances may result in shaping memory associations that link negative affect to the sedative and coping characteristics of substances and consequently elicit compulsive substance use in response to stressful situations (see [40]). Cross-sectional and longitudinal studies have indicated that memory associations about the coping effects of substance use are strong predictors of problematic drinking and drug use [111–114]. In the presence of stress-related cues, memory associations that relate substance use to sedative and coping effects may override the cognitive control system that represents the logical and rational knowledge about the effects of substance use. Difficulties in attention and working memory functions reported in individuals with histories of childhood maltreatment [72–75] may also facilitate the automatic (implicit) retrieval of memory associations. In this line, Barkowsky [115] has indicated that response inhibition and reward sensitivity moderate the relationship between automatic (implicit) memory associations and substance use in adolescents aged 13–14 years old. Difficulties in the cognitive control system in addition to dysfunctional memory associations make individuals susceptible to excessive substance misuse in response to contextual and situational stress. This conclusion is in line with dual system theories that proposed dual processes for explaining human judgement, reasoning and decision-making [112, 116–120]. During adolescence, the developmental gap between early maturity of subcortical circuitry and the late maturation of frontal cortical circuitry and related limited cognitive resources (e.g. lack of adequate top-down control, goal directedness and time perspective) particularly set the stage for increased impulsive decisions and behaviours, such as substance abuse, as a strategy to cope with the dysfunctional memory associations [51].

These processes may actively lead to changes in motives for substance use (i.e. individual's reasons for using substances) [121–124] which will further be discussed in the next section.

### 13.2.2 Negative Reinforcement and Self-Medication Models and Mechanisms

This section highlights the negative reinforcement [45] and self-medication [46] models of substance use which focus on the role of self-regulation and coping processes in the relationship between childhood trauma and SUDs. We also discuss the developmental traumatology model [76, 125, 126] which has studied the trauma-inducing effects of childhood maltreatment on the development of stress response systems as the path from childhood trauma to later SUDs. This section reviews studies that examined the mediating effects of two main sets of explanatory mechanisms related to these models: (1) emotion regulation difficulties and motives and (2) comorbid psychiatric disorders with a focus on PTSD and internalising disorders.

#### 13.2.2.1 Emotion Regulation Difficulties and Motives

Exposure to childhood maltreatment is associated with difficulties in emotion regulation and adaptive responses to stress [47, 127, 128], and increased vulnerability to the effect of later stressful life events, and also predicts continuing exposure to stressful and adverse events and circumstances [129–131]. A history of childhood maltreatment has been related to difficulties in overall emotion regulation [132], and important facets of emotion regulation, including recognition of emotions [133], emotional understanding [134], emotional non-acceptance [135], and higher levels of avoidance strategies [136].

According to both negative reinforcement [45] and self-medication [46] models of substance use, individuals with maltreatment history may use alcohol and drugs to cope with negative emotions, to enhance positive affect, and to cope with stress [137, 138]. In this line, a study on 91 recently abstinent cocaine-dependent adults in an inpatient treatment facility indicated that overall childhood maltreatment severity was significantly associated with higher perceived stress and greater use of avoidance stress-coping strategies [139]. In addition, a recent study of adult patients with concurrent substance use and psychiatric disorders indicated that exposure to childhood maltreatment was associated with higher levels of perceived stress in patients. This association was significant for all types of childhood maltreatment, including emotional and physical neglect, and emotional, physical, and sexual abuse, and remained significant after controlling for demographic factors, diagnoses of psychiatric disorders, and length of stay in the treatment centre [128]. These findings suggest that childhood maltreatment has a long-lasting effect on stress response and regulation, and that the available treatment for general substance use in psychiatric patients may not be sufficient or adequate for patients with histories of childhood maltreatment [43, 140].

Research exploring the motives for substance use have indicated that regulation of emotions and affect is a significant mediator of the relationship between childhood maltreatment and SUDs [37, 38]. Motives reflect individual's reasons for using substances and are important mechanisms that explain the relationship between childhood maltreatment and SUDs and related harm [141]. Several studies have indicated the mediating effect of emotion regulation motives in the path from maltreatment history to SUDs in substance-dependent patients. Importantly, among substance-dependent patients, a history of childhood maltreatment was associated with important dimensions of emotion dysregulation [142]. Drinking motives such as drinking to cope with problems and negative mood, and drinking to enhance positive mood are of great importance [37, 38]. Vilhena [143] indicated that drinking to cope plays a mediating role in the relationship between all types of child maltreatment and alcohol use outcomes among young adults. A study with a sample of adult acute-care psychiatric inpatients indicated that emotion regulation difficulties mediate the relationship between severity of childhood maltreatment and alcohol use severity and problems [144]. Similar mediating effects have been found for the coping motives in the relationship between childhood maltreatment and marijuana problems in a sample of young adults [145]. In addition, emotion dysregulation influenced coping motives and through that increased marijuana problems in maltreated individuals [145]. Another study with a sample of adolescent users of Tramadol, a synthetic opiate, has shown that emotion dysregulation mediated the association of childhood emotional abuse and enhanced motives for substance use [146]. The present findings highlight the role of coping motives and difficulties in emotion regulation as important mechanisms that link the experience of childhood maltreatment to the vulnerability to SUDs. They also highlight the importance of targeted interventions to address the needs of these patients for reducing stress and improving emotional regulation and coping skills [128].

#### 13.2.2.2 Comorbid Psychiatric Disorders

There is evidence that the association between childhood trauma and SUDs is mediated, at least partly, by psychiatric disorders, in particular, post-traumatic stress disorder (PTSD), and anxiety and depressive disorders [8, 29, 32, 48, 49]. For example, in individuals with a lifetime diagnosis of SUDs, mood and anxiety disorders appeared on average 3 years before the first SUD diagnosis and also partially mediated the effect of childhood maltreatment on the risk for SUDs compared to controls. However, childhood maltreatment also has shown an independent strong effect on the development of SUDs, regardless of mood and anxiety disorders [29]. Other studies have indicated that post-traumatic symptoms mediate the relationship between childhood maltreatment and cannabis use in female adolescents (but not males) followed by the child welfare system; the mediating effect of post-traumatic symptoms was significant in the relationship between childhood maltreatment and alcohol use and binge drinking for both genders [147].

PTSD has been found to play a particular mediating role in the relationship between childhood maltreatment and SUDs [32, 147]. PTSD is a mental disorder that may develop after experiencing or witnessing life-threatening or terrifying events, such as natural disaster, sexual assault or car accident, which exhaust individual's capacity to cope. Some post-traumatic symptoms may include flashbacks, severe anxiety, nightmares, invasive uncontrollable thoughts about the event, and emotional numbness. The developmental traumatology model suggests that exposure to chronic childhood trauma has more disruptive physical and psychological effects on developing child compared to an acute or single incidence of trauma, and therefore creates more complex symptomology and PTSD [33, 126, 148]. This model emphasises the traumatic nature of childhood maltreatment and trauma [33, 47, 126], and the consequent symptoms, even when the post-traumatic reactions and symptoms do not meet the diagnostic criteria for PTSD.

The model also suggests that post-traumatic distress, and problem in coping with the trauma experience explain the path from childhood maltreatment to subsequent SUDs [33, 126]. According to this model, exposure to childhood trauma alters individual's stress response systems [33], and increases symptoms of post-traumatic stress [149, 150]. SUDs are the result of a long-term dysfunctional regulation of trauma-related symptoms caused by the maltreatment experience. This dysfunctional regulation of trauma-related distress may increase vulnerability to PTSD and/or affective disorders (e.g. depressive and anxiety disorders), which in turn enhances the risk for substance abuse and dependence [47]. Maltreated individuals may use substances to reduce their negative affect and enhance positive emotions associated with traumatic memories. According to laboratory-based experimental studies on cue-elicited craving, exposure to trauma cues and PTSD symptoms may automatically activate the memory associations and network that connect negative emotions to the coping and sedative effects of alcohol and drugs and thus stimulates craving [101, 151]. This explanation is also in line with the negative reinforcement [45] and self-medication [46] models of substance use.

Other studies have indicated that negative emotions such as shame or guilt, which are prevalent in maltreated individuals, may also stimulate craving in individuals with SUD diagnoses regardless of comorbid PTSD diagnosis [152]. In addition, post-traumatic symptoms and diagnosis may increase the risk of substance abuse and dependence, independent of traumatic experiences [153]. These findings have been replicated in both adolescents [154] and young adults [155]. Externalising disorders and symptoms (e.g. conduct problems, attention-deficit/hyperactivity disorder (ADHD)) are other risk factors that explain the pathway from childhood maltreatment to the onset and development of substance use problems. Findings from the Longitudinal Studies of Child Abuse and Neglect (LONGSCAN) which assessed participants at ages 4, 8, 12, and 18 indicated that a history of childhood maltreatment, and in particular sexual abuse and neglect, predicts the age of onset for alcohol and marijuana use through externalising behaviours (assessed at age 8) [156]. However, internalising behaviours did not mediate the association between childhood maltreatment and onset of substance use in this study [156]. Further research is needed to elucidate the role of comorbid psychiatric disorders in the path from childhood maltreatment to SUDs.

### 13.2.3 Psychosocial Development Models and Mechanisms

Psychosocial models of development also sought to explain the potential underlying pathways linking childhood maltreatment and SUDs. The following section reviews some of the important models and theories that explain how exposure to childhood maltreatment is related to an increased vulnerability to SUDs through reduced self-esteem (e.g. self-dysfunction models [52, 53]; also see the developmental social neuroscience model [51]), relationship difficulties (attachment theory; [54, 157, 158]), and the development of personality risk profiles [50, 86, 159, 160].

#### 13.2.3.1 Attachment and Relationship-Focused Models

Attachment and relationship difficulty models suggest that growing up in adverse and neglectful situations negatively influences child's social and psychological development and the interactions with parents, partners, and peers [51, 54, 157, 161]. In particular, lack of positive caregiver–child relationships and the development of insecure attachment adversely affect the adaptive emotion regulation and behavioural functioning throughout life [162].

Attachment theories particularly suggest that insecure attachment styles (i.e. avoidant and fearful) and subsequent compliant or aggressive behavioural patterns connect the experience of childhood maltreatment to later SUDs [163–166]. Research has indicated that exposure to childhood maltreatment and parental neglect are associated with developing poor mother–child relationship quality and insecure attachment in preschool-aged children [167] which, in turn, increase the risk of substance abuse during adolescence [168]. These dysfunctional attachment styles and related aggressive or compliant behaviours [158] have been connected to problematic relationship with parents (e.g. communication, monitoring, emotional support) and lack of social competency (e.g. conflict with or isolation from peers) during adolescence [168, 169] which, in turn, increase the risk of being involved in deviant peer groups and subsequent substance misuse and related problems [170]. In addition, difficulties in emotion regulation due to the experience of childhood maltreatment and parental neglect may result in the development of maladaptive coping strategies, such as substance abuse [158, 171]. Dunn and colleagues similarly suggested that an insecure attachment style is the mechanism that results in long-lasting use of maladaptive psychosocial functioning linking childhood maltreatment to later substance abuse and SUDs [163].

The relationship with parents and the attachment style also affect the socialisation processes in maltreated individuals. For example, a recent study found that exposure to childhood maltreatment was linked to adolescent and adult marijuana use through parental attachment, which in turn negatively affected the tendency to be involved with peers who approve and use marijuana [172]. These studies support suggestions from social developmental theories [173] and Family Interaction Theory (FIT) [174] that focus on the effect of family and social environment and relationships in linking a history of childhood maltreatment to later substance misuse. For example, within the framework of the FIT model [174], the presence of a close affectional bond, a conflict-free relationship between parent and child, and adolescents' identification with parents' personality, attitudes, and behaviours are critical protective factors in preventing substance abuse [157]. This leads to the formation of psychologically healthy and conventional personality characteristics in adolescence which discourage them from affiliations with peers who use substances and result in adolescent's protection against substance use [157]. Studies based on the FIT developmental model [174] have indicated that a poor parent–child bond was related to the development of vulnerable personality traits (e.g. impulsivity, rebellion), which was related to selection of drug-using peers and partners, and greater risk of SUDs [54]. Similarly, adolescents from neglecting parents (i.e. inadequate monitoring, supervision, communication, and emotional support) were more likely to be influenced by social pressure to drink alcohol and show a higher risk of developing alcohol use disorders compared to non-neglected peers [175]. Most studies on the mediating role of attachment and relationship difficulties are focused on the impact of parental neglect on substance abuse. Additional research is required to precisely examine these developmental models in the context of other types of child maltreatment and trauma. Moreover, it is required to further examine the impact of insecure attachment and mother–child relationship as important mechanisms that link childhood maltreatment and SUDs across the life span.

Socialisation processes and attachment also influence the development of personality traits and self-esteem which, in turn, influence vulnerability to SUDs. The following sections provide more specific studies examining the mediating roles of self-esteem and personality risk factors in the association of childhood maltreatment and SUDs.

#### 13.2.3.2 Self-Esteem

The construct representing an overall evaluation of self-worth (e.g. self-esteem, self-evaluation, self-association, self-processing, self-view) plays an important role in individual's mental health [176]. A negative and unstable self-esteem is associated with higher vulnerability to internalising and externalising problems [177] and substance misuse [178]. The social environment in which individuals develop and interact with other people has a significant impact on the development of self-esteem. Exposure to adverse experiences such as trauma and victimisation has been associated with developing a negative self-esteem [179]. Self-dysfunction models have linked a low self-esteem (i.e. self-derogation) to a higher risk of using substances in order to relieve the emotional pain associated with the experience of childhood maltreatment [52, 180]. Research with incarcerated adolescents has indicated that self-derogation mediated the association of physical and sexual abuse and adolescent substance abuse, regardless of gender [52, 53]. Another study with a group of homeless women indicated a significant association between the experience of child abuse, low self-esteem, depression and substance abuse [181].

Reporting a low level of self-esteem increases the risk of externalising problems, such as delinquency, aggression and antisocial behaviour, in adolescents [182]. This relationship is significant even when the impact of important factors such as achievement-test scores, IQ, supportive parenting, parent–child and peer relationships, narcissism, and socioeconomic status was controlled for [182]. During adolescence, feedback and evaluation of peers become an important source of individual's self-evaluation and self-esteem. Adolescents with a low self-esteem may use substances to cope with the negative affect associated with low self-esteem or as an impression management strategy to gain acceptance and approval in peer groups [51, 183, 184]. Lower levels of self-esteem in adolescence have been related to a history of victimisation (see [179] for meta-analysis and review) and a higher risk of substance misuse during this significant developmental period [51]. A poor emotional regulation and deficient social skills which are prevalent in children and adolescents with histories of maltreatment may affect peer acceptance and rejection resulting in isolation from the mainstream peers [185] or drive them towards deviant peer groups in which behaviours such as violence or substance abuse are central for peer acceptance. This view is in line with the self-derogation/self-enhancement theory of deviant behaviours suggesting that involvement with deviant peer group, and behaviours such as substance abuse may be a way for adolescents to increase their self-esteem [186, 187].

#### 13.2.3.3 Personality Risk Profiles

Exposure to childhood trauma is associated with the development of maladaptive personality profiles [50, 86, 159]. A longitudinal study of 7485 individuals in the age ranges of 20–24, 40–44, and 60–64 years has indicated that higher levels of childhood maltreatment are associated with higher risk of developing maladaptive personality traits, including lower behavioural inhibition and dissocial behaviour, and higher neuroticism and negative affect [87]. Results from cross-sectional [160] and longitudinal [188] studies have indicated that some personality profiles mediate the association between childhood maltreatment and subsequent SUDs. For example, personality traits of impulsivity and sensation seeking mediate the association between childhood maltreatment (e.g. violence, sexual abuse) and substance misuse in community samples of adolescents [160, 189]. In adolescents receiving child welfare services, personality traits of impulsivity, sensation seeking and hopelessness were positively correlated with higher reporting of drinking and more alcohol problems, whereas anxiety sensitivity (i.e. a fear of anxiety-related physical sensations) was associated with difficulties in stopping drinking [190]. Similarly, a study using a community sample of young adults aged 18–25 ($N = 268$) indicated that emotional abuse increases the risk of alcohol use and problems through the personality trait of urgency (i.e. defined as the tendency to act impulsively when distressed) [191].

Personality traits also explain the motivation underlying substance use behaviours in individuals with histories of childhood maltreatment and trauma. For instance, the personality trait of anxiety sensitivity has been related to drinking to conform, whereas hopelessness and impulsivity were linked to drinking to cope with negative emotions in youth receiving child welfare services [192]. These findings suggest that maltreated individuals may use alcohol and drugs to cope with the negative emotions associated with their maladaptive personality traits (i.e. negative reinforcement). Since early relationships and the social and familial environment of a child play a critical role in the development of personality, individuals who grow up in an unfavourable environment (e.g. exposure to abuse and neglect, dysfunctional and hostile caretaking environment) may not be able to develop adaptive personality traits which help them to effectively cope with distressful circumstances and situations [51, 191, 193, 194].

## 13.3 Moderating Effects of Sex/Gender and Race/Ethnicity Differences in the Relationship Between Childhood Trauma and Substance Use Disorders

Factors including gender differences and ethnicity have been suggested to moderate the effect of childhood maltreatment on SUDs. The experience of childhood maltreatment is overrepresented among both men and women with SUDs [8, 195]. However, some gender differences have been reported in previous studies. For example, some studies demonstrated that the prevalence of childhood sexual and physical abuse among substance dependent women is two to three times higher than men and that the association of childhood maltreatment and later SUDs is stronger for women (e.g. [8, 196, 197]). Some studies have indicated that women use more internalising behaviours to cope with the negative affect associated with the experience of trauma and therefore are more prone to use substances to self-medicate, compared to their male counterparts [198]. However, other studies have found opposite or inconsistent findings regarding the effect of gender in the relationship between childhood maltreatment and SUDs [195]. A systematic review of longitudinal studies on the role of gender in the association between childhood maltreatment and substance use outcomes indicated that findings on the effect of gender are inconsistent or mixed [195]. The review found that timing of childhood maltreatment experience and the differences in measurement used to assess substance use outcomes, and the sample composition were potentially important factors for identifying gender effects in the link between childhood maltreatment and SUDs. For example, five of the six papers which indicated gender differences in this review article [195] focused on childhood maltreatment that happened before the age of 12 [199–203], while studies that found no gender effects assessed maltreatments that occurred before the age of 18 (during childhood or adolescence) [204, 205]. Findings from many of these studies suffer from several methodological issues making it difficult to draw a clear conclusion about the presence or absence of gender differences. A further examination of the effect of gender in the relationship between childhood maltreatment and SUDs is needed, in particular, in samples of young adults and in later developmental stages [195]. A recent study using a large sample of adults ($N = 60{,}598$) examined whether race/ethnicity and gender influenced the associations between adverse childhood experiences (ACEs), categorised as household challenges and child abuse, and adult mental health and alcohol outcomes (i.e. heavy drinking and binge drinking) [206]. Findings indicated that gender did not moderate the relationships between ACEs and excessive alcohol use or depression. Race/ethnicity moderated the association between ACEs and heavy drinking, but not other excessive alcohol use outcomes [206]. Hispanics who reported both child abuse and household challenges were more likely to report heavy drinking compared to non-Hispanic whites or blacks who also reported both ACEs. However, the experience of ACEs was associated with higher rates of heavy drinking in all races compared to those with no ACEs exposure [206]. Additional research is required to clearly understand how gender and race/ethnicity interact with the relationship between childhood maltreatment and SUDs.

## 13.4 Clinical Implications

Earlier in this chapter, studies were reviewed which showed that substance dependent patients with histories of childhood trauma have worse treatment outcomes [11, 26] compared with patients without such histories. Other studies indicated that the available treatment for SUDs in general psychiatric patients may not be sufficient or adequate for patients with histories of childhood maltreatment [43, 128, 140]. Unfortunately, there are limited interventions available for substance dependent adults with histories of childhood trauma that impact on both substance use and trauma history. In general, these available interventions are trauma-focused therapies for adults with PTSD and substance use disorders, and not necessarily target histories of childhood trauma [207, 208]. Other line of interventions that address SUDs in individuals with histories of childhood maltreatment target youth in the child welfare system. This section focuses on the limited interventions available for reducing substance use problems in youth involved in the child welfare system. Child welfare systems are designed to receive and investigate reports of possible child abuse and neglect and to provide multi-level services to the maltreated children and adolescents.

Youth involved in the child welfare system are at a high risk of early initiation of substance use and the development of substance use disorders [147, 209–211]. Several factors, such as exposure to childhood maltreatment, parental substance use, multiple placement changes, and lack of family support when transitioning into independent living situations, contribute to the increased risk of SUDs and related problems in youth involved in the child welfare system [212–214]. Edalati and Conrod [215] have reviewed the risk and protective factors for SUDs among youth involved in the child welfare systems and the available interventions for reducing substance use problems in this population. This review suggests that there is still an enormous gap between the needs and availability of the intervention programs specifically targeting substance misuse and related problems in this group [215]. Several interventions have been designed and used to reduce the emotional and behavioural problems, to improve the developmental outcomes, to alleviate the symptoms of complex trauma, and to learn how to cope with negative emotions associated with the traumatic experience in youth involved in the child welfare systems (for a review, see [216, 217]). Although these interventions address the potential pathways from childhood maltreatment to SUDs, such programs do not directly target the risk of substance use initiation and disorders and, in particular, do not address the needs of youth most at risk of transitioning to substance use disorders, and those who have already started using substances [215]. Among few promising approaches to reduce substance use problems in this populations, the Attachment, Self-Regulation and Competency (ARC) intervention [218, 219], the Multidimensional Treatment Foster Care for Adolescents (MTFC-A) [220–223], and the Middle School Success (MSS), a derivative of 'Keeping Foster Parents Trained and Supported' (KEEP) program [224, 225], have shown some limited success in reducing substance misuse and related problems in youth involved in the child welfare systems. Despite these efforts, there are significant limitations in the current research and practice of substance use interventions for these vulnerable populations. Some important limitations of these intervention studies include small effect sizes, short-term effects, limited baseline (pre-intervention) and long-term follow-up data, limited generalisability to other outcome measures and populations, and methodological issues concerning the study design and blindness to study conditions (see [216]).

Adolescence is a critical period for the prevention of SUDs in individuals with childhood trauma experiences. Community- and school-level interventions should extend beyond existing substance use prevention strategies, which target general populations of adolescents (e.g. universal approaches), and aim at providing services for prevention of substance use for high-risk adolescents. It has been suggested that the widespread implementation of selective and targeted intervention programs, such as personality-targeted approach, at the school level offers great advantages over universal approaches, and benefits both youth involved in child welfare systems and those exposed to childhood trauma or living in vulnerable context, but unknown to child welfare services [215, 226]. For children and adolescents who are involved in the child welfare systems, key factors in providing effective substance use intervention include the following: (1) early screening for substance use, (2) development of targeted multi-level treatment interventions, (3) better education and training for the health practitioners, child welfare workers, group home staff, foster parents and school personnel, and (4) efficient communication and better collaboration among these individuals and services. Moreover, trauma-focused interventions have rarely been evaluated for the impact on adolescent substance use outcomes. Such interventions must integrate treatment for both trauma and SUDs using a developmental approach in order to equally impact on substance use and mental health outcomes [215, 227].

In sum, there is a pressing need for the development and implementation of evidence-based targeted substance use prevention strategies that address the special needs and risks of adolescents with truamatic experiences at earlier ages before their vulnerabilities become severe. There is also a great need for the development of targeted interventions that address the specific history of childhood maltreatment in substance-dependent adults to improve their treatment outcomes [43, 140].

## 13.5 Future Directions and Conclusions

This chapter reviewed findings from a number of studies exploring the pathways from childhood trauma to SUDs. Overall, several studies have reported a robust relationship between childhood trauma and vulnerability to SUDs. However, most studies that investigated mediating mechanisms in this relationship were limited in several ways, and the future research is needed to explore unanswered questions in this area. For example, it is not exactly clear whether observed dysfunctions or difficulties are due to the history of childhood trauma or ascribable to other unseen factors (e.g. pre-existing and genetic risk factors, socioeconomic status) that were not controlled in prior study designs. As discussed earlier in this chapter, many of the models and mediating mechanisms reviewed have several overlaps in their concepts and constructs. Moreover, there is a strong interrelationship between some of the reviewed functions and mechanisms. For example, there is a strong relationship between general IQ score and specific cognitive abilities such as memory and executive function [228]. Similarly, maladaptive personality traits not only mediate the association between childhood maltreatment and subsequent SUDs [160, 188], but also explain the motives (e.g. emotion regulation) underlying substance use behaviours [192]. In addition, only few studies have controlled for comorbidity of current and lifetime psychiatric disorders [71] and even the limited number of studies that controlled for some comorbidity produced inconsistent outcomes. Differences in the populations under investigation (e.g. adults vs. adolescents) and small sample sizes contribute to some inconsistencies in findings related to the impact of childhood trauma on mediating mechanisms resulting in SUDs. While the several dysfunctions and difficulties leading to early substance use in maltreated individuals initiate early in childhood, other factors and future events and circumstances may influence the course of the disorder later in life. In particular, findings from most previous studies are based on cross-sectional data and self-reported measures, making it difficult to determine a clear and causal relationship. Future studies are required to separate the effects of childhood maltreatment on mediating mechanisms from the effect of other factors that may influence the risk of SUDs. Yet, there is a great need for large longitudinal studies that precisely investigate the sequence of events in individuals with documented histories of childhood maltreatment.

Most previous studies on the relationship between childhood trauma and SUDs examined substance use problems in individuals reporting histories of childhood maltreatment compared to those without such histories. Childhood maltreatment often is not an isolated one-time event. Important characteristics of the childhood maltreatment which may affect the long-term outcomes, such as severity, co-occurrence of multiple types, chronicity, and age of occurrence, are largely understudied. In addition, further investigation on the impact of neglect, emotional maltreatment, and household dysfunction on SUDs are needed as most prior research has focused on the effects of physical and sexual abuse or the cumulative effect of childhood trauma on the risk of SUDs.

Despite evidence of elevated risk of substance use in youth involved in the child welfare services [147, 209–211], few systematic studies have been performed on the specific patterns of substance use across developmental stages in these vulnerable populations. There is a pressing need to identify unique risk and protective factors of SUDs among youth involved in child welfare systems as the risk and protective factors influencing SUDs in this population might be different from the ones identified in maltreated youth in the community or school-based populations.

Finally, not every individual who is exposed to childhood trauma will develop a problem with SUDs. Few studies, however, have explored resilience in the context of substance use outcomes in maltreated individuals. Another interesting line of research is to compare resilient individuals with those who developed SUDs for possible protective factors. A better understanding of factors underlying resilience to SUDs in populations with traumatic experiences will largely benefit the intervention efforts.

**Acknowledgement** The author of this chapter is supported through a postdoctoral fellowship by the Canadian Institutes of Health Research (CIHR).

*Declarations of competing interest*: The author declares no potential conflicts of interest as regards the authorship and/or publication of this chapter.

## References

1. Anda RF, Whitfield CL, Felitti VJ, Chapman D, Edwards VJ, Dube SR, et al. Adverse childhood experiences, alcoholic parents, and later risk of alcoholism and depression. Psychiatr Serv. 2002;53(8):1001–9.
2. Dube SR, Anda RF, Felitti VJ, Edwards VJ, Croft JB. Adverse childhood experiences and personal alcohol abuse as an adult. Addict Behav. 2002;27(5):713–25.
3. Dube SR, Miller JW, Brown DW, Giles WH, Felitti VJ, Dong M, et al. Adverse childhood experiences and the association with ever using alcohol and initiating alcohol use during adolescence. J Adolesc Health. 2006;38(4):444.e1–10.
4. Edalati H, Barkowsky D, Krank MD. A brief assessment of violence exposure and neglect as a predictor of alcohol use in adolescents: validation and gender differences. Alcohol Clin Exp Res. 2011;35(6):100A.
5. Enoch MA. The role of early life stress as a predictor for alcohol and drug dependence. Psychopharmacology. 2011;214(1):17–31.
6. Langeland W, Hartgers C. Child sexual and physical abuse and alcoholism: a review. J Stud Alcohol. 1998;59(3):336–48.
7. Moran PB, Vuchinich S, Hall NK. Associations between types of maltreatment and substance use during adolescence. Child Abuse Negl. 2004;28(5):565–74.
8. Simpson TL, Miller WR. Concomitance between childhood sexual and physical abuse and substance use problems. A review. Clin Psychol Rev. 2002;22(1):27–77.
9. Tam TW, Zlotnick C, Robertson MJ. Longitudinal perspective: adverse childhood events, substance use, and labor force participation among homeless adults. Am J Drug Alcohol Abuse. 2003;29(4):829–46.
10. Windle M, Windle RC, Scheidt DM, Miller GB. Physical and sexual abuse and associated mental disorders among alcoholic inpatients. Am J Psychiatry. 1995;152(9):1322–8.
11. Westermeyer J, Wahmanholm K, Thuras P. Effects of childhood physical abuse on course and severity of substance abuse. Am J Addict. 2001;10(2):101–10.

12. Nelson EC, Heath AC, Madden PA, Cooper ML, Dinwiddie SH, Bucholz KK, et al. Association between self-reported childhood sexual abuse and adverse psychosocial outcomes: results from a twin study. Arch Gen Psychiatry. 2002;59(2):139–45.
13. Kendler KS, Bulik CM, Silberg J, Hettema JM, Myers J, Prescott CA. Childhood sexual abuse and adult psychiatric and substance use disorders in women: an epidemiological and cotwin control analysis. Arch Gen Psychiatry. 2000;57(10):953–9.
14. Dube SR, Felitti VJ, Dong M, Chapman DP, Giles WH, Anda RF. Childhood abuse, neglect, and household dysfunction and the risk of illicit drug use: the adverse childhood experiences study. Pediatrics. 2003;111(3):564–72.
15. Dunn GE, Ryan JJ, Dunn CE. Trauma symptoms in substance abusers with and without histories of childhood abuse. J Psychoactive Drugs. 1994;26(4):357–60.
16. Tonmyr L, Thornton T, Draca J, Wekerle C. A review of childhood maltreatment and adolescent substance use relationship. Curr Psychiatr Rev. 2010;6(3):223–234(12).
17. Arata CM, Langinrichsen-Rohling J, Bowers D, Farrill-Swails LO. Single versus multi-type maltreatment: an examination of the long-term effects of child abuse. J Aggress Maltreat Trauma. 2005;11:29–52.
18. Pilowsky DJ, Keyes KM, Hasin DS. Adverse childhood events and lifetime alcohol dependence. Am J Public Health. 2009;99(2):258–63.
19. Hamburger ME, Leeb RT, Swahn MH. Childhood maltreatment and early alcohol use among high-risk adolescents. J Stud Alcohol Drugs. 2008;69(2):291–5.
20. Madruga CS, Laranjeira R, Caetano R, Ribeiro W, Zaleski M, Pinsky I, et al. Early life exposure to violence and substance misuse in adulthood—the first Brazilian national survey. Addict Behav. 2011;36(3):251–5.
21. Rothman EF, Edwards EM, Heeren T, Hingson RW. Adverse childhood experiences predict earlier age of drinking onset: results from a representative US sample of current or former drinkers. Pediatrics. 2008;122(2):e298–304.
22. Brems C, Namyniuk L. The relationship of childhood abuse history and substance use in an Alaska sample. Subst Use Misuse. 2002;37(4):473–94.
23. Medrano MA, Hatch JP, Zule WA, Desmond DP. Psychological distress in childhood trauma survivors who abuse drugs. Am J Drug Alcohol Abuse. 2002;28(1):1–13.
24. Evren C, Kural S, Cakmak D. Clinical correlates of childhood abuse and neglect in substance dependents. Addict Behav. 2006;31(3):475–85.
25. Langeland W, Draijer N, van den Brink W. Psychiatric comorbidity in treatment-seeking alcoholics: the role of childhood trauma and perceived parental dysfunction. Alcohol Clin Exp Res. 2004;28(3):441–7.
26. Claus RE, Kindleberger LR. Engaging substance abusers after centralized assessment: predictors of treatment entry and dropout. J Psychoactive Drugs. 2002;34(1):25–31.
27. Evren C, Evren B, Dalbudak E, Ozcelik B, Oncu F. Childhood abuse and neglect as a risk factor for alexithymia in adult male substance dependent inpatients. J Psychoactive Drugs. 2009;41(1):85–92.
28. Ellason JW, Ross CA, Sainton K, Mayran LW. Axis I and II comorbidity and childhood trauma history in chemical dependency. Bull Menn Clin. 1996;60(1):39–51.
29. Douglas KR, Chan G, Gelernter J, Arias AJ, Anton RF, Weiss RD, et al. Adverse childhood events as risk factors for substance dependence: partial mediation by mood and anxiety disorders. Addict Behav. 2010;35(1):7–13.
30. Bernstein DP, Stein JA, Handelsman L. Predicting personality pathology among adult patients with substance use disorders: effects of childhood maltreatment. Addict Behav. 1998;23(6):855–68.
31. Schumacher JA, Coffey SF, Stasiewicz PR. Symptom severity, alcohol craving, and age of trauma onset in childhood and adolescent trauma survivors with comorbid alcohol dependence and posttraumatic stress disorder. Am J Addict. 2006;15(6):422–5.
32. White HR, Widom CS. Three potential mediators of the effects of child abuse and neglect on adulthood substance use among women. J Stud Alcohol Drugs. 2008;69(3):337–47.

33. De Bellis MD, Baum AS, Birmaher B, Keshavan MS, Eccard CH, Boring AM, et al. Developmental traumatology Part I: Biological stress systems. Biol Psychiatry. 1999;45(10):1259–70.
34. Hart H, Rubia K. Neuroimaging of child abuse: a critical review. Front Hum Neurosci. 2012;6:52.
35. Teicher MH, Samson JA. Annual research review: enduring neurobiological effects of childhood abuse and neglect. J Child Psychol Psychiatry. 2016;57(3):241–66.
36. DeWit DJ, MacDonald K, Offord DR. Childhood stress and symptoms of drug dependence in adolescence and early adulthood: social phobia as a mediator. Am J Orthopsychiatry. 1999;69(1):61–72.
37. Goldstein AL, Flett GL, Wekerle C. Child maltreatment, alcohol use and drinking consequences among male and female college students: an examination of drinking motives as mediators. Addict Behav. 2010;35(6):636–9.
38. Grayson CE, Nolen-Hoeksema S. Motives to drink as mediators between childhood sexual assault and alcohol problems in adult women. J Trauma Stress. 2005;18(2):137–45.
39. Hovdestad WE, Tonmyr L, Wekerle C, Thornton T. Why is childhood maltreatment associated with adolescent substance abuse? A critical review of explanatory models. Int J Ment Heal Addict 2011;9(525).
40. Edalati H, Krank MD. Childhood maltreatment and development of substance use disorders: a review and a model of cognitive pathways. Trauma Violence Abuse. 2016;17(5):454–67.
41. Puetz VB, McCrory E. Exploring the relationship between childhood maltreatment and addiction: a review of the neurocognitive evidence. Curr Addict Rep. 2015;2(4):318–25.
42. Teicher MH, Samson JA, Anderson CM, Ohashi K. The effects of childhood maltreatment on brain structure, function and connectivity. Nat Rev Neurosci. 2016;17(10):652–66.
43. Teicher MH, Samson JA. Childhood maltreatment and psychopathology: a case for ecophenotypic variants as clinically and neurobiologically distinct subtypes. Am J Psychiatry. 2013;170(10):1114–33.
44. Teicher MH, Andersen SL, Polcari A, Anderson CM, Navalta CP. Developmental neurobiology of childhood stress and trauma. Psychiatr Clin North Am. 2002;25(2):397–426, vii–viii.
45. Baker TB, Piper ME, McCarthy DE, Majeskie MR, Fiore MC. Addiction motivation reformulated: an affective processing model of negative reinforcement. Psychol Rev. 2004;111(1):33–51.
46. Khantzian EJ. The self-medication hypothesis of substance use disorders: a reconsideration and recent applications. Harv Rev Psychiatry. 1997;4(5):231–44.
47. De Bellis MD. Developmental traumatology: a contributory mechanism for alcohol and substance use disorders. Psychoneuroendocrinology. 2002;27(1–2):155–70.
48. Yoon S, Kobulsky JM, Yoon D, Kim W. Developmental pathways from child maltreatment to adolescent substance use: the roles of posttraumatic stress symptoms and mother-child relationships. Child Youth Serv Rev. 2017;82:271–9.
49. Schuck AM, Widom CS. Childhood victimization and alcohol symptoms in females: causal inferences and hypothesized mediators. Child Abuse Negl. 2001;25(8):1069–92.
50. Kim J, Cicchetti D, Rogosch FA, Manly JT. Child maltreatment and trajectories of personality and behavioral functioning: implications for the development of personality disorder. Dev Psychopathol. 2009;21(3):889–912.
51. Edalati H, Doucet C, Conrod P. A developmental social neuroscience model for understanding pathways to substance use disorders during adolescence. Semin Pediatr Neurol. 2018;27:35–41.
52. Dembo R, Dertke M, La Voie L, Borders S, Washburn M, Schmeidler J. Physical abuse, sexual victimization and illicit drug use: a structural analysis among high risk adolescents. J Adolesc. 1987;10(1):13–34.
53. Dembo R, Williams L, La Voie L, Berry E, Getreu A, Wish ED, et al. Physical abuse, sexual victimization, and illicit drug use: replication of a structural analysis among a new sample of high-risk youths. Violence Vict. 1989;4(2):121–38.

54. Brook JS, Brook DW, Zhang C, Cohen P. Pathways from adolescent parent-child conflict to substance use disorders in the fourth decade of life. Am J Addict. 2009;18(3):235–42.
55. Lupien SJ, McEwen BS, Gunnar MR, Heim C. Effects of stress throughout the lifespan on the brain, behaviour and cognition. Nat Rev Neurosci. 2009;10(6):434–45.
56. Tyrka AR, Price LH, Marsit C, Walters OC, Carpenter LL. Childhood adversity and epigenetic modulation of the leukocyte glucocorticoid receptor: preliminary findings in healthy adults. PLoS One. 2012;7(1):e30148.
57. Andersen SL, Teicher MH. Desperately driven and no brakes: developmental stress exposure and subsequent risk for substance abuse. Neurosci Biobehav Rev. 2009;33(4):516–24.
58. Andersen SL, Tomada A, Vincow ES, Valente E, Polcari A, Teicher MH. Preliminary evidence for sensitive periods in the effect of childhood sexual abuse on regional brain development. J Neuropsychiatry Clin Neurosci. 2008;20(3):292–301.
59. Teicher MH, Andersen SL, Polcari A, Anderson CM, Navalta CP, Kim DM. The neurobiological consequences of early stress and childhood maltreatment. Neurosci Biobehav Rev. 2003;27(1–2):33–44.
60. Anderson CM, Teicher MH, Polcari A, Renshaw PF. Abnormal T2 relaxation time in the cerebellar vermis of adults sexually abused in childhood: potential role of the vermis in stress-enhanced risk for drug abuse. Psychoneuroendocrinology. 2002;27(1–2):231–44.
61. Anderson CM, Rabi K, Lukas SE, Teicher MH. Cerebellar lingula size and experiential risk factors associated with high levels of alcohol and drug use in young adults. Cerebellum. 2010;9(2):198–209.
62. Shin SH, Miller DP, Teicher MH. Exposure to childhood neglect and physical abuse and developmental trajectories of heavy episodic drinking from early adolescence into young adulthood. Drug Alcohol Depend. 2013;127(1–3):31–8.
63. Edmiston EE, Wang F, Mazure CM, Guiney J, Sinha R, Mayes LC, et al. Corticostriatal-limbic gray matter morphology in adolescents with self-reported exposure to childhood maltreatment. Arch Pediatr Adolesc Med. 2011;165(12):1069–77.
64. Paul R, Henry L, Grieve SM, Guilmette TJ, Niaura R, Bryant R, et al. The relationship between early life stress and microstructural integrity of the corpus callosum in a non-clinical population. Neuropsychiatr Dis Treat. 2008;4(1):193–201.
65. Philip NS, Sweet LH, Tyrka AR, Price LH, Carpenter LL, Kuras YI, et al. Early life stress is associated with greater default network deactivation during working memory in healthy controls: a preliminary report. Brain Imaging Behav. 2013;7(2):204–12.
66. Chaney A, Carballedo A, Amico F, Fagan A, Skokauskas N, Meaney J, et al. Effect of childhood maltreatment on brain structure in adult patients with major depressive disorder and healthy participants. J Psychiatry Neurosci. 2014;39(1):50–9.
67. van Harmelen AL, van Tol MJ, van der Wee NJ, Veltman DJ, Aleman A, Spinhoven P, et al. Reduced medial prefrontal cortex volume in adults reporting childhood emotional maltreatment. Biol Psychiatry. 2010;68(9):832–8.
68. van Harmelen AL, van Tol MJ, Dalgleish T, van der Wee NJ, Veltman DJ, Aleman A, et al. Hypoactive medial prefrontal cortex functioning in adults reporting childhood emotional maltreatment. Soc Cogn Affect Neurosci. 2014;9(12):2026–33.
69. Ohashi K, Anderson CM, Bolger EA, Khan A, McGreenery CE, Teicher MH. Susceptibility or resilience to maltreatment can be explained by specific differences in brain network architecture. Biol Psychiatry. 2019;85(8):690–702.
70. McDermott JM, Westerlund A, Zeanah CH, Nelson CA, Fox NA. Early adversity and neural correlates of executive function: implications for academic adjustment. Dev Cogn Neurosci. 2012;2(Suppl 1):S59–66.
71. Majer M, Nater UM, Lin JM, Capuron L, Reeves WC. Association of childhood trauma with cognitive function in healthy adults: a pilot study. BMC Neurol. 2010;10:61.
72. DePrince AP, Weinzierl KM, Combs MD. Executive function performance and trauma exposure in a community sample of children. Child Abuse Negl. 2009;33(6):353–61.

73. Pollak SD, Nelson CA, Schlaak MF, Roeber BJ, Wewerka SS, Wiik KL, et al. Neurodevelopmental effects of early deprivation in postinstitutionalized children. Child Dev. 2010;81(1):224–36.
74. Porter C, Lawson JS, Bigler ED. Neurobehavioral sequelae of child sexual abuse. Child Neuropsychol. 2005;11(2):203–20.
75. Raine A, Park S, Lencz T, Bihrle S, LaCasse L, Widom CS, et al. Reduced right hemisphere activation in severely abused violent offenders during a working memory task: an fMRI study. Aggress Behav. 2001;27:111–29.
76. Beers SR, De Bellis MD. Neuropsychological function in children with maltreatment-related posttraumatic stress disorder. Am J Psychiatry. 2002;159(3):483–6.
77. Perez CM, Widom CS. Childhood victimization and long-term intellectual and academic outcomes. Child Abuse Negl. 1994;18(8):617–33.
78. Prasad MR, Kramer LA, Ewing-Cobbs L. Cognitive and neuroimaging findings in physically abused preschoolers. Arch Dis Child. 2005;90(1):82–5.
79. Loman MM, Wiik KL, Frenn KA, Pollak SD, Gunnar MR. Postinstitutionalized children's development: growth, cognitive, and language outcomes. J Dev Behav Pediatr. 2009;30(5):426–34.
80. Ammerman RT, Cassisi JE, Hersen M, Van Hasselt VB. Consequences of physical abuse and neglect in children. Clin Psychol Rev. 1986;6(4):291–310.
81. Romer D, Betancourt LM, Brodsky NL, Giannetta JM, Yang W, Hurt H. Does adolescent risk taking imply weak executive function? A prospective study of relations between working memory performance, impulsivity, and risk taking in early adolescence. Dev Sci. 2011;14(5):1119–33.
82. Tarter RE, Kirisci L, Mezzich A, Cornelius JR, Pajer K, Vanyukov M, et al. Neurobehavioral disinhibition in childhood predicts early age at onset of substance use disorder. Am J Psychiatry. 2003;160(6):1078–85.
83. Pechtel P, Pizzagalli DA. Effects of early life stress on cognitive and affective function: an integrated review of human literature. Psychopharmacology. 2011;214(1):55–70.
84. Vink M, Zandbelt BB, Gladwin T, Hillegers M, Hoogendam JM, van den Wildenberg WP, et al. Frontostriatal activity and connectivity increase during proactive inhibition across adolescence and early adulthood. Hum Brain Mapp. 2014;35(9):4415–27.
85. Nolin P, Ethier L. Using neuropsychological profiles to classify neglected children with or without physical abuse. Child Abuse Negl. 2007;31(6):631–43.
86. Rutter M. The interplay of nature, nurture, and developmental influences: the challenge ahead for mental health. Arch Gen Psychiatry. 2002;59(11):996–1000.
87. Rosenman S, Rodgers B. Childhood adversity and adult personality. Aust N Z J Psychiatry. 2006;40(5):482–90.
88. Dillon DG, Holmes AJ, Birk JL, Brooks N, Lyons-Ruth K, Pizzagalli DA. Childhood adversity is associated with left basal ganglia dysfunction during reward anticipation in adulthood. Biol Psychiatry. 2009;66(3):206–13.
89. Mueller SC, Maheu FS, Dozier M, Peloso E, Mandell D, Leibenluft E, et al. Early-life stress is associated with impairment in cognitive control in adolescence: an fMRI study. Neuropsychologia. 2010;48(10):3037–44.
90. Navalta CP, Polcari A, Webster DM, Boghossian A, Teicher MH. Effects of childhood sexual abuse on neuropsychological and cognitive function in college women. J Neuropsychiatry Clin Neurosci. 2006;18(1):45–53.
91. Tarter RE, Kirisci L, Habeych M, Reynolds M, Vanyukov M. Neurobehavior disinhibition in childhood predisposes boys to substance use disorder by young adulthood: direct and mediated etiologic pathways. Drug Alcohol Depend. 2004;73(2):121–32.
92. Centers for Disease Control and Prevention. Adverse childhood experiences reported by adults—five states. MMWR Morb Mortal Wkly Rep. 2009;59(49):1609–13.

93. Nigg JT, Wong MM, Martel MM, Jester JM, Puttler LI, Glass JM, et al. Poor response inhibition as a predictor of problem drinking and illicit drug use in adolescents at risk for alcoholism and other substance use disorders. J Am Acad Child Adolesc Psychiatry. 2006;45(4):468–75.
94. Oshri A, Kogan SM, Kwon JA, Wickrama KAS, Vanderbroek L, Palmer AA, et al. Impulsivity as a mechanism linking child abuse and neglect with substance use in adolescence and adulthood. Dev Psychopathol. 2018;30(2):417–35.
95. Walters GD, Espelage DL. Exploring the victimization–early substance misuse relationship: in search of moderating and mediating effects. Child Abuse Negl. 2018;81:354–65.
96. Bremner JD, Randall P, Scott TM, Capelli S, Delaney R, McCarthy G, et al. Deficits in short-term memory in adult survivors of childhood abuse. Psychiatry Res. 1995;59(1–2):97–107.
97. Bremner JD, Vermetten E, Afzal N, Vythilingam M. Deficits in verbal declarative memory function in women with childhood sexual abuse-related posttraumatic stress disorder. J Nerv Ment Dis. 2004;192(10):643–9.
98. Andersen P, Morris R, Amaral D, Bliss T, O'Keefe J. The hippocampus book (Oxford Neuroscience Series). New York, NY: Oxford University Press; 2007.
99. Pitman RK, Orr SP, Shalev AY. Once bitten, twice shy: beyond the conditioning model of PTSD. Biol Psychiatry. 1993;33(3):145–6.
100. van Harmelen AL, de Jong PJ, Glashouwer KA, Spinhoven P, Penninx BW, Elzinga BM. Child abuse and negative explicit and automatic self-associations: the cognitive scars of emotional maltreatment. Behav Res Ther. 2010;48(6):486–94.
101. Potthast N, Neuner F, Catani C. When abuse primes addiction—automatic activation of alcohol concepts by child maltreatment related cues in emotionally abused alcoholics. Addict Behav. 2015;48:62–70.
102. Elwyn L, Smith C. Child maltreatment and adult substance abuse: the role of memory. J Soc Work Pract Addict. 2013;13(3). https://doi.org/10.1080/1533256X.2013.814483.
103. Crews FT, Nixon K. Mechanisms of neurodegeneration and regeneration in alcoholism. Alcohol Alcohol. 2009;44(2):115–27.
104. Conrod PJ, Nikolaou K. Annual research review: on the developmental neuropsychology of substance use disorders. J Child Psychol Psychiatry. 2016;57(3):371–94.
105. Casey BJ, Jones RM. Neurobiology of the adolescent brain and behavior: implications for substance use disorders. J Am Acad Child Adolesc Psychiatry. 2010;49(12):1189–201; quiz 1285
106. Ernst M, Romeo RD, Andersen SL. Neurobiology of the development of motivated behaviors in adolescence: a window into a neural systems model. Pharmacol Biochem Behav. 2009;93(3):199–211.
107. Ernst M, Luciana M. Neuroimaging of the dopamine/reward system in adolescent drug use. CNS Spectr. 2015;20(4):427–41.
108. Guyer AE, Kaufman J, Hodgdon HB, Masten CL, Jazbec S, Pine DS, et al. Behavioral alterations in reward system function: the role of childhood maltreatment and psychopathology. J Am Acad Child Adolesc Psychiatry. 2006;45(9):1059–67.
109. Mahmood OM, Goldenberg D, Thayer R, Migliorini R, Simmons AN, Tapert SF. Adolescents' fMRI activation to a response inhibition task predicts future substance use. Addict Behav. 2013;38(1):1435–41.
110. Wright MO, Crawford E, Del Castillo D. Childhood emotional maltreatment and later psychological distress among college students: the mediating role of maladaptive schemas. Child Abuse Negl. 2009;33(1):59–68.
111. Wiers RW, van Woerden N, Smulders FT, de Jong PJ. Implicit and explicit alcohol-related cognitions in heavy and light drinkers. J Abnorm Psychol. 2002;111(4):648–58.
112. Wiers RW, Stacy AW. Handbook of implicit cognition and addiction. Thousand Oaks, CA: SAGE Publishers; 2006.
113. Krank MD, Schoenfeld T, Frigon AP. Self-coded indirect memory associations and alcohol and marijuana use in college students. Behav Res Methods. 2010;42(3):733–8.
114. Frigon AP, Krank MD. Self-coded indirect memory associations in a brief school-based intervention for substance use suspensions. Psychol Addict Behav. 2009;23(4):736–42.

115. Barkowsky DS. Executive function and future orientation moderate the relationship among substance use associations and outcome expectancies with substance use in adolescents: a pilot study. University of British Columbia; 2013.
116. Sun R, Slusarz P, Terry C. The interaction of the explicit and the implicit in skill learning: a dual-process approach. Psychol Rev. 2005;112(1):159–92.
117. Ricco RB, Overton WF. Dual systems Competence ←-→ Procedural processing: a relational developmental systems approach to reasoning. Dev Rev. 2011;31(2–3):119–50.
118. Nelson C. The ontogeny of human memory: a cognitive neuroscience perspective. Dev Psychol. 1995;31(5):723–38.
119. Gawronski B, Bodenhausen GV. Associative and propositional processes in evaluation: an integrative review of implicit and explicit attitude change. Psychol Bull. 2006;132(5):692–731.
120. Berry DC, Dienes Z. Implicit learning: theoretical and empirical issues. Erlbaum: Hillsdale, NJ; 1993.
121. Kuntsche E, Knibbe R, Gmel G, Engels R. 'I drink spirits to get drunk and block out my problems...' beverage preference, drinking motives and alcohol use in adolescence. Alcohol Alcohol. 2006;41(5):566–73.
122. Cooper ML, Frone MR, Russell M, Mudar P. Drinking to regulate positive and negative emotions: a motivational model of alcohol use. J Pers Soc Psychol. 1995;69(5):990–1005.
123. Prescott CA, Cross RJ, Kuhn JW, Horn JL, Kendler KS. Is risk for alcoholism mediated by individual differences in drinking motivations? Alcohol Clin Exp Res. 2004;28(1):29–39.
124. Grant VV, Stewart SH, O'Connor RM, Blackwell E, Conrod PJ. Psychometric evaluation of the five-factor modified drinking motives questionnaire—revised in undergraduates. Addict Behav. 2007;32(11):2611–32.
125. DE Bellis MD, Hooper SR, Spratt EG, Woolley DP. Neuropsychological findings in childhood neglect and their relationships to pediatric PTSD. J Int Neuropsychol Soc. 2009;15(6):868–78.
126. De Bellis MD, Keshavan MS, Clark DB, Casey BJ, Giedd JN, Boring AM, et al. Developmental traumatology part II: brain development. Biol Psychiatry. 1999;45(10):1271–84.
127. O'Mahen HA, Karl A, Moberly N, Fedock G. The association between childhood maltreatment and emotion regulation: two different mechanisms contributing to depression? J Affect Disord. 2015;174:287–95.
128. Edalati H, Krank M, Schutz CG. Childhood maltreatment and perceived stress in individuals with concurrent psychiatric disorders. J Aggress Maltreat Trauma. 2020;29(1):22–37.
129. Edalati H, Krausz M, Schutz CG. Childhood maltreatment and revictimization in a homeless population. J Interpers Violence. 2016;31(14):2492–512.
130. Edalati H, Nicholls TL. Childhood maltreatment and the risk for criminal justice involvement and victimization among homeless individuals: a systematic review. Trauma Violence Abuse. 2019;20(3):315–30.
131. Uhrlass DJ, Gibb BE. Childhood emotional maltreatment and the stress generation model of depression. J Soc Clin Psychol. 2007;26(1):119–30.
132. Gratz KL, Tull MT, Baruch DE, Bornovalova MA, Lejuez CW. Factors associated with co-occurring borderline personality disorder among inner-city substance users: the roles of childhood maltreatment, negative affect intensity/reactivity, and emotion dysregulation. Compr Psychiatry. 2008;49(6):603–15.
133. Pollak SD, Sinha P. Effects of early experience on children's recognition of facial displays of emotion. Dev Psychol. 2002;38(5):784–91.
134. Shipman K, Zeman J, Penza S, Champion K. Emotion management skills in sexually maltreated and nonmaltreated girls: a developmental psychopathology perspective. Dev Psychopathol. 2000;12(1):47–62.
135. Gratz KL, Bornovalova MA, Delany-Brumsey A, Nick B, Lejuez CW. A laboratory-based study of the relationship between childhood abuse and experiential avoidance among inner-city substance users: the role of emotional nonacceptance. Behav Ther. 2007;38(3):256–68.
136. Batten SV, Follette VM, Aban IB. Experimental avoidance and high-risk sexual behavior in survivors of child sexual abuse. J Child Sex Abus. 2001;10(2):101–20.

137. Jones-Webb R, Jacobs DR Jr, Flack JM, Liu K. Relationships between depressive symptoms, anxiety, alcohol consumption, and blood pressure: results from the CARDIA study. Coronary artery risk development in young adults study. Alcohol Clin Exp Res. 1996;20(3):420–7.
138. Leigh BC. Attitudes and expectancies as predictors of drinking habits: a comparison of three scales. J Stud Alcohol. 1989;50(5):432–40.
139. Hyman SM, Paliwal P, Sinha R. Childhood maltreatment, perceived stress, and stress-related coping in recently abstinent cocaine dependent adults. Psychol Addict Behav. 2007;21(2):233–8.
140. Edalati H, Nicholls TL, Crocker AG, Roy L, Somers JM, Patterson ML. Adverse childhood experiences and the risk of criminal justice involvement and victimization among homeless adults with mental illness. Psychiatr Serv. 2017;68(12):1288–95.
141. Simons J, Correia CJ, Carey KB, Borsari BE. Validating a five-factor marijuana motives measure: relations with use, problems, and alcohol motives. J Couns Psychol. 1998;45(3):265–73.
142. Weiss NH, Tull MT, Lavender J, Gratz KL. Role of emotion dysregulation in the relationship between childhood abuse and probable PTSD in a sample of substance abusers. Child Abuse Negl. 2013;37(11):944–54.
143. Vilhena NA. An examination of the role of motives and emotion regulation in the relationship between child maltreatment and substance use [MA dissertation]. Toronto: Department of Adult and Counselling Psychology, University of Toronto; 2011.
144. Dutcher CD, Vujanovic AA, Paulus DJ, Bartlett BA. Childhood maltreatment severity and alcohol use in adult psychiatric inpatients: the mediating role of emotion regulation difficulties. Gen Hosp Psychiatry. 2017;48:42–50.
145. Vilhena-Churchill N, Goldstein AL. Child maltreatment and marijuana problems in young adults: examining the role of motives and emotion dysregulation. Child Abuse Negl. 2014;38(5):962–72.
146. Barahmand U, Khazaee A, Hashjin GS. Emotion dysregulation mediates between childhood emotional abuse and motives for substance use. Arch Psychiatr Nurs. 2016;30(6):653–9.
147. Wekerle C, Leung E, Goldstein A, Thornton T, Tonmyr L. Substance use among adolescents in child welfare versus adolescents in the general population: a comparison of the Maltreatment and Adolescent Pathways (MAP) Londitudinal Study and the 2005 Ontario Student Drug Use Survey (OSDUS) datasets. https://cwrp.ca/publications/substance-use-among-adolescents-child-welfare-versus-adolescents-general-population. Published 2009.
148. Cloitre M, Stovall-McClough KC, Nooner K, Zorbas P, Cherry S, Jackson CL, et al. Treatment for PTSD related to childhood abuse: a randomized controlled trial. Am J Psychiatry. 2010;167(8):915–24.
149. Pilowsky DJ, Wu LT. Psychiatric symptoms and substance use disorders in a nationally representative sample of American adolescents involved with foster care. J Adolesc Health. 2006;38(4):351–8.
150. Kolko DJ, Hurlburt MS, Zhang J, Barth RP, Leslie LK, Burns BJ. Posttraumatic stress symptoms in children and adolescents referred for child welfare investigation. A national sample of in-home and out-of-home care. Child Maltreat. 2010;15(1):48–63.
151. Stewart SH, Pihl RO, Conrod PJ, Dongier M. Functional associations among trauma, PTSD, and substance-related disorders. Addict Behav. 1998;23(6):797–812.
152. Cooney NL, Litt MD, Morse PA, Bauer LO, Gaupp L. Alcohol cue reactivity, negative-mood reactivity, and relapse in treated alcoholic men. J Abnorm Psychol. 1997;106(2):243–50.
153. Stewart SH. Alcohol abuse in individuals exposed to trauma: a critical review. Psychol Bull. 1996;120(1):83–112.
154. Kilpatrick DG, Acierno R, Saunders B, Resnick HS, Best CL, Schnurr PP. Risk factors for adolescent substance abuse and dependence: data from a national sample. J Consult Clin Psychol. 2000;68(1):19–30.
155. Chilcoat HD, Breslau N. Posttraumatic stress disorder and drug disorders: testing causal pathways. Arch Gen Psychiatry. 1998;55(10):913–7.

156. Proctor LJ, Lewis T, Roesch S, Thompson R, Litrownik AJ, English D, et al. Child maltreatment and age of alcohol and marijuana initiation in high-risk youth. Addict Behav. 2017;75:64–9.
157. Brook JS, Brook DW, Pahl K. The developmental context for adolescent substance abuse intervention. New York, NY: Cambridge University Press; 2006. p. 25–51.
158. Crittenden PM, Claussen AH. Developmental psychopathology perspectives on substance abuse and relationship violence. New York, NY: Brunner-Routledge; 2002. p. 44–63.
159. Nakao K, Takaishi J, Tatsuta K, Katayama H, Iwase M, Yorifuji K, et al. The influences of family environment on personality traits. Psychiatry Clin Neurosci. 2000;54(1):91–5.
160. Edalati H, Krank M. Impulsivity and sensation seeking mediate the relationship between exposure to violence and alcohol use in adolescents and young adults. Alcohol Clin Exp Res. 2015;39:217A.
161. Brook JS, Balka EB, Crossman AM, Dermatis H, Galanter M, Brook DW. The relationship between parental alcohol use, early and late adolescent alcohol use, and young adult psychological symptoms: a longitudinal study. Am J Addict. 2010;19(6):534–42.
162. Sroufe LA. The coherence of individual development: early care, attachment, and subsequent developmental issues. Am Psychol. 1979;34(10):834–41.
163. Dunn MG, Tarter RE, Mezzich AC, Vanyukov M, Kirisci L, Kirillova G. Origins and consequences of child neglect in substance abuse families. Clin Psychol Rev. 2002;22(7):1063–90.
164. Crittenden PM. Children's strategies for coping with adverse home environments: an interpretation using attachment theory. Child Abuse Negl. 1992;16(3):329–43.
165. Flores PJ. Addiction as an attachment disorder: implications for group therapy. Int J Group Psychother. 2001;51(1):63–81.
166. Schindler A, Thomasius R, Petersen K, Sack PM. Heroin as an attachment substitute? Differences in attachment representations between opioid, ecstasy and cannabis abusers. Attach Hum Dev. 2009;11(3):307–30.
167. Stronach EP, Toth SL, Rogosch F, Oshri A, Manly JT, Cicchetti D. Child maltreatment, attachment security, and internal representations of mother and mother-child relationships. Child Maltreat. 2011;16(2):137–45.
168. Ledoux S, Miller P, Choquet M, Plant M. Family structure, parent-child relationships, and alcohol and other drug use among teenagers in France and the United Kingdom. Alcohol Alcohol. 2002;37(1):52–60.
169. Branstetter SA, Furman W, Cottrell L. The influence of representations of attachment, maternal-adolescent relationship quality, and maternal monitoring on adolescent substance use: a 2-year longitudinal examination. Child Dev. 2009;80(5):1448–62.
170. Van Ryzin MJ, Fosco GM, Dishion TJ. Family and peer predictors of substance use from early adolescence to early adulthood: an 11-year prospective analysis. Addict Behav. 2012;37(12):1314–24.
171. Cooper ML, Shaver PR, Collins NL. Attachment styles, emotion regulation, and adjustment in adolescence. J Pers Soc Psychol. 1998;74(5):1380–97.
172. Alex Mason W, Jean Russo M, Chmelka MB, Herrenkohl RC, Herrenkohl TI. Parent and peer pathways linking childhood experiences of abuse with marijuana use in adolescence and adulthood. Addict Behav. 2017;66:70–5.
173. Petraitis J, Flay BR, Miller TQ. Reviewing theories of adolescent substance use: organizing pieces in the puzzle. Psychol Bull. 1995;117(1):67–86.
174. Brook JS, Brook DW, Gordon AS, Whiteman M, Cohen P. The psychosocial etiology of adolescent drug use: a family interactional approach. Genet Soc Gen Psychol Monogr. 1990;116(2):111–267.
175. Clark DB, Thatcher DL, Maisto SA. Adolescent neglect and alcohol use disorders in two-parent families. Child Maltreat. 2004;9(4):357–70.
176. Mann M, Hosman CM, Schaalma HP, de Vries NK. Self-esteem in a broad-spectrum approach for mental health promotion. Health Educ Res. 2004;19(4):357–72.
177. Ybrandt H. The relation between self-concept and social functioning in adolescence. J Adolesc. 2008;31(1):1–16.

178. Carvajal SC, Clair SD, Nash SG, Evans RI. Relating optimism, hope, and self-esteem to social influences in deterring substance use in adolescents. J Soc Clin Psychol. 2017;17(4):443–65.
179. Tsaousis I. The relationship of self-esteem to bullying perpetration and peer victimization among schoolchildren and adolescents: a meta-analytic review. Aggress Violent Behav. 2016;31:186–99.
180. Taylor J, Lloyd DA, Warheit GJ. Self-derogation, peer factors, and drug dependence among a multiethnic sample of young adults. J Child Adolesc Subst Abuse. 2005;15(2):39–51.
181. Stein JA, Leslie MB, Nyamathi A. Relative contributions of parent substance use and childhood maltreatment to chronic homelessness, depression, and substance abuse problems among homeless women: mediating roles of self-esteem and abuse in adulthood. Child Abuse Negl. 2002;26(10):1011–27.
182. Donnellan MB, Trzesniewski KH, Robins RW, Moffitt TE, Caspi A. Low self-esteem is related to aggression, antisocial behavior, and delinquency. Psychol Sci. 2005;16(4):328–35.
183. O'Callaghan F, Doyle J. What is the role of impression management in adolescent cigarette smoking? J Subst Abus. 2001;13(4):459–70.
184. Mcgee R, Williams S. Does low self-esteem predict health compromising behaviours among adolescents? J Adolesc. 2000;23(5):569–82.
185. Kim J, Cicchetti D. Longitudinal pathways linking child maltreatment, emotion regulation, peer relations, and psychopathology. J Child Psychol Psychiatry. 2010;51(6):706–16.
186. Kaplan HB, Martin SS, Robbins C. Pathways to adolescent drug use: self-derogation, peer influence, weakening of social controls, and early substance use. J Health Soc Behav. 1984;25(3):270–89.
187. Mason WA. Self-esteem and delinquency revisited (again): a test of Kaplan's self-derogation theory of delinquency using latent growth curve modeling. J Youth Adolesc. 2001;30(1):83–102.
188. Oshri A, Rogosch FA, Cicchetti D. Child maltreatment and mediating influences of childhood personality types on the development of adolescent psychopathology. J Clin Child Adolesc Psychol. 2013;42(3):287–301.
189. Bailey JA, McCloskey LA. Pathways to adolescent substance use among sexually abused girls. J Abnorm Child Psychol. 2005;33(1):39–53.
190. Stewart SH, McGonnell M, Wekerle C, Adlaf EM, et al. Associations of personality with alcohol use behaviour and alcohol problems in adolescents receiving child welfare services. Int J Ment Heal Addict. 2011;9(5):492–506.
191. Shin SH, Lee S, Jeon SM, Wills TA. Childhood emotional abuse, negative emotion-driven impulsivity, and alcohol use in young adulthood. Child Abuse Negl. 2015;50:94–103.
192. Hudson A, Wekerle C, Stewart SH. Associations between personality and drinking motives in adolescents involved in the child welfare system. Personal Individ Differ. 2015;81:84–9.
193. Johnson JG, Smailes EM, Cohen P, Brown J, Bernstein DP. Associations between four types of childhood neglect and personality disorder symptoms during adolescence and early adulthood: findings of a community-based longitudinal study. J Personal Disord. 2000;14(2):171–87.
194. Laporte L, Paris J, Guttman H, Russell J. Psychopathology, childhood trauma, and personality traits in patients with borderline personality disorder and their sisters. J Personal Disord. 2011;25(4):448–62.
195. Kristman-Valente A, Wells EA. The role of gender in the association between child maltreatment and substance use behavior: a systematic review of longitudinal research from 1995 to 2011. Subst Use Misuse. 2013;48(8):645–60.
196. MacMillan HL, Fleming JE, Streiner DL, Lin E, Boyle MH, Jamieson E, et al. Childhood abuse and lifetime psychopathology in a community sample. Am J Psychiatry. 2001;158(11):1878–83.
197. Najavits LM, Weiss RD, Shaw SR. The link between substance abuse and posttraumatic stress disorder in women. A research review. Am J Addict. 1997;6(4):273–83.
198. Diehl M, Coyle N, Labouvie-Vief G. Age and sex differences in strategies of coping and defense across the life span. Psychol Aging. 1996;11(1):127–39.

199. Lansford JE, Dodge KA, Pettit GS, Bates JE. Does physical abuse in early childhood predict substance use in adolescence and early adulthood? Child Maltreat. 2010;15(2):190–4.
200. Widom CS, Ireland T, Glynn PJ. Alcohol abuse in abused and neglected children followed-up: are they at increased risk? J Stud Alcohol. 1995;56(2):207–17.
201. Widom CS, White HR, Czaja SJ, Marmorstein NR. Long-term effects of child abuse and neglect on alcohol use and excessive drinking in middle adulthood. J Stud Alcohol Drugs. 2007;68(3):317–26.
202. Widom CS, White HR. Problem behaviours in abused and neglected children grown up: prevalence and co-occurrence of substance abuse, crime and violence. Crim Behav Ment Health. 1997;7(4):287–310.
203. Widom CS, Marmorstein NR, White HR. Childhood victimization and illicit drug use in middle adulthood. Psychol Addict Behav. 2006;20(4):394–403.
204. Galaif ER, Stein JA, Newcomb MD, Bernstein DP. Gender differences in the prediction of problem alcohol use in adulthood: exploring the influence of family factors and childhood maltreatment. J Stud Alcohol. 2018;62(4):486–93.
205. Topitzes J, Mersky JP, Reynolds AJ. Child maltreatment and adult cigarette smoking: a long-term developmental model. J Pediatr Psychol. 2010;35(5):484–98.
206. Lee RD, Chen J. Adverse childhood experiences, mental health, and excessive alcohol use: examination of race/ethnicity and sex differences. Child Abuse Negl. 2017;69:40–8.
207. Najavits LM, editor. Seeking safety: a treatment manual for PTSD and substance abuse. New York: Guilford Press; 2002.
208. Najavits LM. Seeking safety: an evidence-based model for substance abuse and trauma/PTSD. In: Witkiewitz KA, Marlatt GA, editors. Therapist's guide to evidence based relapse prevention: practical resources for the mental health professional. San Diego: Elsevier Press; 2007. p. 141–67.
209. Braciszewski JM, Stout RL. Substance use among current and former foster youth: a systematic review. Child Youth Serv Rev. 2012;34(12):2337–44.
210. Traube DE, James S, Zhang J, Landsverk J. A national study of risk and protective factors for substance use among youth in the child welfare system. Addict Behav. 2012;37(5):641–50.
211. Narendorf SC, McMillen JC. Substance use and substance use disorders as foster youth transition to adulthood. Child Youth Serv Rev. 2010;32(1):113–9.
212. Aarons GA, Monn AR, Hazen AL, Connelly CD, Leslie LK, Landsverk JA, et al. Substance involvement among youths in child welfare: the role of common and unique risk factors. Am J Orthopsychiatry. 2008;78(3):340–9.
213. Walsh C, MacMillan HL, Jamieson E. The relationship between parental substance abuse and child maltreatment: findings from the Ontario Health Supplement. Child Abuse Negl. 2003;27(12):1409–25.
214. McCoy H, McMillen JC, Spitznagel EL. Older youth leaving the foster care system: who, what, when, where, and why? Child Youth Serv Rev. 2008;30(7):735–45.
215. Edalati H, Conrod PJ. A review to identify gaps in research and service delivery for substance use prevention among at-risk adolescents involved in child welfare system: the promises of targeted interventions. Int J Child Adolesc Resilience. 2017;5(1):20–39.
216. Leve LD, Harold GT, Chamberlain P, Landsverk JA, Fisher PA, Vostanis P. Practitioner review: children in foster care—vulnerabilities and evidence-based interventions that promote resilience processes. J Child Psychol Psychiatry. 2012;53(12):1197–211.
217. Fratto CM. Trauma-informed care for youth in foster care. Arch Psychiatr Nurs. 2016;30(3):439–46.
218. Hodgdon H, Kinniburgh K, Gabowitz D, Blaustein M, Spinazzola J. Development and implementation of trauma-informed programming in youth residential treatment centers using the ARC framework. J Fam Violence. 2013;28:679–92.
219. Blaustein M, Kinniburgh K. Treating traumatic stress in children and adolescents: how to foster resilience through attachment, self-regulation, and competence. New York, NY: Guilford Press; 2010.

220. Leve LD, Fisher PA, Chamberlain P. Multidimensional treatment foster care as a preventive intervention to promote resiliency among youth in the child welfare system. J Pers. 2009;77(6):1869–902.
221. Kerr DC, Leve LD, Chamberlain P. Pregnancy rates among juvenile justice girls in two randomized controlled trials of multidimensional treatment foster care. J Consult Clin Psychol. 2009;77(3):588–93.
222. Westermark PK, Hansson K, Olsson M. Multidimensional treatment foster care (MTFC): results from an independent replication. J Fam Ther. 2011;33(1):20–41.
223. Smith DK, Chamberlain P, Eddy JM. Preliminary support for multidimensional treatment foster care in reducing substance use in delinquent boys. J Child Adolesc Subst Abuse. 2010;19(4):343–58.
224. Kim HK, Leve LD. Substance use and delinquency among middle school girls in foster care: a three-year follow-up of a randomized controlled trial. J Consult Clin Psychol. 2011;79(6):740–50.
225. Smith DK, Leve LD, Chamberlain P. Preventing internalizing and externalizing problems in girls in foster care as they enter middle school: impact of an intervention. Prev Sci. 2011;12(3):269–77.
226. Edalati H, Conrod P. A review of personality-targeted interventions for prevention of substance misuse and related harm in community samples of adolescents. Front Psychiatry. 2019;9:770.
227. Conrod PJ, Stewart SH. New advancements in the study of co-occurring substance use and psychiatric disorders. J Ment Health. 2006;15(6):615–8.
228. Colom R, Escorial S, Shih PC, Privado J. Fluid intelligence, memory span, and temperament difficulties predict academic performance of young adolescents. Personal Individ Differ. 2007;42(8):1503–14.

# Childhood Trauma in Obsessive-Compulsive Disorder

# 14

Federica Piras and Gianfranco Spalletta

## 14.1 Introduction

The belief that obsessive-compulsive disorder (OCD) is influenced by genetic factors is widely held. Earlier reports on the genetic contribution to OCD are very old, dating back to the beginning of the century, when the relative importance of genetic and environmental factors in predisposing to the disorder was originally tested in twins [1]. This evidence stimulated current neurobiological and cognitive theories of OCD pathophysiology indicating a multimodal model of inheritance, where multiple genetic and biological risk factors act in concert with environmental stressors causing the disease onset [2, 3]. The introduction of the "*stress-diathesis*" concept [4, 5] for explaining how preexisting vulnerability to stressors (the *diathesis* term of the interaction) causes the development of psychopathology led to the investigation of the relationship between stressful life events in childhood and OCD. Indeed, the neurobiological literature identified potential mechanisms whereby either acute and chronic stress may bias an organism toward increased reliance on habits [6–8]. As the process of habit formation is a popular model of compulsivity [9, 10], maladaptive habits have been proposed to play a role in OCD [11], thus identifying a potential mechanism by which stressful/traumatic events may induce OCD symptomatology onset and exacerbation [12].

F. Piras (✉)
Neuropsychiatry Laboratory, IRCCS Santa Lucia Foundation, Rome, Italy
e-mail: federica.piras@hsantalucia.it

G. Spalletta
Neuropsychiatry Laboratory, IRCCS Santa Lucia Foundation, Rome, Italy

Menninger Department of Psychiatry and Behavioural Sciences, Beth K. and Stuart C. Yudofsky Division of Neuropsychiatry, Baylor College of Medicine, Houston, TX, USA
e-mail: g.spalletta@hsantalucia.it



G. Spalletta et al. (eds.), *Childhood Trauma in Mental Disorders*,
https://doi.org/10.1007/978-3-030-49414-8_14

The hallmark signs and symptoms of OCD which affects both cognition and motor behavior are as follows: *obsessions*, thoughts that repeat over and over again, unwanted but insistent; and *compulsions* to act, to repeat fragments of behavior over and over in ritualistic, stereotyped succession [13]. Like habits, which are automatically triggered by stimuli, compulsive behaviors are unproductive and often senseless, rendering behavior insensitive to goal value. OCD patients do have insight of their compulsive actions irrationality; they want to stop but cannot exert control over the urge to act, thus suggesting that a disruption in the balance between goal-directed and habitual behavior is responsible for inducing compulsive actions [9]. The dominant neuroanatomical model of OCD [14] centers on, but is not limited to [15, 16], abnormalities within the regions involved in the balance between goal-directed behavior and habits, i.e., brain areas comprising the fronto-striatal circuits. Concurrently, both preclinical work and human studies suggest that early life stressors have conspicuous effects on cortical and subcortical structures throughout the corticostriatal-limbic circuitry [17]. More specifically, lifetime trauma exposure leads to gray matter loss [18] in key cortical regions implicated in OCD (i.e., the anterior cingulate cortex—ACC and the orbito-frontal cortex—OFC [19]). Thus, as highlighted in Fig. 14.1, the stress/trauma-induced structural and functional changes in corticostriatal-limbic circuitry contributing to the shift in balance between habitual and goal-direct behaviors [20, 21] remarkably resemble the reported neuroanatomical abnormalities associated to the pathogenesis and expression of OCD symptomatology.

The strong overlap between the neurobiological correlates of childhood trauma and OCD [22], and between cognitive models of trauma and OCD symptoms constitute the theoretical framework for investigating the role of stressful/traumatic events in the pathophysiology of OCD. However, when examining the impact of specific and multiple forms of substantiated childhood maltreatment on mental health, several constrains can limit the generalization of findings. The current body of literature on the association between stressful or traumatic life events and OCD is equally hindered by recurring limitations [23]. As a matter of fact, while prospective investigations are rare (e.g., [24]), findings from cross-sectional and case–control studies, which rely on retrospective data, are limited by the potential for recall bias [25]. In fact, OCD onset may occur irrespective of stressful/traumatic events, but nonetheless, be associated with symptoms post hoc, in the effort to make sense of the current experience [22], and retrospective reports of life-course adverse exposure can change over time depending on resilience, recovery [26], and severity of exposures. Moreover, studies generally focus on few specific types of childhood adverse events (for example, sexual assault or abuse, [27–30]) and do not consider the fact that traumatic events usually coexist. That is the case for the few longitudinal studies (e.g., [31]), with the additional limitation of considering clinical rather than general populations. Unfortunately, very often no control is carried out for factors predisposing to both child maltreatment and later adverse mental health outcomes such as familial socio-demographic characteristics, psychopathology, and environmental disadvantages [32]. However, in cross-sectional studies childhood adverse events might be both a cause and an effect of mental illness.

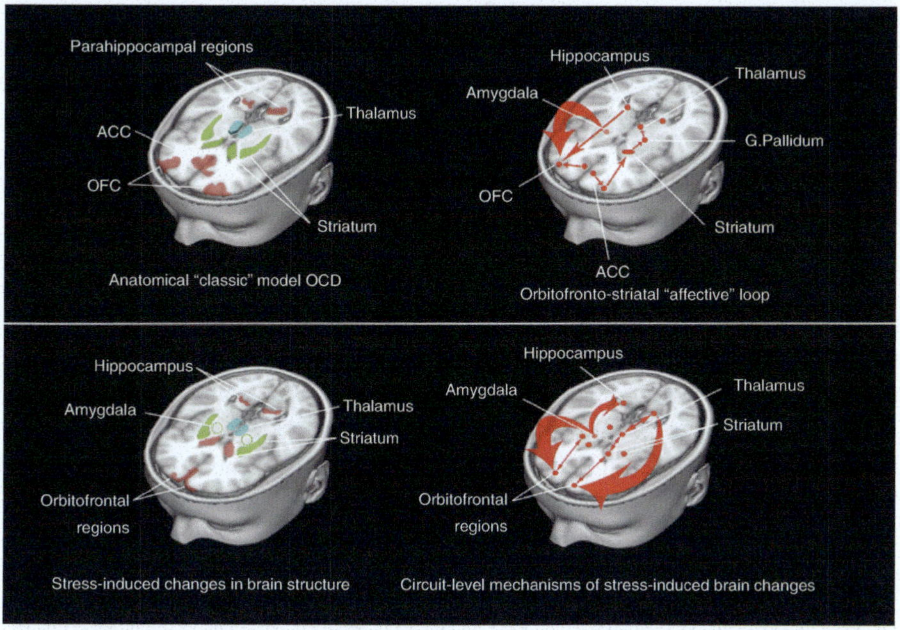


**Fig. 14.1** Neuroanatomical models and circuitry of OCD and trauma exposure/stress-induced changes in the brain. *OCD (upper panel)* is characterized by reduced volume in the ACC and in the medial OFC, and by increased volume in the putamen, striatum, thalamus, and parahippocampal regions. Hypertonic striatum and amygdala generate OCD symptoms by disturbing the fronto-subcortical balance in loops projecting from the prefrontal cortex to the striatum, to the thalamus via the globus pallidus, and back to the cortex. *Trauma exposure and excessive/chronic stress (lower panel)* can result in volume reduction of medial and orbital frontal regions, the hippocampus, and caudate causing hypertrophy in the amygdala and putamen. Acute and protracted stress affects the balance between habits and goal-directed behaviors through amygdala hyperactivity and impairs declarative learning—via atrophy or disrupted neurogenesis of the hippocampus-, and goal-directed behavioral control due to atrophy of frontal cortices

Bearing in mind the abovementioned limitations, different modes of interaction have to be considered in acknowledging the literature documenting an association between stressful/traumatic life events and OCD symptomatology [22]. Indeed, individuals' vulnerability might be either *generic* such that traumas/stressors in childhood lead directly or causally to the development of OCD symptoms, or *specific* as similar traumas may lead some individuals to develop OCD and others to develop different mental disorders. Alternatively, the illness might exist independently, but it may be exacerbated by stressful/traumatic events in childhood, while types and severity of stressful/traumatic experiences may differentially impact the severity, course, and phenotype of the disorder.

In the following sections, we will explore the different modes of the stress-diathesis interaction in OCD considering (1) the possibility that stressful/traumatic events in childhood may be a mechanism in the development of OCD, as substantiated by preclinical research and longitudinal studies; (2) the existence in OCD clinical samples, of a specific antecedent vulnerability to the pathogenic effects of traumas/stressors, as investigated by studies comparing the incidence and effects of stressful/traumatic experiences in OCD to those in other clinical conditions; and (3) the potentiality that OCD exists independently from exposure to traumatic events in childhood, but is exacerbated by the latters, as suggested by studies hypothesizing a distinct post-traumatic subtype of OCD [33] or an OCD phenotype increasing the risk for developing post-traumatic stress disorder.

## 14.2 Early Life Stress or Trauma as Mechanisms of Obsessive-Compulsive (OC) Symptoms

Considering the methodological limitations of the current literature, prospective population-based or longitudinal studies are most informative for investigating the potential causal association between stressful/traumatic events in childhood and OCD. Indeed, in clinical populations, especially when retrospective data are used, the propensity to recall stressful or traumatic life events might be biased by OCD severity [34], thus misrepresenting the role of childhood trauma in the pathogenesis of the disorder. However, the tremendous overlap between symptoms of OCD and of post-traumatic stress disorder (PTSD) [35], which are both characterized by recurrent and intrusive thoughts experienced as anxiety/fear inducing, and their unexpectedly frequent co-occurrence [35] are the foundation of emerging research examining the potential unique relation between trauma exposure and OC symptoms. Evidence exists that obsessions or intrusive thoughts are also experienced by 80–90% of the general population [36], and numerous different models have been put forward to clarify causes of transition from normal intrusive thoughts into clinical obsession.

### 14.2.1 Population-Based and Preclinical Research

Published studies investigated the relationship between adverse early experience (also defined as Stressful Life Experience, SLEs, or Potentially Traumatic Events, PTEs) and the degree to which individuals exhibited OC symptoms in the general population. The mediating role of specific cognitive, personological, and experiential factors was also considered.

In exploring the potential causal association between traumatic experiences and the development and maintenance of OC symptoms, the majority of studies referred to contemporary cognitive models of OCD [37] positing that symptoms arise from particular kinds of dysfunctional beliefs, where the strength of beliefs influences the development and severity of OCD symptoms. Three intercorrelated types of beliefs have been theoretically and empirically linked to OCD symptoms: (1) perfectionism and intolerance of uncertainty; (2) overimportance of thoughts and the need to control them; and (3) inflated responsibility/overestimation of threat [38]. These theories propose that adverse early experiences lead to assumptions and general beliefs about personal responsibility, which then drive maladaptive interpretations of normally occurring intrusive thoughts thus determining the transition to clinical obsessions [39]. Indeed, in the psychological, cognitive, and physiological struggle to integrate and manage a traumatic experience [40], learned assumptions and beliefs initially form as adaptive ways of coping with problematic aspects of early experience [41]. Over time, with repeated exposure to trauma, coping resources may be depleted, and assumptions and beliefs, particularly when activated by critical incidents, may be transformed from protective to maladaptive, triggering an obsessional disorder [41]. Actually, in a fairly vast ($N = 313$) sample of young (mean age ~28 years) Australian participants from the general community, the association between adverse childhood experience and OC symptoms was indirect, via the relationship with OCD dysfunctional beliefs (responsibility and threat estimation, perfectionism and intolerance for uncertainty, and importance and control of though) and with anxiety and depression [42].

Since PTEs, and more in general, SLEs can be argued to involve different degrees of controllability, if uncontrollable life events are combined with OCD dysfunctional beliefs, such as the excessive need for control, then the disorder could be a predictable outcome. Indeed, in the general population (269 Australian students, 18–51 years), a low perceived capacity to control a recent SLE, coupled with strong efforts to suppress thoughts, was associated with increased OC symptoms [43]. Therefore, it could be that people who experience SLEs, which are perceived as being difficult to control, might attempt to employ more mental control (i.e., increased thought suppression), as to compensate for a less controllable external environment. Such avoidance strategy might inadvertently lead to an OCD-type cycle since subjects fail to habituate to the relevant fear stimuli [43, 44]. Additionally, specific personality characteristics that have been proposed as vulnerability traits for OCD symptoms, particularly conscientiousness, play an indirect role in the relationship between childhood trauma and the development of clinically significant, maladaptive levels of OC symptoms, even in a nonclinical samples (more than 900 American undergraduate students) [45]. However, in the cited study conscientiousness was not associated with probable OCD. This would suggest that the interaction between consciousness and trauma exposure is more important in the development of moderate, rather than severe OC symptoms, and likely associated with adaptive, rather than maladaptive, behavior patterns [45]. Increased levels of OC symptoms have also been associated with a variety of maladaptive or dysfunctional parenting styles that are related to emotional abuse and neglect, leading to insecure attachment style and attachment anxiety [46]. The association may be driven by the effect that inadequate parental care, and the consequent development of distorted attachment styles, have on assumptions and beliefs about the world and the self. Researchers have proposed several pathways that can lead to obsessive beliefs, including authoritarian and neglectful parenting, which can give rise to maladaptive interpersonal perfectionism [47, 48], and to increased threat appraisal and overestimation since significant others are viewed as unwilling or unable to provide support in times of need [49]. As a matter of fact, in a sample of 338 American students, attachment anxiety partially mediated the association between parent–child relationships and obsessive beliefs, thus demonstrating that perceived inadequate parental care is related to distorted cognitions about the self and others, which then function as cognitive vulnerabilities for OCD [50].

An additional mechanism that may help in understanding the role that trauma exposure plays in the transition from normal intrusive thoughts into clinical obsession is experiential avoidance. This can be characterized as an unwillingness or inability to remain in contact with internal experiences (thoughts, memories, emotions, and/or bodily sensations), or any attempt to alter or escape the experience. Such avoidant coping strategy in response to trauma has been identified as an important mechanism leading to psychopathology manifestation in adolescents [51] and young adults. It was also proved to be a powerful mediator between trauma exposure and OC symptoms in both high-risk (53 American students referred for academic failure, substance abuse, risk for dropout, or psychosocial difficulties) and nonclinical (400 college students) samples [52]. Avoidance of trauma-related memories, thoughts, and emotions may be one of the most common coping strategies among individuals exposed to trauma [53]. However, when it becomes a disordered process, such that enormous time, effort, and energy is devoted to managing, controlling, or struggling with unwanted private events [54], it might lead to increased salience of the unwanted stimulus in the long term [55, 56]. Chronic avoidance might therefore be the cause of the *rebound effect* (i.e., the increased frequency of the suppressed thoughts in subjects who display a strong tendency to suppress them) [57] responsible for the maintenance of OC symptoms in trauma sufferers [52].

Risk factors for subclinical OCD and transition probabilities to either the OCD or the healthy subjects group have been examined in a two-stage epidemiological study, originally designed to investigate depression between 1987 and 1989 in the USA [36]. Although it is unclear whether subclinical OCD (i.e., a weak manifestation of the disorder not severe enough to meet clinical OCD criteria) [58] represents a stage of development, a precursor of clinical disorder, or a level of severity on a continuum, it does not necessarily indicate progression to OCD. Authors showed that a weak manifestation of OCD was not predictive of OCD diagnosis 1 year later, while predictors for subclinical OCD were similar to that for OCD, and included more baseline undesirable life events, fewer baseline desirable life events, and medium to high socioeconomic status. The strong association between desirable and undesirable life events and the onset of the disorder, with a significant protective association of desirable life events with OCD and subclinical OCD, suggests that undesirable life events act in the opposite manner by exacerbating symptomatology in potentially susceptible individuals [36].

### 14.2.2 Longitudinal Studies

Although longitudinal studies offer more robust evidence than cross-sectional investigations, very few ones [24, 59, 60] have evaluated prospective psychosocial risk factors for OCD and OC symptoms. Moreover, most of them relied on retrospectively collected data about childhood trauma exposure and only in one [59] stressful/traumatic events exposure was questioned at first visit in a sample of children and adolescents who experienced a specific trauma (two major earthquakes) [59].

Potential risk factors for symptoms or diagnosis of OCD in adulthood, and for specific adult OC symptom dimensions were examined in the Dunedin Study [24], a longitudinal investigation on health and behavior of a complete cohort of children born during a 1-year period in 1972–1973 in Dunedin, New Zealand. Here, the presence of obsessions and compulsions and other psychological dimensions was investigated at ages 26, when the potential exposure to childhood physical and sexual abuse was assessed retrospectively, and 32 years ($N = 959$). Neurodevelopmental factors, childhood temperament, and behavior were evaluated from age 3, and personality factors at age 18 years. Results showed that cohort members who retrospectively reported a history of childhood physical or sexual abuse were more likely diagnosed with an anxiety disorder in adulthood. Those who reported physical abuse had seven times greater odds of developing OCD relative to healthy controls, and almost four times increased odds of developing the disorder relative to anxious controls. When risk factors were examined for adult OC symptoms rather than clinical diagnosis, the majority of the childhood risk factors were significant predictors, thus highlighting the importance of including subclinical cases in examining vulnerability for OC symptomatology. Indeed, each of the neurodevelopmental risk factors, several of the temperamental and behavioral characteristics, all of the personality dimensions, and most of the childhood stressors significantly predicted OC symptoms at age 26 or 32. Furthermore, several significant associations emerged between stressful/traumatic events exposure and specific OCD symptom dimensions. Sexual abuse was linked to increased risk for each symptom dimensions (except contamination/washing), physical abuse to the shameful thoughts dimension, loss of a parent to both the harm/checking and shameful thoughts dimensions and number of residence changes to each OCD symptom dimensions. These findings support the proposition that specific characteristics of a traumatic event may play a role in determining the symptom profile of OCD. Conversely, exposure to natural disasters, which have been associated in survivors with a range of psychopathologies including PTSD, was not associated with OCD (1 subject, 0.22%) in more than 400 Chinese children and adolescents who survived the 2013 Ya'an earthquake and were assessed via a face-to-face structured interview 12 months after, and followed-up ($N = 153$) at 30 months via telephone. Post-traumatic stress reactions were quite common at 12 months (43.9%) and decreased significantly from 12 to 30 months after the earthquake (15.7%), while depression was the second most prevalent disorder at 12 and 30 months (20.9 and 21.6%, respectively). Losing a family member and witnessing someone's death during the earthquake were risk factors for PTSD and depressive disorder, and not for OCD, given the very low incidence of the illness in the present sample.

Turning to a symptomatic population, the association of PTSD symptoms and the potential development of anxiety disorders (including OCD) was investigated in an American sample ($N = 34$) of children and adolescents victimized by interpersonal trauma and followed-up 12–18 months after the first assessment [40]. Among those who received the diagnosis of anxiety disorders at T2, no subject manifested OCD or OC symptoms, suggesting that PTSD symptomatology does not represent a high risk for OCD. Indeed, youth meeting full PTSD criteria at first visit were 25 times more likely to have a non-PTSD anxiety disorder (specific and social phobia, separation anxiety, and General Anxiety Disorder (GAD)) at T2. In addition, two specific PTSD clusters (i.e., avoidance/numbing and hyperarousal) prospectively predicted diagnosis of anxiety disorders, but not OCD diagnosis, and this association emerged only longitudinally. This observation suggests that chronic avoidance and a failure in decreasing levels of physiological and psychological arousal associated with trauma exposure predispose children to the development of an anxiety disorder, excluding OCD.

Thus, we might assert that adverse and potentially traumatic early experiences have a substantial role in the development of clinically significant maladaptive levels of OC symptoms in the general population. This relationship is both direct, as exposure to trauma causes maladaptive ways of coping, and dysfunctional responses that lead to the emergence of OC symptoms, and indirect, since the association seems to be mediated by specific personality characteristics that have been recognized as vulnerability traits for OC symptoms. Indeed, in the struggle to integrate and manage experiences that are perceived as difficult to control [43], the repeated exposure to multiple types of interpersonal trauma [61] leads children and adolescents to develop strategies such as thought suppression [43] and experiential avoidance [52]. This may inadvertently cause an OCD-type cycle as considerable energy is expended in avoiding memories or situations related to trauma, and/or numbing or constricting emotions [40]. The frequent and persistent presence of trauma-related intrusive thoughts can also determine, in abused children, the overestimation of thought importance, and the misinterpretation of mental contents that are characteristic of OCD. In the process of recovery, the persistence of trauma-related intrusive thoughts might convince abused children that these thoughts are important, that they reflect their true evil nature, and even that thinking about a bad event makes it more likely to happen. This contributes to constitute the inflated sense of responsibility that is accountable for the transition into clinical obsessions [37, 41]. Repeated exposure to uncontrollable traumatic events has also a critical effect on children's learned assumptions and beliefs about the world (such as threat estimation, intolerance for uncertainty and unpredictability, or excessive need for control), which initially form as adaptive ways of coping, and may be transformed from protective to maladaptive as coping resources are depleted over time, thus triggering an obsessional disorder [41].

Finally, some personality facets like conscientiousness [45] may indirectly mediate the relationship between stressful/traumatic events exposure and the development of clinically significant, maladaptive levels of OCD symptoms.

The considered studies provide further support for the concept of post-traumatic OCD, suggesting that some individuals may experience traumatic events that predispose them to OCD symptoms. They also help to further elucidate specific targets for clinical and preventive interventions.

## 14.3 Specific Antecedent Vulnerability in OCD to the Pathogenic Effects of Traumas/Stressors

In the previous paragraph, we revised evidence favoring the *general quantitative theory* of the relationship between stressful/traumatic events and some psychiatric disorders [62] stating that clustering of social or life events achieves etiological significance as a necessary, but insufficient cause of illness. Particularly, we showed that in OCD symptoms such relationship is partially mediated by some personality facets, and that repeated exposure to traumatic events may deplete coping resources increasing the risk of maladaptive functioning and/or cognitive distortions leading to the emergence of OCD symptoms [63]. In the current paragraph we will present studies supporting the *general qualitative theory* [64] emphasizing the undesirability or threatening quality of life events, and suggesting that there may be specific events that are important for specific disorders. We will therefore describe investigations comparing the effect of stressful/traumatic experiences in OCD to those in other clinical conditions.

The impact of childhood exposure to multiple types of PTEs compared to a single type, on prevalence of psychiatric disorders and somatic discomfort, was investigated in more than 6000 (18–74 years) Korean adults [61]. This community-based epidemiological study demonstrated that childhood trauma exposure is closely linked to increased risk of developing psychiatric disorders, including alcohol and nicotine addiction, mood, anxiety, eating, and psychotic disorders. Subjects who were exposed to multiple types of PTEs showed increased risk for developing psychiatric disorders. In particular, OCD, GAD, and somatoform disorder were associated with exposure to multiple types of PTEs, and not with exposure to a single type. Specifically, OCD was the most probable outcome (with a four times increased risk of developing the disorder) after repeated traumas. Since all three disorders (i.e., OCD, GAD, and somatoform disorder) are associated with a high level of trait anxiety, such observation reinforces the widely reported relationship between childhood abuse or trauma and the subsequent development of anxiety disorders. The evidence that exposure to multiple types of PTEs specifically predisposed to OCD strengthens the hypothesis that adverse childhood experiences are associated with maladaptive beliefs and lead to OCD symptoms.

Additional sparse research essentially investigated the role of stressful/traumatic events in the emergence of OCD symptoms, as opposed to the development of disorders previously deemed [65] as lying on the same psychopathological spectrum (non-OCD anxiety disorders) [30, 66, 67], or characterized by overlapping phenomenology and psychobiology such as trichotillomania (TTM) [68], Tourette Syndrome (TS) [69, 70], and PTSD [71]. Indeed, although studies point on the high co-occurrence of OCD and PTSD following traumatic events, possible factors (either demographic or trauma-related) explaining why some PTSD patients also develop OCD have been rarely investigated. Evidence showed that following an exposure to combat or terror-related trauma, roughly 60% of Israeli subjects ($N = 44$) developed OC symptoms, and for the majority of these (>70%) symptoms were severe enough to meet OCD diagnostic criteria [71]. However, demographic or trauma-dependent differences were not observed between PTSD and PTSD-OCD groups, thus suggesting that post-traumatic obsessions are not related to these variables. Equally, when the association between life events, personality factors, and anxiety-related pathology was investigated in children and adolescents with a primary diagnosis of OCD ($N = 28$), or other anxiety disorders ($N = 28$; GAD, social and specific phobias, and panic disorder) [66] the two groups had significantly more total life events and negative life events (both lifetime, and 1 year prior to onset of the disorder) than healthy control subjects. However, no significant difference was observed between OCD and AD patients. The only event that significantly differentiated among the three groups was having experienced a major illness or injury of a relative, which was more common in the OCD and the AD groups, compared to normal controls. Moreover, a temperamental factor (harm avoidance, i.e., a tendency to react with avoidance and inhibition to aversive stimuli) [72] distinguished OCD and AD from healthy subjects. This factor positively correlated with negative life events and with the self-rated actual amount of stress resulting from having experienced specific events. It is therefore possible that genetic or environmental factors induce harm avoidance tendencies in children. When a stressful/traumatic event occurs, these children tend to perceive it more negatively, such that pathological anxious mechanisms are activated, leading to the emergence of AD (especially OCD). This finding replicates in patient samples, the reported indirect effect (mediated by personality and temperamental facets) [45, 52] that traumatic events in childhood exert on the development of maladaptive levels of OC symptoms in the general population.

Equally, a significantly higher frequency of childhood sexual abuse involving physical contact was found among 30 patients diagnosed with OCD (53.3%) and 17 diagnosed with panic disorder (PD) (52.9%) as compared to 26 non-psychiatric rheumatic patients (23.1%) [30]. This suggests that sexual abuse in childhood might be a contributing factor in the etiology of anxiety disorders, OCD and PD in particular, as it impinges upon normal development. Alternatively, a familial, environmental or personality structure may generate the risk for childhood sexual abuse, as well as for anxiety disorders.

However, when a history of trauma was compared between OCD patients and subjects meeting DSM-IV diagnostic criteria for social anxiety disorder (SAD) [67], which, contrary to AD and PTSD, is not trauma-related, OCD patients showed significantly lower rates of exposure to traumatic events. Authors speculated on this negative finding assuming that as the obsessional system is a psychological process primarily designed to generate harm avoidance behavior in response to dangers, the disorder might be analogous to a mental autoimmune disease, i.e., a protective response that goes beyond the point of usefulness, and becomes self-destructive [73].

The role of childhood trauma was also investigated across TTM and OCD [68], two disorders that in the DSM-5 lie on the same psychopathological spectrum [74] and are characterized by overlapping phenomenology. Both OCD ($N = 74$) and TTM ($N = 36$) patients showed a significantly greater severity of childhood trauma compared to normal controls, differing specifically for the emotional neglect (with no difference between patients) and physical abuse subscale (with higher scores in TTM patients).

Based on previous studies on TS and early onset OCD, which consistently suggested that these etiologically related disorders are sensitive to psychosocial stress (e.g., [75, 76]), two more recent investigations [69, 70] explored in children and adolescents (7–18 years) the role of non-shared environmental factors in the pathogenesis of TS and OCD. In the first study [69], the risk of developing TS was less likely to be influenced by stressful/traumatic events than in OCD, as the latter group ($N = 28$) showed more SLEs either lifetime, and in the year prior the disease onset. Injury to a family member was the most common event experienced lifetime, and changing residence and birth of a sibling the most frequent SLE in the year preceding the onset of symptoms. Moreover, OCD patients rated SLEs impact significantly higher than control subjects. In another study, the global impact of major life events was measured using objective threat ratings (made independently from the subjects' own ratings and referring to the degree of stress most people would experience under similar contextual circumstances) in children and adolescents with TS ($N = 28$), OCD ($N = 23$), or both ($N = 18$) [70]. Either group reported a higher number of daily stressful events and major life stressors, particularly of a chronic, negative nature, although no cross-sectional relationship was observed between global stress and severity of TS and OCD symptoms. The fact that measures of daily life stressors consistently showed significant associations with tic and OCD symptom severity ratings suggests that these are more closely related to the severity of stress for daily events, rather than for major life events.

As summarized in Table 14.1, the reported studies, although quite limited, suggest that specific (i.e., negative and potentially traumatic) life events have a different role in activating pathological mechanisms in anxiety disorders and OCD. Indeed, although previously subsumed within the same psychopathological spectrum, the unique differences between AD and disorders with OC features justified a separate category for OCD and related disorders [77]. Such differences also include the potential pathogenic effect of traumas as in OCD the triggering effect may be mediated by personality and temperamental facets such as a proneness to confer more impact to life events, and a tendency to harm avoidance [66, 69]. This observation would favor the relational-cognitive-orientation approach—which emphasizes the meaning attributed to life events by the individual [78], for investigating the relationship of stressful/traumatic events to OCD onset.

## 14.4 Specific Latent Vulnerability to OCD and the Exacerbating Effect of Childhood Trauma

The third mode of the stress-diathesis interaction posits that an illness may exist independently, but it may be exacerbated (either permanently or episodically) by repeated stressful/traumatic events. The type and severity of stressors may be of importance, and specific types of stressors may differentially affect vulnerabilities and manifestations of symptom. In this last paragraph, we will first revise studies investigating the importance of traumatic incidence in shaping OCD etiology and pathophysiology, and will enquire whether victimization has an independent

**Table 14.1** Summary of studies (and the relative strength of the evidence) testing the hypothesis that trauma exposure and cognitive factors combine to produce vulnerability for the development of OC symptoms

| Direct and indirect relationship between trauma exposure and the emergence of OC symptoms | | |
|---|---|---|
| | Type of evidence | Strength of evidence |
| **(a) Typical OCD dysfunctional beliefs mediating the relationship between trauma and OC-OCD symptoms** | | |
| Inflated personal responsibility [42, 50, 81] | Mostly nonclinical samples | Strong mediating effect |
| Intolerance to uncertainty [42, 50] | Nonclinical samples only | Strong mediating effect |
| **(b) Maladaptive ways of coping with trauma leading to the emergence of OC-OCD symptoms** | | |
| Thought suppression [42, 43] | Mostly nonclinical samples | Medium mediating effect |
| Experiential avoidance [40, 42, 52, 67] | Nonclinical and at-risk samples; clinical samples | Moderate to medium mediating effect in nonclinical samples; strong effect in at-risk subjects; moderate effect in clinical samples |
| Excessive need for control [43] | Nonclinical and clinical sample | Small to medium mediating effect |
| **(c) OC personality traits interacting with stress and trauma in determining the emergence of OC-OCD symptoms** | | |
| Conscientiousness [45] | Nonclinical sample | Small interactive effect |
| Harm avoidance [66, 69] | Clinical samples | Small to medium interactive effect in OCD, TS, and AD |

The limited evidence demonstrates that trauma can directly cause typical OCD dysfunctional beliefs (a) and maladaptive ways of coping (b) leading to the emergence of OC symptoms. Such association may also be indirect (c) and mediated by personality facets, which are recognized as vulnerability traits for OCD

Abbreviations: *AD* anxiety disorders; *OC* obsessive-compulsive; *OCD* obsessive-compulsive disorder; *TS* Tourette syndrome

association with the disorder, apart from its association with PTSD. The complex interplay of experience, environment, and genetics in the disorder will be explored discussing studies on the interaction between candidate genes for OCD susceptibility and childhood trauma. The effect of "non-shared" environmental factors [79] in the etiology and course of the disorder, as investigated in twins studies, will be also considered. Finally, we will explore the importance of history of trauma, and the potential relationship with types of traumatic life events, in shaping the severity, course, and phenotypic presentation of OCD.

### 14.4.1 Etiological Importance of Trauma Incidence in OCD

The frequency of life events was explored in two studies [80, 81] assuming that several psychosocial factors may contribute toward vulnerability to OCD, and that traumatic events may act as stressors causing increased susceptibility [82]. Significantly more lifetime events were reported by the OCD group, and this excess spanned the 6 months prior to the onset of illness, with a peak at 1 month [81]. A dose–response relationship between the number of presumptive SLEs and OCD symptoms severity, and a cumulative effect over a lifetime on severity of obsessions were evidenced [80]. However, a negative correlation was observed between premorbid personality traits (obsessional, anxious, and most markedly, self-conscious) and SLEs (including serious illness, arguments, childbirth, and traumatic brain injury) [81]. This suggests that subjects with abnormal traits develop OCD symptoms without experiencing an excess of stressful life events, while those with normal premorbid personality require a significant excess of events before developing such symptoms. Once again, personality facets seem to mediate the relationship between OCD onset and trauma exposure.

To further clarify the link between traumatic events and OCD, several studies [33, 35, 83–86] investigated whether traumatic stress exposure impacts OCD independently from its relation to PTSD, possibly examining the effect of the seemingly unique relationship between the two disorders on treatment response [84, 85, 87]. When the relation between victimization status, crime factors, PTSD, and several other psychological disorders was explored among a community sample of 391 American women (18 years or older) [86], results showed that at the univariate level, victims were more likely than non-victims to currently suffer from several Axis I disorders, including OCD (4.1% of the whole sample). Comorbidity with PTSD was quite high in the OCD sub-sample (27%), and demographics, victimization status, and crime factors (e.g., life threat, injury) were largely associated with the development of non-PTSD disorders. However, rape remained a significant predictor of both OCD (with an almost fourfold increased risk for developing the disorder) and social phobia. Such observation implies that in the victimization-psychopathology relation for OCD, comorbid PTSD is not the only mediating factor.

Likewise, in a large sample of children with OCD, PTSD rates and evidence of psychological trauma were higher than in a comparable non-OCD cohort, and those affected by concurrent disorders had more severe OCD symptoms [83]. PTSD diagnosis (either full or subthreshold) resulted in a 14 times increased risk of developing OCD, while trauma exposure determined a ninefold higher risk of onset, thus highlighting the overall salience of potential environmental triggers in pediatric OCD.

The unique relation between OCD and PTSD thus far outlined is supposed to result in negative treatment outcome for OCD, and several case studies (e.g., [35]) have documented the possible development of treatment-resistant OCD after various types of trauma. Indeed, in a naturalistic retrospective chart review of 104 individuals diagnosed with treatment-resistant OCD, 82% reported a history of trauma, while roughly 40% of the whole sample met diagnostic criteria for PTSD, i.e., half of those who experienced adulthood or childhood interpersonal violence [84]. The negative prediction afforded by PTSD on treatment outcome for OCD was mediated by the incidence of comorbid disorders such as borderline personality disorder or major depressive disorder. In fact, the presence of history of trauma or PTSD in patients with OCD ($N = 215$) did not negatively affect the response to treatment (either cognitive behavioral therapy or pharmacological treatment). Conversely, a greater reduction in OCD and anxiety scores was observed in the comorbid OCD + PTSD group, which represented the 60% of OCD patients exposed to traumatic experiences [85].

Nevertheless, the PTSD-OCD comorbidity picture remains quite blurred as there are several possible explanations for how comorbid conditions might develop. Temporally, three possibilities exist: the OCD diagnosis may precede PTSD, follow development of PTSD, or begin at the same time as PTSD. The last hypothesis was tested in 210 cases with OCD [88] and 133 sex- and age-matched controls from the adult general population. Results demonstrated that the lifetime prevalence rates of traumatization, PTSD, and acute stress disorder were not different between OCD patients and healthy controls. Given the low rate of trauma-related disorders occurring before (2.9%) or within (1.5%) the same year as the onset of OCD, authors concluded that other factors than severe traumatic events determine the onset of the disorder in most of the cases.

Two further studies [89, 90] compared the clinical features of OCD with and without stressful life events preceding it, and examined the relationship between type of SLEs and OCD symptom dimensions in 412 and 329 OCD patients, respectively. Those with OCD onset close to an SLE showed a distinct clinical pattern with a later onset, less family history of the disorder, and the presence of contamination/cleaning symptoms of OCD [89]. At least one event preceded the onset of OCD in more than 60% of patients, and this was significantly associated with female gender, abrupt onset, and somatic obsessions. Moreover, three specific traumatic events, i.e., hospitalization of a family member, major personal physical illness, and loss of personally valuable object, were significantly associated with prominent symmetry/ordering symptoms [90], thus suggesting that at least the contamination/cleaning and symmetry/ordering symptoms are more related to environmental factors.

Finally, the temporal relationship between trauma and OCD onset was explored in 106 patients developing OCD after PTSD (post-traumatic OCD), 41 patients developing OCD before PTSD (pre-traumatic OCD), and 810 OCD patients without any history of PTSD (non-traumatic OCD) [33]. Again, PTSD prevalence was quite high (19%) and OCD occurring after or in close proximity to the onset of PTSD was associated with distinct clinical features, particularly when patients did not present a history of pre-traumatic OCD symptoms. These results suggest that there might be a specific pre-traumatic OCD phenotype that increases the risk of developing comorbid PTSD. Conversely, post-traumatic OCD patients experienced traumatic events at earlier age had a more severe clinical picture and a later age at onset suggesting that late-onset OCD may be more likely to be precipitated by a trauma in childhood/adolescence, rather than by genetic factors. Based on the reported evidence authors suggested a working model encompassing different OCD phenotypes and their relationship to trauma [33]. Figure 14.2 depicts the etiological overlap of the post-traumatic OCD phenotype showing that the presence of OCD symptoms predating trauma exposure defines a group of patients more etiologically related to OCD, as opposed to those without pre-traumatic OCD symptoms in whom the disorder is more akin to PTSD.

Considering the high prevalence of traumatic experiences in OCD, and the consequent high OCD-PTSD comorbidity, questions arise regarding the nature of the supposed preexisting vulnerability required to interact with childhood trauma to increase risk for the disorder. In some recent studies [91, 92], gene-by-environment

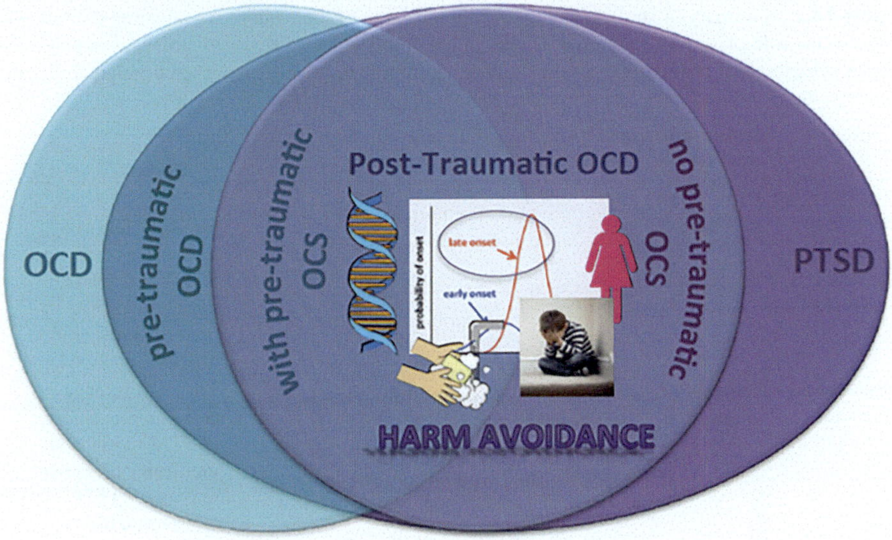


**Fig. 14.2** Post-traumatic OCD. When OCD (pale green circle) and PTSD (purple oval) are comorbid, two different OCD phenotypes are observable: (1) patients with OCD onset before PTSD (*the pre-traumatic OCD phenotype*, etiologically related to primary OCD, and predisposing to the development of comorbid PTSD), and (2) patients developing OCD after PTSD onset (*post-traumatic OCD*). The latter is associated to a specific personality trait (i.e., harm avoidance) [66, 69], and characterized by a definite genotype [91, 92], later onset [33, 89], predominance of females [90] and of contamination/washing symptoms [33, 89] and by more experienced childhood traumas [33]. When OCS symptoms are present before PTSD onset, the disorder is etiologically related to primary OCD, while in cases not showing pre-traumatic OCS, the disorder is more akin to PTSD. (Adapted from Ref. [33]). *OCD* obsessive-compulsive disorder; *OCS* obsessive-compulsive symptoms; *PTSD* post-traumatic stress disorder

interaction analyses investigated the role that environmental factors (i.e., traumatic experiences in childhood) play in the development of OCD, and how genetic constitution could modify this association. Results showed that both a polymorphism of a key neurotrophic factor (BDNF, playing a crucial role in neurodevelopment and synaptic plasticity and moderating the effect of stressful life events and the subsequent risk for psychopathology) [91] and polymorphisms within genes encoding serotonergic and dopaminergic systems (COMT, already associated with increased susceptibility for OCD, and MAOA-B involved in stress-related pathogenicity) [92] interacted with childhood trauma (specifically emotional and sexual abuse) in increasing the risk for OCD. This suggests that post-traumatic OCD is characterized by a specific genotype moderating the effect of traumatic events in increasing the risk for the disorder.

Taken together the reported evidence (summarized in Table 14.2) suggests that traumatic events do cause increased susceptibility to OCD by interacting with specific personality traits, and genes moderating the effect of stressful life events and the subsequent risk for psychopathology (the post-traumatic OCD phenotype). The

**Table 14.2** Summary of studies (and the relative strength of the evidence) evaluating the following: (a) the role of stressful/traumatic events in the emergence of OCD symptoms, as opposed to the development of anxiety and trauma-related disorders; and (b) the antecedent vulnerability in anxiety, trauma-related, and OC disorders to the pathogenic effect of specific traumas

| Differential effect of traumas in triggering OCD and non-OCD anxiety disorders | | | | |
|---|---|---|---|---|
| | AD | PTSD | OCD | Difference |
| **(a) Prevalence (%) of disorders in traumatized samples** | | | | |
| Ref. [61] General population, multiple PTEs | **28.2** | 12.3 | 0.7 | **Small difference between prevalence of AD and OCD**; trivial difference between prevalence of PTSD and OCD |
| Ref. [86] General population | 1.8 | **6**<br>**PTSD + OCD 27.3** | **4.1** | Trivial difference between prevalence of PTSD and OCD; **medium difference between PTSD and comorbid PTSD-OCD; small difference between prevalence of AD and OCD** |
| Ref. [84] Clinical sample (treatment-resistant OCD) | | **PTSD + OCD 49.4** | **50.6** | Trivial difference between prevalence of treatment-resistant OCD and comorbid OCD + PTSD |
| Ref. [33] Clinical sample | | PTSD+OCD 19.1 | **Pre-trauma OCD 4.1**<br>**Post-trauma OCD 10.6**<br>**No trauma OCD 80** | Trivial difference between prevalence of pre- and post-traumatic OCD and between prevalence of trauma-related and non-traumatic OCD |
| Ref. [90] Clinical sample | **Comorbid in OCD without SLEs 20.2**<br>**Comorbid in OCD with SLEs 43** | | **OCD without SLEs 39.2**<br>**OCD with SLEs 60.8** | Significant, but trivial difference between prevalence of OCD following SLEs and OCD without SLEs. Trivial difference in AD comorbidity according to presence/absence of SLEs |
| **(b) Prevalence (%) of TLEs in psychiatric samples** | | | | |
| Ref. [66] Negative life events 1 yr prior onset | **3.0** | | **3.8** | Significant, but trivial difference between incidence of negative life events 1 yr prior onset in AD and OCD, and between prevalence of relative's major illness or injury |
| Major illness or injury of a relative | **18** | | **32** | |

**Table 14.2** (continued)

| Differential effect of traumas in triggering OCD and non-OCD anxiety disorders | | | | |
|---|---|---|---|---|
| | AD | PTSD | OCD | Difference |
| Ref. [67] Attempted or actual breaking into home | **36.7** | | **11.8** | Significant, but trivial difference between prevalence of TLEs in AD and OCD |
| Exposure to dead bodies | **70.0** | | **35.3** | |
| News of serious injury, illness, or unexpected death of someone close | **86.7** | | **64.7** | |
| Ref. [24] Clinical sample CSA | **52.9** | | **53** | Non-significant difference between prevalence of CSA between AD and OCD and between prevalence of AD and OCD after CSA (compared to no CSA) |
| No CSA | **47.1** | | **43** | |

Abbreviations: *AD* anxiety disorders; *CSA* childhood sexual abuse; *yr* year; *OC* obsessive-compulsive; *OCD* obsessive-compulsive disorder; *PTSD* post-traumatic disorder; *SLE* stressful life events; *TLE* traumatic life event

co-occurrence of PTSD is not therefore the only mediating factor in this pathogenic relation, and does not affect treatment outcome, unless additional comorbid disorders are present [85].

### 14.4.2 Separating the Genetic and Environmental Sources of Increased Risk for OCD. Twin Studies Considering Traumas as Predisposing Environmental Factors

Twin studies of OCD have a long history, starting back in 1929 [1], and evolving from single case reports to large epidemiological studies. Indeed, genetically informative studies, in particular those employing the discordant monozygotic twin design (MZ, sharing 100% of their genetic background and growing up largely in the same environment), are better suited to test whether the association between an environmental measure and an observed phenotype is likely to be consistent with a causal effect, as they provide excellent control of both genetic and shared environmental effects. Several studies (e.g., [79, 93–95]) demonstrated that OCD disorder and OC symptoms exhibit a familial pattern of transmission being influenced by genetic factors (approximately 55%), and also by unique environmental factors (approximately 45%). Few studies [96, 97] evaluated in twins the specific contribution of traumatic/stressful life events on the presence and severity of OC symptoms. In a large Swedish population-based cohort of

22.084 twins [96] within a wide range of retrospectively reported stressful life events, a significant association was found between certain types of SLEs, i.e., history of physical abuse, neglect and family disruption, and OC symptoms severity. Because this association remained significant when genetic factors shared environmental factors and co-occurring depression were controlled for, these aversive events play a modest (about 3% of the variance), but significant role in the severity of OC symptoms. When the MZ-difference method was used to isolate the non-shared environment effect (i.e., controlling for genetic and shared environment effects), only SLEs related to abuse and family disruption were associated to differences in OC symptoms severity. Although these factors contain items that are by definition shared environment, it is important to note that technically they could be classified as non-shared (e.g., parenting being experienced as different among siblings). This finding is reminiscent of the literature on parenting styles in OCD [50] and suggests that some dysfunctional styles could be associated to the severity of OC symptoms.

In order to investigate environmental factors that protect against or exacerbate OC symptoms, 25 MZ twin pairs discordant, 17 MZ twin pairs concordant high and 34 MZ pairs concordant low on OC symptoms were selected from a large longitudinal Dutch sample of adult twin pairs and their family members [97]. Unique environmental factors were studied using within-discordant MZ pair comparisons, whereas comparisons between concordant MZ pair were used to study environmental factors shared by the twins of an MZ pair. The high-scoring MZ twins of the discordant group reported more life events (especially sexual abuse) than their low-scoring twin siblings, thus suggesting that this non-shared environmental factor determined more severe OC symptoms. Interestingly, although they showed similar OC symptoms, no sexual assault was reported by the concordant high-scoring MZ twin pairs, suggesting that OC symptoms in the high-scoring twins of the discordant pairs are associated more with environmental stressors (i.e., sexual assault) than the OC symptoms in the concordant high-scoring pairs. As a matter of fact, concordance between MZ twin pairs on OC behavior is supposed to result from genetic similarity between the twins of a pair, thus suggesting a stronger inheritance of symptomatology.

Despite the strong evidence here reported, it is clear that prospective longitudinal studies are needed before potentially traumatic/stressful life events can be considered truly causal.

### 14.4.3 On the Relationship Between Types of Traumatic Life Events and OCD Severity, Course, and Phenotype

Another variable that needs to be considered when conceptualizing the role of trauma in the genesis of OCD symptoms is type of trauma. Interpersonal trauma (such as sexual assault or physical violence), when compared to non-interpersonal trauma, has been found to affect more severely an individual, especially in the development of psychological symptoms.

Using a meta-analytic approach, the strength of the association between trauma exposure and symptoms within the obsessive-compulsive spectrum (OCS) disorders symptoms [74] was explored to better understand relevant trauma sequelae outside of PTSD symptoms [98]. Gender and relationship status were used as moderators since the association between trauma and OCS symptoms may be greater for women than men. Also the buffering effect that social support has on traumatized individuals by providing needed interpersonal resources, reducing negative appraisals of the event, and encouraging coping skills, was considered. Apart from examining the overall effect of past trauma on OCS symptoms, effect sizes were calculated separately for each of four types of interpersonal trauma exposure, i.e., violence, emotional abuse, sexual abuse, and neglect, and two dimensions of OCS symptoms (obsessions and compulsions). The overall relation between past trauma exposure and severity of OCS symptoms for the total sample (24 included studies comprising more than 4500 patients) was small, but statistically significant. This association was larger for females and was not weakened in the context of committed relationships. A closer examination of effect sizes for specific types of interpersonal traumas revealed that the four types were each associated with more severe OCS symptoms, and effects were similar in magnitude, implying that the important component of trauma is trauma severity, rather than type of trauma. The association between past trauma and OCS symptoms was present for compulsions, but not for obsessions, as compulsions may represent a more behavioral response to trauma, while strength of obsessions may be more innate in origin [98].

The reported evidence reinforces the role of environmental risk factors on the occurrence of OCD, while it is still controversial whether these influences have a clear predictive value on its clinical course (i.e., the probability for chronicity). The hypothesis that the interactions between familiarity, environmental exposure to SLEs before OCD onset, and gender are related to variations in the chances of presenting a chronic course of the disorder was tested in more than 400 OCD patients [99]. Results indicated that familiarity alone does not reliably predict the course of OCD. Rather, the predictive value of familial loading on course seems to be significantly altered by both exposure to SLE before onset and gender as the former was associated with increased odds of a chronic course at low levels of familial loading. Equally, male gender was associated with an increased probability of presenting a chronic course at low levels of familiarity. Thus, the lack of a clear mode of inheritance and the strong impact of lifetime experiences on OCD indicate that the course of the disorder may be influenced by multiple factors. This additive moderation model is of clinical importance and suggests that the information regarding familial background must be complemented with other patients' characteristics (i.e., SLE and gender) in order to reliably predict the course of OCD.

A last question that needs to be addressed is whether individual differences in the experience of stressors can affect the expression of the clinical condition in trauma-related OCD. In addition to a predisposition/diathesis-stress approach, the pathoplastic model proposes that regardless of the role that stress may play in the development of OCD, exposure to stressful/traumatic events may modify the manifestation of the disorder. The majority of studies specifically addressed the relationship between PTE/SLEs and the onset and severity of hoarding symptoms [100–104] thus adding further evidence to the already cited studies suggesting that the contamination/cleaning and symmetry/ordering symptoms are more related to environmental stressful factors [33, 90]. Based on the assumption that psychiatric disorders mediate the relation of childhood trauma to suicidality, two studies also sought to determine the association of traumatic/stressful events exposure with the probability of suicide in OCD [105, 106].

Hoarding disorder (HD) is typically defined as the acquisition of, and failure to discard, a large number of possessions, which results in debilitating clutter and subsequent distress or impairment [107]. The supposed specific relationship between hoarding behaviors and stress lies in the hypothesis that possessions may acquire an association with safety and security in response to a threatening environment, and in the provisional support from current literature for a putative non-genetic risk factor for compulsive HD [101]. Most studies examined the onset and course of HD in relation to PTE/SLEs exposure either in a vast population of subjects with self-reported HD ($N = 751$) [104], and in OCD patients ($N = 180$) meeting criteria for HD (24%) respect to OCD patients with no HD [101], or in small samples of hoarders ($N = 26$), as compared to healthy subjects [103]. The considered studies found a robust relationship between PTEs and HD onset, while exposure to traumatic events was disproportionately associated in time with periods of symptoms exacerbation [104]. The relationship with symptom severity was independent from general OCD symptomatology, age of OCD onset, current age, or mood, anxiety, and depression comorbidity [101]. Accidents [101, 103] and interpersonal violence [104] were the most frequent traumatic events reported, thus suggesting that such events might lead individuals to perceive the world as unpredictable and uncontrollable, predisposing them to seek comfort and security in personal possessions through which they can exert some degree of control [103].

When rates of specific traumas were compared between those with OCD symptoms alone ($N = 112$), HD alone ($N = 42$), and comorbid OCD and HD ($N = 53$) [102], HD severity was positively associated with the number of PTEs before onset (while no difference was observed between groups in the total number of experienced traumatic events lifetime), thus supporting the notion of cumulative trauma in hoarding. Indeed, HD patients (either with comorbid OCD and with no OCD symptoms) reported a greater exposure to a range of PTE/SLEs compared to patients without hoarding symptoms [100]. Specifically, subjects with HD alone had higher rates of physical assault and transportation accidents prior to symptom onset compared to those with OCD symptoms alone, while patients with comorbid HD and OCD reported higher rates of sexual assault prior to symptom onset than those with OCD alone [102].

Overall, although data on traumas were acquired retrospectively, which may encourage participants to draw connections between traumatic events and symptom onset, the reported studies indicate that the hoarding dimension of OCD is strongly related to repeated PTEs exposure before symptoms manifestation. They also suggest a potential link between hoarding symptoms and specific types of traumatic events, such as physical and sexual assault and transportation accidents.

Finally, both studies investigating the relationship between childhood trauma and the probability of suicide in OCD [105, 106] found that the latter was increased (independently form comorbid depression and anxiety) [105] in the traumatized group. Specifically, history of sexual abuse and the OCD symptom of unacceptable thoughts explained suicide ideation [106]. This bears important implications for the assessment and treatment of suicide risk in OCD patients with lifetime suicide attempts.

## 14.5 General Conclusions

Here we discussed the current body of literature on putative environmental risk factors for OCD as to ascertain whether stressful/traumatic life events (particularly those experienced in childhood and adolescence) may act as an etiological factor in the development of the disorder. We also examined the potential role of types of traumatic life events in shaping the severity, course, and phenotypic presentation of OCD. Although several constrains limit the generalization of findings, results demonstrate that clustering of traumatic events achieves etiological significance as a necessary but insufficient mechanism of illness. Particularly, we showed that in OCD such relationship is partially mediated by some personality facets, and that repeated exposure to traumatic events may deplete coping resources determining a tendency toward maladaptive functioning and/or cognitive distortions leading to the emergence of OCD symptoms. As for a potential specific antecedent vulnerability in OCD to the pathogenic effects of traumas/stressors, the current literature demonstrates that negative and potentially traumatic life events have a role in activating pathological anxious mechanisms leading to the onset of anxiety, stressor-related, and OCD disorders. More specifically in the latters, the triggering effect may be mediated by a proneness to confer more impact to life events, and a tendency to harm avoidance. The putative post-traumatic OCD phenotype seems to be characterized by contamination/cleaning, symmetry/ordering, and hoarding symptoms suggesting that these OCD symptoms are more related to environmental stressful factors and to the cumulative effect of traumas. However, evidences regarding the existence of a specific post-traumatic OCD phenotype are for now scanty therefore impeding firm conclusions.

Based on the discussed evidence we therefore envision a multifactorial model of inheritance [90] in which especially interpersonal traumatic events in childhood and adolescence cause increased susceptibility to OCD by interacting with specific personality traits and genes moderating the subsequent risk for psychopathology. Since cognitive and behavioral treatment strategies have proved effective on the content of OCD dysfunctional beliefs, and in reducing experiential avoidance, the cognitive distortions that mediate the effect of repeated trauma exposure on the emergence of OCD should be targeted as to prevent the development of the disorder in traumatized youth..

## References


1. Lange J. Leistungen der Zwillingpathologie für die Psychiatrie [The importance of twin pathology for psychiatry]. Allg Zeitschrift fur Psychiatr und Psych Med. 1929;90:122–42.
2. Grisham JR, Anderson TM, Sachdev PS. Genetic and environmental influences on obsessive-compulsive disorder. Eur Arch Psychiatry Clin Neurosci. 2008;258:107–16.
3. Hettema JM, Neale MC, Kendler KS. A review and meta-analysis of the genetic epidemiology of anxiety disorders. Am J Psychiatry. 2001;158:1568–78.
4. Zuckerman M. Vulnerability to psychopathology: a biosocial model. American Psychological Association; 1999.
5. Ingram RE, Luxton DD. Vulnerability-stress models. In: Development of psychopathology: a vulnerability-stress perspect. Thousand Oaks, CA: Sage Publications; 2005. p. 32–46.
6. Wirz L, Bogdanov M, Schwabe L. Habits under stress: mechanistic insights across different types of learning. Curr Opin Behav Sci. 2018;20:9–16.
7. Schwabe L, Wolf OT. Stress prompts habit behavior in humans. J Neurosci. 2009;29:7191–8.
8. Ferragud A, Haro A, Sylvain A, Velázquez-Sánchez C, Hernández-Rabaza V, Canales JJ. Enhanced habit-based learning and decreased neurogenesis in the adult hippocampus in a murine model of chronic social stress. Behav Brain Res. 2010;210:134–9.
9. Gillan CM, Papmeyer M, Morein-Zamir S, Sahakian BJ, Fineberg NA, Robbins TW, et al. Disruption in the balance between goal-directed behavior and habit learning in obsessive-compulsive disorder. Am J Psychiatry. 2011;168:718–26.
10. Boulougouris V, Chamberlain SR, Robbins TW. Cross-species models of OCD spectrum disorders. Psychiatry Res. 2009;170:15–21.
11. Gillan CM, Robbins TW. Goal-directed learning and obsessive–compulsive disorder. Philos Trans R Soc B Biol Sci. 2014;369:20130475.
12. Gillan CM, Robbins TW, Sahakian BJ, van den Heuvel OA, van Wingen G. The role of habit in compulsivity. Eur Neuropsychopharmacol. 2016;26:828–40.
13. Graybiel AM, Rauch SL. Toward a neurobiology of obsessive-compulsive disorder. Neuron. 2000;28:343–7.
14. Alexander GE, DeLong MR, Strick PL. Parallel organization of functionally segregated circuits linking basal ganglia and cortex. Annu Rev Neurosci. 1986;9:357–81.
15. Boedhoe PSW, Schmaal L, Abe Y, et al. Cortical abnormalities associated with pediatric and adult obsessive-compulsive disorder: findings from the ENIGMA Obsessive-Compulsive Disorder Working Group. Am J Psychiatry. 2017;175(5):453–62.
16. Piras F, Piras F, Chiapponi C, Girardi P, Caltagirone C, Spalletta G. Widespread structural brain changes in OCD: a systematic review of voxel-based morphometry studies. Cortex. 2015;62:89–108. https://doi.org/10.1016/j.cortex.2013.01.016.
17. Edmiston EE, Wang F, Mazure CM, Guiney J, Sinha R, Mayes LC, et al. Corticostriatal-limbic gray matter morphology in adolescents with self-reported exposure to childhood maltreatment. Arch Pediatr Adolesc Med. 2011;165:1069–77.
18. Ansell EB, Rando K, Tuit K, Guarnaccia J, Sinha R. Cumulative adversity and smaller gray matter volume in medial prefrontal, anterior cingulate, and insula regions. Biol Psychiatry. 2012;72:57–64.
19. Saxena S, Rauch SL. Functional neuroimaging and the neuroanatomy of obsessive-compulsive disorder. Psychiatr Clin North Am. 2000;23:563–86.
20. Goodman J, Leong KC, Packard MG. Emotional modulation of multiple memory systems: implications for the neurobiology of post-traumatic stress disorder. Rev Neurosci. 2012;23:627–43.
21. Dias-Ferreira E, Sousa JC, Melo I, Morgado P, Mesquita AR, Cerqueira JJ, et al. Chronic stress causes frontostriatal reorganization and affects decision-making. Science. 2009;325(80):621–5.
22. Adams TG, Kelmendi B, Brake CA, Gruner P, Badour CL, Pittenger C. The role of stress in the pathogenesis and maintenance of obsessive-compulsive disorder. Chronic Stress. 2018;2:247054701875804.

23. Brander G, Pérez-Vigil A, Larsson H, Mataix-Cols D. Systematic review of environmental risk factors for obsessive-compulsive disorder: a proposed roadmap from association to causation. Neurosci Biobehav Rev. 2016;65:36–62.
24. Grisham JR, Fullana MA, Mataix-Cols D, Moffitt TE, Caspi A, Poulton R. Risk factors prospectively associated with adult obsessive-compulsive symptom dimensions and obsessive-compulsive disorder. Psychol Med. 2011;41:2495–506.
25. Fergusson DM, Horwood LJ, Woodward LJ. The stability of child abuse reports: a longitudinal study of the reporting behaviour of young adults. Psychol Med. 2000;30:529–44.
26. Dekel S, Bonanno GA. Changes in trauma memory and patterns of posttraumatic stress. Psychol Trauma Theory Res Pract Policy. 2013;5:26–34.
27. Burnam MA, Stein JA, Golding JM, Siegel JM, Sorenson SB, Forsythe AB, et al. Sexual assault and mental disorders in a community population. J Consult Clin Psychol. 1988;56:843–50.
28. Saunders BE, Villeponteaux LA, Lipovsky JA, Kilpatrick DG, Veronen LJ. Child sexual assault as a risk factor for mental disorders among women: a community survey. J Interpers Violence. 1992;7:189–204.
29. Chou KL. Childhood sexual abuse and psychiatric disorders in middle-aged and older adults: evidence from the 2007 adult psychiatric morbidity survey. J Clin Psychiatry. 2012;73:e1365–71.
30. Caspi A, Vishne T, Sasson Y, Gross R, Livne A, Zohar J. Relationship between childhood sexual abuse and obsessive-compulsive disorder: case control study. Isr J Psychiatry Relat Sci. 2008;45:177–82.
31. Lin H, Katsovich L, Ghebremichael M, Findley DB, Grantz H, Lombroso PJ, et al. Psychosocial stress predicts future symptom severities in children and adolescents with Tourette syndrome and/or obsessive-compulsive disorder. J Child Psychol Psychiatry. 2007;48:157–66.
32. Kisely S, Abajobir AA, Mills R, Strathearn L, Clavarino A, Najman JM. Child maltreatment and mental health problems in adulthood: birth cohort study. Br J Psychiatry. 2018;213:1–6.
33. Fontenelle LF, Cocchi L, Harrison BJ, et al. Towards a post-traumatic subtype of obsessive-compulsive disorder. J Anxiety Disord. 2012;26:377–83.
34. Maina G, Albert U, Bogetto F, Vaschetto P, Ravizza L. Recent life events and obsessive-compulsive disorder (OCD): the role of pregnancy/delivery. Psychiatry Res. 1999;89:49–58.
35. Gershuny BS, Baer L, Radomsky AS, Wilson KA, Jenike MA. Connections among symptoms of obsessive-compulsive disorder and posttraumatic stress disorder: a case series. Behav Res Ther. 2003;41:1029–41.
36. Valleni-Basile LA, Garrison CZ, Waller JL, Addy CL, McKeown RE, Jackson KL, et al. Incidence of obsessive-compulsive disorder in a community sample of young adolescents. J Am Acad Child Adolesc Psychiatry. 1996;35:898–906.
37. Salkovskis PM. Obsessional-compulsive problems: a cognitive-behavioural analysis. Behav Res Ther. 1985;23:571–83.
38. Olatunji BO, Taylor S, Wu KD, Tolin DF, Carmin C, Abramowitz JS, et al. How are dysfunctional beliefs related to obsessive-compulsive symptoms? J Cogn Psychother. 2010;24:165–76.
39. Salkovskis PM, McGuire J. Cognitive-behavioural theory of OCD. Obs Compuls Disord Theory Res Treat. 2003;37:783. https://doi.org/10.1111/j.1440-1614.2003.01289.x.
40. Cortes AM, Saltzman KM, Weems CF, Regnault HP, Reiss AL, Carrion VG. Development of anxiety disorders in a traumatized pediatric population: a preliminary longitudinal evaluation. Child Abus Negl. 2005;29:905–14.
41. Salkovskis PM, Forrester E. Responsibility. In: Cognitive approaches to obsessions and compulsions. Pergamon; 2002. p. 45–61.
42. Briggs ES, Price IR. The relationship between adverse childhood experience and obsessive-compulsive symptoms and beliefs: the role of anxiety, depression, and experiential avoidance. J Anxiety Disord. 2009;23:1037–46.
43. McLaren S, Crowe SF. The contribution of perceived control of stressful life events and thought suppression to the symptoms of obsessive-compulsive disorder in both non-clinical and clinical samples. J Anxiety Disord. 2003;17:389–403.

44. Wegner DM, Zanakos S. Chronic thought suppression. J Pers. 1994;62:615–40.
45. Mathews CA, Kaur N, Stein MB. Childhood trauma and obsessive-compulsive symptoms. Depress Anxiety. 2008;25:742–51.
46. Bowlby J. Attachment and loss. In: Key concepts in family studies. Basic Books; 1982. p. 9–12.
47. Ulu I, Tezer E. Adaptive and maladaptive perfectionism, adult attachment, and big five personality traits. J Psychol Interdiscip Appl. 2010;144:327–40.
48. Wei M, Mallinckrodt B, Russell DW, Abraham WT. Maladaptive perfectionism as a mediator and moderator between adult attachment and depressive mood. J Couns Psychol. 2004;51:201–12.
49. Doron G, Moulding R, Kyrios M, Nedeljkovic M, Mikulincer M. Adult attachment insecurities are related to obsessive compulsive phenomena. J Soc Clin Psychol. 2009;28:1022–49.
50. Yarbro J, Mahaffey B, Abramowitz J, Kashdan TB. Recollections of parent-child relationships, attachment insecurity, and obsessive-compulsive beliefs. Pers Individ Dif. 2013;54:355–60.
51. Venta A, Sharp C, Hart J. The relation between anxiety disorder and experiential avoidance in inpatient adolescents. Psychol Assess. 2012;24:240–8.
52. Kroska EB, Miller ML, Roche AI, Kroska SK, O'Hara MW. Effects of traumatic experiences on obsessive-compulsive and internalizing symptoms: the role of avoidance and mindfulness. J Affect Disord. 2018;225:326–36.
53. Tull MT, Gratz KL, Salters K, Roemer L. The role of experiential avoidance in posttraumatic stress symptoms and symptoms of depression, anxiety, and somatization. J Nerv Ment Dis. 2004;192:754–61.
54. Kashdan TB, Barrios V, Forsyth JP, Steger MF. Experiential avoidance as a generalized psychological vulnerability: comparisons with coping and emotion regulation strategies. Behav Res Ther. 2006;44:1301–20.
55. Chawla N, Ostafin B. Experiential avoidance as a functional dimensional approach to psychopathology: an empirical review. J Clin Psychol. 2007;63:871–90.
56. Thompson RW, Arnkoff DB, Glass CR. Conceptualizing mindfulness and acceptance as components of psychological resilience to trauma. Trauma Violence Abuse. 2011;12:220–35.
57. Clark DA, Purdon CL. The assessment of unwanted intrusive thoughts: a review and critique of the literature. Behav Res Ther. 1995;33:967–76.
58. Thomsen PH. Obsessive-compulsive disorder in children and adolescents. A 6-22 year follow-up study of social outcome. Eur Child Adolesc Psychiatry. 1995;4:112–22.
59. Tang W, Zhao J, Lu Y, Yan T, Wang L, Zhang J, et al. Mental health problems among children and adolescents experiencing two major earthquakes in remote mountainous regions: a longitudinal study. Compr Psychiatry. 2017;72:66–73.
60. Cortes AM, Saltzman KM, Weems CF, Regnault HP, Reiss AL, Carrion VG. Development of anxiety disorders in a traumatized pediatric population: a preliminary longitudinal evaluation. Child Abuse Negl. 2005;29:905–14.
61. Park S, Hong JP, Bae JN, et al. Impact of childhood exposure to psychological trauma on the risk of psychiatric disorders and somatic discomfort: single vs. multiple types of psychological trauma. Psychiatry Res. 2014;219:443–9.
62. Holmes TH, Rahe RH. The social readjustment rating scale. J Psychosom Res. 1967;11:213–8.
63. Dykshoorn KL. Trauma-related obsessive–compulsive disorder: a review. Heal Psychol Behav Med. 2014;2:517–28.
64. Sarason IG, Sarason BR, Potter EH, Antoni MH. Life events, social support, and illness. Psychosom Med. 1985;47:156–63.
65. Guze SB. Diagnostic and statistical manual of mental disorders, 4th ed. (DSM-IV). Am J Psychiatry. 1995;152:1228.
66. Gothelf D, Aharonovsky O, Horesh N, Carty T, Apter A. Life events and personality factors in children and adolescents with obsessive-compulsive disorder and other anxiety disorders. Compr Psychiatry. 2004;45:192–8.
67. Fontenelle LF, Domingues AM, Souza WF, Mendlowicz MV, De Menezes GB, Figueira IL, et al. History of trauma and dissociative symptoms among patients with obsessive-compulsive disorder and social anxiety disorder. Psychiatry Q. 2007;78:241–50.

68. Lochner C, du Toit PL, Zungu-Dirwayi N, Marais A, van Kradenburg J, Seedat S, et al. Childhood trauma in obsessive-compulsive disorder, trichotillomania, and controls. Depress Anxiety. 2002;15:66–8.
69. Horesh N, Zimmerman S, Steinberg T, Yagan H, Apter A. Is onset of Tourette syndrome influenced by life events? J Neural Transm. 2008;115:787–93.
70. Findley DB, Leckman JF, Katsovich L, Lin H, Zhang H, Grantz H, et al. Development of the Yale Children's Global Stress Index (YCGSI) and its application in children and adolescents with Tourette's syndrome and obsessive-compulsive disorder. J Am Acad Child Adolesc Psychiatry. 2003;42:450–7.
71. Nacasch N, Fostick L, Zohar J. High prevalence of obsessive-compulsive disorder among posttraumatic stress disorder patients. Eur Neuropsychopharmacol. 2011;21:876–9.
72. Cloninger CR, Bayon C, Svrakic DM. Measurement of temperament and character in mood disorders: a model of fundamental states as personality types. J Affect Disord. 1998;51:21–32.
73. Abed RT, de Pauw KW. An evolutionary hypothesis for obsessive compulsive disorder: a~psychological immune system? Behav Neurol. 1998;11:245–50.
74. American Psychiatric Association. Diagnostic and statistical manual of mental disorders, 5th ed. 2013. https://doi.org/10.1176/appi.books.9780890425596
75. Munoz DM, Barickman J, Friedhoff AJ. Course of obsessive-compulsive disorder in children and adolescents: a prospective follow-up study of 23 Danish cases. J Child Psychol Psychiatry. 1995;36:305–12.
76. Munoz DM, Barickman J, Friedhoff AJ. Environmental factors and related fluctuation of symptoms in children and adolescents with Tourette's disorder. J Child Psychol Psychiatry. 1995;36:305–12.
77. American Psychiatric Association. Diagnostic and statistical manual of mental disorders. 2013. https://doi.org/10.1176/appi.books.9780890425596
78. Lazarus RS, DeLongis A, Folkman S, Gruen R. Stress and adaptational outcomes. The problem of confounded measures. Am Psychol. 1985;40:770–9.
79. Van Grootheest DS, Cath DC, Beekman AT, Boomsma DI. Genetic and environmental influences on obsessive-compulsive symptoms in adults: a population-based twin-family study. Psychol Med. 2007;37:1635–44.
80. Sarkhel S, Praharaj SK, Sinha VK. Role of life events in obsessive compulsive disorder. Isr J Psychiatry Relat Sci. 2011;48:182–5.
81. McKeon J, Roa B, Mann A. Life events and personality traits in obsessive-compulsive neurosis. Br J Psychiatry. 1984;144:185–9.
82. Selye H. The stress of life. New York, NY: McGraw-Hill; 1956.
83. Lafleur DL, Petty C, Mancuso E, McCarthy K, Biederman J, Faro A, et al. Traumatic events and obsessive compulsive disorder in children and adolescents: is there a link? J Anxiety Disord. 2011;25:513–9.
84. Gershuny BS, Baer L, Parker H, Gentes EL, Infield AL, Jenike MA. Trauma and posttraumatic stress disorder in treatment-resistant obsessive-compulsive disorder. Depress Anxiety. 2008;25:69–71.
85. Shavitt RG, Valério C, Fossaluza V, Da Silva EM, Cordeiro Q, Diniz JB, et al. The impact of trauma and post-traumatic stress disorder on the treatment response of patients with obsessive-compulsive disorder. Eur Arch Psychiatry Clin Neurosci. 2010;260:91–9.
86. Boudreaux E, Kilpatrick G, Resnick HS, Best CL, Saunders BE. Criminal victimization, posttraumatic stress disorder, and comorbid psychopathology among a community sample of women. J Trauma Stress. 1998;11:665–78.
87. Gershuny BS, Baer L, Jenike MA, Minichiello WE, Wilhelm S. Comorbid posttraumatic stress disorder: impact on treatment outcome for obsessive-compulsive disorder. Am J Psychiatry. 2002;159:852–4.
88. Grabe HJ, Ruhrmann S, Spitzer C, et al. Obsessive-compulsive disorder and posttraumatic stress disorder. Psychopathology. 2008;41:129–34.
89. Real E, Labad J, Alonso P, Segalàs C, Jiménez-Murcia S, Bueno B, et al. Stressful life events at onset of obsessive-compulsive disorder are associated with a distinct clinical pattern. Depress Anxiety. 2011;28:367–76.

90. Rosso G, Albert U, Asinari GF, Bogetto F, Maina G. Stressful life events and obsessive-compulsive disorder: clinical features and symptom dimensions. Psychiatry Res. 2012;197:259–64.
91. Hemmings SMJ, Lochner C, van der Merwe L, Cath DC, Seedat S, Stein DJ. BDNF Val66Met modifies the risk of childhood trauma on obsessive-compulsive disorder. J Psychiatr Res. 2013;47:1857–63.
92. McGregor NW, Lochner C, Calmarza-Font I, Hemmings SMJ, Erdman L, Stein DJ. Modification of the association between early adversity and obsessive-compulsive disorder by polymorphisms in the MAOA, MAOB and COMT genes. Psychiatry Res. 2016;246:527–32.
93. Hudziak JJ, van Beijsterveldt CEM, Althoff RR, Stanger C, Rettew DC, Nelson EC, et al. Genetic and environmental contributions to the child behavior checklistobsessive-compulsive scale. Arch Gen Psychiatry. 2004;61:608.
94. Taylor S, Jang KL. Biopsychosocial etiology of obsessions and compulsions: an integrated behavioral-genetic and cognitive-behavioral analysis. J Abnorm Psychol. 2011;120:174–86.
95. Jonnal AH, Gardner CO, Prescott CA, Kendler KS. Obsessive and compulsive symptoms in a general population sample of female twins. Am J Med Genet Neuropsychiatr Genet. 2000;96:791–6.
96. Vidal-Ribas P, Stringaris A, Rück C, Serlachius E, Lichtenstein P, Mataix-Cols D. Are stressful life events causally related to the severity of obsessive-compulsive symptoms? A monozygotic twin difference study. Eur Psychiatry. 2015;30:309–16.
97. Cath DC, Van Grootheest DS, Willemsen G, Van Oppen P, Boomsma DI. Environmental factors in obsessive-compulsive behavior: evidence from discordant and concordant monozygotic twins. Behav Genet. 2008;38:108–20.
98. Miller ML, Brock RL. The effect of trauma on the severity of obsessive-compulsive spectrum symptoms: a meta-analysis. J Anxiety Disord. 2017;47:29–44.
99. Goldberg X, Soriano-Mas C, Alonso P, et al. Predictive value of familiality, stressful life events and gender on the course of obsessive-compulsive disorder. J Affect Disord. 2015;185:129–34.
100. Landau D, Iervolino AC, Pertusa A, Santo S, Singh S, Mataix-Cols D. Stressful life events and material deprivation in hoarding disorder. J Anxiety Disord. 2011;25:192–202.
101. Cromer KR, Schmidt NB, Murphy DL. Do traumatic events influence the clinical expression of compulsive hoarding? Behav Res Ther. 2007;45:2581–92.
102. Przeworski A, Cain N, Dunbeck K. Traumatic life events in individuals with hoarding symptoms, obsessive-compulsive symptoms, and comorbid obsessive-compulsive and hoarding symptoms. J Obsessive Compuls Relat Disord. 2014;3:52–9.
103. Hartl TL, Duffany SR, Allen GJ, Steketee G, Frost RO. Relationships among compulsive hoarding, trauma, and attention-deficit/hyperactivity disorder. Behav Res Ther. 2005;43:269–76.
104. Tolin DF, Meunier SA, Frost RO, Steketee G. Course of compulsive hoarding and its relationship to life events. Depress Anxiety. 2010;27:829–38.
105. Ay R, Erbay LG. Relationship between childhood trauma and suicide probability in obsessive-compulsive disorder. Psychiatry Res. 2018;261:132–6.
106. Khosravani V, Kamali Z, Jamaati Ardakani R, Samimi Ardestani M. The relation of childhood trauma to suicide ideation in patients suffering from obsessive-compulsive disorder with lifetime suicide attempts. Psychiatry Res. 2017;255:139–45.
107. Frost RO, Hartl TL. A cognitive-behavioral model of compulsive hoarding. Behav Res Ther. 1996;34:341–50.

# Childhood Trauma in Eating Disorders 15

Carolina Anaya, Seeba Anam, Hana F. Zickgraf,
Shannon M. O'Connor, and Jennifer E. Wildes

## 15.1 Introduction

Eating disorders are serious psychiatric illnesses characterized by maladaptive thoughts and behaviors related to eating or weight control. The most recent edition of the Diagnostic and Statistical Manual of Mental Disorders [1] includes three primary eating disorders: anorexia nervosa (AN), bulimia nervosa (BN), and binge-eating disorder (BED), as well as a residual category of Other Specified Feeding or Eating Disorder (OSFED) (Table 15.1). Eating disorders typically onset between the ages of 18 and 21 [2, 3] and are associated with significant impairment, morbidity, and mortality [2, 4, 5]. Lifetime prevalence of AN, BN, or BED has been estimated at 1.01% in a meta-analysis of 15 general population samples spanning the United States, Latin America, Western Europe, New Zealand, South Korea, and several other countries [6]. Importantly, the studies included in this meta-analysis used DSM-IV criteria for eating disorders, and there is evidence that changes in diagnostic criteria from the DSM-IV to DSM-V resulted in an increased prevalence of AN and BN and a decreased prevalence of OSFED [7].

Additionally, three feeding disorders: Avoidant Restrictive Food Intake Disorder (ARFID), Rumination Disorder, and Pica, were grouped with the eating disorders in DSM-5 (Table 15.1). Feeding disorders tend to start in young childhood [8] and though studies are limited, data from a population-based sample of 8- to 13-year-old school children in Switzerland suggest a prevalence of 3.2% for ARFID, 3.8% for Pica, and 1.7% for Rumination [9]. Most of the current literature studying trauma within the eating disorder population focuses on AN, BN, and BED. Therefore, the remainder of this chapter focuses on these disorders.

C. Anaya · S. Anam · H. F. Zickgraf · S. M. O'Connor · J. E. Wildes (✉)
Department of Psychiatry and Behavioral Neuroscience, Billings Hospital, The University of Chicago, Chicago, IL, USA
e-mail: anayac@uchicago.edu; sanam@yoda.bsd.uchicago.edu; hzickgraf@bsd.uchicago.edu; Soconnor4@bsd.uchicago.edu; jwildes@uchicago.edu



G. Spalletta et al. (eds.), *Childhood Trauma in Mental Disorders*,
https://doi.org/10.1007/978-3-030-49414-8_15

**Table 15.1** DSM-5 Feeding and Eating Disorders

| Diagnosis | Brief description | Lifetime prevalence | Age of onset (median) |
|---|---|---|---|
| Anorexia nervosa | Characterized by restriction of food intake, leading to low weight, and a fear of gaining weight | 0.6% [2] | 18 years [2] |
| Bulimia nervosa | Characterized by recurrent binge eating, defined as eating large amounts of food while experiencing loss of control over eating, and engaging in compensatory behaviors such as self-induced vomiting, fasting, laxative or diuretic abuse, or excessive exercise | 0.8–1% [2] | 18 years [2] |
| Binge-Eating Disorder | Characterized by recurrent binge eating in the absence of compensatory behaviors | 1.4–2.8% [2] | 21 years [2] |
| Other Specified Feeding or Eating Disorder (OSFED) | Characterized by significant eating disturbance and impairment with symptoms that fall below severity or frequency thresholds for AN, BN, or BED, and thus do not meet DSM criteria | 2–4.4% [82] | 18–21 years [82] |
| Avoidant Restrictive Food Intake Disorder (ARFID) | Characterized by avoidance and/or restriction of food intake in the absence of shape or weight concerns | ~3.2% [8] | Early childhood [8] |
| Rumination disorder | Characterized by repeated regurgitation of food (i.e. spitting, re-chewing, re-swallowing) | ~1.7% [9] | Early childhood [9] |
| Pica | Characterized by persistent eating of nonnutritive, non-food substances | ~3.8% [9] | Early childhood [9] |

Note: Lifetime prevalence and age of onset for AN, BN, and BED from Hudson et al. [2]; Lifetime prevalence for OSFED from Smink et al. [82]; Lifetime prevalence and age of onset for ARFID, Rumination Disorder and Pica estimated from one population-based sample of 8- to 13-year-old school children in Switzerland [9], and one Swiss sample of 8- to 13-year-old school children [8]

The aim of this chapter is to summarize existing data on the association between eating disorders and childhood trauma. Relations between trauma and eating disorder severity and response to treatment also are discussed. Finally, potential mechanisms underlying the association between childhood trauma and eating disorders are described, and clinical implications are proposed.

## 15.2 Prevalence of Eating Disorders in Populations with Childhood Trauma

In a sample of 36,309 adult men and women in the United States—the largest single sample in which this relationship has been explored—Afifi and colleagues [10] examined the prevalence of eating disorders across different forms of maltreatment experienced during childhood. Childhood maltreatment was categorized broadly as "harsh physical punishment," "physical abuse," "sexual abuse," "emotional abuse," "emotional neglect," "physical neglect," and "exposure to intimate partner violence". Prevalence estimates for childhood maltreatment across the sample were quite high, likely due to the use of broad definitions of maltreatment (e.g., physical neglect was defined as any response other than "never" to "*having ever been left alone or unsupervised before the age of 10 years or going without necessary clothing, school supplies, food, or medical treatment*"). Nevertheless, the proportion of individuals reporting any form of childhood maltreatment was significantly higher in the eating disorder group (70.6% for men, 67.6% for women) than in the control group (49.3% for men; 48% for women).

Adjusting for gender, reporting any type of child maltreatment was associated with increased lifetime risk for all DSM-5 eating disorders, including AN (OR = 1.78, $p \leq .001$), BN (OR = 2.05, $p \leq .05$), and BED (OR = 2.98, $p \leq .001$) [10]. Odds were even higher for those who experienced multiple types of childhood maltreatment (OR's = 2.46–3.82, *p*'s < .001). When all types of childhood abuse were entered simultaneously into models predicting AN, BN, or BED diagnosis, childhood physical abuse had significant independent associations with AN (OR = 1.91, $p \leq .001$) and BED (OR = 2.83, $p \leq .001$). Sexual abuse was significantly associated with AN (OR = 2.11, $p \leq .001$), BN (OR = 2.43, $p \leq .05$), and BED (OR = 3.05, $p \leq .001$). Emotional abuse was significantly associated with AN (OR = 2.25, $p \leq .001$), BN (OR = 2.97, $p \leq .001$), and BED (OR = 3.67, $p \leq .001$). Emotional neglect was significantly associated with BN (OR = 3.18, $p \leq .001$) and BED (OR = 1.99, $p \leq .001$). Physical neglect was significantly associated with AN (OR = 1.6, $p \leq .001$) and BED (OR = 2.92, $p \leq .001$). Finally, the authors found that child sexual abuse and physical neglect had the strongest association with eating disorders in men (sexual abuse: OR = 2.92, $p \leq .001$; physical neglect: OR = 2.19, $p \leq .001$), and child sexual abuse and emotional abuse had the strongest association with eating disorders in women (sexual abuse: OR = 2.27, $p \leq .001$; emotional abuse: OR = 2.48, $p \leq .001$) [10].

## 15.3 Prevalence of Childhood Trauma in Populations with Eating Disorders

Though childhood maltreatment is considered a relatively nonspecific risk factor for psychopathology [11], meta-analytic data suggest that individuals with eating disorders may be more likely than those with substance use or depressive disorders to report childhood maltreatment. Specifically, in a meta-analysis of 82 studies, including 13,059 participants of all ages meeting DSM-IV criteria for AN, BN, BED and Eating Disorder Not Otherwise Specified (EDNOS; similar to OSFED in DSM-5) and 15,092 individuals diagnosed with other psychiatric disorders (e.g., depression, substance dependence), Molendijk and colleagues [12] found higher rates of childhood maltreatment (i.e., sexual, physical, and emotional maltreatment) in the eating disorder samples (21–59%) compared to the psychiatric reference samples (5–46%) and healthy control samples ($N = 7736$, 1–35%). Overall, the prevalence of any history of childhood maltreatment was higher in patients with eating disorders than in those with other psychiatric illnesses (OR = 1.31, $p < .001$).

When compared to healthy controls, the association between childhood maltreatment and eating disorders was even higher (overall childhood maltreatment OR = 2.47, $p < .001$), and there was more evidence of variability among types of childhood maltreatment and eating disorder diagnosis [12]. Within the AN group, 26% reported a history of sexual childhood maltreatment (OR = 1.98 vs. healthy controls, $p < .001$), 17% reported physical childhood maltreatment (OR = 2.42, $p < .001$), and 34% reported emotional childhood maltreatment (OR = 3.81, $p < .001$). Within the BN group, 35% reported a history of sexual childhood maltreatment (OR = 2.57, $p < .001$), 33% reported physical childhood maltreatment (OR = 3.78, $p < .001$), and 81% reported emotional childhood maltreatment (OR = 5.13, $p < .001$). Within the BED group, 24% reported a history of sexual childhood maltreatment (OR = 1.88, $p < .001$), 23% reported physical childhood maltreatment (OR = 2.57, $p < .001$), and 59% reported emotional childhood maltreatment (OR = 2.44, $p < .001$).

There may be differences in the strength of associations between childhood maltreatment across the eating disorder diagnoses. In particular, the connection between binge/purge behaviors and trauma history seems especially strong. For example, several studies have shown that patients with AN–binge-eating/purging subtype are more likely to report a history of trauma than patients with AN-restricting type [12–14]. These findings suggest a potential mechanism underlying the association between binge-eating/purging symptoms and childhood trauma, which will be elaborated upon later in this chapter.

Previous studies, including many from population-based samples, have explored the relationship between eating disorder risk and specific types of childhood maltreatment, most often childhood physical and sexual abuse [15, 16]. More recent evidence from large population-based samples (e.g., Afifi et al. 2017) [10] and meta-analyses (e.g., Molendijk et al. 2017) [12] has supported the hypothesis that childhood sexual and physical abuse are strongly linked to eating disorder risk, but also suggests that broadly defined childhood maltreatment (i.e., including physical neglect and emotional abuse and neglect) is a risk factor for eating disorders. Nevertheless, as noted in a recent meta-analysis of 23 studies exploring the relationship between adult eating disorders and childhood emotional abuse, emotional neglect, and exposure to interpersonal violence, this literature is relatively limited compared to the larger and older literature focusing on childhood sexual and physical abuse [17]. Moreover, few studies have explored the relationship between eating disorders and categories of trauma other than childhood maltreatment, including food insecurity, exposure to community violence, war, terrorism, natural disasters, medical traumas, and accidents [17, 18].

## 15.4 Clinical Correlates

Table 15.2 summarizes research documenting the clinical correlates of trauma and eating disorders. Of note, the majority of studies have explored the correlates of trauma across the life span rather than being limited to traumatic events during childhood. As such, when not specified, "trauma" described in the following studies includes, but does not refer exclusively, to childhood trauma.

**Table 15.2** Clinical correlates of trauma and eating disorders

| Study | Sample | Eating disorder assessment | Trauma assessment | Measure of clinical correlate | Key findings |
|---|---|---|---|---|---|
| *Eating disorder severity* | | | | | |
| Backholm et al. [19] | 4524 Swedish adults from clinical sample meeting DSM-IV criteria for AN, BN, BED or EDNOS | SCID-I until February 2008; SEDI | PTSD section of the SCID-I | EDE-Q; CIA | Patients with trauma history:<br>↑ EDE-Q subscales (restraint, eating concern, weight concern, shape concern)<br>↑ CIA Total |
| Kong and Bernstein [20] | 73 South Korean ED patients aged 14–36 years | Clinical interview | CTQ | EDI-2 | Physical neglect ↑ correlated with drive for thinness scale<br>Sexual abuse ↑correlated with impulse regulation and perfectionism scales |
| White et al. [21] | 182 US ED patients aged 12–22 | Clinical interview | Yes/no format, experience of: CPA, CSA, exposure to domestic violence, significant bullying, motor vehicle accident, neighborhood violence, war, terrorism, medical procedures, national disaster, and significant death/loss | BMI; EBW | Childhood trauma exposure ↑ correlated with body weight |

(continued)

**Table 15.2** (continued)

| Study | Sample | Eating disorder assessment | Trauma assessment | Measure of clinical correlate | Key findings |
|---|---|---|---|---|---|
| *Response to treatment* | | | | | |
| Carter et al. [14] | 77 Canadian patients with AN admitted to an inpatient ED unit | Clinical interview | Investigator-based interview developed by Welch and Fairburn (1994) | EDE and EDE-Q | ↑ likelihood of dropout in AN-binge/purge type patients with exposure to CSA |
| Mahon et al. [22] | 114 English female patients of an outpatient ED clinic with BN or atypical BN | Clinical interview using ICD-10 | Assessors ask about sexual and physical abuse, and parental losses informally | Drop out defined as ceasing contact with clinic against advice of therapist before tenth session | ↑ in childhood traumatic events correlated with ↑ proportion of patients who dropped out |
| Rodriguez et al. [23] | 160 Columbian women aged 12–49 with AN, BN, or BED who had completed treatment for at least 4 months | SCID-I | Clinical interview | Poor treatment response defined by weight progress, reduction in binge/purge frequencies, or reductions on self-report measures of disordered eating | Patients with sexual abuse and exposure to other violent acts had the highest probability of poor outcome<br>Exposure increased the risk of drop out or relapse |
| Vrabel et al. [24] | 74 Adult patients with AN, BN, or EDNOS in an inpatient ED program in Norway | SCID-I and EDE | CSA assessed retrospectively using medical charts and personal interviews | Global EDE score | CSA predicted global EDE and improvement in scores, and mediated influence of ASPD on EDE score |
| Calugi et al. [25] | 81 Patients aged 16–45 years with AN admitted to Italian inpatient program | Italian version of the EDE | Investigator-based interview developed by Welch and Fairburn (1994) | Changes in BMI; EDE; BSI; WSAS | No differences on measures between patients with and without a history of CSA |

| | | | | | |
|---|---|---|---|---|---|
| *Comorbid psychiatric disorders* | | | | | |
| Hudson et al. [2] | 9282 English-speaking adults ages 18 and older in the US | CIDI | CIDI | PTSD diagnosed using CIDI | ↑ likelihood of patients with BN or BED having co-morbid PTSD |
| Inniss et al. [26] | 78 Canadian adult women with BN and 61 adult women with no ED | EDE; EAT-26 | SCID-I | SCID-I | Threshold and subthreshold PTSD more common in individuals with EDs, as compared to individuals without EDs |
| Mitchell et al. [27] | 5702 Adult men and women in United States from part II of the National Comorbidity Survey-Replication (NCS-R) | CIDI | Participants were asked whether they had ever been exposed to a variety of traumatic events | PTSD diagnosed using clinical interviews derived from DSM-IV | Majority of participants with AN, BN, and BED reported a history of interpersonal trauma; rates of PTSD were significantly higher among those with BN and BED |
| Reyes-Rodriguez et al. [28] | 753 Women aged 16+ with DSM-IV AN and PTSD from multiple countries (e.g., United States, Germany, England, Canada) | Expanded modified version of module H of the SCID-I; SIAB; YBC-EDS | SCID-I | SCID-I | ↑ odds of having PTSD in patients with AN purging type versus AN restricting type |
| Tagay et al. [29] | 103 German female patients aged 18+ with AN or BN | EDI-2 | ETI | ETI using clinical cutoffs | No significant difference between AN and BN patients with regard to the lifetime prevalence of trauma; PTSD prevalence = 23.1% (AN) and 25.6% (BN) |

(continued)

**Table 15.2** (continued)

| Study | Sample | Eating disorder assessment | Trauma assessment | Measure of clinical correlate | Key findings |
|---|---|---|---|---|---|
| Isomaa et al. [30] | 843 Swedish adult ED patients who reported a trauma | SCID-I until February 2008; SEDI | PTSD section of the SCID-I; LES | EDE-Q; CPRS-S-A | 24.1% of ED patients had a lifetime diagnosis of PTSD; PTSD had an impact on ED severity, which was mediated by psychological distress |
| Mitchell at al [31]. | 65 US women with a history of childhood or adult sexual or physical assault | SCID-I patient edition | Standardized trauma interview | PDS; EDI-2 | ↓ PTSD symptom scores significantly associated with ↓ in several EDI-2 subscales: Impulse regulation, interoceptive awareness, interpersonal distrust, ineffectiveness, and maturity fears |
| Mayes et al. [32] | 90 US children ages 7–18 with BN or AN | Clinical interview using DSM-5 criteria | PBS | Mothers rated their children on PBS | Suicidal ideation and attempts more prevalent among patients with BN as compared to patients with AN; correlates of suicide attempts in BN patients include CPA and CSA |
| Backholm et al. [19] | 4524 Swedish adults from a clinical sample meeting DSM-IV criteria for AN, BN, BED or EDNOS | SCID-I until February 2008; SEDI | PTSD section of the SCID-I | CPRS-S-A | ↑ depression, anxiety, and obsessive-compulsive symptoms in patients with a history of trauma |

| | | | | | |
|---|---|---|---|---|---|
| Kong and Bernstein [20] | 73 South Korean ED patients aged 14–36 years | Clinical interview | CTQ | BDI | Depression-mediated association between some forms of childhood trauma and ED psychopathology |
| Richardson et al. [33] | 89 Canadian women aged 17–49 years with DSM-IV bulimia-spectrum disorders | SCID-I | SCID-I | SCID-I | ↑ likelihood of comorbid depression, substance use, and/or anxiety in patients with trauma exposure |
| Carter et al. [14] | 77 Canadian patients with AN admitted to an inpatient ED unit | Clinical interview | Investigator-based interview developed by Welch and Fairburn (1994) | BDI-II; RSES; PI; BSI | ↑ depression, anxiety, and obsessive-compulsive symptoms, and ↓ self-esteem in patients with a history of CSA |
| Utzinger et al. [34] | 133 US women with DSM-IV BN | SCID-I, patient edition | CTQ | SCID; DIB-R | History of sexual trauma or polytrauma significantly correlated with ↑ DIB-R scores |

Note: *AN* anorexia nervosa; *ASPD* antisocial personality disorder; *BDI* Beck Depression Inventory; *BDI-II* Beck Depression Inventory II; *BED* binge-eating disorder; *BMI* body mass index; *BN* bulimia nervosa; *BSI* Brief Symptom Inventory; *CIA* Clinical Impairment Assessment; *CIDI* World Health Organization Composite International Diagnostic Interview; *CPA* childhood physical abuse; *CPRS-S-A* Comprehensive Psychiatric Rating Scale, self-rated version of the affective subscales; *CSA* childhood sexual abuse; *CTQ* Childhood Trauma Questionnaire; *DIB-R* Diagnostic Interview for Borderlines-Revised; *EAT-26* Eating Attitudes Test; *EBW* expected body weight; *ED* eating disorder; *EDE* Eating Disorder Examination interview; *EDE-Q* Eating Disorder Examination Questionnaire; *EDI-2* Eating Disorders Inventory-2; *EDNOS* eating disorder not otherwise specified; *ETI* Essen Trauma–Inventory; *ICD-10* International Statistical Classification of Diseases; *LES* Life Events Checklist; *PBS* Pediatric Behavior Scale; *PDS* Post-traumatic Stress Diagnostic Scale; *PI* Padua Inventory; *PTSD* Post-traumatic stress disorder; *RSES* Rosenberg Self-esteem Scale; *SCID-I* Structured Clinical Interview for DSM-IV Axis I Disorders; *SEDI* Structured Eating Disorders Interview; *SIAB* Structured Interview for Anorexia Nervosa and Bulimic Syndromes; *WSAS* Work and Social Adjustment Scale; *YBC-EDS* Yale–Brown–Cornell Eating Disorder Scale

### 15.4.1 Eating Disorder Severity

In their dose–response meta-analysis of 82 studies, Molendijk and colleagues found that childhood maltreatment within eating disorder populations may be linked to earlier onset, higher likelihood of binge/purge behaviors, and higher overall symptom severity [12]. For example, in a large clinical adult sample in Sweden ($N = 4524$), eating disorder patients with a trauma history had higher scores on self-reported symptoms, with small but significant effects on the Eating Disorder Examination Questionnaire (EDE-Q) subscales: Restraint ($n^2_p = .01$), Eating Concern ($n^2_p = .01$), Weight Concern ($n^2_p = .01$), and Shape Concern ($n^2_p = .01$) [19]; trauma exposure was also associated with greater secondary psychosocial impairment ($n^2_p = .01$) [19]. Similarly, in a small sample of 73 treatment-seeking South Korean girls and women ages 14–36 years, physical neglect predicted drive for thinness, whereas sexual abuse predicted impulse regulation and perfectionism [20].

Interestingly, in a single study in an adolescent treatment-seeking sample (ages 12–22 years), body mass index (BMI) and expected body weight (EBW) were shown to be influenced by exposure to trauma during childhood, such that trauma exposure was linked to higher body weight [21]. This finding should be interpreted cautiously until further studies are conducted, as patients who endorsed a trauma history were more likely to have BN, which by definition is associated with higher body weight than AN.

### 15.4.2 Response to Treatment

Several studies have explored eating disorder treatment response among adult patients in trauma-exposed populations. Treatment response generally has been measured as the rate of premature dropout from treatment as well as changes in eating disorder severity. Some studies have found that a history of childhood trauma, including sexual abuse and physical abuse, increased the likelihood of dropping out of treatment in patients with binge–purge symptoms [14, 22], where an increase in the number of traumatic events may further increase the likelihood of dropout compared to patients who had no history of trauma [22]. Other studies have found that patients who have been exposed to trauma are more likely to have a poor treatment outcome (e.g., not making necessary weight progress, smaller reduction in binge/purge frequencies, or smaller reductions on self-report measures of disordered eating) [23, 24]. Alternatively, a recent study by Calugi et al. [25] found that a history of childhood sexual abuse in patients with AN did not make a difference in treatment outcomes using inpatient-enhanced cognitive-behavioral therapy. Of note, research documenting a significant relationship between trauma history and treatment outcome included samples consisting of patients with AN, BN, BED, or EDNOS, whereas the study by Calugi et al. only included one type of trauma and one type of eating disorder. This highlights the need for future research exploring the impact of different types of trauma, eating disorder diagnoses, and treatment settings. It may also be beneficial to compare changes in symptom severity over time between patients with and without a history of trauma, rather than focusing on one data point.

### 15.4.3 Comorbid Psychiatric Disorders

A history of childhood trauma also has been shown to increase the likelihood of having a comorbid psychiatric diagnosis in patients with eating disorders. Many studies have focused on comorbid post-traumatic stress disorder (PTSD) within this group. Patients with BN and a trauma history have been found to have a significantly higher likelihood of having PTSD or subthreshold PTSD compared to other eating disorder patients [2, 26, 27]; similar findings have been reported in patients with the binge/purge subtype of AN, BED, or subclinical binge eating [2, 28]. A few studies have found that patients with AN and a history of trauma are similar to those with BN in risk for comorbid PTSD, but findings have been mixed and subtypes of AN were not included in the original papers [29, 30]. Of note, comorbid PTSD within trauma-exposed eating disorder populations has been consistently linked to increased eating disorder severity, but the degree to which this effect is explained by elevated overall psychological distress [30] versus shared symptoms between eating disorders and PTSD, including emotion dysregulation and impulsivity [31], is unclear.

Although most studies have focused on PTSD, several recent investigations have documented other comorbidities in individuals with trauma and eating disorders. For example, patients with eating disorders and a history of childhood trauma have been shown to have a higher rate of suicidal ideation and attempts, as compared to patients without a trauma history [12]; comorbid suicidality is also more prevalent within the BN spectrum than the restricting spectrum [32]. Comorbid depression has been linked to all types of childhood trauma in individuals with eating disorders, and there is evidence that depressive symptoms may mediate the impact of childhood trauma on eating disorder risk [19, 20]. This finding has been documented in both bulimic-spectrum disorders [33] and AN [14]. Within BN, a history of trauma has been associated with borderline personality and substance-use disorders [33, 34]. Across all eating disorder diagnoses, trauma history has been associated with higher rates of anxiety [14, 19, 33], obsessive-compulsive symptoms [14, 19], and lower self-esteem [14, 19].

## 15.5 Potential Mechanisms

Multiple mechanisms have been proposed to mediate the relationship between traumatic exposures and the development or maintenance of eating disorders, though the precise pathway remains unclear. Both psychological and biological constructs have been identified as potential models linking environmental trauma, especially childhood emotional and sexual abuse and maltreatment, to eating-related disorders. Several proposed mechanisms integrate both biological and psychological factors as putative agents in the development of AN, BN, and BED.

### 15.5.1 Psychological Mediators and Mechanisms

In considering psychological mechanisms, emotion dysregulation, negative maladaptive cognitions, dissociation, and impulsivity associated with exposure to childhood adverse events have been linked to eating-related pathology [35]. Childhood maltreatment confers elevated risk of emotion dysregulation, which in turn may elevate the risk for disordered eating patterns. Several studies implicate emotion dysregulation as a contributing factor in both the development and maintenance of eating disordered behaviors, as well as associated symptoms of self-harm in patients exposed to childhood abuse, particularly emotional and sexual abuse [36–40]. Emotional dysregulation, subsequent to traumatic exposure, has been proposed as a stable trait culminating in the greater complexity of the psychopathology and clinical presentation of patients with a history of childhood abuse [41].

In addition to emotion dysregulation, negative affect has been implicated as a potential mediator in the relationship between childhood adverse experiences or trauma and eating disorders. Negative affect, related to depression, suicidality, anxiety, and irritability, has been identified as a potential contributing factor in the relationship between binge eating and purging and childhood trauma [12, 42]. Maladaptive core beliefs of emotional inhibition (i.e. the belief that emotional expression is undesirable or unacceptable) and defectiveness (i.e. the belief that one is inwardly defective or fundamentally unlovable) have been associated with symptoms of BN, and self-criticism has been linked to binge-eating behaviors [43, 44]. The affect regulation theory of binge eating proposed by Dingemans, Danner, and Parks (2017) hypothesizes that binge eating functions, in part, to help regulate emotional distress in patients exposed to trauma [45]. Other mediators between childhood emotional and sexual abuse and eating disorder pathology include maladaptive core beliefs about mistrust/abuse and abandonment and beliefs about ineffectiveness [46].

Dissociation also has been tied to eating disorder pathology, generally linked with other mediating factors. Moulton et al. (2015) found both emotion dysregulation and dissociation to be significant mediators between childhood trauma and eating psychopathology [37]. Several other studies replicated the mediating effect of both dissociation and emotion dysregulation in the relationship between childhood trauma and eating disorder psychopathology [37] or binge-eating symptoms [47, 48]. In a sample of 133 patients with AN and BN, Castellini et al. (2018) found that greater impulsivity, as well as higher levels of depersonalization, was related to body uneasiness in patients with history of abuse compared with their non-abused counterparts [41]. Other studies showed that body dissatisfaction and dissociation were linked specifically in patients with histories of childhood sexual abuse [12, 49–52]. Finally, a recent study found that patients with eating disorders ($n = 86$) demonstrated higher levels of self-reported childhood trauma and dissociation compared to healthy controls ($n = 86$). Individuals with BN and AN binge-eating/purging subtype exhibited the highest levels of dissociation, and dissociation and childhood trauma both predicted the severity of binge-eating symptoms [53].

Dissociation is often related to traumatic exposure and is linked intrinsically to post-traumatic stress disorder (PTSD), being recently described as a subtype of PTSD in the nosology [1]. A number of studies demonstrate that symptoms of PTSD and trauma-associated behaviors mediate the relationship between eating disorders and potentially traumatic exposures. Furthermore, several studies posit that exposure to potentially traumatic events precipitates the onset of disordered eating, while other studies indicate that PTSD symptoms may carry greater impact on the relationship between eating disorders and potentially traumatic exposures than the events themselves [54–56]. Exposure to potentially traumatic events has been associated with binge eating and purging more than other eating disordered behaviors, even in patients with AN [28, 29, 42, 55]. Trottier et al. [35] propose that PTSD and eating disorder behaviors share a bidirectional and functional relationship, in that behaviors such as severe food restriction, binge eating, and purging may attenuate PTSD symptoms by dampening hyperarousal and reinforcing avoidance symptoms of PTSD, although the behavioral pattern is responsible for the maintenance of both disorders. This bidirectional and reinforcing relationship between the maintenance of PTSD and eating disorders is supported by several studies [35, 57, 58]. Specific factors associated with PTSD and the development of eating disorder psychopathology include body dissatisfaction, disinhibition, and impulsivity [29, 42, 55]. Brewerton, Cotton, and Kilpatrick [59] identified elevated rates of sensation-seeking behaviors and disinhibition in women with binge-type eating compared to their peers without BN or BED.

Identification of specific psychological mediators and their impact on the onset and maintenance of eating disorder psychopathology are critical to understanding the nature of the relationship and the mechanistic steps paving the path between eating disorders and exposure to potentially traumatic events. Given that the bulk of the extant literature examining the mediators between eating disorders and trauma is cross sectional in nature, temporal relationships are not delineated, and true mechanistic causal pathways between the two constructs have yet to be clearly defined.

### 15.5.2 Neurobiological Mediators and Mechanisms

Modulation of the hypothalamus–pituitary–adrenal (HPA) axis, which regulates stress reactivity, has been identified as a likely mechanism transposing the effects of early traumatic exposures by elevating risks of long-term detrimental sequelae on physical and mental health [60]. Both HPA axis hyper- and hypo-reactivity have been associated with exposure to trauma [61]. HPA axis modulation may mediate the relation between childhood trauma and the development of adult eating disorders. Monteleone et al. have found several implications of early trauma exposure on the HPA axis functioning of patients with eating disorders including impaired cortisol awakening response in patients with AN and BN, decreased salivary cortisol response to Trier Social Stress Test (TSST), and increased anxiety levels before and after the TSST in patients with AN and childhood maltreatment compared to their counterparts without a history of maltreatment [62–64]. Their studies indicate that childhood trauma exposure has persistent detrimental effects on HPA axis reactivity to a psychosocial stressor in adults with AN [64]. Developmental stage at time of traumatic exposure, as well as number of exposures at developmentally sensitive ages may differentially impact the modulation of HPA axis, where earlier exposure may more significantly impact developmental trajectories and regulation of stress response in patients with eating disorders [41].

The examination of epigenetics mechanisms in eating disorders has focused predominantly on DNA methylation, in which environmental influences may alter future gene expression and function by addition of methyl groups to specific genomic regions. Currently identified environmental determinants shaping DNA methylation include early exposure to adverse childhood experiences, dietary factors, and perinatal insults [65–68].

Several studies have examined DNA methylation in patients with BN and comorbid borderline personality disorder (BPD) or suicidality, who have also experienced childhood abuse. Thaler et al. [69] examined the association between BN and variations in brain-derived neurotrophic factor (BDNF) and observed that women with BN, especially when comorbid with BPD or history of childhood abuse, demonstrated elevated rates of methylation at specific BDNF promoter region sites. They propose that hypermethylation of the BDNF gene may be related to eating disorder status, developmental stress exposure, and comorbid psychopathology [69]. In addition, women with BN and comorbid BPD or history of suicidality demonstrated hypermethylation of specific GR exon 1C promoter sites, though in that study childhood abuse did not demonstrate parallel effects [70]. Women with bulimia spectrum disorder (BSD) and a reported history of childhood sexual abuse demonstrated a trend-level elevation of DRD2 methylation compared with women with no eating disorder (NED), though BSD and NED groups did not differ with respect to mean percent DRD2 promoter methylation. This finding may indicate that epigenetic modification reflects the impact of childhood trauma more than the impact of the eating disorder [43]. Gene × environment mechanisms have also been implicated in the transcription of the (S) allele of the serotonin transporter promoter polymorphism 5-HTTLPR in patients with BN. Carriers of the 5-HTTLPR S allele with both BN and a reported history of severe childhood abuse displayed more pronounced dissocial behavior, characterized by novelty seeking, recklessness, or hostility [71]. Additionally, carriers of the variant glucocorticoid receptor (GR) BCL1 C allele with a history of reported childhood abuse demonstrated elevated rates of BN relative to counterparts who did not experience abuse and those with other allelic variations [72]. Though future studies will likely reveal a greater impact of epigenetic mechanisms on eating disorder psychopathology and exposure to potentially traumatic events, at present, epigenetic studies focusing on AN and BN are limited in number and sample size. Moreover, as reviewed by Hubel et al. [73], studies show extensive sample overlap such that one-third of the available studies recruited participants from the same two eating disorder clinics.

The investigation of neurobiological structures in patients with both eating disorders and history of childhood maltreatment also is in its infancy. Monteleone et al. [73] detected reduced white matter integrity and reduction of gray matter volume in maltreated patients with AN and BN in brain structures modulating processes that typically play a role in eating disorder pathology, such as reward, taste, and body image perception. The neuroimaging profile of patients with BN and BED remains a subject of ongoing and future investigation, and the neurobiology of the intersection among trauma, binge eating, and BN is yet to be determined [74].

## 15.6 Clinical Implications

Given the evidence of an overlap between individuals with eating disorders and trauma exposure, treatment providers working with either group should be aware of this comorbidity and prepared to screen and assess patients for both trauma-related symptoms and eating disorders. It is important to note that not all trauma exposures lead to trauma-related symptomatology, and not all the sequelae of trauma require psychosocial treatment. However, in patients with comorbid eating disorders and PTSD, severe underweight should be addressed before beginning trauma work, given the evidence that malnutrition interferes with learning [55]. There is less agreement regarding the management of comorbid eating disorders and PTSD in patients who are more nutritionally stable. Some argue that untreated symptoms of PTSD can interfere with adherence to eating disorder-focused treatment and increase the likelihood that patients will drop out of eating disorder treatment [55]. The developers of enhanced cognitive behavior therapy (CBT-E) for eating disorders suggest focusing on one treatment at a time, either for disordered eating or PTSD, and giving patients the choice of which symptoms to address first [75]. Others have suggested integrated treatment focusing on both trauma-related symptoms and eating disorders due to the high co-occurrence and shared mechanisms between the two [35].

One way to target both trauma symptoms and eating disorder symptoms is by using treatments that address shared mechanisms, such as emotion dysregulation and maladaptive beliefs. Cognitive processing therapy (CPT), historically focused on treating PTSD symptoms, is a promising example of a combined treatment, with preliminary evidence from a single study ($n = 65$) suggesting improvement in symptoms shared by PTSD and eating disorders [31]. The drawback is that CPT has not been shown to improve symptoms specific to eating disorders, such as low weight, shape and weight concerns, and the frequency of binge/purge episodes [31].

Dialectical behavior therapy also has been suggested as a combined treatment for trauma symptoms and disordered eating, as it involves learning ways to regulate emotions and manage psychological distress [35, 76]. Eating disorder symptoms, including binge eating, purging, and dietary restriction, are conceptualized as maladaptive coping techniques, which patients are taught to replace with other, more adaptive methods for managing distress. There is some evidence for the efficacy of dialectical behavior therapy for promoting abstinence from binge eating in BED [77], and for overall eating disorder symptom reduction in women with comorbid BN and borderline personality disorder [78], although these studies are small and lack active control groups. Dialectical behavior therapy interventions have also been integrated into family-based therapy for eating disorders, with positive results from pilot studies showing reductions in binge/purge behaviors [79].

Evidence-based treatments for PTSD, which may also be useful for individuals with subthreshold trauma symptoms, include CBT with prolonged exposure, eye movement desensitization and reprocessing (EMDR), and pharmacotherapy [55]. Evidence-based treatments for EDs include CBT-E, the treatment with the strongest evidence, as well as interpersonal psychotherapy (IPT), and family-based treatment for children and adolescents (FBT). Unlike combined treatment modalities, there is robust evidence for the efficacy of these treatments in their intended clinical populations (e.g. DBT for borderline personality disorder; CBT-E for eating disorders). However, there is little evidence for their effectiveness in individuals with comorbid eating disorders and trauma-related conditions such as PTSD and borderline personality disorder.

Ultimately, screening for either eating disorder or trauma in these populations is an important first step to determine the best treatment options. The Childhood Trauma Questionnaire (CTQ) is a recommended screening assessment using retrospective self-report [80]. When screening for trauma, it is important also to screen for trauma-related symptoms, and whether they point or not to a potential PTSD diagnosis, as a history of trauma alone does not mean trauma-focused treatment is necessary or appropriate. Eating disorders can be screened using the EDE-Q, which includes subscales of restraint, eating concerns, shape concerns, and weight concerns [81]. Patients with BN are most likely to report childhood trauma, especially sexual abuse and emotional abuse. Health providers should be particularly vigilant for binge eating and inappropriate compensatory behaviors in patients with a history of childhood trauma.

## 15.7 Conclusion

Overall, the research literature on the relationship between childhood trauma and eating disorders shows a definite association, where the eating disorder population reported a history of childhood trauma at a higher rate than the general population. Of note, patients with BN are most likely to endorse childhood trauma, in particular childhood sexual abuse and emotional abuse. Cognitive, physiological, and epigenetic mediators include emotion dysregulation, negative affect, modulation of the HPA axis, and DNA methylation. Although studies have shown that there is a clinically significant relationship between childhood trauma and eating disorder psychopathology [35], further research needs to be conducted to determine whether childhood trauma is a specific risk factor for eating disorders or a general risk factor for psychopathology. Future research should also include the study of individuals with diagnoses such as ARFID, Pica, and Rumination Disorder, and expand upon the types of trauma experienced by the patients and when it occurred. Regardless of causality, there is a clear need to consider the potential impact of trauma on eating disorder treatment.

## References

1. Association AP. Diagnostic and statistical manual of mental disorders (DSM-5). American Psychiatric Pub; 2013.
2. Hudson JI, Hiripi E, Pope HG Jr, Kessler RC. The prevalence and correlates of eating disorders in the National Comorbidity Survey Replication. Biol Psychiatry. 2007;61(3):348–58.

3. Volpe U, Tortorella A, Manchia M, Monteleone AM, Albert U, Monteleone P. Eating disorders: what age at onset? Psychiatry Res. 2016;238:225–7.
4. Arcelus J, Mitchell AJ, Wales J, Nielsen S. Mortality rates in patients with anorexia nervosa and other eating disorders: a meta-analysis of 36 studies. Arch Gen Psychiatry. 2011;68(7):724–31.
5. Mehler PS, Brown C. Anorexia nervosa–medical complications. J Eat Disord. 2015;3(1):11.
6. Qian J, Hu Q, Wan Y, Li T, Wu M, Ren Z, et al. Prevalence of eating disorders in the general population: a systematic review. Shanghai Arch Psychiatry. 2013;25(4):212–23.
7. Dahlgren CL, Wisting L, Rø Ø. Feeding and eating disorders in the DSM-5 era: a systematic review of prevalence rates in non-clinical male and female samples. J Eat Disord. 2017;5:56.
8. Kurz S, Van Dyck Z, Dremmel D, Munsch S, Hilbert A. Early-onset restrictive eating disturbances in primary school boys and girls. Eur Child Adolesc Psychiatry. 2015;24(7):779–85.
9. Murray HB, Thomas JJ, Hinz A, Munsch S, Hilbert A. Prevalence in primary school youth of pica and rumination behavior: the understudied feeding disorders. Int J Eat Disord. 2018;51(8):994–8.
10. Afifi TO, Sareen J, Fortier J, Taillieu T, Turner S, Cheung K, et al. Child maltreatment and eating disorders among men and women in adulthood: results from a nationally representative United States sample. Int J Eat Disord. 2017;50(11):1281–96.
11. Carr CP, Martins CMS, Stingel AM, Lemgruber VB, Juruena MF. The role of early life stress in adult psychiatric disorders: a systematic review according to childhood trauma subtypes. J Nerv Mental Dis. 2013;201(12):1007–20.
12. Molendijk M, Hoek H, Brewerton T, Elzinga B. Childhood maltreatment and eating disorder pathology: a systematic review and dose-response meta-analysis. Psychol Med. 2017;47(8):1402–16.
13. Jaite C, Schneider N, Hilbert A, Pfeiffer E, Lehmkuhl U, Salbach-Andrae H. Etiological role of childhood emotional trauma and neglect in adolescent anorexia nervosa: a cross-sectional questionnaire analysis. Psychopathology. 2012;45(1):61–6.
14. Carter JC, Bewell C, Blackmore E, Woodside DB. The impact of childhood sexual abuse in anorexia nervosa. Child Abuse Negl. 2006;30(3):257–69.
15. Smolak L, Murnen SK. A meta-analytic examination of the relationship between child sexual abuse and eating disorders. Int J Eat Disord. 2002;31(2):136–50.
16. Romans SE, Gendall KA, Martin JL, Mullen PE. Child sexual abuse and later disordered eating: a New Zealand epidemiological study. Int J Eat Disord. 2001;29(4):380–92.
17. Kimber M, McTavish JR, Couturier J, Boven A, Gill S, Dimitropoulos G, et al. Consequences of child emotional abuse, emotional neglect and exposure to intimate partner violence for eating disorders: a systematic critical review. BMC Psychol. 2017;5(1):33.
18. Pignatelli AM, Wampers M, Loriedo C, Biondi M, Vanderlinden J. Childhood neglect in eating disorders: a systematic review and meta-analysis. J Trauma Dissociation. 2017;18(1):100–15.
19. Backholm K, Isomaa R, Birgegård A. The prevalence and impact of trauma history in eating disorder patients. Eur J Psychotraumatol. 2013;4:10. https://doi.org/10.3402/ejpt.v4i0.22482.
20. Kong S, Bernstein K. Childhood trauma as a predictor of eating psychopathology and its mediating variables in patients with eating disorders. J Clin Nurs. 2009;18(13):1897–907.
21. White AAH, Pratt KJ, Cottrill C. The relationship between trauma and weight status among adolescents in eating disorder treatment. Appetite. 2018;129:62–9.
22. Mahon J, Bradley SN, Harvey PK, Winston AP, Palmer RL. Childhood trauma has dose-effect relationship with dropping out from psychotherapeutic treatment for bulimia nervosa: a replication. Int J Eat Disord. 2001;30(2):138–48.
23. Rodríguez M, Pérez V, García Y. Impact of traumatic experiences and violent acts upon response to treatment of a sample of Colombian women with eating disorders. Int J Eat Disord. 2005;37(4):299–306.
24. Vrabel KR, Hoffart A, Rø Ø, Martinsen EW, Rosenvinge JH. Co-occurrence of avoidant personality disorder and child sexual abuse predicts poor outcome in long-standing eating disorder. J Abnorm Psychol. 2010;119(3):623–9.
25. Calugi S, Franchini C, Pivari S, Conti M, El Ghoch M, Dalle GR. Anorexia nervosa and childhood sexual abuse: treatment outcomes of intensive enhanced cognitive behavioural therapy. Psychiatry Res. 2018;262:477–81.

26. Inniss D, Steiger H, Bruce K. Threshold and subthreshold post-traumatic stress disorder in bulimic patients: Prevalences and clinical correlates. Eating Weight Disorders. 2011;16(1):30–6.
27. Mitchell KS, Mazzeo SE, Schlesinger MR, Brewerton TD, Smith BN. Comorbidity of partial and subthreshold PTSD among men and women with eating disorders in the national comorbidity survey-replication study. Int J Eat Disord. 2012;45(3):307–15.
28. Reyes-Rodríguez ML, Ann Von Holle T, Thornton LM, Klump KL, Brandt H, Crawford S, et al. Post traumatic stress disorder in anorexia nervosa. Psychosom Med. 2011;73(6):491–7.
29. Tagay S, Schlottbohm E, Reyes-Rodriguez ML, Repic N, Senf W. Eating disorders, trauma, PTSD, and psychosocial resources. Eat Disord. 2014;22(1):33–49.
30. Isomaa R, Backholm K, Birgegård A. Posttraumatic stress disorder in eating disorder patients: the roles of psychological distress and timing of trauma. Psychiatry Res. 2015;230(2):506–10.
31. Mitchell KS, Wells SY, Mendes A, Resick PA. Treatment improves symptoms shared by PTSD and disordered eating. J Trauma Stress. 2012;25(5):535–42.
32. Mayes SD, Fernandez-Mendoza J, Baweja R, Calhoun S, Mahr F, Aggarwal R, et al. Correlates of suicide ideation and attempts in children and adolescents with eating disorders. Eat Disord. 2014;22(4):352–66.
33. Richardson J, Steiger H, Schmitz N, Joober R, Bruce KR, Israel M, et al. Relevance of the 5-HTTLPR polymorphism and childhood abuse to increased psychiatric comorbidity in women with bulimia-spectrum disorders. J Clin Psychiatry. 2008;69(6):981–90.
34. Utzinger LM, Haukebo JE, Simonich H, Wonderlich SA, Cao L, Lavender JM, et al. A latent profile analysis of childhood trauma in women with bulimia nervosa: associations with borderline personality disorder psychopathology. Int J Eat Disord. 2016;49(7):689–94.
35. Trottier K, MacDonald DE. Update on psychological trauma, other severe adverse experiences and eating disorders: state of the research and future research directions. Curr Psychiatry Rep. 2017;19(8):45.
36. Michopoulos V, Powers A, Moore C, Villarreal S, Ressler KJ, Bradley B. The mediating role of emotion dysregulation and depression on the relationship between childhood trauma exposure and emotional eating. Appetite. 2015;91:129–36.
37. Moulton SJ, Newman E, Power K, Swanson V, Day K. Childhood trauma and eating psychopathology: a mediating role for dissociation and emotion dysregulation? Child Abuse Negl. 2015;39:167–74.
38. Mills P, Newman EF, Cossar J, Murray G. Emotional maltreatment and disordered eating in adolescents: testing the mediating role of emotion regulation. Child Abuse Negl. 2015;39:156–66.
39. Gordon KH, Simonich H, Wonderlich SA, Dhankikar S, Crosby RD, Cao L, et al. Emotion dysregulation and affective intensity mediate the relationship between childhood abuse and suicide-related behaviors among women with bulimia nervosa. Suicide Life Threat Behav. 2016;46(1):79–87.
40. Racine SE, Wildes JE. Emotion dysregulation and anorexia nervosa: an exploration of the role of childhood abuse. Int J Eat Disord. 2015;48(1):55–8.
41. Castellini G, Lelli L, Cassioli E, Ciampi E, Zamponi F, Campone B, et al. Different outcomes, psychopathological features, and comorbidities in patients with eating disorders reporting childhood abuse: a 3-year follow-up study. Eur Eat Disord Rev. 2018;26(3):217–29.
42. Vanzhula IA, Calebs B, Fewell L, Levinson CA. Illness pathways between eating disorder and post-traumatic stress disorder symptoms: understanding comorbidity with network analysis. Eur Eat Disord Rev. 2019;27(2):147–60.
43. Groleau P, Joober R, Israel M, Zeramdini N, DeGuzman R, Steiger H. Methylation of the dopamine D2 receptor (DRD2) gene promoter in women with a bulimia-spectrum disorder: associations with borderline personality disorder and exposure to childhood abuse. J Psychiatr Res. 2014;48(1):121–7.
44. Feinson MC, Hornik-Lurie T. Binge eating & childhood emotional abuse: the mediating role of anger. Appetite. 2016;105:487–93.
45. Dingemans A, Danner U, Parks M. Emotion regulation in binge eating disorder: a review. Nutrients. 2017;9(11).

46. Jenkins PE, Meyer C, Blissett JM. Childhood abuse and eating psychopathology: the mediating role of core beliefs. J Aggress Maltreat Trauma. 2013;22(3):248–61.
47. Kent A, Waller G, Dagnan D. A greater role of emotional than physical or sexual abuse in predicting disordered eating attitudes: the role of mediating variables. Int J Eat Disord. 1999;25(2):159–67.
48. Rodriguez-Srednicki O. Childhood sexual abuse, dissociation, and adult self-destructive behavior. J Child Sex Abus. 2002;10(3):75–89.
49. Dunkley DM, Masheb RM, Grilo CM. Childhood maltreatment, depressive symptoms, and body dissatisfaction in patients with binge eating disorder: the mediating role of self-criticism. Int J Eat Disord. 2010;43(3):274–81.
50. Muehlenkamp JJ, Claes L, Smits D, Peat CM, Vandereycken W. Non-suicidal self-injury in eating disordered patients: a test of a conceptual model. Psychiatry Res. 2011;188(1):102–8.
51. Duarte C, Ferreira C, Pinto-Gouveia J. At the core of eating disorders: overvaluation, social rank, self-criticism and shame in anorexia, bulimia and binge eating disorder. Compr Psychiatry. 2016;66:123–31.
52. Preti A, Incani E, Camboni MV, Petretto DR, Masala C. Sexual abuse and eating disorder symptoms: the mediator role of bodily dissatisfaction. Compr Psychiatry. 2006;47(6):475–81.
53. Palmisano GL, Innamorati M, Susca G, Traetta D, Sarracino D, Vanderlinden J. Childhood traumatic experiences and dissociative phenomena in eating disorders: level and association with the severity of binge eating symptoms. J Trauma Dissociation. 2018;19(1):88–107.
54. Collins B, Fischer S, Stojek M, Becker K. The relationship of thought suppression and recent rape to disordered eating in emerging adulthood. J Adolesc. 2014;37(2):113–21.
55. Brewerton TD. Eating disorders, trauma, and comorbidity: focus on PTSD. Eat Disord. 2007;15(4):285–304.
56. Holzer SR, Uppala S, Wonderlich SA, Crosby RD, Simonich H. Mediational significance of PTSD in the relationship of sexual trauma and eating disorders. Child Abuse Negl. 2008;32(5):561–6.
57. Mitchell K, Porter B, Boyko E, Field A. Longitudinal associations among posttraumatic stress disorder, disordered eating, and weight gain in military men and women. Am J Epidemiol. 2016;184(1):33–47.
58. Trottier K, Wonderlich SA, Monson CM, Crosby RD, Olmsted MP. Investigating posttraumatic stress disorder as a psychological maintaining factor of eating disorders. Int J Eat Disord. 2016;49(5):455–7.
59. Brewerton TD, Cotton BD, Kilpatrick DG. Sensation seeking, binge-type eating disorders, victimization, and PTSD in the National Women's study. Eat Behav. 2018;30:120–4.
60. Cicchetti D, Toth SL. Child maltreatment. Annu Rev Clin Psychol. 2005;1:409–38.
61. Binder EB, Holsboer F. Low cortisol and risk and resilience to stress-related psychiatric disorders. Biol Psychiatry. 2012;71(4):282–3.
62. Monteleone AM, Monteleone P, Serino I, Scognamiglio P, Di Genio M, Maj M. Childhood trauma and cortisol awakening response in symptomatic patients with anorexia nervosa and bulimia nervosa. Int J Eat Disord. 2015;48(6):615–21.
63. Monteleone A, Monteleone P, Volpe U, De Riso F, Fico G, Giugliano R, et al. Impaired cortisol awakening response in eating disorder women with childhood trauma exposure: evidence for a dose-dependent effect of the traumatic load. Psychol Med. 2018;48(6):952–60.
64. Monteleone AM, Patriciello G, Ruzzi V, Cimino M, Del Giorno C, Steardo L, et al. Deranged emotional and cortisol responses to a psychosocial stressor in anorexia nervosa women with childhood trauma exposure: evidence for a "maltreated ecophenotype"? J Psychiatr Res. 2018;104:39–45.
65. Dauncey M. Genomic and epigenomic insights into nutrition and brain disorders. Nutrients. 2013;5(3):887–914.
66. McGowan PO, Meaney MJ, Szyf M. Diet and the epigenetic (re) programming of phenotypic differences in behavior. Brain Res. 2008;1237:12–24.
67. Crider KS, Yang TP, Berry RJ, Bailey LB. Folate and DNA methylation: a review of molecular mechanisms and the evidence for folate's role. Adv Nutr. 2012;3(1):21–38.

68. Thaler L, Steiger H. Eating disorders and epigenetics. In: Neuroepigenomics in aging and disease. Springer; 2017. p. 93–103.
69. Thaler L, Gauvin L, Joober R, Groleau P, de Guzman R, Ambalavanan A, et al. Methylation of BDNF in women with bulimic eating syndromes: associations with childhood abuse and borderline personality disorder. Prog Neuro-Psychopharmacol Biol Psychiatry. 2014;54:43–9.
70. Steiger H, Labonté B, Groleau P, Turecki G, Israel M. Methylation of the glucocorticoid receptor gene promoter in bulimic women: associations with borderline personality disorder, suicidality, and exposure to childhood abuse. Int J Eat Disord. 2013;46(3):246–55.
71. Steiger H, Richardson J, Joober R, Israel M, Bruce KR, Ng Ying Kin N, et al. Dissocial behavior, the 5HTTLPR polymorphism, and maltreatment in women with bulimic syndromes. Am J Med Genet. 2008;147(1):128–30.
72. van Eekelen JAM, Ellis JA, Pennell CE, Craig J, Saffery R, Mattes E, et al. Stress-sensitive neurosignalling in depression: an integrated network biology approach to candidate gene selection for genetic association analysis. Ment Illn. 2012;4(2):e21.
73. Monteleone AM, Monteleone P, Esposito F, Prinster A, Ruzzi V, Canna A, et al. The effects of childhood maltreatment on brain structure in adults with eating disorders. World J Biol Psychiatry. 2019;20(4):301–9.
74. Donnelly B, Touyz S, Hay P, Burton A, Russell J, Caterson I. Neuroimaging in bulimia nervosa and binge eating disorder: a systematic review. J Eat Disord. 2018;6:3. https://doi.org/10.1186/s40337-018-0187-1.
75. Fairburn CG. Cognitive behavior therapy and eating disorders. Guilford Publications; 2008.
76. Linehan MM, Chen EY. Dialectical behavior therapy for eating disorders. In: Freeman A, editor. Encyclopedia of cognitive behavior therapy. New York: Springer Science & Business Media; 2006. p. 168–71.
77. Telch CF, Agras WS, Linehan MM. Dialectical behavior therapy for binge eating disorder. J Consulting Clin Psychol. 2001;69(6):1061–5.
78. Chen EY, Matthews L, Allen C, Kuo JR, Linehan MM. Dialectical behavior therapy for clients with binge-eating disorder or bulimia nervosa and borderline personality disorder. Int J Eat Disord. 2008;41(6):505–12.
79. Murray SB, Anderson LK, Cusack A, Nakamura T, Rockwell R, Griffiths S, et al. Integrating family-based treatment and dialectical behavior therapy for adolescent bulimia nervosa: preliminary outcomes of an open pilot trial. Eat Disord. 2015;23(4):336–44.
80. Bernstein DP, Fink L, Handelsman L, Foote J, Lovejoy M, Wenzel K, et al. Initial reliability and validity of a new retrospective measure of child abuse and neglect. Am J Psychiatry. 1994;151(8):1132–6.
81. Fairburn CG, Beglin SJ. Assessment of eating disorders: interview or self-report questionnaire? Int J Eat Disord. 1994;16(4):363–70.
82. Smink FRE, van Hoeken D, Hoek HW. Epidemiology, course, and outcome of eating disorders. Curr Opin Psychiatry. 2013;26(6):543–8. https://doi.org/10.1097/YCO.0b013e328365a24f.

# 16 Childhood Trauma and Dissociative Disorders

Vedat Şar

## 16.1 Introduction

Dissociation is characterized by a disruption or discontinuity in usually integrated psychological functions such as memory, consciousness, perception, sense of self and agency, or sensorimotor abilities [1]. Thus, dissociation may affect any psychological faculty, albeit reversibly. In its most dramatic and chronic form, such discontinuity takes the form of marked identity disruptions typically observed in dissociative identity disorder (DID). Other types of dissociative disorders represent either partial representations of chronic dissociation (i.e., limited to a smaller number of symptoms), or an acute and/or transient dissociative reaction to a stressful event (either of the mono- and poly-symptomatic type).

## 16.2 Epidemiology of Dissociative Disorders

Studies conducted in various countries led to a consensus about the prevalence of dissociative disorders which is slightly over 10% in clinical settings, where relatively severe types of the disorders predominate [2]. The prevalence rates seem to be higher in the general population due to the preponderance of milder or partial types of dissociative disorders [3, 4]. There are special populations where the prevalence of dissociative disorders exceed these rates; i.e., adolescent psychiatric outpatients, patients who are admitted to psychiatric emergency ward, and dependents of chemical drugs (Table 16.1).

While women predominate in clinical settings [5, 6], one study in the general population [3] based on a standardized diagnostic clinical interview yielded no significant difference in gender distribution of dissociative disorders. A further study in

V. Şar (✉)
Department of Psychiatry, Koç University School of Medicine (KUSOM), Istanbul, Turkey
e-mail: vsar@ku.edu.tr



G. Spalletta et al. (eds.), *Childhood Trauma in Mental Disorders*,
https://doi.org/10.1007/978-3-030-49414-8_16

**Table 16.1** Prevalence of dissociative disorders in various settings (adapted from [2])

| Setting | Country | n | Dissociative identity disorder (%) | All dissociative disorders (%) |
|---|---|---|---|---|
| *Psychiatric inpatient unit* | | | | |
| Tutkun et al. [6] | Turkey | 166 | 5.4 | 10.2 |
| Modestin et al. [165] | Switzerland | 207 | 0.4 | 5.0 |
| Gast et al. [166] | Germany | 115 | 0.9 | 4.3 |
| Friedl et al. [167] | Netherlands | 122 | 2.0 | 8.0 |
| Ginzburg et al. [168] | Israel | 120 | 0.8 | 12.0 |
| Saxe et al. [169] | Germany | 172 | 4.0 | 13.0 |
| Ross et al. [170] | USA | 484 | 5.4 | 20.7 |
| Lipsanen et al. [171] | Norway | 39 | – | 21.0 |
| Ross et al. [172] | USA | 407 | 7.5 | 40.8 |
| *Psychiatric outpatient unit* | | | | |
| Şar et al. [5] | Turkey | 150 | 2.0 | 12.0 |
| Şar et al. [173] | Turkey | 240 | 2.5 | 13.8 |
| Foote et al. [174] | USA | 82 | 6.0 | 29.0 |
| Lipsanen et al. [171] | Norway | 39 | – | 14.0 |
| *Emergency psychiatric ward* | | | | |
| Şar et al. [14] | Turkey | 43 | 14.0 | 34.9 |
| *General population* | | | | |
| Şar et al. [4], women | Turkey | 628 | 1.1 | 18.3 |
| Johnson et al. [3] | USA | 658 | 1.5 | 8.6 |
| Ross [175, 170] | USA | 454 | 3.1 | 11.2 |
| *Substance dependents (chemical)* | | | | |
| Tamar-Gürol et al. [176] | Turkey | 104 | 26.0 | 5.8 |
| Ross et al. [177] | USA | 100 | – | 39.0 |
| Dunn et al. [178] | USA | 100 | – | 15.0 |
| *Substance dependents (alcohol)* | | | | |
| Evren et al. [179] | Turkey | 54 | 9.0 | 0.0 |
| *Women in prostitution* | | | | |
| Yargic et al. [180] | Turkey | 50 | 18.0 | – |
| Ross et al. [181] | USA | 20 | 5.0 | 55.0 |
| *Exotic dancers* | | | | |
| Ross et al. [182] | USA | 20 | 35.0 | 80.0 |

the general population based on self-report evaluation documented that twice women as much as men had a dissociative experiences score above the cutoff level although there was no significant difference on average scores [7]. Apparently, women seem to be more vulnerable in development of a relatively severe clinical condition; i.e., they are overrepresented in clinical settings and among groups with more symptoms. This may be due to the differences in traumatic antecedents, gender role characteristics, help-seeking behavior, and last but not least, ways of expressing distress. For example, men with dissociative disorders are known to hide their symptoms more readily.

A study among adolescent psychiatric outpatients [8], however, which was based on a standardized clinical diagnostic interview, did not yield a gender difference in the prevalence of dissociative disorders. It is possible that the dissociative disorders of adolescence may improve or may turn to another diagnosis in a less vulnerable subgroup of the cases over time which could explain the change in gender distribution toward adulthood. Study characteristics including differences in assessment instruments may also affect the reported prevalence rates [9]. For example, differences in self-report and clinical interview may have systematic reasons [10–12]. Diminished self-awareness about dissociative experiences may possibly occur due to amnesias which may interfere with self-report in particular. Fear and shame, on the other hand, may interfere with reporting in an interpersonal setting.

Both clinical as well as general population studies screen rather chronic dissociative disorders. Recently, the DSM-5 [1] introduced acute dissociative reaction to a stressful event as a new diagnostic category which is listed among other specific dissociative disorders. Per definition, this category covers conditions that last less than 1 month. There is surprisingly little information about prevalence of such reactions which cover a large spectrum of severity, reaching even the scope of a brief psychotic disorder in its most extreme form. Acute reactions may also be superposed to a chronic dissociative disorder such as DID or its partial forms which makes an additional diagnosis redundant [13]. Applications to the emergency psychiatric ward due to such acute transient conditions ("nervous breakdown") usually serve as a diagnostic window to the clinician when core symptoms of a chronic dissociative disorder remain dormant until a stressful event triggers a more prominent manifestation [14].

## 16.3 Childhood Trauma: The Evidence

Among all psychiatric conditions, dissociative disorders are the diagnostic group which is associated with the highest frequencies of childhood trauma reports [15]. DID is the most studied type of dissociative disorders regarding a traumatic etiology, namely clinical studies in the United States, Australia, Turkey, Puerto Rico, the Netherlands, Germany, and Canada have consistently found that DID is linked to chronic abusive experiences in childhood, typically at the hands of an attachment figure [16–20].

By using corroborating documentation from hospital, police and child protection agencies or witnesses, several studies have confirmed histories of severe abuse in DID [18, 21, 22]. There are patients who remember certain traumatic memories during treatment which were previously covered by dissociative amnesia. However, such "repressed" or "recovered memories" do not have any effect on the prevalence rates of traumatic life histories, namely in most clinical series, childhood abuse and/or neglect is reported by 90–100% of the patients during the study examination [23, 24], i.e., before any specific treatment intervention.

For example, to eliminate any specific treatment effect and to demonstrate the accuracy of trauma reports, a screening study was conducted in a rather uninformed population of an Anatolian town (Turkey) in the 1990s [7]. A large representative sample of women from the general population (N = 994) was evaluated in three stages: completion of a self-report measure of dissociation, administration of a standardized diagnostic interview, and clinical evaluation by a study psychiatrist. In particular, two groups of participants with high and low dissociative scores in the first stage of the study, were evaluated using a structured diagnostic clinical interview by a researcher blind to the dissociation scores. Those participants who got the diagnosis of a dissociative disorder and another group of non-dissociative controls were evaluated by a study psychiatrist blind to the diagnoses and group membership. There was 100% agreement between clinicians of the second and third step in dissociative disorder diagnoses. Four cases of DID were identified and all of them reported childhood abuse and/or neglect.

Nevertheless, beside developmental traumatization, socio-cognitive sequalae, trauma-generated neurobiological responses, biologically derived traits, and epigenetic mechanisms may also contribute to the emergence of dissociative psychopathology [25]. There are also genetic links to dissociation in general, and in relation to childhood adversity ([26, 27]). High hypnotizability—itself a non-pathological, genetically derived capacity—has also been proposed to be a necessary diathesis for dissociative disorders [28]. Although patients with dissociative disorders have higher hypnotizability than those with other mental disorders, this characteristic is also found in patients with chronic refractory post-traumatic states in general [29–31]. In conclusion, dissociative disorders may be seen as an exemplary disease model of the *biopsychosocial paradigm* in psychiatry [25].

## 16.4 Psychopathogenesis of Mental Fragmentation

To go beyond limitations of prevalence studies, the association between childhood trauma and dissociative disorders needs to be elaborated in terms of overall conceptualization and psychopathogenesis as well. First of all, rather than being merely an anxiety-dominated response to a single traumatic event, the body of evidence drives clinicians and researchers to conceptualize trauma-related disorders also in terms of a long-lasting and multi-dimensional consequence of chronic, early, and interpersonal (developmental) traumatization [32]. Developmental trauma refers to a type of stressful event that occurs repeatedly and cumulatively, usually over a period of time, and within specific relationships and contexts [33]. Childhood abuse (sexual, emotional, and physical) and neglect (physical and emotional) constitute typical forms of chronic traumatization. The distinction between acute and chronic stress has neurobiological repercussions as well. For example, unlike claims about consequences of chronic traumatization, a recent meta-analysis on "simplex" PTSD revealed no significant changes in gray matter volume [34].

Dissociative disorders are currently understood as post-traumatic developmental conditions where adverse experiences usually begin in early childhood, i.e., before puberty. For example, the identity alterations observed in DID and its subthreshold forms may be considered as an elaborated version of trauma-related mental *intrusions* and *avoidance* which corresponds to the basic mechanism of PTSD [12, 35, 36]. While the dynamics are similar, there are discrete identities with their own first-person perspective in DID, as well as breaks in consciousness between these identities, which does not occur in PTSD.

Nevertheless, these discrete identities are usually linked to certain traumatic experiences (i.e., they may carry memories, cognitions, and emotions associated with the experience). However, such relationship may also occur in a disguised form, i.e., a distinct personality state may carry compensatory cognitive schemas or psychological defences, such as denial, projection, or rationalization, to cope with traumatic experience. The mental organization on the basis of discrete identities allows long-term maintenance of an internal system which is composed of incompatible pieces of mental content. Distinct identities may also be formed by narcissistic-identificatory-fusionary processes, i.e., through copying or imitation of "others" including creation of new and modified versions of pre-existing distinct entities of the internal system upon further stressful experiences [37]. Growing up in a neglectful environment and the already started dissociative psychopathology increases these tendencies. The trauma-related interpersonal relationship pattern of "attachment to the perpetrator" (also known as "identification with the aggressor" or "Stockholm Syndrome") is an attempt to cope with the abuse through such fusionary tendencies.

On the other hand, developmental trauma may start even in disturbances of early interpersonal attachment with caregivers. Deficiencies in mirroring is one example of the mismatch between the infant and the caregiver [38]. Bowlby [39] proposed that such experiences lead the infant to develop multiple internal representations of self and attachment figures which he called internal working models (IWMs). One IWM becomes dominant in regulating interpersonal relationships in a certain context, while the other IWMs remain separated from mainstream conscious experience. The latter surface in stressful situations to regulate emotions and cognitions in a way that may be perceived as alien to the person's usual sense of self. Thus, later traumatizations such as abuse and neglect or being exposed to an overtly or covertly dysfunctional family may further accentuate such fragmentation. Sense of self and agency is influenced by these alterations.

Inspired by Janet, the model of structural dissociation of personality [40] is based on the idea of an underlying division of personality, in which each dissociative subsystem has its own first-person perspective. While the "apparently normal" part is oriented in daily functions and avoids the traumatic mental content, the "emotional" part of personality is fixed in traumatic experiences. Traumatized individuals alternate between these different parts which may be activated in a sequence or in parallel. Each part of personality is composed of multiple actions systems devoted to survival of the individual (e.g., fight-flight, submission, freezing, cry for help) or the species.

## 16.5 Psychosomatic Implications of Dissociation

Putnam [41] proposes that trauma-related discrete behavioral states do not only differ in their manifest behavior (e.g., sleeping, feeding, socializing, escaping danger) but also in all other psychophysiological dimensions, including arousal level, heart rate, motivation, affective tone, thought patterns and content, appraisals, and brain areas activation. Conversely, among infants and young children, such psychophysiological differences may constitute the precursors of later pathological dissociative symptoms, and in the most extreme case, dissociative identities, namely he proposes that, initially through biological decrees (need to eat, sleep) and then through experience (e.g., exposure to traumatic stress), different behavioral states emerge to support survival and promote adaptation to the environment [42]. Instead of providing the origin of an integrated self-identity, emerging discrete behavioral states may be elaborated over time while being further utilized as coping mechanisms during subsequent traumatic experiences.

Although bodily aspects of dissociation may be subsumed under the rubric of "somatoform dissociation" [43], a particular type of manifestations should be differentiated from other somatic symptoms in this spectrum: conversion (functional neurological) symptoms [44]. Per definition, conversion symptoms affect voluntary organ systems (e.g., vomiting, blindness, paralysis of the limbs, fainting, pseudoseizure). While chronic somatoform dissociation may point to a hidden or difficult to recognize trauma history such as childhood neglect [45, 46], conversion symptoms usually have an acute quality which constitutes a medical emergency [14]. Acute and transient conversion symptoms may be accompanied by clear-cut dissociative symptoms (e.g., depersonalization, amnesia). In particular, severe and persistent conversion symptoms may be an indicator of an unrecognized chronic dissociative disorder or PTSD.

In conclusion, dissociative disorders are composed of both negative (e.g., amnesia, loss of sensorimotor functions such as anesthesia) and positive (e.g., hallucinations, flashbacks, pseudoseizures) symptoms, i.e., symptom clusters of either the intrusion or the omission type. Those symptoms which affect sensorimotor functions (e.g., functional neurological symptoms such as pseudoseizure and dissociative blindness) may appear as presentation of a bodily (i.e., usually neurological) illness, which require general medical attention.

To considerate the autonomous nervous system functions may be helpful in understanding these psychosomatic repercussions of dissociative disorders. Given the insights of the polyvagal theory [47] and of the empirical research on heart rate variability in PTSD, the sympathetic and parasympathetic response types to threat and the role of the nervous vagus in connecting "higher" mental functions with organs as well as psychobiological action systems (i.e., fight-flight, submission, freezing) serving survival in front of threat, places the autonomous nervous system at the centre of psychosomatic phenomena related to dissociation.

## 16.6 Dissociation in Trauma: Confounder or Common Denominator?

As a way of coping with unbearable pain, dissociation is increasingly recognized as a common feature of trauma-related conditions [2, 48]. For example, both peritraumatic and persistent dissociation have been considered as components of PTSD with variability between individuals [49]. There are ongoing debates on whether dissociation is the common denominator [50] or a confounder of the entire trauma spectrum [32]. The latter stance is represented by the newly introduced dissociative subtype of PTSD [1] which is a construct defined solely by negative symptoms of dissociation: depersonalization and derealization (Table 16.2).

Specifically, while providing neurobiological underpinnings of a dissociative subtype of PTSD, a group of researchers [51] described two types of reaction to traumatic stress: overmodulation (inhibition) and undermodulation (arousal) of emotions. The former response types were taken as the basis of dissociation. This view is challenged by supporters of the structural dissociation theory [40] which proposes that dissociation also appears in positive symptoms (e.g., flashbacks, intrusive memories). While assuming dissociation as the central mechanism rather than a marginal feature of PTSD, these authors propose that complex PTSD involves a more complex structural dissociation than simplex PTSD [52].

From a clinical point of view, dissociation covers a diagnostic spectrum from the mildest to the most severe, covering acute dissociative reactions to stressful events, acute stress disorder, Simplex PTSD, dissociative subtype of PTSD, complex PTSD, and chronic dissociative disorders such as DID, respectively [2]. Besides

**Table 16.2** Symptoms of complex post-traumatic stress disorder (PTSD) and dissociative subtype of PTSD (Şar, 2011)

| Features | Complex PTSD | Dissociative subtype of PTSD |
|---|---|---|
| *General characteristics* | | |
| Core symptoms | "Simplex" PTSD | "Simplex" PTSD |
| Traumatic stress | Cumulative trauma | Developmental trauma |
| Psychiatric comorbidity | Almost always | Less common |
| *Specific characteristics* | | |
| Additional symptoms | Affect dysregulation, somatization, dissociation (disturbances of sense of self), impulsive, and/or self-destructive behavior | Depersonalization and/or derealization |
| Dissociation | Not required | Required |
| "Borderline" features | Common | Less common |
| *Potential correlates* | | |
| Sense of agency | Identity confusion and/or alteration | Possession |
| Sensorimotor dissociation | Somatoform dissociation | Functional neurological symptoms |

constituting a disorder on its own, dissociation may accompany several other psychiatric disorders which are known to have etiologies other than psychological trauma, e.g., schizophrenic disorder [53, 54]. When it does, concurrent dissociation is usually linked to a history of chronic developmental trauma, independent from the accompanying psychiatric disorder [55].

Conversely, such "true" comorbidity should be differentiated from "syndromal façades" of dissociative disorders on the clinical surface [56, 57], namely the clinical surface may imitate or dissimulate such comorbidity through a "phenocopy" created by the symptoms which are common to both dissociative and other psychiatric disorders, e.g., hallucinations, passive influence experiences, "borderline" features, and mood disturbances. "True" comorbidity requires specific treatment which cannot be addressed by solely the treatment of dissociation. Syndromal "façades," on the other hand, can only be treated by addressing the dissociative disorder. One of these conditions is the "dissociative depression" [2] which is usually resistant to "treatment-as-usual" of depressive disorders [57]. The latter is an intermediate condition between true comorbidity and façade (Table 16.3).

The recently proposed "*trimodal model*" of trauma and dissociation [58, 59] provides an integrative solution between the "structural theory" and the conceptualizing of the dissociative subtype of PTSD as a condition characterized by overmodulation of emotions only. Actually, the trimodal model assumes that each of the three modes (i.e., acute reaction, chronic trauma illness, and alienation) of response to trauma operates between *two poles*. Each mode is characterized by an interaction between trauma-related *intrusions* and processes of *controlling* the psychological pain initiated by the former, in order to keep the tension inside a "window of tolerance" [60]. Dissociative amnesia may operate within all modes ([58, 59]; see also [61]). In addition to amnesia, denial, avoidance, and alienation are phenomena which dampen the pain of mental intrusions. Each of these experiences of omission represent overmodulation of emotions in context of one of the three modes: denial is covered by the first mode only, avoidance by the second mode, and alienation is considered as the main feature of the third mode. This understanding of coping assumes that amnesia, denial, avoidance, and alienation (the "four horsemen") are different mechanisms despite their resemblance in the surface. The main difference between them seems to be in "realization" (in fact, "non-realization") of the experience by modified "ownership" (personalization) and deficient presentification (detemporalization) (Table 16.4).

**Table 16.3** Clinical features of "dissociative depression" (Şar 2011; Şar et al. 2013)

| Depressive symptoms | Overall condition | Identity fragmentation |
|---|---|---|
| Suicidal ideas | Suicide attempts | Associated symptoms of DID (voices hearing, etc.) |
| Appetite and weight changes | More severe depression, more psychiatric comorbidity | Borderline personality disorder criteria |
| Thoughts of guilt and worthlessness | Younger age | Experiences of possession and extrasensory perception |
| Diminished concentration and indecisiveness | Polytraumatization in childhood | Schneiderian symptoms |

**Table 16.4** The "trimodal response" to trauma [59]

| Components of psychopathology | Mode 1 "Acute Inflammation" | Mode 2 "The Trauma Illness" | Mode 3 "Alienation" |
|---|---|---|---|
| *Undermodulation* | | | |
| Cognition (knowing and not knowing) | Current and/or lifetime re-experiencing | – | – |
| Hyperarousal | Lifetime hyperarousal | Current hyperarousal | – |
| Impaired sense of agency (intruding entities) | – | Passive influence, possession, "borderline" phenomena | Dissociative identities (switching) |
| *Overmodulation* | | | |
| Cognition (knowing and not knowing) | Denial, dissociative amnesia | Lifetime and current avoidance, dissociative amnesia | "Not me" experience, dissociative amnesia |
| Hypoarousal | – | (Dissociative) depression | Depersonalization, derealization |
| Impaired sense of agency (avoidant entities) | – | Absorption trance | Dissociative identities (switching) |

The tertiary mode (alienation) may represent both state and trait dissociation. In a recent study [62], state dissociation during psychotherapy sessions predicted improvement after dialectical behavior therapy (DBT) for PTSD: patients with a low dissociative state during treatment had a higher chance to show substantial improvement. This relation consistently emerged across subgroups of PTSD patients with and without borderline personality disorder (BPD). Trait dissociation was not a significant predictor in either direction. Probably, state dissociation more readily interferes with interpersonal communication due to its impact on cognitive abilities. DBT, on the other hand, may have addressed trait dissociation rather successfully. However, the latter requires to be shown empirically.

## 16.7 Alienation as the Basis of Dissociation

Alienation (estrangement to oneself and/or the environment) is the core of dissociation. This is represented in the clinical phenomenon of depersonalization which connotes an impairment of personalization [63]; i.e., the experience that all psychological faculties (perception, memory, imagination, thought, feeling, etc.) belong to oneself. This is the basis of the sense of self and agency. Loss of time perception (detemporalization) may also accompany such depersonalization. Janet considered both depersonalization and detemporalization as the core impairment which undermines the experience of realization [64], i.e., the true understanding of the meaning of an experience which is necessary for integration.

While alienation in the form of depersonalization is common to all dissociative disorders, switching to or co-existence of alternate personality states is more typical of cases with identity fragmentation. On the other hand, a clinical condition predominantly characterized by or limited to depersonalization and derealization may turn to one of the identity fragmentations over time. This usually occurs due to better expression of previously suppressed structural dissociation.

In an empirical study [65], alienation was the only cognitive appraisal variable to differentiate DID from PTSD. While the groups had similar appraisals of shame, betrayal, self-blame, anger, and fear, the DID participants had higher appraisal of themselves as experiencing alienation. Indeed, these patients experience depersonalization and derealization which may go back to their childhood [10]. To put it simply, they feel alone, disconnected, and different. They often feel very isolated/lonely experiencing themselves as the only one in the universe who is "different" from others. Even they have difficulties in understanding themselves.

This condition affects the individual twofold. First of all, abuse and neglect may activate feelings of alienation, isolation, and aloneness. On the other hand, relational support is necessary to constructively process a specific abuse. If this is not available, the child cannot make sense of this experience through narrative that the related affective states are contained. This hinders the integration of the abuse with other autobiographical experiences. In fact, such integrative "metabolization" of the traumatic experience has an interpersonal aspect as psychosocial validation to achieve meaning. Consequently, the representations of abuse/neglect experiences remain mentally isolated. With further incidences and isolation, the child's ability to develop an ordinary sense of self-in-relation-to-others, based on a coherent narrative which includes the abuse experiences, is impeded and dissociative identities may begin to form [25]. This is the "downward spiral" of trauma.

Not only the perception of the subject about the trauma experience but also the perception of the subject about oneself is affected by this process. The traumatized subject evaluates oneself from the perspective of multiple versions of reality [66]. The experience may differ after each repetition therefore the affected person develops *isolated subjectivities* [67]. With the contribution of dissociative amnesias, such perceptual fragmentations lead to depersonalization, derealization, and identity alterations [10–12] because multiple perceptions of reality may destroy personalization, i.e., one's experience that all psychological faculties (perception, body perception, memory retrieval, imagination, thought, feeling, etc.) belong to oneself [63]. Depersonalization is the core element of clinical categories which are considered to be trauma-related conditions, e.g., dissociative, borderline personality, conversion, and certain types of depressive disorders [10, 68–70].

Thus, dissociation is a non-interactive solution [71] and, as such, may lead to biased processing of social inclusion [72]. The development of an internal "ghetto" [73] of "alternate identities" leads to the emergence and dominance of relationships in the individual's internal world. Interpersonal and internal phobias (e.g., phobias of other dissociative identities) then interfere with change, integration, and growth [74].

## 16.8 Dissociation and Altered Consciousness

Dissociative individuals may also suffer from alterations of consciousness, which are common among patients with trauma histories [75]. Specifically, in their 4-D model of consciousness, Frewen and Lanius [76] differentiated trauma-related altered state of consciousness (TRASC) from normal waking consciousness (NWC). These two poles of experiencing can be observed in consciousness of time-memory (flashbacks versus intrusive recall and distressful reminders), thought (voice-hearing versus negative self-referential thinking), body (disembodied versus embodied experiences of distress), and emotions (numbing and affective shutdown versus non-dissociative forms of negative emotionality). In fact, all these four dimensions reflect cognitive-emotional and bodily aspects of depersonalization; i.e., alienation, self-detachment, or estrangement [10].

Inspired by Pierre Janet, the structural theory of dissociation underlines a "*division of personality*" ("doublement") or "fragmentation" as the main characteristic of dissociation, while absorption experiences are considered as non-pathological [77, 78]. A recent study by Schimmenti and Sar [61] proposed that absorption trance was itself also a dissociative phenomenon with strong relationship to hypnotic phenomena [28]. In fact, absorption trance is based on the narrowing of consciousness which does not differ from a "division" in the final analysis.

Hence, the definition of "*normative dissociation*" [79, 80] should be based on lack of "clinical" symptoms rather than on preponderance of certain allegedly "benign" or "non-pathological" dissociative experiences such as absorption. Moreover, structural theory of dissociation may be applicable to "normative dissociative phenomena" which support the maintenance of the "apparently normal" deficient adjustment to daily life. Indeed, such adjustment may remain unperceived by the subject and his social circle, unless the distinct "action systems" operate in a disorganized or desynchronized manner.

## 16.9 Traumatic Memory as an Internal Driver

The vast proliferation (inflation) of options for mental operations in the aftermath of a traumatic experience facilitates cognitive alienation as well. Such options are usually based on representations of inadequate operations in other past problematical experiences. They are transferred to inactive memory during these repetitions, either partially or totally. From the perspective of time dimension that remain in the past. As perception is embedded in time, while processing the trauma experience, the subject concentrates on the past experience while being in the present [64]. This de-doublement of time leads to *detemporalization* [81] because the subject's contact with the present time weakens.

The repetitions of the representations of these operations in the active memory are attempts to solve the trauma. However, solution scenarios for recurrent traumatic experiences and repeated cognitions detach from each other rather than achieving a convergence. They become autonomous and reveal separate domains.

Sar and Ozturk [81] propose that the excluded operations may lay the foundation for the immediate or future development of *distinct mental states* [41] or *parallel-distinct mental structures* [35] of dissociative individuals. Mental operations, which are excluded from current processing and are formed to distinct personality states, are then activated as solution in further domains of life problems [81].

The hallmark of trauma resolution is the ability and opportunity of the subject to respond to a traumatic experience adequately. Escape, partial denial, or processing the situation until it is resolved is possible. Inadequate processing of the traumatic experience causes the fact that past trauma is then repeatedly handled in the context of present time in the person's active memory [81]. That is why, in their "preliminary" publication on traumatized dissociative patients, Breuer and Freud [82] stated as follows: "*the hysteric suffers mainly from reminiscences*." The need to match new information with inner models based on older information, and the revision of both until they agree, is called a *completion tendency* [83]. The completion principle summarizes the human mind's intrinsic ability to continue to process new information in order to update the inner schemata of the self and the world.

By definition, any change in the perception of traumatic experience leads to the emergence of a new internal and external world, i.e., a change in reality and its perception [81]. The organism inquiries ways of adaptation to the changes in the real world in the aftermath of the traumatic experience. Trauma is, however and per definition, a threatening experience which turns an adaptive process to a *maladaptive* one [81]. This is the condition when upsetting and unpredicted situational and/or enduring factors interrupt the psycho-sociological experiencing significantly, and interfere, for a certain amount of time, with the coping capacity of the person.

Namely, although being disruptive on the perceived continuity of both internal and external "realities," a traumatic experience is per definition an "event" that contains a message about the future [84] on the basis of a new horizon of possibilities generated by the event. Carl Gustav Jung stated once that the individual is programmed for *uniqueness* in order to perceive oneself as a living entity [85]. Paradoxically, living organisms are also evolved to a capacity of coping by responding to stressors with adaptation [86], i.e., change. Thus, while survival requires a personal update in the aftermath of major changes, it is expected that one keeps his or her unique identity more or less existing [35]. This dilemma remains temporarily unresolved in dissociative disorders, i.e., adoption of a new and adapted identity seems to be postponed until a definitive processing of the disruption becomes available [59].

Pain is a signal of the threat to the homeostasis in the context of the supreme mission of survival. Psychological trauma creates *mental pain* which is related to memories, sensations, emotions, and thoughts about the stressful experience. With their painful quality, traumatic memories seem to be the main driver of the "trauma response." One natural reaction of the organism to pain is avoidance. Hence, the individual is concerned with ways of keeping pain within bearable limits, while preserving its signaling function [59].

The *trimodal response model* [59] is concerned with a proposed model of response to complex psychological trauma and dissociation. Rather than consecutive phases, the response of the individual to developmental trauma is described in three modes which can co-occur: Acute reaction, chronic process, alienation. Each mode operates in a window of overmodulation and undermodulation of emotions. The trimodal model resembles medical conceptualizations of injury, response, and illness as they occur to the body. Psychotherapeutic intervention to trauma-related conditions has to consider the possible copresence of the three modes. The model tries to cover the mental striving of the traumatized individual to deal with unbearable pain, while fighting for overall survival. Nevertheless, survival has not only physical but also psychological aspects which converge on the maintenance of one's unique self-identity.

## 16.10 Dissociative Amnesia, Reenactment, and Identity

The concept of identity has a rich history in psychology and social sciences; however, its implications for psychiatry have been relatively little considered. For example, in DSM-5, only two disorders are based on an identity disturbance: DID and borderline personality disorder (BPD). Interestingly, both diagnostic categories are related to childhood adversities [87]. Moreover, a large descriptive overlap between two disorders has been repeatedly shown [68].

While the illumination of the true nature of this descriptive overlap requires further studies, the problem has conceptual aspects as well, namely the definition of the boundary between a "personality disorder" and a "dissociative disorder," or even "any psychiatric disorder" needs to be revisited [88]. According to the DSM-5, diagnosis of personality disorders should not be made if the condition can be better explained by an other disorder. This rule becomes imperative for the "dissociation-borderline" realm, i.e., for patients who demonstrate the phenomenologies of both disorders. However, there are several options to explain this overlap, which is too common to be a coincidence. Both conditions may be two faces of the same coin, i.e., a trauma-based psychopathology or one condition may lead to the other. In any case, the relationship to "personality" remains obscure if there is any.

Among many other aspects, identity confusion and alteration may be represented by the so-called Schneiderian symptoms among non-psychotic individuals. Rather than being delusions, they constitute passive influence experiences in dissociative disorders, i.e., disturbances of sense of self and agency. Notwithstanding the large descriptive overlap between two conditions, Schneiderian experiences represent mainly mental intrusions from "within," while features of BPD seem to constitute representations of a dissociative "inner space" in the external (i.e., interpersonal) world. Thus, subjects with BPD are known to "divide" external world by "splitting and projective identification" (in fact, they are mechanisms of strong dissociative quality), while dissociative individuals perceive themselves as "divided." Little is known about the role of absorption trance in these dynamics [61]. A recent study conducted on a Chinese college population demonstrated that dissociative, BPD, and Schneiderian symptoms were different, but highly inter-related dimensions of psychopathology [89]. Nevertheless, both conditions may still represent two faces of the same coin.

A series of studies conducted in a college population [10–12, 68] propose that both dissociative amnesia and diminished awareness (e.g., denial of the experience or idealization of the perpetrator) about childhood trauma may affect the way patients express their unresolved mental processing on the borderline-dissociation spectrum. In these studies, discrepancies between self-report and clinical assessment led to important insights about relationships between childhood trauma, disturbances of memory, and core dimensions of dissociative psychopathology. Amnesia to symptoms and/or childhood trauma or perceptual alterations seemed to explain these discrepancies. Self-report measures were more sensitive than clinician assessment except for patients with dissociative disorders who had "amnesia to amnesia" in their self-report assessment. The latter phenomenon describes the lack of awareness of dissociative individuals about their amnesias. Unlike dissociative disorders and alongside clinical assessment, BPD was associated with self-reported amnesia as well [11]. Thus, the main difference between the two diagnostic categories was in "awareness about amnesia" (Table 16.5).

In self-report, both disorders were associated with "cognitive-emotional self-detachment" but only BPD was associated with "detachment from reality" [10]; i.e., beside awareness of amnesia, and being a dimension of derealization, detachment from reality seemed to discriminate BPD from dissociative disorders. It was correlated with total childhood trauma as well. However, such correlation with childhood trauma was not observed for clinician-assessed derealization. Similarly to detachment from reality, self-reported identity alteration was correlated with all childhood trauma types; however, in clinical assessment, it was correlated with childhood sexual abuse only [12]. Thus, self-report instruments seemed to be more sensitive than clinician-administered assessment in demonstrating the relationship between childhood trauma, derealization, and identity alteration. This may be due to the basically subjective quality of experiences of derealization and identity alteration. Additionally, the presence and intervention of an interviewer may blockade the "flow" of expression upon questions due to factors such as shame and basic mistrust. Childhood sexual abuse seemed to be the most harmful type of childhood trauma in terms of its relationship to identity disturbance on BPD-dissociation spectrum [90].

**Table 16.5** Differences between borderline personality disorder and dissociative disorder in self-report and standardized clinical interview

| Components of assessment | Specific to borderline personality disorder | Common to both disorders | Specific to dissociative disorder |
|---|---|---|---|
| Detachment | Detachment from reality in self-report (correlated with total childhood trauma score) | Cognitive-emotional self-detachment in self-report | |
| Identity alteration | | Present in self-report | Present in clinical assessment (correlated with childhood sexual abuse report) |
| Amnesia | Present in self-report | Present in clinical assessment | Not present in self-report (amnesia to amnesia) |

## 16.11 Disturbed Interpersonal Attachment, Betrayal, and Denial

Another study demonstrated that BPD criteria as seen among patients with DID were culture sensitive. Specifically, in a comparison of Turkish and Dutch patients with DID, large differences existed between the two groups in meeting BPD criteria [16, 24]. Indeed, Dutch patients reported frequent mood swings, physically self-damaging acts, identity confusion, and impulsive and unpredictable behavior more frequently than Turkish patients. These phenomena pointed to the preponderance of affect dysregulation and disturbances of sense of self and agency, and to a possible role of abusive experiences. In turn, Turkish patients reported intense anger and lack of control of this emotion, chronic feelings of emptiness and boredom, efforts to avoid abandonment, and intense but unstable relationships more frequently than Dutch patients. These differences pointed to the predominance of attachment disturbance among Turkish patients and possible role of childhood neglect (Table 16.6).

This is why (with its sensitivity to rejection) BPD appears to be an attachment disorder. As much as culture, differences in traumatic antecedents may have also played a role in explaining the different prevalence of these distinct patterns in the two cultures. In another study, and possibly as an indicator of the importance of relational issues in the local culture, Turkish adolescent outpatients with dissociative disorders differed from non-dissociative psychiatric outpatients in respect to the increased prevalence of concurrent separation anxiety disorder [8].

Some data and theories suggest that disorganized attachment style may facilitate the development of dissociative disorders [91–97]. Bowlby [39] proposed that inadequate care-seeking interactions with primary caregivers could lead the infant to develop multiple internal representations of the self and attachment figures (which he called Internal Working Models; IWM). Contradictory IWMs develop to represent the caregiver as dangerous and safe at the same time. Early onset abuse and/or neglect by a relational figure is associated with disorganized attachment [94]. Main and Hesse [98] identified disorganized attachment developing from a relational context where the child, who is seeking for safety and comfort, is frightened by the caregiver. The child may also frighten the insecure caregiver which may impede connection.

**Table 16.6** Borderline personality disorder features that dominate the clinical condition of Turkish and Dutch patients with dissociative identity disorder: a comparison (adapted from [24])

| | Turkish | Dutch |
|---|---|---|
| Affect | Intense anger and lack of control of anger | Frequent mood swings |
| Sense of self | Chronic feelings of emptiness and boredom | Identity confusion |
| Emotion regulation | Efforts to avoid abandonment | Physically self-damaging acts |
| Behavioral regulation | Intense but unstable relationships | Impulsive or unpredictable behavior |

"Betrayal trauma" [99] is the trauma which is perpetrated by someone the victim relies on, e.g., by a primary caregiver. Betrayal trauma theory suggests that dissociative amnesia is an adaptive response to childhood abuse that allows for survival by enabling the child to maintain attachment to an abusive figure who is also vital to his or her development. A recent study by Kaehler and Freyd [100] found that higher betrayal traumas are associated with greater "borderline" characteristics which are common in DID as secondary features which do not necessarily point to an underlying personality disorder [10–12, 68]. Betrayal trauma is common in family systems which are characterized by secrets and denial [101].

Not only the pain related to the traumatic experience itself, but also amnesia for experience or its denial or minimization are also part of the clinical problem. Such diminished awareness, as depicted by the Janetian concept of "lack of realization," may lead to either "return of the dissociated," or to diminished sense of self and agency as represented by identity confusion and alteration (e.g., Schneiderian passive influence phenomena), depersonalization-derealization, and absorption trance.

Clinical studies have demonstrated that at least a subgroup of patients tend to minimize the traumatic quality of their childhood [102]. Patients with conversion and dissociative disorders [68, 69], mothers of children with masturbatory behavior [103], and individuals who report fear of happiness [104] are among them. This observation may be related to a relatively blank response (e.g., dissociative amnesia and absorption) due to the "betrayal" [99] in the ongoing attachment [105].

In a neuroimaging study conducted on adolescents with PTSD due to childhood sexual abuse [58], earlier age and a more severe type of childhood sexual abuse (i.e., involving coitus) was associated to a larger left anterior cingulate, while the opposite was observed for sexual abuse by a perpetrator in a closer relationship with the victim. Those adolescents who were sexually abused by their biological father or brother reported more dissociative amnesia and absorption, compared to that of the victims of other perpetrators. However, there was no correlation between specific or total childhood trauma scores and brain volume. An earlier study could not determine the relative importance of specific types of events in neurobiological variables [106]. Hence, these neurobiological and psychological phenomena may serve for stress alleviation by facilitating the "attachment to the perpetrator" who was also a "caretaker" [107]. However, such alterations in perception of trauma lead to ponder on principles of reality regulation in stressful conditions.

## 16.12 Earliest Developmental Traumatization

Regulation of reality perception requires consideration of the mutuality between the internal and the external world [35]. Childhood abuse, neglect, and insecure attachment disrupt this balance such that internal reality becomes more compelling. From a developmental point of view, in order to establish a balance between the external and the internal world, the caregiver's adequate *mirroring* is necessary. That means, the caregiver's responses should accurately match the infant's mental state (Table 16.7).

**Table 16.7** Types of developmental traumatization

| Years of age | Intrusion | Omission | Psychopathology |
|---|---|---|---|
| *Cognitive-emotional trauma* | | | |
| 0–2 | Unmarked mirroring | Lack of mirroring | Isolation, insecure attachment, cognitive-emotional depersonalization |
| 2–10 | Emotional abuse | Emotional neglect | Disorganized attachment |
| *Bodily trauma* | | | |
| 0–2 | Embodied intrusion | Embodied detachment | Bodily depersonalization |
| 2–10 | Sexual and physical abuse | Physical neglect | Somatic dissociation |

The equation of internal and external world which typifies toddlers' and preschoolers' way of thinking is called *psychic equivalence* [108]. This mode does not allow consideration of alternative perspectives on reality. Hence, a fantasy may be experienced as potentially real. This is why the acquisition of a sense of pretend in relation to mental states is essential. In *pretend mode,* thoughts and feelings can be envisioned and talked about, but they do not correspond to real. Otherwise, the subject is bound with a black and white type of perception of reality as observed in teleological mode. This restricts the symbolic-associative thinking and even undermines sense of humor in certain conditions, namely the *teleological mode* (i.e., the opposite of the pretend mode) is based on imputing intention from what is physically apparent.

Experiencing internal reality both in psychic equivalent and pretend modes is typical for dissociation as seen in all types of dialectical mental operations, e.g., the "dissociation paradox" [109], namely compared to other non-psychotic psychiatric disorders, individuals with a dissociative disorder have elevated self-certainty which is a reason of delusional thinking in psychotic disorders if combined with diminished self-reflection. As the latter is not disturbed in dissociative disorders, the increased self-certainty does not lead to loss of cognitive insight which is empirically defined as the difference between self-reflection and self-certainty.

*Mentalization* is a construct which provides hints about the developmental and interpersonal origins of perception of reality [38]. It is defined as the ability to understand the mental state of oneself or others, which underlies overt behavior. There is a relationship between development of mentalization capacity and experiences of mirroring with the caregiver. For example, if the caregiver is not able to express an affect while indicating she is not expressing her own feelings (*unmarked* mirroring), the child would perceive the response of the caregiver as the mirroring of his or her (i.e., the child's) affect. This would mean that the caregiver's expression may seem to externalize the infant's experience and may overwhelm the infant. Such a breach of the window of tolerance would make the response of the caregiver contagious and would lead to escalation of the affect rather than to regulate the child's state [110]. Moreover, a predisposition of experiencing emotions through other people might be established by this early interpersonal template [108]. This is the first step leading to emotional dysregulation which further affects perception of reality.

In a recent study [111], child mentalization partially mediated the relationship between childhood sexual abuse and depressive symptoms. The effects of childhood sexual abuse on externalizing symptoms and sexualized behavior difficulties were sequentially mediated through mentalization and dissociation. Not rarely, families with dysfunctionalities, for instance, affect dysregulation among family members, may also be developmentally traumatizing for the offsprings [112]. Based on the assumption that very early traumatization has a critical role in dissociative disorders, a prospective study documented that childhood neglect is a significant predictor of dissociation in early adulthood [113].

## 16.13 Covert Trauma and "Apparently Normal" Families

Recent research underlines the importance of the context (e.g., family) where specific types of abuse and neglect occur [112]. Moreover, dissociative disorders may be associated with traumatization that is covert, such as enduring dysfunctional communication and relationship styles in family members, including subtle forms of emotional neglect. Krüger and Fletcher [23] demonstrated that self-reported emotional neglect by biological parents or siblings in childhood was the strongest individual predictor of an adult diagnosis of a dissociative disorder in psychiatric patients (out of all other combinations of abuse type and abuser-abused relational bonds).

Dysfunctions in the family may partly originate from parents' own traumatic antecedents which lead to inter-generational transmission of developmental stress as reported in the context of the "apparently normal (dissociative) family" by Öztürk and Şar [112]. In their empirical study, family members of patients with DID and related dissociative disorders reported frequent mood swings, intense anger and inability to control anger, transient dissociative experiences or paranoid ideas, and identity confusion more frequently than controls. Some of these features were correlated with certain types of childhood trauma in this group. For example, frequent mood swings were associated with all types of childhood trauma except sexual abuse (probable role of dissociative amnesia to the latter?) and identity confusion was correlated with emotional abuse.

A study conducted on a large group of college students [68] demonstrated that, not only emotional neglect but also *minimization (denial) of childhood trauma* predicted a dissociative disorder diagnosis. Systemic denial of multigenerational childhood trauma and betrayal may be important characteristics of "apparently normal" families. One type of hidden childhood trauma is overcontrol which may be disguised by the "overprotection" type of parenting [114]. While some of the children and adolescents are capable to escape from such oppressive practices, the seemingly positive (i.e., "excess" of love, the intention of protection against perceived "threats") motivation behind this attitude may make the realization of the "danger" by the offspring rather difficult. Additionally, such "intrusive" practices usually lead to "overcontrol" which is sometimes combined with omissions (i.e., neglect) in other contexts which is perceived as inconsistency (i.e., betrayal) by the offspring. Undermining the development of a healthy psychological autonomy and growth, a symbiotic coupling between generations may settle over time which may extend to the adulthood of the offspring. Such developments may be built on attachment disturbances.

In such families with subclinical dissociative characteristics, individuals can interchange their social roles over time, alternating between being a victim, the abuser, and the rescuer [107]. Depending on their own traumatic past, or on their current interaction between each other and with their children, the parents may maintain trust and present themselves in a positive role ("angelic," affectionate/compassionate parent), but they can turn to an abusive parenting style (angry, agressive, insistent) at any time. The changing attitudes of their parents and the marital discord will often cause contradictory feelings within the children. Family members often feel trapped, first being unable to leave in the midst of a crisis as it is not safe. Then, they do not leave the family when the crisis is over and the need to escape has vanished, as the atmosphere becomes less threatening and more settled. Third, in an environment of neglect, chaos may be an opportunity for making contact with others in the unit [112]. Upon direct traumatization early in life, the ever-changing roles in an enduring family system continue to push children and adolescents toward a dissociative adaptation style in a period sensitive to the establishment of a stable identity.

## 16.14 Dissociation in Childhood and Adolescence

Adolescents are known to be dissociation prone as children are. Moreover, they are directly faced with the developmental task of solving their normative identity crisis through integration ([42, 115–120]). Ironically, studies on dissociative disorders in children and adolescents are in their infancy; i.e., they lag behind those on adult survivors of childhood adversities [41]. In fact, one of the rare diagnostic screening studies shows that the highest prevalence of dissociative disorders is seen among adolescents [8]. Another Turkish study on adolescents showed that each type of trauma and dissociation contributed to suicide attempts and self-mutilation [121]. Dissociation was the most powerful predictor. A recent meta-analysis documented that childhood maltreatment—except for emotional neglect—predicted self-mutilative behavior [122].

Children and adolescents are disadvantaged compared to adults in term of conceiving and reporting their dissociative experiences in an understandable way as they usually lack the required armamentarium of communication. Nowadays, increasing accessibility to internet resources may started to open the ways of correct understanding of their suffering for adolescents. Anecdotal experiences on children and adolescents reveal that certain types of alternate personality states are common which are relatively difficult to detect due to their resemblance to normative behavior of child and adolescent. They may also imitate diagnostic categories which keep dissociative psychopathology hidden behind a clinical surface, e.g., major depressive disorder, disruptive dysphorical mood disorder, attention deficit hyperactivity disorder, reactive attachment disorder, or oppositional defiant disorder [8, 123, 124].

Among such types of alternate personality states, those characterized by an emotion, an experience of possession, personality states with identical age and/or name, or imaginary companion are common among adolescents. Switching experiences of children and adolescents may also remain unrecognized by general label of regressive and angry behavior. Children with alternate personalities which are restricted to certain emotions (e.g., anger) may get the diagnosis of disruptive dysregulated mood disorder or bipolar mood disorder in adolescents which may lead to inaccurate and ineffective pharmacotherapeutic strategies. Upon newly prescribed antidepressant and/or antipsychotic medication, immediate switching to a cheerful personality state may resemble a hypomanic response to pharmacological intervention which may have occurred in the aftermath of a "dissociative depression" [2]. Delineation of a new category of "dissociative mood disorder" may be considered for future research among children and adolescents in particular.

As seen in Japanese hikikomori (i.e., apparently a chronic dissociative disorder usually with an onset in adolescence), placement of the dissociative psychopathology in culture-bound domain by rupturing its association with general trauma-related psychopathology may hinder the access to proper treatment [125]. This condition characterized by social withdrawal of the offspring of "apparently normal" middle and middle-upper class families is characterized by a history of emotional neglect. These cases seem to have "dual personality" type of DID covering an "unemotional" and "apparently normal" host personality state occasionally switching to an "emotional" and "angry" personality state with persecution of the possibly "hard-working" and "neglectful" parents. Childhood sexual abuse histories typical for many other cultures are lacking in these cases which underline the equal importance of other developmental adversities in dissociative disorders.

## 16.15 Society and Culture as Origin or Context of Trauma

Most clinicians agree on the importance of contextual factors which operate in the environment of the subject exposed to traumatic experiences. Such factors either accentuate the impact of the experience and/or constitute a social network which fails in providing the support the subject needs to overcome the threat. Thus, the controversy between the so-called socio-cognitive and trauma theories of dissociation is useless because these perspectives are two faces of the same coin. In fact, trauma emerges in the context of socio-cognitive factors [126]. One culture-sensitive aspect of dissociation is the dissociative somatic phenomena and indeed their prevalence varies in various parts of the world [36, 127].

Although childhood abuse and neglect require the presence and actions of "perpetrators," they can occur only in a suitable environment. This environment is characterized by denial, boundary violations, reality distortions, paranoia, narcissism, and dramatic posturing which usually serve the purpose of maintaining the family structure. These features and dynamics may derive from psychological, relational, and economic needs of one or both of the parents, as well as oppressive traditions which do not allow a dissolution of marriage, and other contextual issues in the family [128].

Cultural processes influence the development and phenomenology of dissociative disorders [129, 130]. The role of culture may be divided in two components: as the origin of trauma and as a modifier of the disorder expression. To explain the interface between society and the individual, Sar and Ozturk proposed the concept of "*sociological self*" [79]. Krüger, Sokudela, Motlana, Mataboge, and Dikobe [131] underlined the inevitability of dissociation to live in an oppressive and constantly traumatizing community. Gold [132] depicts dissociogenic aspects of contemporary societies, and Sar and Öztürk [80] describe how they are misused as a tool of oppression. These theories and models provide the underpinnings to the concept of "normative dissociation."

Last but not least, DSM-5 recognizes conditions related to organizational abuse such as consequences of coercive persuation, cult attendance, terror organizations as identity disturbances. They are listed among other specific dissociative disorders. Empirical research is very limited on this neglected area of dissociation, while the sociocultural burden of such aberrations is devastating in many societies. Not only participants of cult-type organizations but also individuals showing sudden violence ("individual cult") while living in seemingly good psychosocial adjustment or those who "deliberately" join such malignant organizations ("Stockholm Syndrome") are suspect of such dissociative psychopathology, although they may not show overt clinical presentations [133, 134]. According to the theory of "functional dissociation of the self," this is an appearance of a malignantly hypertrophied sociological self detached from one's psychological self which is underdeveloped upon such detachment beginning from early years of life on.

## 16.16 Neurobiology of Dissociation: Orbitofrontal Hypothesis

A structural MRI study established that DID patients have smaller hippocampi and amygdalae compared to normal controls [135]. Ehling et al. [136] also found reduced volumes in the parahippocampal gyrus of individuals with DID and strong correlations between reduction of parahippocampal volume (as compared to healthy subjects) and both cognitive-emotional and sensorimotor dissociation. In two SPECT studies, DID patients in "host" identities exhibited orbitofrontal hypoperfusion in comparison to normal controls [137, 138]. There were no significant differences between perfusions obtained when the patient was in control of different alter personality states [137]. Bilaterally increased perfusion in prefrontal regions and occipital areas was also observed in one of these studies [138]. In the other one, increased perfusion in the left (dominant hemisphere) lateral temporal region was shown compared to healthy controls [137]. This lateralization was not replicated in a follow-up study [138].

Notwithstanding the possible effect of psychiatric comorbidity as a confounding factor (hence, it cannot be considered as specific to DID), the findings concerning orbitofrontal hypoperfusion do not seem to be at odds with the theoretical understanding of developmental neurobiology. Longitudinal neuroimaging studies suggest that the orbitofrontal cortex is one of the last regions in the brain to fully develop in humans [139]. For example, a tensor-based morphometry investigation indicated that orbitofrontal cortex volumes were smaller in children who have suffered early aberrant parental care in the form of physical abuse, and that these volumetric alterations were associated with difficulties children experience in various aspects of their social lives [140]. The orbitofrontal cortex is a key component of a circuit that facilitates adaptation to changing environmental contingencies and plays an important role in the control of emotion and motivational states. In this regard, Schore [141] reported that there is a relationship between the development of the orbitofrontal cortex, emotion regulation, and attachment.

In accordance with these observations and based on a neurodevelopmental approach, Forrest [142] proposed an "*orbitofrontal model*" for DID which integrates and elaborates on theory and research from four domains: the neurobiology of the orbitofrontal cortex and its protective inhibitory role in the temporal organization of the behavior, the development of emotion regulation, the development of the self, and experience-dependent maturation of the orbitofrontal cortex. This model hypothesizes that the orbitofrontal cortex plays a critical role in the development of distinct mental states (i.e., dissociative identities) due to its inhibitory functions.

## 16.17 Lateralization and Connectivity

Possibly in concert with right hippocampus and anterior cingulate in "remembering" traumatic memories, *right* amygdala seems to be the main driver of the post-traumatic process, i.e., re-experiencing, avoidance, and hyperarousal (Table 16.3). This pattern clearly represents the basic phenomenology of simplex PTSD. Representing operations of control, denial of trauma was associated with thinner *right* prefrontal cortex and larger *right* thalamus which may dampen the perception of psychological pain [58]. The size of the *right* amygdala was also correlated with depression, passive influence experiences, absorption trance, and negative affect intrusions [58, 143] which seemed to represent complex PTSD and "dissociative depression" [70]. This pattern suggested the dominant role of *right* hemisphere in post-traumatic process, i.e., the presence of a sort of *lateralization*.

Shore [144] stated that "the right brain is fundamentally involved in an avoidant-defensive mechanism for coping with emotional stress, including the passive survival strategy of dissociation." Mutluer et al. [58] pointed to the bilateral but asymmetrical impact of PTSD on the brain with a predominant role of the right hemisphere in primary and secondary modes of post-traumatic reaction, i.e., acute and chronic response of simplex of complex PTSD type. Left prefrontal cortex was involved with symptoms representing dissociative subtype of PTSD or DID. Thus, unlike proposed by Shore, *core symptoms of dissociation were proposed to be related to the left brain hemisphere,* and, in particular, to the *left prefrontal cortex.* In contrast to the general volume decrease in other brain structures, thickness of the *left prefrontal cortex* was correlated with dissociative phenomena, suggesting a possible neuro-protective phenomenon.

*Denial* (thinner right prefrontal cortex), *avoidance* (smaller right amygdala), and *alienation* (thicker left prefrontal cortex) seem to have different neurobiological associations [58]. Representing the distinctness of the components, volumetric abnormalities in these regions were not correlated. Interestingly, both right and left *prefrontal cortex* were involved with *altered awareness* of traumatic experiences but not with symptoms of PTSD (see also [145]). Subcortical structures seemed to be more involved with re-evaluation of reality as triggers; denial seemed to represent the worst scenario and was related to a thinner right prefrontal cortex.

Some of the studies on neurobiological effects of childhood adversities suggested diminished connectivity between the two hemispheres or different areas of the brain. For example, decreased right/left cortical integration has been proposed as associated with childhood sexual abuse and/or physical abuse [146]. Corpus callosum is the major neural pathway that connects homologous cortical areas of the two cerebral hemispheres both in an excitatory and inhibitory role [147]. The total corpus callosum area of the abused/neglected patients was smaller than in controls and psychiatric patients who had not been abused or neglected [148]. Sexual abuse was the strongest factor associated with reduced corpus callosum size in girls. In a diffusion tensor imaging (DTI) study, adolescents with childhood sexual abuse-related PTSD showed decreased fractional anisotropy (i.e., white matter integrity) in the corpus callosum [149]. Abnormalities in the integrity of the corpus callosum were related to anger. Another DTI study documented significantly decreased fractional anisotropy in right anterior corona radiata of dissociative patients [150]. An association between bad paternal relationships and lower fractional anisotropy in the genu of the corpus callosum was shown in female patients who were maltreated by their fathers.

These findings on the neurobiological consequences of childhood trauma may have implications for dissociative disorders. For example, Farina, Speranza, Dittoni, Gnoni, Trentini, Vergano et al. [151] demonstrated that, compared to controls, dissociative individuals did not show an increase in EEG connectivity after administration of an interview triggering memories of early attachment; i.e., the brain's overall response lacked the integrative reaction shown in healthy controls. Accordingly, a recent study [152] also demonstrated decreased EEG connectivity in dissociative absorption, which was considered as a type of trait dissociation.

Considering both findings on lateralization and connectivity, Mutluer et al. [58] study led to the speculation that diminished connectivity may be part of the "protective" response among traumatized adolescents to "quarantine" the left hemisphere while the right hemisphere was operating in "frontline" [59, 145], at least through adolescence. Such lateralization seems also to point to the central role of the "right brain" in processing the interrupted trauma resolution with particular emphasis on memory and emotions. Although thicker left prefrontal cortex is not an absolute neurobiological marker of mental health, the obvious relationship between psychopathology and the downsizing of all evaluated brain regions in PTSD supports this proposal. On the other hand, a particular role of the left brain in dissociation may also point to a difference between PTSD and dissociative disorders in this respect, namely lateralization seems to be a phenomenon related to PTSD while dissociative disorders may be accompanied by bilateral response of the brain. Further research would shed light to potential accuracy of these speculations.

## 16.18 Integration as Healing: Conclusive Remarks

Considering the perspective provided by the elaborations in this paper, a definition of dissociation may be as follows: traumatic experiences and consequently altered self-perceptions contribute to the impairment of the mutuality between internal world and external reality [35, 59]. This is accompanied by a renewed perception of the self in the context of a different reality, accompanied by altered vigilance, awareness, control (agency), and concentration. Depersonalization is the core clinical element of this condition.

Understanding the etiology of dissociative disorders requires integration of trauma-exposure, coping, cognitive, neurobiological, systemic, and developmental factors. These include not only traumatic experiences but also family dynamics, child development, and attachment [99, 119, 153]. Dissociative disorder develops when a child is exposed to chaos, coercion, and overt severe physical and/or sexual abuse, or alternatively, to "apparently normal" dissociative families often with subtle neglect, disorganized attachment to caregivers, emotion dysregulation, and misattuned communication styles [130]. Overwhelmed by intense conflicting needs and emotions, the child is unable to integrate discrete behavioral and emotional states into a coherent or relatively integrated self according to the appropriate sociocultural construction of self [48, 119]. While the role of the child's biological capacity to dissociate to an extreme level is yet unclear, there is evidence demonstrating the neurobiological impact of developmental stress. The latter converges around an impairment of connectivity in the central nervous system in affected individuals.

The possibility of successful treatment ("restitutio ad integrum") of dissociative disorders by means of psychotherapy [154] even at a later time in life and the probable positive natural course of dissociative disorders in a subgroup of adolescents [8] support the possible role of dissociation in mental survival [59, 79, 155]. Traumatic memories do not only seem to be drivers in the psychopathogenesis of post-traumatic conditions, but their processing also seem to serve to re-establish the sense of self and agency [37, 156–158]. Even disturbances of attachment cannot be reparated solely by a good therapist–patient relationship, unless some trauma processing occurs in this context [48].

Researchers agree on the significant role of connection between amygdala and the frontal lobes in PTSD [159, 160]. Individuals with childhood trauma have inhibitory failure and frontal lobe dysfunction in regions related to Nogo-P3 in EEG [161]. Hence, interactions between frontal lobe and amygdala seem to be crucial in the establishment of mental integration. However, this connection seems to be more complex than a simple balance between and excitatory and inhibitory functions [162]. For example, with its role as a "hub" embedded in numerous structures of the limbic system alongside its contribution to the integration of emotion, perception, and cognition (including memories of past autobiographical events), amygdala does not only play a role in intrusive phenomena, but it also forges the establishment and maintenance of an integrated self [163].

Both studies on connectivity and lateralization inspire treatment methods such as Eye Movement Desensitization and Reprocessing (EMDR), which covers bilateral stimulation of the brain [164] and neurobiologically informed mindfulness therapies addressing inter-hemispheric balance [60]. However, those and other methods should always be embedded in the larger context of trauma psychotherapy, e.g., phase-oriented based on stabilization, trauma work, and integration [40]. Maintenance of a good balance between overmodulation and undermodulation of emotions throughout treatment [51], and consideration of alterations of consciousness to manage perceptual alterations [76] is essential to fit the requirements of an orchestration as proposed by the mode-oriented approach [59]. Last but not least, the illumination of the role of autonomous nervous system dysfunction in dissociation may open ways to better psychotherapy and even pharmacotherapeutic management of these conditions.

These considerations would not be complete without a renewed definition of integration as ultimate goal of psychotherapy in dissociative disorders. Thus, integration takes place by *letting the individual perceive oneself as oneself in the face of each of the diverse psychological realities, while developing sociopsychological connections between each of these psychological realities and kernels of the self* [35, 66].

## References


1. American Psychiatric Association. Diagnostic and statistical manual of mental disorders (DSM-5). 5th ed. Washington, DC: American Psychiatric Association; 2013.
2. Şar V. Epidemiology of dissociative disorders: an overview. Epidemiol Res Int. 2011a;2011:404538. https://doi.org/10.1155/2011/404538.
3. Johnson JG, Cohen P, Kasen S, Brook JS. Dissociative disorders among adults in the community, impaired functioning, and axis I and II comorbidity. J Psychiatr Res. 2006;40(2):131–40.
4. Şar V, Akyüz G, Dogan O. Prevalence of dissociative disorders among women in the general population. Psychiatry Res. 2007;149(1–3):169–76.
5. Şar V, Tutkun H, Alyanak B, Bakım B, Baral I. Frequency of dissociative disorders among psychiatric outpatients in Turkey. Compr Psychiatry. 2000;41:216–22.
6. Tutkun H, Şar V, Yargıç Lİ, Özpulat T, Yanık M, Kızıltan E. Frequency of dissociative disorders among psychiatric inpatients in a Turkish university clinic. Am J Psychiatr. 1998;155:800–5.
7. Akyüz G, Doğan O, Şar V, Yargıç LI, Tutkun H. Frequency of dissociative identity disorder in the general population in Turkey. Compr Psychiatry. 1999;40:151–9.
8. Şar V, Önder C, Kılınçaslan A, Zoroğlu SS, Alyanak B. Dissociative identity disorder among adolescents: prevalence in a university psychiatric outpatient unit. J Trauma Dissociation. 2014;15(4):402–19.

9. Friedl MC, Draijer N, de Jonge P. Prevalence of dissociative disorders in psychiatric inpatients: the impact of study characteristics. Acta Psychiatr Scand. 2000;102:423–8.
10. Şar V, Alioğlu F, Akyüz G. Depersonalization and derealization in self-report and clinical interview: the spectrum of borderline personality disorder, dissociative disorders, and healthy controls. J Trauma Dissociation. 2017;18(4):490–506.
11. Şar V, Alioğlu F, Akyüz G, Karabulut S. Dissociative amnesia in dissociative disorders and borderline personality disorder: self-rating assessment in a college population. J Trauma Dissociation. 2014;15(4):477–93.
12. Şar V, Alioğlu F, Akyüz G, Tayakısı E, Öğülmüş EF, Sönmez D. Awareness of identity alteration and diagnostic preference between borderline personality disorder and dissociative disorders. J Trauma Dissociation. 2017;18(5):693–709.
13. Şar V, Öztürk E. Psychotic symptoms in dissociative disorders. In: Moskowitz A, Schaefer I, Dorahy M, editors. Psychosis, trauma and dissociation: emerging perspectives on severe psychopathology. 2nd ed. New York: Wiley Press; 2019. p. 195–206.
14. Şar V, Koyuncu A, Öztürk E, Yargic LI, Kundakci T, Yazici A, et al. Dissociative disorders in psychiatric emergency ward. Gen Hosp Psychiatry. 2007;29(1):45–50.
15. Dorahy MJ, Brand BL, Şar V, Kruger C, Stavropoulos P, Martinez-Taboas A, et al. Dissociative identity disorder: an empirical overview. Aust N Z J Psychiatry. 2014;48:402–17.
16. Boon S, Draijer N. Multiple personality disorder in The Netherlands: a clinical investigation of 71 patients. Am J Psychiatr. 1993;150:489–94.
17. Dorahy MJ, Middleton W, Seager L, Williams M, Chambers R. Child abuse and neglect in complex dissociative disorder, abuse-related chronic PTSD and mixed psychiatric samples. J Trauma Dissociation. 2016;17(2):223–36.
18. Martinez-Taboas A. Multiple personality in Puerto Rico: analysis of fifteen cases. Dissociation. 1991;4:189–92.
19. Middleton W, Butler J. Dissociative identity disorder: an Australian series. Aust. N. Z. J. Psychiatry. 1998;32(6):786–804.
20. Ross CA, Norton GR, Wozney K. Multiple personality disorder: an analysis of 236 cases. Can J Psychiatr. 1989;34(5):413–8.
21. Coons PM. Confirmation of childhood abuse in child and adolescent cases of multiple personality disorder and dissociative disorder not otherwise specified. J Nerv Ment Dis. 1994;182(8):461–4.
22. Lewis DO, Yeager CA, Swica Y, Pincus JH, Lewis M. Objective documentation of child abuse and dissociation in 12 murderers with dissociative identity disorder. Am J Psychiatr. 1997;154(12):1703–10.
23. Krüger C, Fletcher L. Predicting a dissociative disorder from type of childhood maltreatment and abuser-abused relational tie. J Trauma Dissociation. 2017;18(3):356–72.
24. Şar V, Yargic LI, Tutkun H. Structured interview data on 35 cases of dissociative identity disorder in Turkey. Am J Psychiatr. 1996;153:1329–33.
25. Şar V, Dorahy MJ, Krüger C. Revisiting the etiological aspects of dissociative identity disorder: a biopsychosocial perspective. Psychol Res Behav Manage. 2017;10:137–46.
26. Wolf EJ, Rasmusson AM, Mitchell KS, Logue MW, Baldwin CT, Miller MW. A genome-wide association study of clinical symptoms of dissociation in a trauma-exposed sample. Depress Anxiety. 2014;31:352–60.
27. Lochner C, Seedat S, Hemmings SMJ, Kinnear CJ, Corfield VA, Niehaus DJ, et al. Dissociative experiences in obsessive-compulsive disorder and trichotillomania: clinical and genetic findings. Compr Psychiatry. 2004;45(5):384–91.
28. Dell PF. Is high hypnotizability a necessary diathesis for pathological dissociation? J Trauma Dissociation. 2017;18(1):58–87.
29. Frischholz E, Lipman L, Braun B, Sachs R. Psychopathology, hypnotizability, and dissociation. Am J Psychiatr. 1992;149:1521–5.
30. Spiegel D, Hunt T, Dondershine HE. Dissociation and hypnotizability in posttraumatic stress disorder. Am J Psychiatr. 1988;145:301–5.

31. Stutman RK, Bliss EL. Post-traumatic stress disorder, hypnotizability and imagery. Am J Psychiatr. 1985;142:741–2.
32. Şar V. Developmental trauma, complex PTSD and the current proposal of DSM-5. Eur J Psychotraumatol. 2011b;2:5662. https://doi.org/10.3402/ejpt.v2i0.5622.
33. Courtois CA. Complex trauma, complex reactions: assessment and treatment. Psychother Theory Res Pract Train. 2004;41(4):412–25.
34. Tench CR, Tanasescu R, Jethwa KD, Constantinescu CS. Coordinate based meta-analysis of whole-brain voxel-based morphometry studies does not show evidence of grey matter loss specific to PTSD. Biorxiv Preprint. 2018; https://doi.org/10.1101/265496.
35. Şar V. Parallel-distinct structures of internal world and external reality: disavowing and reclaiming the self-identity in the aftermath of trauma- generated dissociation. Front Psychol. 2017;8:216. https://doi.org/10.3389/fpsyg.2017.00216.
36. Şar V. Identity revised: a clinician's perspective on how an identity-based model of mind would look like. In: Sinnott J, editor. Identity flexibility during adulthood: perspectives in adult development. New York: Springer; 2018. p. 265–88.
37. Öztürk E, Şar V. Formation and functions of alter personalities in dissociative identity disorder: a theoretical and clinical elaboration. J Psychol Clin Psychiatry. 2016a;6(6):00385. https://doi.org/10.15406/jpcpy.2016.06.00385.
38. Fonagy P, Allison E. What is mentalization ? The concept and its foundations in developmental research. In: Midgley N, Vrouva I, editors. Minding the child: mentalization-based interventions with children, young people, and their families. New York: Routledge Press; 2012. p. 11–34.
39. Bowlby J. Attachment and loss. New York: Basic Books; 1969/1982.
40. Van der Hart O, Nijenhuis ERS, Steele K. The haunted self: structural dissociation and the treatment of chronic traumatization. New York: Norton; 2006.
41. Putnam FW. Dissociation in children and adolescents: a developmental perspective. New York: Guilford Press; 1997.
42. Putnam FW. The way we are. How states of mind influence our identities, personality and potential for change. Los Gatos, CA: International Psychoanalytic Books; 2016.
43. Nijenhuis ERS, Spinhoven P, Van Dyck R, Van der Hart O, Vanderlinden J. Degree of somatoform and psychological dissociation in dissociative disorder is correlated with reported trauma. J Trauma Stress. 1998;11:711–30.
44. Şar V, Akyüz G, Doğan O, Öztürk E. The prevalence of conversion symptoms in women from a general Turkish population. Psychosomatics. 2009;50(1):50–8.
45. Kılıç Ö, Şar V, Taycan O, Aksoy-Poyraz C, Erol TC, Tecer Ö, et al. Dissociative depression among women with fibromyalgia or rheumatoid arthritis. J Trauma Dissociation. 2014;15(3):285–302.
46. Taycan O, Şar V, Çelik C, Erdoğan-Taycan S. Trauma-related psychiatric comorbidity of somatization disorder among women in eastern Turkey. Compr Psychiatry. 2014;55(8):1827–46.
47. Porges SW. The polyvagal theory. Neuropsychological foundations of emotions, attachment, communication, & self-regulation. New York: Norton; 2011.
48. Schimmenti A, Caretti V. Linking the overwhelming with the unbearable: developmental trauma, dissociation, and the disconnected self. Psychoanal Psychol. 2016;33(1):106–28.
49. Briere J, Scott C, Weathers F. Peritraumatic and persistent dissociation in the presumed etiology of PTSD. Am J Psychiatr. 2005;162:2295–301.
50. Nijenhuis ERS. Ten reasons for conceiving and classifying posttraumatic stress disorder as a dissociative disorder. Eur J Trauma Dissociation. 2017;1:47–61.
51. Lanius RA, Vermetten E, Loewenstein RJ, Brand B, Schmahl C, Bremner JD, et al. Emotion modulation in PTSD: clinical and neurobiological evidence for a dissociative subtype. Am J Psychiatr. 2010;167(6):640–7.
52. Van der Hart O, Nijenhuis ERS, Steele K. Dissociation: an insufficiently recognized major feature of complex posttraumatic stress disorder. J Trauma Stress. 2005;18(5):413–23.

53. Lyssenko L, Schmahl C, Bockhacker L, Vonderlin R, Bohus M, Kleindienst N. Dissociation in psychiatric disorders: a meta-analysis of studies using the Dissociative Experiences Scale. Am J Psychiatr. 2018;175(1):37–46.
54. Şar V, Taycan O, Bolat N, Özmen M, Duran A, Öztürk E, et al. Childhood trauma and dissociation in schizophrenia. Psychopathology. 2010;43:33–40.
55. Şar V, Ross CA. Dissociation as a confounding factor in psychiatric research. Psychiatr Clin N Am. 2006;29:129–44.
56. Şar V. The many faces of dissociation: opportunities for innovative research in psychiatry. Clin Psychopharmacol Neurosc. 2014;12(3):171–9.
57. Şar V. The psychiatric comorbidity of dissociative identity disorder: an integrated look. In: van der Merwe AP, Sinason V, editors. Shattered but unbroken: voices of triumph and testimony. London: Karnac Press; 2016. p. 181–210.
58. Mutluer T, Şar V, Kose-Demiray C, Arslan H, Tamer S, Ünal S, et al. Lateralization of neurobiological response in adolescents with PTSD related to severe childhood sexual abuse: the tri-modal reaction (T-MR) model of protection. J Trauma Dissociation. 2018;19(1):108–25.
59. Şar V. The tri-modal reaction (T-MR) model of complex trauma and dissociation: a proposal. Quaderni of Psicoterapia Cognitiva. 2019
60. Siegel D. The developing mind. New York: Guilford; 1999.
61. Schimmenti A, Şar V. A correlation network analysis of dissociative experiences. J Trauma Dissociation. 2019;20(4):402–19.
62. Kleindienst N, Priebe K, Görg N, Dyer A, Steil R, Lyssenko L, et al. State dissociation moderates response to dialectical behavior therapy for posttraumatic stress disorder in women with and without borderline personality disorder. Eur J Psychotraumatol. 2016;7:30375. https://doi.org/10.3402/ejpt.v7.30375.
63. Jaspers K. Allgemeine psychopathologie (general psychopathology). Berlin: Springer Verlag; 1913.
64. Van der Kolk BA, Van der Hart O. The intrusive past: the flexibility of memory and the engraving of trauma. Am Imago. 1991;48(4):425–252.
65. DePrince AP, Huntjens RJC, Dorahy MJ. Alienation appraisals distinguish adults diagnosed with DID from PTSD. Psychol Trauma Theory Res Pract Policy. 2015;7:578–82.
66. Bromberg PM. Standing in the spaces. Essays on clinical process trauma & dissociation. Hilsdale, NJ: The Analytic Press; 1998.
67. Chefetz AR, Bromberg PM. Talking with 'me' and 'not-me.' A dialogue. Comtemporary Psychoanal. 2004;40:409–64.
68. Şar V, Akyüz G, Kuğu N, Ozturk E, Ertem-Vehid H. Axis-I dissociative disorder comorbidity of borderline personality disorder and childhood trauma reports. J Clin Psychiatry. 2006;67(10):1583–90.
69. Şar V, Akyüz G, Kundakcı T, Kızıltan E, Dogan O. Childhood trauma, dissociation and psychiatric comorbidity in patients with conversion disorder. Am J Psychiatr. 2004;161:2271–6.
70. Şar V, Akyüz G, Öztürk E, Alioğlu F. Dissociative depression among women in the community. J Trauma Dissociation. 2013;14(4):423–38.
71. Crandell J, Morrison R, Willis K. Using psychomotor to treat dissociative identity disorder. J Trauma Dissociation. 2002;3:57–80.
72. Weinbrecht A, Niedeggen M, Roepke S, Renneberg B. Feeling excluded no matter what? Bias in the processing of social participation in borderline personality disorder. Neuroimage: Clinical. 2018;19:343–50.
73. Frankel AS, O'Hearn TC. Similarities in responses to extreme and unremitting stress: cultures of communities under siege. Psychother Theory Res Pract Train. 1996;33:485–502.
74. Steele K, van der Hart O, Nijenhuis ERS. Dependency in the treatment of complex posttraumatic stress disorder and dissociative disorders. J Trauma Dissociation. 2001;2(4):79–116.
75. Ross CA, Browning E. Altered states of consciousness among inpatients in a trauma program. J Trauma Dissociation. 2018;19(5):596–606.
76. Frewen PA, Lanius R. Trauma-related altered states of consciousness (TRASC): exploring the 4-D model. J Trauma Dissociation. 2014;15(4):436–56.

77. Nijenhuis ERS, Van der Hart O. Dissociation in trauma: a new definition and comparison with previous formulations. J Trauma Dissociation. 2011;12(4):416–45.
78. Van der Hart O, Nijenhuis ERS, Steele K, Brown D. Trauma-related dissociation: conceptual clarity lost and found. Aust N Z J Psychiatry. 2004;38:906–14.
79. Şar V, Öztürk E. Functional dissociation of the self: a sociocognitive approach to trauma and dissociation. J Trauma Dissociation. 2007;8(4):69–89.
80. Şar V, Öztürk E. Stimulus deprivation and overstimulation as dissociogenic agents in post-modern oppressive societies. J Trauma Dissociation. 2013;14(2):198–212.
81. Şar V, Öztürk E. What is trauma and dissociation? J Trauma Pract. 2005;4(1–2):7–20.
82. Breuer J, Freud S. 1895. On the psychical mechanism of hysterical phenomena: preliminary communication. In: Strachey J, editor. The standard edition of the complete psychological works of Sigmund Freud, volume II (1893-1895): Studies on hysteria. New York: Wintage Publs; 1893/1999. p. 1–17.
83. Horowitz MJ. Stress response syndromes. 2nd ed. Northwale NJ: Jason Aronson Inc.; 1986.
84. Badiou A. Being and event. New York, NY: Continuum; 2005.
85. Singer JK. Boundaries of the soul: the practice of Jung's psychology. New York: Anchor Books; 1994.
86. Dhabbar FS. The short-term stress response-mother nature's mechanism for enhancing protection and performance under conditions of threat, challenge, and opportunity. Front Neuroendocrinol. 2018;49:175–92. https://doi.org/10.1016/j.yfrne.2018.03.004.
87. Brand B, Şar V, Korzekwa M, Kruger C, Martinez-Taboas A, Stavropoulos P, et al. Separating fact from fiction: an empirical examination of six myths about disociative identity disorder. Harv Rev Psychiatry. 2016;24(4):257–70.
88. MacIntosh HB, Godbout N, Dubash N. Borderline personality disorder: disorder of trauma or personality, a review of the empirical literature. Can J Psychol. 2015;56(2):227–41.
89. Fung HW, Ling HW-H, Ross CA, Tse JW-L, Liu KW. Dissociative, Schneiderian and borderline personality symptoms in a non- clinical sample in Hong Kong: a preliminary report. Eur J Trauma Dissociation. 2019;
90. de Aquino Ferreira LF, Pereira FHQ, Benevides AMLN, Melo MCA. Borderline personality disorder and sexual abuse: a systematic review. Psychiatry Res. 2018;262:70–7.
91. Barach PM. Multiple personality disorder as an attachment disorder. Dissociation. 1991;4(3):117–23.
92. Blizard RA. Disorganized attachment, development of dissociated self states, and a relational approach to treatment. J Trauma Dissociation. 2003;4(3):27–50.
93. Byun S, Brumariu LE, Lyons-Ruth K. Disorganized attachment in young adulthood as a partial mediator of relations between severity of childhood abuse and dissociation. J Trauma Dissociation. 2016;17(4):460–79.
94. Liotti G. Trauma, dissociation, and disorganized attachment: three strands of a single braid. Psychother Theory Res Pract Train. 2004;41(4):472–86.
95. Liotti G. A model of dissociation based on attachment theory and research. J Trauma Dissociation. 2006;7(4):55–73.
96. Lyons-Ruth K, Dutra L, Schuder MR, Bianchi I. From infant attachment disorganization to adult dissociation: relational adaptations or traumatic experiences? Psychiatr Clin N Am. 2006;29(1):63–86.
97. Sachs A. Through the lens of attachment relationship: stable DID, active DID and other trauma-based mental disorders. J Trauma Dissociation. 2017;18(3):319–39.
98. Main M, Hesse E. Parents' unresolved traumatic experiences are related to infant disorganized attachment status: is frightened or frightening parental behavior the linking mechanism? In: Greenberg M, Cicchetti C, E.M. Cummings EM., editors. Attachment in the preschool years. Chicago, IL: University of Chicago Press; 1990. p. 161–82.
99. Freyd JJ. Betrayal trauma: traumatic amnesia as an adaptive response to childhood abuse. Ethics Behav. 1994;4(4):307–29.
100. Kaehler LA, Freyd JJ. Borderline personality disorder: a betrayal trauma approach. Psychol Trauma Theory Res Pract Policy. 2009;1(4):261–8.

101. Goldsmith RE, Freyd JJ, DePrince AP. Betrayal trauma: associations with psychological and physical symptoms in young adults. J Interpers Violence. 2012;27(3):547–67.
102. Mac Donald K, Thomas M, Sciolla A, Schneider B, Pappas K, Bleijenberg G, et al. Minimization of childhood maltreatment is common and consequential: results from a large, multinational sample using the Childhood Trauma Questionnaire. PLoS One. 2016;11(1):e0146058. https://doi.org/10.1371/journal.pone.0146058.
103. Mutluer T, Necef I, Eray Ş, Kaçar AŞ, Şar V. Mothers of children with masturbatory behaviour: does the child mirror mother's unresolved childhood trauma? (submitted). n.d.
104. Şar V, Türk T, Öztürk E. Fear of happiness among college students: the role of childhood trauma and dissociation. Indian J Psychiatry. 2019;61(4):389.
105. Freyd JJ, Deprince AP, Zurbriggen EL. Self-reported memory for abuse depends upon victim-perpetrator relationship. J Trauma Dissociation. 2001;2(3):5–15.
106. Cohen RA, Grieve S, Hoth KF, Paul RH, Sweet L, Tate D, et al. Early life stress and morphometry of the adult anterior cingulate cortex and caudate nuclei. Biol Psychiatry. 2006;59(10):975–82.
107. Ross CA. Dissociative identity disorder: diagnosis, clinical features, and treatment of multiple personality. 2nd ed. New York: Wiley; 1997.
108. Fonagy P, Gergely G, Jurist E, Target M. Affect regulation, mentalization, and the development of the self. New York: Other Press; 2002.
109. Şar V, Öztürk E, Islam S, Küçükgöncü S, Yumbul Ç, Ertem-Vehid H. Zwischen Selbstreflexion und Selbst-Überszeugtsein: kognitive Einsicht bei dissoziativen und schizophrenen Störungen und das Dissoziationsparadox. In: Özkan I, Sachsse IU, Streeck-Fischer A, editors. Zeit heilt nicht alle Wunden: Kompendium zur Psychotraumatologie. Göttingen: Vandenhoeck Ruprecht; 2012. p. 161–72.
110. Gergely G, Watson J. Early social development: contingency perception and the social biofeedback model. In: Rochat P, editor. Early social cognition: understanding others in the first months of life. Hillsdale, NJ: Erlbaum; 1999. p. 101–37.
111. Ensink K, Bégin M, Normandin L, Godbout N, Fonagy P. Mentalization and dissociation in the context of trauma: implications for child psychopathology. J Trauma Dissociation. 2017;18(1):11–30.
112. Öztürk E, Şar V. "Apparently normal" family: a contemporary agent of transgenerational trauma and dissociation. J Trauma Pract. 2005;4(3–4):287–303.
113. Ogawa JR, Sroufe LA, Weinfield NS, Carlson EA, Egeland B. Development and the fragmented self: longitudinal study of dissociative symptomatology in a nonclinical sample. Dev Psychopathol. 1997;9:855–79.
114. Hernandez A, Gallardo-Pujol D, Pereda N, Arntz A, Bernstein DP, Gaviria AM, et al. Initial validation of the Spanish Childhood Trauma Questionnaire-Short Form: factor structure, reliability and association with parenting. J Interpers Violence. 2013;28(7):1498–518.
115. Alexander PC, Schaeffer CM. A typology of incestuous families based on cluster analysis. J Fam Psychol. 1994;8:458–70.
116. Carlson EB, Dalenberg C, Armstrong J, Daniels JW, Loewenstein R, Roth D. Multivariate prediction of posttraumatic symptoms in psychiatric inpatients. J Trauma Stress. 2001;14:549–67.
117. Dalenberg CJ, Brand BL, Gleaves DH, Dorahy MJ, Loewenstein RJ. Evaluation of the evidence for the trauma and fantasy models of dissociation. Psychol Bull. 2012;138:550–88.
118. Howell EF, Blizard RA. Chronic relational trauma disorder: a new diagnostic scheme for borderline personality and the spectrum of dissociative disorders. In: Dell PF, O'Neil JA, editors. Dissociation and the dissociative disorders: DSM-V and beyond. New York: Routledge; 2009. p. 495–510.
119. Putnam FW. Dissociative disorders. In: Cicchetti D, Cohen DJ, editors. Developmental psychopathology, vol. 2. New York: Wiley; 2006. p. 657–95.
120. Allen JG, Fultz J, Huntoon J, Brethour JR Jr. Pathological dissociative taxon membership, absorption, and reported childhood trauma in women with trauma-related disorders. J Trauma Dissociation. 2002;3:89–110.

121. Zoroğlu SS, Tüzün Ü, Şar V, Tutkun H, Savaş HA, Öztürk M, et al. Suicide attempt and self-mutilation among Turkish high-school students in relation with abuse, neglect and dissociation. Psychiatry Clin Neurosci. 2003;57(1):119–26.
122. Liu RT, Scopelliti KM, Pittman SK, Zamora AS. Childhood maltreatment and non-suicidal self-injury: a systematic review and meta-analysis. Lancet. 2018;5:51–64.
123. Bozkurt H, Düzman-Mutluer T, Kose C, Zoroglu S. High psychiatric comorbidity in adolescents with dissociative disorders. Psychiatry Clin Neurosci. 2014;69(6):369–74.
124. Fujisawa TX, Shimada K, Takiguchi S, Mizushima S, Kosaka H, Teicher MH, et al. Type and timing of childhood maltreatment and reduced visual cortex volume in children and adolescents with reactive attachment disorder. Neuroimage: Clin. 2018;20:216–21.
125. Hattori Y. Social withdrawal in Japanese youth: a case study of thirty-five hikikomori clients. J Trauma Pract. 2006;4(3–4):181–201.
126. Şar V, Krüger C, Martinez-Taboas A, Middleton W, Dorahy MJ. Sociocognitive and post-traumatic models are not opposed. J Nerv Ment Dis. 2013;201(5):439–40. https://doi.org/10.3402/ejpt.v6.28213.
127. Martinez-Taboas A, Lewis-Fernandez R, Şar V. Cultural aspects of psychogenic non-epileptic seizures. In: Schachter SC, La France C, editors. Gates & Rowan's non-epileptic seizures. 4th ed. New York: Cambridge University Press; 2018. pp. 137–149.
128. Gold SN. Not trauma alone: therapy for child abuse survivors in family and social context. Philadelphia, PA: Brunner & Routledge; 2000.
129. Dorahy MJ. Culture, cognition and dissociative identity disorder. In: Schumaker JF, Ward T, editors. Culture, cognition and psychopathology. Westport, CT: Praeger; 2001. p. 157–69.
130. Krüger C. Variations in identity alteration—a qualitative study of experiences of psychiatric patients with dissociative identity disorder. In: van der Merwe AP, Sinason V, editors. Shattered but unbroken: voices of triumph and testimony. London: Karnac Books; 2016. p. 133–61.
131. Krüger C, Sokudela BF, Motlana LM, Mataboge CK, Dikobe AM. Dissociation: a preliminary contextual model. S Afr J Psychiatry. 2007;13(1):13–21.
132. Gold SN. Fight club: a depiction of contemporary society as dissociogenic. J Trauma Dissociation. 2004;5(2):13–34.
133. Şar V. Dissociative depression is resistant to treatment-as-usual. J Psychol Clin Psychiatry. 2015a;3(2):00128. https://doi.org/10.15406/jpcpy.2015.03.00128.
134. Şar V. Establishing the common ground in European psychotraumatology. Eur J Psychotraumatol. 2015b;6:28213. https://doi.org/10.3402/ejpt.v6.28213.
135. Vermetten E, Schmahl C, Lindner S, Loewenstein R, Bremner JD. Hippocampal and amygdalar volumes in dissociative identity disorder. Am J Psychiatr. 2006;163(4):630–6.
136. Ehling T, Nijenhuis ERS, Krikke AP. Volume of discrete brain structures in complex dissociative disorders: preliminary findings. In: de Kloet ER, Oitzl MS, Vermetten E, editors. Prog Brain Res. 2007;167:307–10.
137. Şar V, Ünal SN, Kızıltan E, Kundakçı T, Öztürk E. HMPAO SPECT study of cerebral perfusion in dissociative identity disorder. J Trauma Dissociation. 2001;2(2):5–25.
138. Şar V, Ünal SN, Öztürk E. Frontal and occipital perfusion changes in dissociative identity disorder. Psychiatry Res Neuroimaging. 2007;156(3):217–23.
139. Shore AN. The experience-dependent maturation of a regulatory system in the orbital prefrontal cortex and the origin of developmental psychopathology. Development & Psychopathology, 1996;8:54–87.
140. Hanson JL, Chung MK, Avants BB, Shirtcliff EA, Gee JC, Davidson RJ, et al. Early stress is associated with alterations in the orbitofrontal cortex: a tensor-based morphometry investigation of brain structure and behavioral risk. J Neurosci. 2010;30(22):7466–72.
141. Schore AN. The experience-dependent maturation of a regulatory system in the orbital prefrontal cortex and the origin of developmental psychopathology. Dev Psychopathol. 1996;8:54–87.
142. Forrest KA. Toward an etiology of dissociative identity disorder: a neurodevelopmental approach. Conscious Cogn. 2001;10(3):259–93.

143. DePierro J, D'Andrea W, Frewen P, Todman M. Alterations in positive affect: relationship to symptoms, traumatic experiences, and affect ratings. Psychol Trauma Theory Res Pract Policy. 2018;10(5):585–95.
144. Shore AN. Relational trauma and the developing right brain an interface of psychoanalytic self psychology and neuroscience. Self and systems. Ann N Y Acad Sci. 2009;1159:189–203.
145. Depue BE, Curran T, Banich MT. Prefrontal regions orchestrate suppression of emotional memories via a two-phase process. Science. 2007;317(5835):215–9.
146. Teicher M, Ito Y, Glod C, Schiffer F, Gelbard H. Early abuse, limbic system dysfunction, and borderline personality disorder. In: Silk K, editor. Biological and neurobehavioral studies of borderline personality disorder. Washington, DC: American Psychiatric Association Press; 1994. p. 177–207.
147. Bloom JS, Hynd GW. The role of the corpus callosum in interhemispheric transfer of information: excitation or inhibition? Neuropsychol Rev. 2005;15(2):59–71.
148. Teicher MH, Dumont NL, Ito Y, Vaituzis C, Giedd JN, Andersen SL. Childhood neglect is associated with reduced corpus callosum area. Biol Psychiatry. 2004;56:80–5.
149. Rinne-Albers MAW, Van Der Werff SJA, Van Hoof M-J, Van Lang ND, Lamers-Winkelman F, Rombouts SA, et al. Abnormalities of white matter integrity in the corpus callosum of adolescents with PTSD after childhood sexual abuse: a DTI study. Eur Child Adolesc Psychiatry. 2016;25(8):869–78.
150. Basmacı-Kandemir S, Bayazıt H, Selek S, Kılıçaslan N, Kandemir H, Karababa IF, et al. Tracking down the footprints of bad paternal relationship in dissociative disorders: a DTI study. J Trauma Dissociation. 2016;17(3):371–81.
151. Farina B, Speranza AM, Dittoni S, Gnoni V, Trentini C, Vergano CM, et al. Memories of attachment hamper EEG cortical connectivity in dissociative patients. Eur Arch Psychiatry Clin Neurosci. 2014;264(5):449–58.
152. Soffer-Dudek N, Todder D, Shelef L, Deutsch I, Gordron S. A neural correlate for common trait dissociation: decreased EEG connectivity is related to dissociative absorption. J Pers. 2019;87(2):295–309.
153. Kluft RP. Multiple personality disorder. In: Spiegel D, editor. Dissociative disorders: a clinical review. Lutherville, MD: Sidran Press; 1993. p. 17–44.
154. Brand BL, Classen CC, McNary SW, Zaveri P. A review of dissociative disorders treatment studies. J Nerv Ment Dis. 2009;197:646–54.
155. Ross CA, Goode C, Schroeder E. Hippocampal volumes in a sample of trauma patients: a possible neuro-protective effect of dissociation. Open Psychiatry J. 2015;9(1):7–10.
156. Conway MA. Memory and the self. J Memory Lang. 2005;53:594–628.
157. Öztürk E, Şar V. The trauma-self and its resistances in psychotherapy. J Psychol Clin Psychiatry. 2016b;6:00386. https://doi.org/10.15406/jpcpy.2016.06.00386.
158. Prebble SC, Addis DR, Tippett LJ. Autobiographical memory and sense of self. Psychol Bull. 2013;139:815–40.
159. Gard AM, Waller R, Swartz JR, Shaw DS, Forbes EE, Hyde LW. Amygdala functional connectivity during socioemotional processing prospectively predicts increases in internalizing symptoms in a sample of low-income, urban, young men. NeuroImage. 2018;178:562–73.
160. Lobo I, de Oliveira L, David IA, Pereira MG, Volchan E, Rocha-Rego V, et al. The neurobiology of posttraumatic stress disorder: dysfunction in the prefrontal-amygdala circuit? Psychol Neurosci. 2011;4(2):191–203.
161. Kim S, Kim JS, Jin MJ, Im C-H, Lee S-H. Dysfunctional frontal lobe activity during inhibitory tasks in individuals with childhood trauma: an event-related potential study. Neuroimage: Clin. 2017;17:935–42.
162. Solms M, Panksepp J. The "Id" knows more than the "Ego" admits: neuropsychoanalytic and primal consciousness perspectives on the interface between affective and cognitive neuroscience. Brain Sci. 2012;2:147–75.
163. Markowitsch HJ, Staniloiu A. Amygdala in action: relaying biological and social significance to autobiographical memory. Neuropsychologia. 2011;49:718–33.

164. Laugharne J, Kullack C, Lee CW, McGuire T, Brockman S, Drummond PD, et al. Amygdala volumetric change following psychotherapy for posttraumatic stress disorder. J Neuropsychiatr Clin Neurosci. 2016;28(4):312–8.
165. Modestin J, Ebner G, Junghan M, Erni T. Dissociative experiences and psychiatric disorders in acute psychiatric inpatients. Compr Psychiatry. 1996;37:355–61.
166. Gast U, Rodewald F, Nickel V, Emrich HM. Prevalence of dissociative disorders among psychiatric inpatients in a German university clinic. J Nerv Ment Dis. 2001;189:249–57.
167. Friedl MC, Draijer N. Dissociative disorders in Dutch psychiatric inpatients. Am J Psychiatr. 2000;157:1012–3.
168. Ginzburg K, Somer E, Tamarkin G, Kramer L. Clandestine psychopathology: unrecognized dissociative disorders in inpatient psychiatry. J Nerv Ment Dis. 2010;198:378–81.
169. Saxe GN, van der Kolk BA, Berkowitz R, Chinman G, Hall K, Lieberg G, et al. Dissociative disorders in psychiatric inpatients. Am J Psychiatr. 1993;150:1037–42.
170. Ross CA, Anderson G, Fleisher WP, Norton GR. The frequency of multiple personality disorder among psychiatric inpatients. Am J Psychiatr. 1991;148:1717–20.
171. Lipsanen T, Korkeila J, Peltola P, Jarvinen J, Langen K, Lauerma H. Dissociative disorders among psychiatric patients. Comparison with a nonclinical sample. Eur Psychiatry. 2004;19(1):53–5.
172. Ross CA, Duffy C, Ellason JW. Prevalence, reliability and validity of dissociative disorders in an inpatient setting. J Trauma Dissociation. 2002;3:7–17.
173. Şar V, Kundakçi T, Kızıltan E, Yargıç IL, Tutkun H, Bakım B, Aydıner O, Özpulat T, Keser V, Özdemir Ö. Axis I dissociative disorder comorbidity of borderline personality disorder among psychiatric outpatients. J Trauma Dissociation. 2003;4(1):119–36.
174. Foote B, Smolin Y, Kaplan M, Legatt ME, Lipschitz D. Prevalence of dissociative disorders in psychiatric outpatients. Am J Psychiatr. 2006;163(4):623–9.
175. Ross CA. Epidemiology of multiple personality disorder and dissociation. Psychiatr Clin N Am. 1991;14:503–17.
176. Tamar-Gürol D, Sar V, Karadag F, Evren C, Karagoz M. Childhood emotional abuse, dissociation and suicidality among patients with drug dependency in Turkey. Psychiatry Clin Neurosci. 2008;62(5):540–7.
177. Ross CA, Kronson J, Koensgen S, Barkman K, Clark P, Rockman G. Dissociative disorder comorbidity in 100 chemically dependent patients. Hosp Community Psychiatry. 1992;43:840–2.
178. Dunn GE, Ryan JJ, Paolo AM, van Fleet JN. Comorbidity of dissociative disorders among patients with substance use disorders. Psychiatr Serv. 1995;46:153–6.
179. Evren C, Sar V, Karadag F, Tamar-Gurol D, Karagoz M. Dissociative disorders among alcohol-dependent inpatients. Psychiatry Res. 2007;152:233–41.
180. Yargic LI, Sevim M, Arabul G, Ozden SY. Childhood trauma histories and dissociative disorders among prostitutes in Turkey. Paper presented at the Annual Conference of the International Society for the Study of Dissociation. San Antonio, TX. 2000.
181. Ross CA, Farley M, Schwartz HL. Dissociation among women in prostitution. J Trauma Pract. 2004;2(3):199–212.
182. Ross CA, Anderson G, Heber S, Norton GR. Dissociation and abuseamong multiple personality patients, prostitutes, and exotic dancers. Hosp Community Psychiatry. 1990;41:328–30.

# Trauma in Children with Neurodevelopmental Disorders: Autism, Intellectual Disability, and Attention-Deficit/Hyperactivity Disorder

17

Daniel W. Hoover

## 17.1 Introduction

Children with neurodevelopmental disorders (NDD) are frequently exposed to violence and maltreatment with potentially far-reaching negative effects on their overall adjustment and mental health [1, 2]. The causes, symptoms, and effects of trauma and NDD are intimately intertwined, complicating diagnosis and treatment. Growing evidence suggests that the shape and course of NDD may be strongly impacted by traumatic experiences, especially when the trauma occurs early and often during key developmental periods [3, 4]. Trauma symptoms and NDD-related impairments, such as overarousal, inattention, sleep, and emotional and behavioral regulation difficulties, overlap to a considerable degree, making it difficult to discriminate whether behaviors are primarily a result of trauma, NDD, or both [5]. Further, NDD-related immature or externalizing childhood behavior itself may increase the risk of physical or sexual assault [6]. Population-based twin studies point to a family-based genetic relationship between the neurological immaturity of children with NDD and vulnerability to violence and maltreatment [6–8]. All of this leads to the oft-cited conclusion that children with various types of NDD are two to seven times more likely to experience maltreatment or other violence than their typically developing peers [2].

Defined as "a group of heterogeneous conditions characterized by a delay or disturbance in the acquisition of skills in a variety of developmental domains, including motor, language, and cognition (p. 690)," [9] neurodevelopmental disorders in the DSM-5 include autism spectrum disorder (ASD), intellectual developmental disorder (ID), and attention-deficit/hyperactivity disorder (ADHD) as well as communication, motor, and specific learning disorders. The NDDs are usually

D. W. Hoover (✉)
Center for Child and Family Traumatic Stress, Kennedy Krieger Institute, Johns Hopkins School of Medicine, Baltimore, MD, USA
e-mail: hooverd@kennedykrieger.org



G. Spalletta et al. (eds.), *Childhood Trauma in Mental Disorders*,
https://doi.org/10.1007/978-3-030-49414-8_17

first diagnosed in early childhood but may not be identified until later in life [10]. While NDD is generally thought to originate in brain-based functions affected by genetics, prenatal or perinatal insults, the unfolding of these disorders during development can be worsened or ameliorated by environmental and psychosocial events [9, 11].

Psychological trauma occurs at a high rate among children and youth in the United States. The majority have been exposed to traumatic situations with estimates as high as 68% lifetime prevalence before age 17 [12–14]. While most children are resilient to isolated exposures, repeated or severe trauma can have devastating effects on mental and emotional well-being and physical health well into adulthood [12, 15, 16]. The DSM-5 defines traumatic experience as "exposure to actual or threatened death, serious injury, or sexual violence (p. 271)" [10] and includes witnessing traumatic events and learning that they happened to caregivers or other close family members. Consistent with current definitions of trauma in childhood [17], it is broadly defined in this chapter to include the various forms of maltreatment, exposure to community and family violence, and severe verbal and physical victimization by peers.

This chapter aims to clarify what is currently known about the prevalence and unique effects of trauma in the population of children with autism spectrum disorder, intellectual disability, and ADHD. Clinical implications for evaluating and treating co-occurring childhood NDD and trauma-related disorders will be discussed.

## 17.2 Case Example: Trauma and Autism Spectrum Disorder

Maya is an 8-year-old Latina girl who has been diagnosed with autism as well as a moderate range intellectual disability. She has significant speech articulation impairments making verbal communication almost impossible for most people who try to engage her in conversation. Her mother is able to understand her and often serves as an "interpreter." Maya has witnessed several instances of violence in her community. In one episode, she was present when a close family member was robbed and shot. Maya has always displayed considerable anxiety and tended to avoid communicating with people outside of her family. Since the shooting, she has shown increased separation anxiety, seems always to be on the lookout for danger, has stopped speaking except to her mother, and has returned to sleeping in her mother's bed. She avoids all reminders of the traumatic events by refusing to go out or attend school.

This case illustrates the often thorny diagnostic and treatment dilemmas that arise when a child with ASD and intellectual disorder encounters potentially traumatic events. In the presence of known traumas (witnessing community violence, assault, and murder of a family member), a child with preexisting communication, language, and anxiety challenges may show regression in communication and self-regulation perhaps putting her at risk for further victimization. Behavioral manifestations of autism such as difficulty discriminating between safe and unsafe individuals might increase the risk of victimization and maltreatment. Trauma symptoms such as avoidance, overarousal, and negative emotion may lead to further regression in her language or other developmental skills [6, 18, 19]. The diagnostic understanding (i.e., an emphasis on her NDD and associated behavioral challenges, vs. a trauma-informed formulation) would influence the choice of treatment approach. A purely behavioral treatment regimen may focus on improving verbal skills, ability to sleep independently, and perhaps reducing separation anxiety. A trauma-focused treatment regimen would provide psycho-education about trauma, enhanced coping skills, and gradual exposure through narration of the trauma, to the degree possible, given the extent of the child's and parent's ability to engage in this type of treatment.

## 17.3 Prevalence of Trauma in Neurodevelopmental Disorders

The extant literature suggests that, broadly speaking, children with disabilities are more often traumatized than those without disabilities. The results of meta-analyses demonstrate that children with a range of broadly defined intellectual and physical disabilities are more likely to be victims of violence than nondisabled age-mates [20] and large-scale population studies reveal greater risk of maltreatment among children with various kinds of disability [2, 21]. However, this conclusion is based on few studies with heterogeneous definitions of violence and disability and limited evidence regarding causation [20]. Conclusions about the increased odds of traumatization in children with NDD and other disabilities are greatly influenced by the types of trauma studied, the ways in which disability is defined and identified, and the methodological quality of the study [22, 23].

Whether or not trauma exposure and NDD are causally related, and the direction of any such causation is difficult to ascertain. Evidence suggests that having an NDD such as autism spectrum disorder (ASD), attention-deficit/hyperactivity disorder (ADHD), and intellectual developmental disorder (ID) increases the odds that a child will be exposed to maltreatment such as physical and sexual abuse [6, 21, 24]. Other studies suggest that early trauma, such as physical neglect or exposure to violence during early development, may lead to the development of ADHD or ASD symptoms [3, 25]. A group of recent twin studies looking at the associations between early NDD diagnoses and maltreatment suggest that tendencies for both NDD and child abuse run in families as a set of genetic traits [7, 8]. For example, Dinkler and colleagues [7] assessed monozygotic and dizygotic twins in a cross-sectional sample from a population-based Swedish twin study of 8192 nine-year-old twins born in Sweden between 1997 and 2005. Of the twins studied, 26.7% were monozygotic; 374 of them had a history of maltreatment. The prevalence of neurodevelopmental disorders, including ADHD, ASD, learning disorders, and tic disorders, was assessed. While the findings revealed that maltreated children were up to seven times more likely to be diagnosed with one or more NDDs, cross-twin analysis suggested the effect was mainly due to family genetic factors, a finding that has been replicated in similar studies [8]. Thus, it appears that child maltreatment and NDD tend to run together in families across generations. As a result of all of these "chicken and egg" causal conceptions and the degree of covariance between trauma exposure and NDD, researchers and clinicians recommend that childhood victims of abuse and trauma be evaluated for NDDs, and vice versa.

### 17.3.1 Current Barriers in Assessing Trauma in NDD

Research in the field of trauma and NDD is in its infancy but is expanding rapidly. Major barriers have included: (a) diagnostic overshadowing; (b) gaps in the availability of measurement tools for assessing traumatic experience in children with developmental and language delays; and (c) underdeveloped research methodologies that have over-relied upon convenience samples and tend not to have typically developing control groups [22, 23]. Diagnostic overshadowing occurs when symptoms are attributed to the developmental disability rather than to mental or physical health problems that arise from other sources. This potential bias leads clinicians and researchers to consistently underestimate the impact of conditions other than the disability (e.g., trauma) and to overemphasize core biological or behavioral features of the NDD as explanations of the child's symptoms [26–28].

Also, when attempting to assess the effects of traumatic events on children with NDD, clinicians have tended to rely solely on parent or other adult reports with the assumption that affected children could not adequately report on their experiences [29]. This leaves out self-report data that make a unique contribution to the overall diagnostic picture [30]. Survey research has looked at large samples collected from educational and child protection records without having precise definitions of what constitutes traumatic events, and different types of NDD have not always been differentiated from each other or compared directly with non-NDD populations [22]. These methodological inadequacies have led to possible exaggeration and unsound conclusions regarding causal associations between NDD and trauma.

## 17.4 Autism Spectrum Disorder and Trauma

Autism spectrum disorder is a highly prevalent and increasingly identified neurodevelopmental disorder. Recent estimates suggest that 1–1.2 million children and adolescents in the United States under age 21 have been diagnosed [31]. The increasing incidence of ASD diagnosis in children is a trend that gained momentum with the labeling of a wider range of symptoms pertaining to autism in the DSM-IV [32]. Recent estimates in the United States suggest a prevalence rate of 1 in 59 children being identified with ASD based on national survey data, which is consistent with international estimates employing equivalent diagnostic criteria with school-aged children [33, 34]. ASD is characterized by pervasive developmental deficits in social communication and interaction as well as rigid, repetitive patterns of behavior, interests, or activities. The DSM-5 groups those diagnosed into three severity levels based on functional behavior [10].

There are contradictory research findings regarding the extent of maltreatment in children with autism. Several population-based studies suggest a lower risk of maltreatment compared with children who have intellectual disabilities [2, 21], with abuse rates similar to those of typically developing children [35], thus calling into question the conclusion that they are more often maltreated than typically developing children [36]. However, methodological weaknesses described above may account for some of the contradictory findings. In an effort to address these methodological concerns, a recent large-scale study merged social services records and expert-derived diagnostic data from an Autism and Developmental Disabilities Monitoring Network (ADDM) site for 4988 children. The findings demonstrated increased risk of both reported and substantiated maltreatment for those with ASD, ASD + ID, and ID only [24]. In this study, children identified as having ASD both with and without ID were more likely to experience substantiated physical neglect than a comparison group of nondisabled peers. They also experienced more documented social services reports of physical abuse, sexual abuse, and emotional abuse than the control group. Those who were reported to experience maltreatment showed significantly increased rates of aggression, hyperactivity, inattention, and temper tantrums.

Other studies of clinically referred children with ASD and exposure to traumatic events have identified increased behavioral and emotional problems in clinical settings [37, 38]. Abused children with ASD are more likely to be placed in foster care [39], to have more family mental health problems, and to require more social services than children without ASD who are similarly reported to have been maltreated [35].

Children with ASD show high rates of comorbid anxiety and other emotional and behavioral problems, which may be worsened by trauma [5, 40]. As emotional regulation is a particular vulnerability for children on the autism spectrum, they may have more severe emotional reactions to traumatic events with a tendency to utilize fewer formal coping strategies [41]. Some studies have found regression in previously attained adaptive functioning in children following potentially traumatizing experiences, including reduced functional communication and a return of previously extinguished stereotypies, toileting, sleeping, eating, and other adaptive skills [18, 19].

As illustrated by the case study highlighted in this chapter, diagnostic complexity is amplified by the fact that PTSD and ASD have similar and overlapping symptoms. Behaviors such as sleep problems, overactivity and arousal, emotion regulation problems, temper tantrums, irritability, and attention problems are common to both syndromes [10]. Further, the effects of early neglect and trauma may mimic autism, leading to attachment irregularities and disrupted social skills, detached and flat affect, and perseverative behavior characteristic of autism. While this neglect syndrome has been termed "quasi-autism," and is distinguishable by the absence of developmental characteristics central to autism, recent findings suggest that an autism diagnosis may be warranted in many cases [25, 42].

### 17.4.1 Peer Victimization

Bullying is the most studied and possibly most prevalent form of traumatic occurrences in children with autism [1]. Children with ASD are bullied more often than peers with other disabilities, more often than nondisabled peers, those with intellectual disabilities alone, and their typically developing siblings [43–45]. A recent meta-analysis of well-conducted studies estimates that children with ASD are bullied at a rate three times that of typically developing children. Physical, verbal, and relational school bullying were reported in 33, 50, and 31% of ASD students, respectively [46]. Bullying of children with ASD has been shown to have serious negative impacts on children's academic and social functioning [47].

Mayes and colleagues [48] found that youth with ASD who were teased were three times more likely than those who were not teased to report suicidal ideation or to make a suicide attempt.

### 17.4.2 Adverse Childhood Experiences (ACEs)

A growing number of studies demonstrate that adverse childhood events (ACEs) such as homelessness, mental illness of a family member, and the experiences of racism and poverty contribute to later negative medical and mental health outcomes decades after the original events. Adults who report a range of adverse experiences in childhood have been shown to have poorer long-term health and mental health outcomes [15, 49].

Families with a child on the autism spectrum as compared to those without ASD have significantly greater odds of experiencing divorce, the death of a parent, and having a family member with mental illness [49, 50]. The combination of income insufficiency and comorbid conditions, not including intellectual disability, may account more for the association between ASD and ACEs than the autism itself [51]. These findings also suggest that childhood adversity contributes to the high rate of mental health problems experienced by individuals with ASD [50]. Additionally, those with more ACEs have been shown to have later ASD diagnosis than those with fewer ACEs, perhaps because family stress interferes with access to services, and thus delays early intervention efforts [52]. Figure 17.1 summarizes the main findings regarding trauma and adversity in children with an autism diagnosis.

- Maltreatment prevalence findings vary by methodological quality of study
- Abuse victims more often placed in foster care, receive more social services
- Bullying is common with broad negative effects
- Regression in functioning is often seen
- Unusually high levels of childhood adversity (e.g., poverty, family divorce death, and mental illness)

**Fig. 17.1** Key points: Autism and trauma

## 17.5 Intellectual Disabilities and Trauma

Intellectual disability is defined in the DSM-5 [10] as an early-onset developmental disorder combining deficits in intellectual functions and adaptive functioning. The main causes are chromosomal abnormalities, prenatal and postnatal infections, and exposure to toxic substances such as alcohol in utero [52]. Severity levels range from mild to profound impacts on conceptual, social, and practical functions. The incidence of ID in childhood is estimated to be 1–2% internationally [53]. Compared to children with ASD alone, currently available research more strongly supports claims of disproportionate vulnerability to maltreatment and other adverse events in those with intellectual disability.

### 17.5.1 Prevalence of Maltreatment

Recent population-based prevalence studies [2, 21, 24, 54] and meta-analytic data [20] provide consistent evidence that children with ID are on average two to four times more likely to be victims of various types of maltreatment and violence than their nondisabled peers, though the specific types and extent of maltreatment vary by study. Meta-analysis of a small group of studies that included children with ID compared with nondisabled peers found children with ID to be three times more likely to be victims of physical violence, nearly five times more likely to encounter sexual violence, and four times more likely to experience emotional abuse [20].

The first large-scale study that made use of educational records merged with police records, child protective services, and foster care records [2] reported a 9% prevalence rate of reported maltreatment in children without an educationally identified disability and a 31% prevalence rate among those with any form of disability classification. Relative risk ratios for children with ID compared with nondisabled youth were as follows: neglect (3.8), physical abuse (3.8), emotional abuse (3.8), and sexual abuse (4.0). In this study, children with behavior disorders and speech/language disorders were more likely to be maltreated than children with ID, and children with other physical disabilities had the same level of maltreatment as those with ID. While maltreated youth were significantly more likely to come from geographical areas in which socioeconomic status was lower on average, individual data were not available and not controlled for in the overall analysis, leaving an unresolved confound between poverty and ID in affecting maltreatment risk.

In a more recent population-based analysis of children born in Western Australia between 1990 and 2010, children with disabilities were identified by using data obtained from early identification registries, public disabilities agencies, and inpatient and outpatient mental health records [21]. Risk of maltreatment was determined through child protective services records. While children with all types of disabilities made up approximately 11% of the overall sample, they were overrepresented among those with maltreatment allegations (25.9%) and substantiated abuse (29%); children with ID made up 6.7% of allegations. The degree of risk for maltreatment varied by severity of the intellectual disability. More substantially impaired children had approximately the same risk as nondisabled peers and those with borderline range/mild-level intellectual disabilities had nearly three times the number of maltreatment allegations. After adjusting for SES and other family risk factors, children with ID were more likely to have neglect allegations (33%) compared with all disabilities (25%) and were twice as likely as nondisabled children in the sample to have maltreatment allegations. These results suggest that allegations of abuse are increased in children with ID.

In a study analyzing the 2008 Canadian Incidence Study of children and youth with substantiated child maltreatment, approximately 1 in 10 children was identified with ID based on either a "known" medical diagnosis or "suspected" diagnosis based on caseworkers' records review. Among that group, emotional abuse (25.5% vs. 16.5) and neglect (59.1 vs. 39.1%) were more frequently identified than in nondisabled children. Additionally, children with ID more often experienced multiple substantiated incidents (73.7 vs. 59.3%). These results are perhaps compromised by the lack of a consistent and reliable approach to identifying ID. Among the ID group, caregivers were more likely to have their own cognitive impairments, cases were more frequently previously opened, and more often required ongoing services and supports including referral to other child supporting agencies and children with ID were more often placed in out-of-home care than those without ID [53]. Residential and foster home placements can carry greater risk of sexual victimization for children with mild ID compared to those without ID [55].

McDonnell and colleagues [24] conducted the most methodologically defensible population-based study to date, employing the South Carolina Autism and Developmental Disabilities Monitoring Network (SC ADDM) for expert assessment of co-occurring and separate diagnoses of ASD and ID. Department of Social Services records of both alleged and substantiated abuse formed the basis of maltreatment data. After assessing for the potential impact of SES, findings demonstrated significantly higher risk as indicated by odds ratio comparisons with nondisabled peers for reported (OR 2.19) and substantiated physical abuse (OR 1.88), reported (OR 2.76) and substantiated sexual abuse (OR 2.40), and reported (OR 2.08) and substantiated emotional abuse (OR 2.23) in children with ID. As with their counterparts diagnosed with co-occurring ASD, those with ID had higher risk of reported (OR 2.51) and substantiated physical neglect (OR 2.67).

### 17.5.2 Adverse Childhood Experiences

There are indications that adults with ID are more likely to have experienced various forms of maltreatment and other adverse events in childhood. The frequency and severity of child abuse and exposure to intimate partner violence have been shown to be associated with self-reported depression, further incidents of adverse events and abuse, and PTSD in adults [56, 57]. Children with ID have been shown to have more adverse events such family financial crises, parental marital separation, parents involved in the legal system, serious illness requiring hospitalization,

- On average two to four times more likely to be victims of maltreatment and violence
- Studies possess confounds of SES, identification,definitions of ID and trauma
- ID associated with greater childhood adversity (e.g., poverty, divorce, loss, illness)

**Fig. 17.2** Key points: Intellectual disability and trauma

and traumatic loss of family members. It appears that there is a significant association of these adverse events with income insufficiency that may account for both circumstances leading to the intellectual and developmental delays and the adverse experiences [58]. Figure 17.2 presents summary points drawn from the literature on intellectual disability, trauma, and adversity in children.

## 17.6 Trauma in Attention-Deficit/Hyperactivity Disorder

Attention-deficit/hyperactivity disorder is the most commonly diagnosed childhood NDD and is characterized by core features of inattention, overactivity, and impulsivity typically first observed in early childhood [10]. International prevalence estimates vary widely between 4% and 12% or more [59]. ADHD is often associated with later adolescent and adult problems such as conduct and substance abuse disorders [6, 60]. The causes of the disorder are genetically influenced and mediated by an impaired or immature nervous system, but symptoms may also be shaped by childhood environmental factors such as disorganized family functioning, childhood stress, abuse, and other adverse experiences [9, 11]. Since the disorder is thought to be largely biological in nature and most often treated with medication, clinicians may overlook key psychosocial factors affecting development, severity, and expression of ADHD [4]. For example, foster children diagnosed with ADHD at a high rate (up to 14%) have been shown to have significant reductions in core symptoms with stable placement settings and higher parental warmth [61].

A growing number of studies document strong associations between childhood ADHD diagnosis, exposure to abuse and neglect, and posttraumatic stress disorder (PTSD) diagnosis [3, 6, 62–65]. There are several apparent explanations for this. First, as with other NDDs, diagnostic criteria for ADHD and PTSD overlap considerably with symptoms of overarousal in the form of hyperactivity, inattention, and distractibility. Second, early neglect and abuse during key developmental windows are known to impact the formation of neural circuitry in brain areas responsible for reasoning, planning, judgment, inhibition, and emotional control, thus setting the stage for neurodevelopmental disorders such as ADHD [3, 4].

Third, behavior related to ADHD symptoms may be seen as annoying or difficult to manage, leading to mistreatment by peers [65–67] and adults [6]. Children with ADHD report being bullied significantly more often than those without ADHD and the difference may be more notable for girls than for boys [67]. Early childhood features of ADHD are associated with later odds of being a victim or a combination victim bully in later childhood [66] and at least one study indicates they are victimized more than children with all other types of disability [68].

**Fig. 17.3** Key points: ADHD and trauma

- Strong associations between ADHD, trauma exposure, and trauma symptoms
- ADHD and PTSD symptoms overlap considerably
- Early abuse and neglect affect development of ADHD
- ADHD symptoms increase risk of victimization
- ADHD and trauma may be genetically linked

Finally, recent population-based twin studies suggest genetic covariance between ADHD and exposure to various forms maltreatment [7, 8, 69]. Population-based twin studies of child participants demonstrate that while sexual victimization among children with ASD and ADHD exceeds the rate of sexual victimization of nondisabled peers, this is more accounted for by family genetic effects related to the broad NDD phenotype rather than behaviors specific to either ASD or ADHD [8]. These authors note that the recognition of "difference" in general may set children up for victimization by peers. Further, twin studies of adults diagnosed with ADHD as children queried retrospectively about abuse and maltreatment experiences, reveal evidence of familial genetic associations but causal inferences about the direction of causation between childhood trauma and ADHD are suspect in research involving retrospective accounts [62, 69]. Figure 17.3 summarizes the key points currently established in the literature on trauma in children with ADHD.

## 17.7 Assessment and Treatment of Trauma in NDD

### 17.7.1 Assessment

Because of the demonstrated associations between NDD and traumatic experiences, it is important to carefully assess trauma exposure and symptoms in evaluations of children with NDD [7]. Trauma-informed evaluation and therapy with children who have NDD are commonly undertaken by clinicians but require adaptation, especially for those children with ASD and ID or other significant limitations in language or cognition [70]. There is a paucity of research literature due in part to the nonexistence of trauma measures designed and validated for ASD or other developmental disorders [1].

Traditional paper–pencil measures most often comprise rating scales completed by caregivers and youth regarding the child's exposure to potentially traumatic events and symptoms of posttraumatic stress disorder [71]. Several are well-validated and commonly used with typically developing children [72, 73]. Other measures are designed to be more "child-friendly" with age-appropriate language and graphically illustrated formats and response options [74, 75]. However, none of the paper–pencil or child-friendly scales has been specifically tested with children who have ASD or other developmental disorders.

Likely because of core deficits in communication, emotional understanding, and expression, children with ASD show differences from non-ASD diagnosed children when asked to report psychological symptoms on behavior-rating scales [6, 29]. On self-report measures, they tend to self-enhance relative to typically developing peers on behavior-rating scales, fail to complete measures, or interpret items in an overly literal manner, though there is evidence that higher functioning youth with ASD show no less self-awareness than typically developing youth [76]. Children's self-reports are needed to complete the diagnostic picture and add significantly to assessment in ways that are unique and not accounted for by reports of adult caregivers [30, 77].

Screening and assessment measures for children with ASD and ID would optimally be normed specifically for these populations and capable of addressing a range of functional levels both for self- and caregiver- reporting. A solution to this need may be through the development of web-based, touchscreen applications. Tablet-based "apps" are widely used and familiar for many children and provide flexible and compelling platforms similar to what is being done in the assessment and rehabilitation of autism [78, 79]. While no such format currently exists for assessing trauma, a web-based touchscreen assessment tool is being developed, showing initial evidence of psychometric validity and reliability [80].

### 17.7.2 Trauma-Informed Treatment

The developmental challenges of many children with NDD may present obstacles for trauma treatment including language and conceptual delays, cognitive rigidity, and difficulty generalizing skills learned in treatment to real-world settings. Given the high heritability of NDD, parents often have delays in these same areas, requiring adaptation of therapies to work with both parents and children who have developmental differences.

There are a number of evidence-based treatments with demonstrated effectiveness that may be adapted for use with this population. Solid evidence is accumulating in the area of general anxiety treatment for children with ASD, showing that many respond well to cognitive-behavioral therapies [81]. Summarizing the active ingredients of several of these approaches, Moree and Davis [82] specify: (a) the creation of disorder-specific hierarchies of goals and treatment, (b) the use of concrete, visual tactics, (c) incorporating the child's "special interests" or preoccupations, and (d) parent involvement.

In this author's clinical experience, the need for rigid routines and "sameness" in children who have ASD and ID may in some cases be helpful qualities for making use of treatment. Once children overcome initial resistance to a new setting or therapist, they build treatment into their routine and participate regularly. They relatively easily go along with a pattern of providing "updates" about the previous week, move into skill-building and gradual exposure activities, and then spend a few minutes at the end of the session engaging with the therapist in activities related to their special interests. Special interests and preoccupations can be used to make therapy concepts come alive for children and can reinforce their involvement. This pattern lies at the core of such interventions as the Power Cards [83] and Real Life Heroes [84]. Another treatment-related strength in many children with ASD is their natural tendency to take seriously and adhere strongly to written rules and schedules. We often find that children with ASD stick rigidly to expectations and carry out "homework" assignments more readily than do many typically developing youth.

Recommendations have been made for applying Trauma-Focused Cognitive Behavior Therapy (TF-CBT), one of the most widely disseminated and research-supported interventions, for use with children who have developmental disabilities. Grosso [85] highlights the need for sensitive assessment procedures using drawings, a "Rain Cloud Likert Scale", and completing standard scales such as the UCLA-PTSD Reaction Index [72]. The TF-CBT treatment steps are discussed in detail with accompanying suggestions for adapting them for use with children who have a variety of developmental differences.

The initial steps of TF-CBT or "PRAC" skills (Psychoeducation and Parent Training, Relaxation, Affect Regulation, Cognitive Coping) address basic coping and regulation techniques that are well disposed for use with children who have neurodevelopmental disorders. We recommend that at each stage of TF-CBT, clinicians trained in the model first assess core features of these disorders that commonly complicate treatment efforts: verbal language comprehension problems, delays in conceptual understanding, sensory differences, motivational issues, and difficulty with generalization of skills. Modifications can be made at each step to accommodate the child's and family's particular weaknesses and strengths in these areas. For example, a child with autism being raised by a parent who also has characteristics of ASD may have a strong aversion to beginning therapy to address fears and anxieties related to the trauma, which may be reinforced by the parent. In such a case, careful and perhaps longer-than-typical psycho-education of the parent is needed to form an alliance and understanding of the unique ways in which the trauma affects the child's behavior. A reinforcement schedule and charted practice sessions at home with clinician review at each session may help both child and parent make the association between coping skills learned in the clinician's office and their use when encountering real-world stressors or reminders of the trauma.

## 17.8 Conclusions

Maltreatment, bullying, childhood adversities, and other potentially traumatic events are not unusual for children with NDD and have far-reaching impacts on children's well-being and future adjustment. While for many years there has been a public interest in the disability/trauma connection, researchers are only recently refining methods to specify prevalence, causality, and the unique effects of trauma on this population of children. The limited array of assessment tools, diagnostic overshadowing bias, and other conceptual and definitional barriers remain obstacles to the interpretation of findings and require further attention in order to inform research and intervention efforts. Children with autism and other forms of NDD can be treated with evidence-based models that have proven effective with other populations, but with appropriate modifications.

## References

1. Hoover DW. The effects of psychological trauma on children with autism spectrum disorders: a research review. Rev J Autism Dev Disord. 2015;2(3):287–99. https://doi.org/10.1007/s40489-015-0052-y.
2. Sullivan PM, Knutson JF. Maltreatment and disabilities: a population-based epidemiological study. Child Abuse Negl. 2000;24(10):1257–73.
3. Lewis T, Schwebel DC, Elliott MN, Visser SN, Toomey SL, McLaughlin KA, et al. The association between youth violence exposure and attention-deficit/hyperactivity disorder (ADHD) symptoms in a sample of fifth-graders. Am J Orthopsychiatry. 2015;85(5):504–13.
4. Richards L. It is time for a more integrated bio-psycho-social approach to ADHD. Clin Child Psychol Psychiatry. 2012;18:483–503.
5. Kerns C, Newschaffer C, Berkowitz S. Traumatic childhood events and autism spectrum disorder. J Autism Dev Disord. 2015;45(11):3475–86.
6. Stern A, Agnew-Blais J, Danese A, Fisher HL, Jaffee SR, Matthews T, et al. Associations between abuse/neglect and ADHD from childhood to young adulthood: a prospective nationally-representative twin study. Child Abuse Negl. 2018;81:274–85.
7. Dinkler L, Lundstrom S, Gajwani R, Lichtenstein P, Gillberg C, Minnis H. Maltreatment-associated neurodevelopmental disorders: a co-twin control analysis. J Child Psychol Psychiatry. 2017;58(6):691–701.
8. Ohlsson Gotby V, Lichtenstein P, Långström N, Pettersson E. Childhood neurodevelopmental disorders and risk of coercive sexual victimization in childhood and adolescence - a population-based prospective twin study. J Child Psychol Psychiatry. 2018;59(9):957–65.
9. Jeste SS. Neurodevelopmental behavioral and cognitive disorders. Continuum (Minneap Minn). 2015;21(3):690–714.
10. American Psychiatric Association. Diagnostic and statistical manual of mental disorders. 5th ed. Arlington, VA: Author; 2013.
11. Davis AS, Hoover KL, Mion AM. In: Butcher JN, Kendall PC, editors. Understanding and treating children and adolescents with neurodevelopmental disorders. Washington, DC: American Psychological Association; 2018. p. 279–315.
12. Copeland WE, Keeler G, Angold A, Costello EJ. Traumatic events and posttraumatic stress in childhood. Arch Gen Psychiatry. 2007;64(5):577–84.
13. Kessler RC, Avenevoli S, McLaughlin KA, Green JG, Lakoma MD, Petukhova M, et al. Lifetime co-morbidity of DSM-IV disorders in the US National Comorbidity Survey Replication Adolescent Supplement (NCS-A). Psychol Med. 2012;42(9):1997–2010.
14. McLaughlin KA, Koenen KC, Hill ED, Petukhova M, Sampson NA, Zaslavsky AM, et al. Trauma exposure and posttraumatic stress disorder in a national sample of adolescents. J Am Acad Child Adolesc Psychiatry. 2013;52(8):815–30.
15. Felitti VJ, Anda RF, Nordenberg D, Williamson DF, Spitz AM, Edwards V, et al. Relationship of childhood abuse and household dysfunction to many of the leading causes of death in adults. The Adverse Childhood Experiences (ACE) Study. Am J Prev Med. 1998;14(4):245–58.
16. Layne CM, Greeson JKP, Ostrowski SA, Kim S, Reading S, Vivrette RL, et al. Cumulative trauma exposure and high risk behavior in adolescence: findings from the National Child Traumatic Stress Network Core Data Set. Psychol Trauma. 2014;6:S40–9.
17. American Psychological Association Presidential Task Force on Posttraumatic Stress Disorder and Trauma in Children and Adolescents. Children and Trauma: Update for Professionals. 2008.
18. Mehtar M, Mukaddes NM. Posttraumatic stress disorder in individuals with diagnosis of autistic spectrum disorders. Res Autism Spectr Disord. 2011;5(1):539–46.
19. Valenti M, Di Giovanni C, Mariano M, Pino MC, Sconci V, Mazza M. Autism after an earthquake: the experience of L'Aquila (Central Italy) as a basis for an operative guideline. Epidemiol Prev. 2016;40(2 Suppl 1):49–52.
20. Jones L, Bellis MA, Wood S, Hughes K, McCoy E, Eckley L, et al. Prevalence and risk of violence against children with disabilities: a systematic review and meta-analysis of observational studies. Lancet. 2012;380(9845):899–907.

21. Maclean MJ, Sims S, Bower C, Leonard H, Stanley FJ, O'Donnell M. Maltreatment risk among children with disabilities. Pediatrics. 2017;139(4):e20161817. https://doi.org/10.1542/peds.2016-1817.
22. Leeb RT, Bitsko RH, Merrick MT, Armour BS. Does childhood disability increase risk for child abuse and neglect? J Mental Health Res Intellect Disabil. 2012;5(1):4–31.
23. Sullivan PM. Violence exposure among children with disabilities. Clin Child Fam Psychol Rev. 2009;12(2):196–216.
24. McDonnell CG, Boan AD, Bradley CC, Seay KD, Charles JM, Carpenter LA. Child maltreatment in autism spectrum disorder and intellectual disability: results from a population-based sample. J Child Psychol Psychiatry. 2019;60(5):576–84.
25. Green J, Leadbitter K, Kay C, Sharma K. Autism spectrum disorder in children adopted after early care breakdown. J Autism Dev Disord. 2016;46(4):1392–402.
26. Kerns CM, Kendall PC, Zickgraf H, Franklin ME, Miller J, Herrington J. Not to be overshadowed or overlooked: functional impairments associated with comorbid anxiety disorders in youth with ASD. Behav Ther. 2015;46(1):29–39.
27. Levitan GW, Reiss S. Generality of diagnostic overshadowing across disciplines. Appl Res Ment Retard. 1983;4(1):59–64.
28. Meera SS, Kaipa R, Thomas J, Shivashankar N. Brief report: An unusual manifestation of diagnostic overshadowing of pervasive developmental disorder—not otherwise specified: a five year longitudinal case study. J Autism Dev Disord. 2013;43(6):1491–4.
29. Mahan S, Matson JL. Children and adolescents with autism spectrum disorders compared to typically developing controls on the Behavioral Assessment System for Children, Second Edition (BASC-2). Res Autism Spectr Disord. 2011;5(1):119–25.
30. Adams RE, Fredstrom BK, Duncan AW, Holleb LJ, Bishop SL. Using self- and parent-reports to test the association between peer victimization and internalizing symptoms in verbally fluent adolescents with ASD. J Autism Dev Disord. 2014;44(4):861–72.
31. Centers for Disease Control and Prevention. Prevalence of autism spectrum disorder among children aged 8 years—autism and developmental disabilities monitoring network, 11 sites, United States, 2010. Morb Mortal Wkly Rep Surveill Summ. 2014;63(2):1–13.
32. American Psychiatric Association. Diagnostic and statistical manual of mental disorders. 4th ed. Arlington, VA: Author; 1994.
33. Baio J, Wiggins L, Christensen DL, Maenner MJ, Daniels J, Warren Z, et al. Prevalence of autism spectrum disorder among children aged 8 years—autism and developmental disabilities monitoring network, 11 sites, United States, 2014. MMWR Surveill Summ. 2018;67:1–23.
34. Schendel DE, Thorsteinsson E. Cumulative incidence of autism into adulthood for birth cohorts in Denmark, 1980-2012. JAMA. 2018;320(17):1811–3.
35. Hall-Lande J, Hewitt A, Mishra S, Piescher K, LaLiberte T. Involvement of children with autism spectrum disorder (ASD) in the child protection system. Focus Autism Other Dev Disabl. 2015;30(4):237–48.
36. Hoover DW, Kaufman J. Adverse childhood experiences in children with autism spectrum disorder. Curr Opin Psychiatry. 2018;31(2):128–32.
37. Brenner J, Pan Z, Mazefsky C, Smith K, Gabriels R. Behavioral symptoms of reported abuse in children and adolescents with autism Spectrum disorder in inpatient settings. J Autism Dev Disord. 2017;48(11):3727–35.
38. Mandell DS, Walrath CM, Manteuffel B, Sgro G, Pinto-Martin J. The prevalence and correlates of abuse among children with autism served in comprehensive community-based mental health settings. Child Abuse Negl. 2005;29(12):1359–72.
39. Cidav Z, Xie M, Mandell DS. Foster care involvement among Medicaid-enrolled children with autism. J Autism Dev Disord. 2018;48(1):176–83.
40. Konst MJ, Matson JL. Comorbid psychopathology symptom rates in infants and toddlers with Autism Spectrum Disorders. Res Autism Spectr Disord. 2014;8(2):147–55.
41. Mazefsky CA, Borue X, Day TN, Minshew NJ. Emotion regulation patterns in adolescents with high-functioning autism spectrum disorder: comparison to typically developing adolescents and association with psychiatric symptoms. Autism Res. 2014;7(3):344–54.

42. Rutter M, Andersen-Wood L, Beckett C, Bredenkamp D, Castle J, Groothues C, et al. Quasi-autistic patterns following severe early global privation. J Child Psychol Psychiatry. 1999;40(4):537–49.
43. Nowell KP, Brewton CM, Goin-Kochel R. A multi-rater study on being teased among children/adolescents with autism spectrum disorder (ASD) and their typically developing siblings: associations with ASD symptoms. Focus Autism Other Dev Disabl. 2014;29(4): 195–205.
44. Sreckovic MA, Hume K, Able H. Examining the efficacy of peer network interventions on the social interactions of high school students with autism spectrum disorder. J Autism Dev Disord. 2017;47(8):2556–74.
45. Zeedyk SM, Rodriguez G, Tipton LA, Baker BL, Blacher J. Bullying of youth with autism spectrum disorder, intellectual disability, or typical development: victim and parent perspectives. Res Autism Spectr Disord. 2014;8(9):1173–83.
46. Maiano C, Normand CL, Salvas MC, Moullec G, Aime A. Prevalence of school bullying among youth with autism spectrum disorders: a systematic review and meta-analysis. Autism Res. 2016;9(6):601–15.
47. Adams R, Taylor J, Duncan A, Bishop S. Peer Victimization and educational outcomes in mainstreamed adolescents with autism spectrum disorder (ASD). J Autism Dev Disord. 2016;46(11):3557–66.
48. Mayes SD, Gorman AA, Hillwig-Garcia J, Syed E. Suicide ideation and attempts in children with Autism. Res Autism Spectr Disord. 2013;7(1):109–19.
49. Berg KL, Shiu CS, Acharya K, Stolbach BC, Msall ME. Disparities in adversity among children with autism spectrum disorder: a population-based study. Dev Med Child Neurol. 2016;58(11):1124–31.
50. Rigles B. The relationship between adverse childhood events, resiliency and health among children with autism. J Autism Dev Disord. 2017;47(1):187–202.
51. Kerns CM, Newschaffer CJ, Berkowitz S, Lee BK. Brief report: examining the association of autism and adverse childhood experiences in the national survey of children's health: the important role of income and co-occurring mental health conditions. J Autism Dev Disord. 2017;47(7):2275–81.
52. Berg KL, Acharya K, Shiu CS, Msall ME. Delayed diagnosis and treatment among children with autism who experience adversity. J Autism Dev Disord. 2018;48(1):45–54.
53. Bhaumik S, Kiani R, Michael DM, Gangavati S, Khan S, Torales J, et al. World Psychiatric Association (WPA) report on mental health issues in people with intellectual disability: Paper 1: intellectual disability and mental health: an overview. Int J Cult Ment Health. 2016;9(4):417–29.
54. Dion J, Paquette G, Tremblay KN, Collin-Vezina D, Chabot M. Child maltreatment among children with intellectual disability in the Canadian Incidence Study. Am J Intellect Dev Disabil. 2018;123(2):176–88.
55. Euser S, Alink LR, Tharner A, van IJzendoorn MH, Bakermans-Kranenburg MJ. The prevalence of child sexual abuse in out-of-home care: increased risk for children with a mild intellectual disability. J Appl Res Intellect Disabil. 2016;29(1):83–92.
56. Catani C, Sossalla IM. Child abuse predicts adult PTSD symptoms among individuals diagnosed with intellectual disabilities. Front Psychol. 2015;6:1600.
57. Santoro A, Shear S, Haber A. Childhood adversity, health and quality of life in adults with intellectual and developmental disabilities. J Intellect Disabil Res. 2018;62(10):854–63.
58. Hatton C, Emerson E. The relationship between life events and psychopathology amongst children with intellectual disabilities. J Appl Res Intellect Disabil. 2004;17(2):109–17.
59. Brown RT, Freeman WS, Perrin JM, Stein MT, Amler RW, Feldman HM, et al. Prevalence and assessment of attention-deficit/hyperactivity disorder in primary care settings. Pediatrics. 2001;107(3):E43.
60. Erskine HE, Norman RE, Ferrari AJ, Chan GCK, Copeland WE, Whiteford HA, et al. Long-term outcomes of attention-deficit/hyperactivity disorder and conduct disorder: a systematic review and meta-analysis. J Am Acad Child Adolesc Psychiatry. 2016;55(10):841–50.

61. Linares LO, Li M, Shrout PE, Ramirez-Gaite M, Hope S, Albert A, et al. The course of inattention and hyperactivity/impulsivity symptoms after foster placement. Pediatrics. 2010;125(3):e489–98.
62. Fuller-Thomson E, Lewis DA. The relationship between early adversities and attention-deficit/hyperactivity disorder. Child Abuse Negl. 2015;47:94–101.
63. Ouyang L, Fang X, Mercy J, Perou R, Grosse SD. Attention-deficit/hyperactivity disorder symptoms and child maltreatment: a population-based study. J Pediatr. 2008;153(6):851–6.
64. Cromer KD, Villodas MT. Post-traumatic stress as a pathway to psychopathology among adolescents at high-risk for victimization. Child Abuse Negl. 2017;67:182–92.
65. McQuade J, Breslend N, Groff D. Experiences of physical and relational victimization in children with ADHD: the role of social problems and aggression. Aggress Behav. 2018;44(4):416–25.
66. Verlinden M, Jansen PW, Veenstra R, Jaddoe VW, Hofman A, Verhulst FC, et al. Preschool attention-deficit/hyperactivity and oppositional defiant problems as antecedents of school bullying. J Am Acad Child Adolesc Psychiatry. 2015;54(7):571–9.
67. Wiener J, Mak M. Peer victimization in children with attention-deficit/hyperactivity disorder. Psychol Sch. 2009;46(2):116–31.
68. Blake JJ, Kim ES, Lund EM, Zhou Q, Kwok O, Benz MR. Predictors of bully victimization in students with disabilities: a longitudinal examination using a national data set. J Disabil Policy Stud. 2016;26(4):199–208.
69. Capusan AJ, Kuja-Halkola R, Bendtsen P, Viding E, McCrory E, Marteinsdottir I, et al. Childhood maltreatment and attention deficit hyperactivity disorder symptoms in adults: a large twin study. Psychol Med. 2016;46(12):2637–46.
70. Mazefsky CA, Kao J, Oswald DP. Preliminary evidence suggesting caution in the use of psychiatric self-report measures with adolescents with high-functioning autism spectrum disorders. Res Autism Spectr Disord. 2011;5(1):164–74.
71. Eklund K, Rossen E, Koriakin T, Chafouleas SM, Resnick C. A systematic review of trauma screening measures for children and adolescents. Sch Psychol Q. 2018;33(1):30–43.
72. Briere J. Trauma symptom checklist for children. Odessa: Psychological Assessment Resources, Inc.; 1996.
73. Steinberg AM, Brymer MJ, Kim S, Briggs EC, Ippen CG, Ostrowski SA, et al. Psychometric properties of the UCLA PTSD reaction index: Part I. J Trauma Stress. 2013;26(1):1–9.
74. Geller PA, Neugebauer R, Possemato AK, Walter P, Dummit ES, Silva RR. Psychometric properties of Darryl, a cartoon based measure to assess community violence-related PTSD in children. Psychiatric Q. 2007;78(2):157–68.
75. Praver F, DiGiuseppe R, Pelcovitz D, Mandel FS, Gaines R. A preliminary study of a cartoon measure for children's reactions to chronic trauma. Child Maltreat. 2000;5(3):273–85.
76. Schriber RA, Robins RW, Solomon M. Personality and self-insight in individuals with autism spectrum disorder. J Pers Soc Psychol. 2014;106(1):112–30.
77. Cohen JA, Bukstein O, Walter H, Benson RS, Chrisman A, Farchione TR, et al. Practice parameter for the assessment and treatment of children and adolescent with posttraumatic stress disorder. J Am Acad Child Adolesc Psychiatry. 2010;49(4):414–30.
78. Brooks BA, Haynes K, Smith J, McFadden T, Robins DL. Implementation of web-based autism screening in an urban clinic. Clin Pediatr. 2016;55(10):927–34.
79. Kobak KA, Stone WL, Wallace E, Warren Z, Swanson A, Robson K. A Web-based tutorial for parents of young children with autism: results from a pilot study. Telemed J E-Health. 2011;17(10):804–8.
80. Hoover DW, Romero EMG. The Interactive Trauma Scale: a web-based measure of psychological trauma in children with autism. J Autism Dev Disord. 2019;49:1686. https://doi.org/10.1007/s10803-018-03864-3.
81. Attwood T, Scarpa A. In: Scarpa A, Williams White S, Attwood T, editors. Modifications of cognitive-behavioral therapy for children and adolescents with high-functioning ASD and their common difficulties. New York, NY: Guilford Press; 2013. p. 27–44.

82. Moree BN, Davis TEI II. Cognitive-behavioral therapy for anxiety in children diagnosed with autism spectrum disorders: modification trends. Res Autism Spectr Disord. 2010;4(3):346–54.
83. Gagnon E, Myles BS. The power card strategy: an evidence-based practice. Shawnee, KS: AAPC Publishing; 2016.
84. Kagan R, Henry J, Richardson M, Trinkle J, LaFrenier A. Evaluation of Real Life Heroes treatment for children with complex PTSD. Psychol Trauma. 2014;6(5):588–96.
85. Grosso CA. Children with developmental disabilities. In: Cohen JA, Mannarino AP, Deblinger E, editors. Trauma-focused CBT for children and adolescents: treatment applications. New York, NY: Guilford Press; 2012. p. 149–74.

# 18 Neurobiological Basis of Childhood Trauma and the Risk for Neurological Deficits Later in Life

Francesca Assogna, Federica Piras,
and Gianfranco Spalletta

## 18.1 Introduction

Childhood trauma (CT) is a comprehensive concept of encompassing experiences of stressful or traumatic life events that occur during the first 18 years of life, such as emotional, physical, or sexual abuse, emotional or physical neglect, or other forms of family dysfunctions, or pervasive and significant public health problems [1–3]. These traumatic events in the early years of childhood are not lost but kept for life as an enduring experience. In fact, time does not cure the wounds that occur in early years of life, time just hides them [4, 5] as they are not misplaced and become part of the body in the form of disorders (Fig. 18.1).

Only in the last few decades, the problem of trauma-related damage during human development has been recognized and its importance considered. A real comprehension of the impact that CT can have on health and illness allows us to understand how we become the people that we are, not only as biological entities but also as real human beings with an external personality and an inner soul.

In the last two decades, there has been an explosion of knowledge about how experience shapes the central nervous system and the formation of the self. Advances in neuroscience and studies on the relationship between attachment styles and brain information processing show how experience shapes brain function and that life itself can continuously influence perceptive and biological levels. Overwhelming experiences alter the capacity for self-regulation, attention, and memory processing due to changes occurring at the neuronal level. Traces of trauma imprinted in memory are

F. Assogna (✉) · F. Piras
Laboratory of Neuropsychiatry, IRCCS Santa Lucia Foundation, Rome, Italy
e-mail: f.assogna@hsantalucia.it

G. Spalletta
Laboratory of Neuropsychiatry, IRCCS Santa Lucia Foundation, Rome, Italy

Menninger Department of Psychiatry and Behavioral Sciences, Baylor College of Medicine, Houston, TX, USA



G. Spalletta et al. (eds.), *Childhood Trauma in Mental Disorders*,
https://doi.org/10.1007/978-3-030-49414-8_18

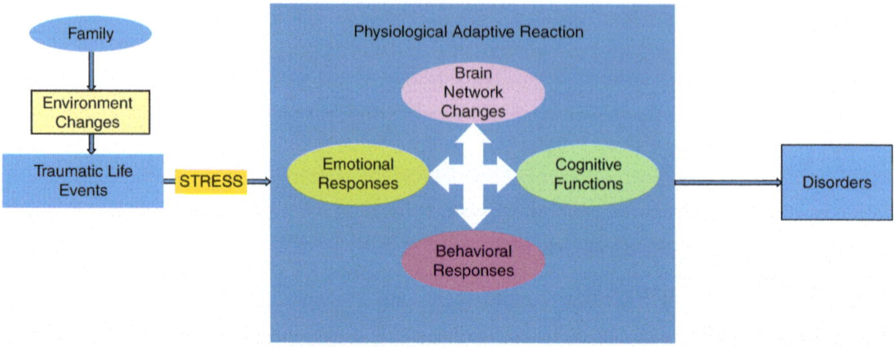


**Fig. 18.1** Traumatic life events cause stress due to environmental changes that destabilize the family. Child has physiological adaptive reactions to stress, characterized by brain network modifications and cognitive, emotional, and behavioral responses. If stressor events are prolonged in time, children may have disorders related to maladaptive responses

preserved as body states and patterns of physical action. As a result, the entire organism automatically reacts to current experiences as if they are repetitions of the past.

A failure to cope with environmental changes defines stress as potentially traumatic. Stress becomes traumatic for the subject when health and well-being are threatened, making the individual impotent in front of danger, and by violating the basic assumptions of survival and highlighting the impossibility of controlling and predicting events. Accordingly, the coping skills of the individual are overcome from a physical, emotional, and cognitive point of view, giving rise to physical and mental illnesses.

Early traumatic relational experience alters the physiology and structure of the brain and has an impact on the selection of information processes during the early stages of development. Therefore, traumatic stress affects the development of the brain, altering the regulation of emotions, the ability to control impulses, reasoning, and memory. In stress situations, especially if prolonged, the neurotoxic action of steroid hormones, which are released in the bloodstream in response to stress or danger, affects the functioning of both hippocampus and amygdala. This effect explains why many details and circumstances of the trauma can be forgotten, while the fear and anxiety associated with it can be triggered more easily by any circumstance that presents evocative signals.

All these alterations of the brain network structure and biochemistry have long-term effects, well beyond the traumatic event and its resolution, as a sort of neuronal physical trace of the damage suffered. In other words, as noted by Wordsworth [6], "*the child is father of the man*," and disease is an integral part of this observation.

## 18.2 Neurobiological Basis of Trauma

When we are exposed to signals of danger, we feel threatened and are naturally led to prepare an adaptive response. The perception of fear and the consequent behavioral responses are crucial for the adaptation to the environment and for the survival of species. The activation of the nervous system to the perception of danger determines a reaction of *fight/attack*, or *escape/freezing*, that are all physiological adaptive reactions to stress. The involved brain regions are part of a complex system, the limbic system, a phylogenetically old system, which should not be considered an anatomical entity, but rather a neurophysiological system [7]. The limbic system intervenes in the selection of behaviors that are correlated with the survival of the species, elaborates the emotions and related vegetative manifestations, and is involved in the memorization processes. In particular, the amygdala plays a central role in this system related to the perception of fear and the elaboration of the resulting reactions. The amygdala is considered a sort of gateway for the emotions that are here recorded, triggering adaptive physiological reactions involving thalamus, sensory circuits, hippocampus, some deep nuclei of the medulla oblongata, and predominantly frontal cortical regions.

The organization and functional capacity of the human brain depends upon an extraordinary set and sequence of developmental and environmental experiences that influence the expression of the genome [7–9]. The psychological trauma that comes from being exposed to an event that is perceived as potentially dangerous for one's own life or that of others, or potentially capable of generating serious physical injury to oneself or others, may alter the way genes express themselves. Specifically, extreme and repetitive trauma during critical or circumscribed periods of brain development in childhood can determine experience-driven chemical modifications of genes, which can compromise, often permanently, the activity of major neuroregulatory systems, with enduring and profound neurobiological and behavioral consequences [9–15]. In the biology of trauma, the interaction between negative environmental stimulation, responses/reactions and pathology onset must be evaluated on the basis of the interaction between the nervous and the endocrine systems within specific brain circuits. Specifically, the macro-neurobiological systems that regulate responses to stress are as follows: (1) the neuroendocrine pathway; (2) neurotransmitters; and (3) a network, both conscious and unconscious, connecting different deep and cortical brain regions. Although the major growth of brain occurs during the pre- and early postnatal periods, key maturation processes, marked by myelination and an increasing complexity of connectivity, continue through late adolescence. This prolonged period provides high opportunity for various insults to interfere with brain development although compromised function may not be expressed until later life [16].

### 18.2.1 The Neuroendocrine Pathway: The Role of the Hypothalamic–Pituitary–Adrenal Axis in Trauma

The stress/trauma circuit includes the activation of neuroendocrine and sensitive brain areas such as the hypothalamic–pituitary–adrenal (HPA) axis. The HPA axis is a complex network of direct influences and feedback interactions among three components: the hypothalamus, the pituitary gland, and the adrenal glands. The HPA axis is activated under stress conditions [7, 8, 15] in order to respond very quickly to potential threats according to an overlearned sequence. Some anatomical

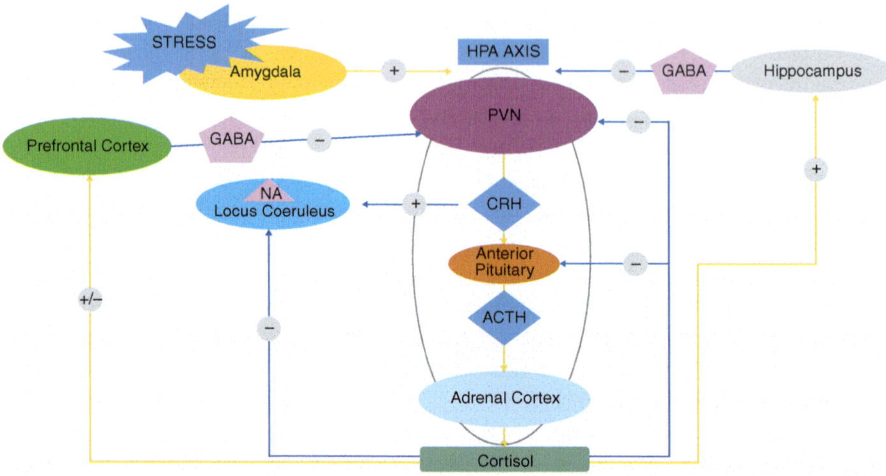


**Fig. 18.2** In response to stress in the environment, the amygdala actives the hypothalamus–pituitary–adrenal (HPA) axis detecting and sending inputs to the paraventricular nucleus (PVN) of the hypothalamus that responds producing the corticotropin-releasing hormone (CRH). CRH, through the anterior pituitary gland, determines the release of the adrenocorticotropic hormone (ACTH) that stimulates cortisol from the adrenal cortex. Cortisol auto-regulates the HPA axis, inhibiting the PVN of the hypothalamus and pituitary gland's activities. Oversecretion of stress hormones is also regulated, through negative feedback, by the hippocampus. Moreover, high levels of cortisol deactivate the locus coeruleus and modulate prefrontal cortex activity. The prefrontal cortex also regulates the stress reactions, decreasing activity of HPA axis

brain areas such as the amygdala, hippocampus, and prefrontal cortex modulate the activation of the HPA axis. The amygdala, in particular, detects stress in the environment and sends inputs to the paraventricular nucleus of the hypothalamus (PVN). Following inputs from amygdala, the PVN produces the corticotropin-releasing hormone (CRH), which is a key mediator of the stress response [17] (Fig. 18.2).

CRH reaches the anterior pituitary gland and promotes the release of the adrenocorticotropic hormone (ACTH). The latter stimulates glucocorticoids release from the adrenal cortex. Glucocorticoids activate the conversion of proteins and lipids into carbohydrates, making the body ready for immediate reactions of highly adaptive value for survival [18] (Fig. 18.3).

When absolute or relative stressors chronically activate the stress reaction, the effects are negative. These fundamental stress mediators cannot distinguish between real and objective threats and threats subjectively perceived dangers as actual. This determines a chronicity of the response with repeated secretion of glucocorticoids in time, even when no objectively traumatic condition is present.

### 18.2.2 The Role of Neurotransmitters in Trauma

The neurochemical core of the trauma lies in the abnormal regulation of catecholamines, serotonin, some amino acids, some peptides, and opioid neurotransmitters, each of which plays a role as neuromediator in the brain circuits that regulate/

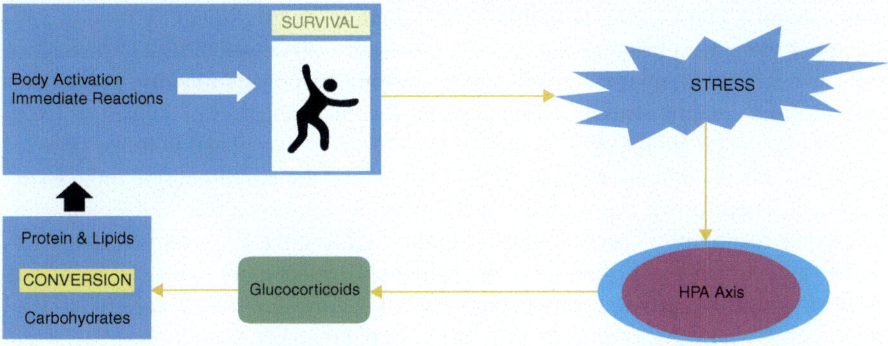


**Fig. 18.3** The activation of the hypothalamic-pituitary-adrenal (HPA) axis leads to production of glucocorticoid hormones (mainly cortisol). Glucocorticoids stimulate the conversion of protein and lipids to carbohydrate for activating an immediate reaction to stress

integrate the response to stress and fear [19]. Catecholamines (adrenaline, noradrenaline, and dopamine) are hormones and neurotransmitters derived from tyrosine and released from the body in stressful situations [20, 21]. They are produced by the adrenal glands as hormones released in the bloodstream, and as neurotransmitters in the presynaptic ganglia.

Adrenaline is the main neurotransmitter of the sympathetic nervous system, involved in the body's preparation for the "*attack-or-flight*" response. Many studies showed that patients with chronic post-traumatic stress disorder (PTSD) have a high activation of the vegetative nervous system and of the alert state, as evidenced by increased heart rate, blood pressure, and skin conductance and other augmented psychophysiological measures, correlated with a dysregulation of the noradrenergic pathway [22–24].

The main site for the noradrenaline (NA) synthesis in the brain is the locus coeruleus (LC) also called the "blue spot" due to its color tending to blue as a consequence of melanin granules [18]. LC is a nucleus of medium-sized neurons, located in the brainstem, between the midbrain and the pons; neural connections from this nucleus reach the spinal cord, the brainstem, the cerebellum, the hypothalamus, the thalamus, the amygdala, the base of the telencephalon, and the cerebral cortex. Through its connections with the frontal and the temporal cortices, the thalamus and the hypothalamus, the LC is involved in the regulation of attention, sleep–wake cycle, mood and appetite, anxiety/stress mechanisms, pain perception, learning processes. It has been observed that a massive secretion of noradrenergic hormones, at the time of trauma, determines a long-term potentiation (LTP) in hippocampal area CA1 and, therefore, the overconsolidation of traumatic memories [25–28]. LTP is an activity-dependent increase in synaptic transmission [29, 30] that correlates with particular forms of hippocampus-dependent learning and memory [28, 31, 32]. More specifically, forms of LTP induced by multiple trains of high-frequency stimulation correlate with hippocampus-dependent long-term memory. Noradrenergic afferents extensively innervate the hippocampus [33], and b-adrenergic receptors activation can recruit intracellular signaling pathways that are important for hippocampal LTP [34, 35]. Noradrenergic signaling can significantly modify hippocampal synaptic plasticity and enhance memory processes [36, 37] through a sustained enhancement of the postsynaptic neuronal response following a particular pattern of presynaptic stimuli. LTP is crucial for the evaluation of subsequent sensory inputs, which are matched to the memory "traces" associated with the primary event, and this phenomenon seems to be mediated by the noradrenergic input to the amygdala. Therefore, it is possible that NA plays an important role in the characteristic hyper-reactivity of people who have suffered trauma. Exposure to stress following a severe trauma, therefore, accompanied by an increased release of NA could sensitize the subject, thus generating a kind of amplified reactivity toward subsequent stimuli. As a consequence, intrusive phenomena, such as flashbacks and nightmares in patients with PTSD, could be correlated with increased noradrenergic innervation of the connections between LC, hippocampus, amygdala, and temporal neocortex. It is important to underline that although the high release of NA at the time of trauma causes excessive consolidation of traumatic memories, other hormones, such as oxytocin and endogenous opioids, can contribute to the creation of amnesia, which is a common correlate of PTSD.

As to dopamine, although the mechanism involved in trauma is still not clear, it may be linked to glucocorticoid secretion with subsequent increases in the dopaminergic activity in some brain regions [38, 39] including the mesolimbic system [40]. Animal studies showed how the administration of substances capable of suppressing glucocorticoid secretion also reduces the release of dopamine, in similar way to antipsychotics [41], while the administration of agents that implement dopaminergic activity also increases the activity of the HPA axis and cortisol secretion in humans. In acute stress, we observe a potent activation of dopaminergic neurons, particularly those projecting from the ventral tegmental area and from the substantia nigra to the limbic system and medial prefrontal cortex, with consequent cognitive and behavioral modifications [42]. More specifically, the mesocortical dopaminergic system includes dopaminergic neurons in the ventral tegmental area that receives excitatory inputs and sends them to the medial prefrontal cortex. A study by Butts and collaborators [43] hypothesized that glucocorticoid receptors in the medial prefrontal cortex modulate the activity of glutaminergic input descending to the ventral tegmental area that occurs during an acute stress. Using microdialysis in vivo, they demonstrated that blockade of glucocorticoids receptors in the medial prefrontal cortex of rats, but not in the ventral tegmental area, attenuates the mesocortical dopamine efflux in response to acute stress. Acute stress leads to a significant increase in glutamate release in the ventral tegmental area, while blockade of prefrontal glucocorticoid receptors can attenuate such stress-evoked dopamine efflux in the medial prefrontal cortex. The functional impact of the increase in the efflux of mesocortical dopamine resulting from acute stress has been shown by evaluating performance in working memory and set-shifting task [42, 44, 45]. In particular, impairment induced by stress in working memory is attenuated by receptors blockage in the medial prefrontal cortex. These data suggest that glucocorticoids act locally in the prefrontal cortex by modulation of dopaminergic mesocortical flow and glutamate release in the ventral tegmental area that lead to executive functions impairment observed during acute stress.

In chronic stress, adaptive reduction in the number of dopaminergic neurons is observed with probable alteration of the response to subsequent stressors [46]. The existence of a hypothalamic factor, called corticotropin release-inhibitory factor (CRIF), has also been postulated, which would inhibit ACTH secretion. This substance could be involved in the control of hypercortisolism from activation of the HPA axis [47].

Serotonin and endogenous opioids also play an active role in the stress response. Serotonin (5-HT) is a monoamine neurotransmitter, synthesized by serotonergic neurons in the central nervous system (CNS), as well as enterochromaffin cells in the gastrointestinal tract. It is an important CNS transmitter and is present in high concentrations in specific areas of the midbrain. In the CNS, serotonin plays a crucial role in regulating mood, sleep, body temperature, sexuality, and appetite. Although the precise consequences of serotonergic dysregulation are not known, many studies indicate that it is implicated in PTSD and in other psychopathological disorders [18, 48–50]. Two serotonergic pathways were identified as important in the development of PTSD symptoms: the first interconnects the dorsal raphe (a nucleus of the brain stem) with the amygdala, involving the postsynaptic serotonergic receptors that mediate conditional avoidance behaviors; and the second path interconnects the median raphe and the hippocampus and is involved in the ability to recover and adapt to stress.

In summary, the dysfunctions in the serotonergic system [51], induced by a condition of high chronic stress, could damage the functioning of the behavioral inhibitory system, resulting in different symptoms encountered in individuals suffering from post-traumatic reactions, such as aggressive explosions, hypervigilance, impulsivity, and intrusive memories.

Endorphins are a group of substances produced in the pituitary anterior lobe, they are classifiable as neurotransmitters, and have analgesic and physiological properties similar to those of morphine and opium, but with a wider effect. They are present in the tissues of highly advanced animals and are released in particular conditions of physical fatigue and stress; a strong emotion can also stimulate endorphins release. The endorphinic system determines analgesia during the stress response through the release of endogenous opioids that inhibit pain and reduce panic. Endogenous opioids exert inhibitory influences on the HPA axis attenuating the emotional response to environmental stressors; their increase is responsible for emotional analgesia when confronted with severe stress and trauma remainders. Thus, endogenous opioids potentially reduce overwhelming psychological distress and prevent impulsive and dysfunctional behavioral responses. For instance, in line with what was saying, the responses of stiffening and stunning in animals, as well as human dissociative reactions in response to trauma, could allow the body to "not consciously experience" or to not remember situations of oppressive stress, thus determining the inability to learn experience.

Also, glucocorticoids have a crucial role in the stress. In fact, the brain areas involved in the stress circuit, the frontal lobes, the amygdala, and the hippocampus, contain glucocorticoid receptors. Glucocorticoids with their lipophilic properties can pass the blood–brain barrier, and bind to different types of receptors, thus influencing brain function and behavior. Moreover, cortisol, a steroid hormone that belongs to the glucocorticoid class of hormones, represents the most important stress hormone that exerts direct effects on genes and, through this action, chronic stress permanently inhibits production of granular cells—types of neurons with very small cell bodies within the granular layer of hippocampus—thus altering hippocampal function. The reduced activity of the HPA axis in maltreated children has an adaptive value, protecting them from the consequences of chronic hypercortisolism, at the cost of a reduced social competence and an increase in anxious and worried internalizing behavior. Hypercortisolism was found in children who had suffered multiple maltreatments, while hypocortisolism was found in children who had suffered physical abuse. Hypercortisolism characterized the group of children with internalizing symptomatology and was a consequence of prolonged exposure to infantile traumatic conditions [18]. An overwhelming stressful event when lasting over time, from birth to adolescence, involves considerable risks for the evolving brain architecture. Amygdala and hippocampus volume reductions are the consequence of traumas, especially when experienced early in life and repeated. Substantial deprivation and negligence may cause morphological abnormalities such as brain cortical atrophy and subsequent brain ventricular enlargement. The glucocorticoids increase, associated with alterations of other neurotransmitters (serotonin), is responsible for the inhibition of neurogenesis and the apoptosis of already formed neurons (see Table 18.1).

### 18.2.3 The Implications of Trauma on Cortical and Subcortical Structures

The ability of the brain and body to adapt to acute and chronic stress is well-known. The brain is the central organ for perception and response to stressors, and it is a target of allostatic load/overload along with the rest of the body. Biological embedding of early experiences interacts with physical environment. The individual adapts to particular environments and experiences to achieve reproductive success; however, these adaptations to one context may be maladaptive to another environment and, as a result, they may predispose the individual to greater allostatic load/overload. Traumatic events that affect the mental and psychophysiological functioning of the subject also determine detectable modifications in brain [7, 18, 51], which are more serious if they occur at earlier developmental stages.

The most vulnerable brain regions include those that slowly develop during the postnatal period, have a high density of glucocorticoid receptors, and continue to generate new neurons after birth. They are as follows: (1) HIPPOCAMPUS: encodes the memory of the experience lived in a space-time dimension and its functioning is necessary for explicit, declarative memory [52]; (2) AMYGDALA: plays a pivotal role in the evaluation of the emotional meaning of the afferent stimuli [53]; (3) THE PREFRONTAL CORTICAL REGION: has direct connections with the amygdala and is involved in the activation of stress and fear circuits [51]; (4) CORPUS CALLOSUM: potentially vulnerable to the impact of early exposure to excessive

**Table 18.1** The role of brain areas and neurotransmitters in stress

| Brain area | Neurotransmitter | Stress role | Function |
|---|---|---|---|
| Hippocampus | GABA | Inhibition of HPA axis | Memory function (recovery of episodic memories and the ability to remember when and where an event occurred), learning |
| Locus coeruleus | NA | Interaction with the limbic system | Learning, attention, concentration, cognitive vigilance, evaluation of sensorial input, autonomic responses, anxiety/stress mechanisms, brain reactivity, speed of information processing in the sensory and motor systems |
| Prefrontal cortex | GABA | Connection with the amygdala | Memory of experience, cognitive and behavioral responses to stress stimuli (exerts inhibitory control on the stress responses and reactions to emotions) |
| Hypothalamus–pituitary–adrenal cortex | CRH<br>ACTH<br>Glucocorticoids | Stress core circuit | Increased attention and body activation for immediate reactions |
| Nuclei of the dorsal raphe | 5-HT | Connection with the amygdala | Avoidant behaviors |
| Nuclei of the medial raphe | 5-HT | Connection with the hippocampus | Adaptive behaviors |
| Anterior pituitary | Endorphins | Inhibition of ACTH production | Analgesic and physiological action, inhibit pain and reduce panic |
| Amygdala | Glu, NA | Activation of HPA axis | Alert, emotional evaluation, and responses of stress |
| Substantia nigra<br>Ventral tegmental area | DA | Activation of the adrenal cortex for increasing adrenaline in peripheral nervous system | Fight/flight reactions, emotional processing of pleasure and reward, physiology of psychological reinforcement and learning processes |

*GABA* gamma-amino-butyric acid, *HPA* hypothalamus–pituitary–adrenal axis, *NA* noradrenaline, *CRH* corticotropin-releasing hormone, *ACTH* adrenocorticotropic hormone, *5-HT* serotonin, *Glu* glutamic acid, *DA* dopamine

levels of stress hormones, which suppress glial cell division critical for myelination [51]; (5) CEREBELLAR VERMIS: a brain region that is extraordinarily sensitive to the effects of early maltreatment [51].

The cerebral cortex is involved in high-level cognitive functions (attention; working memory, reasoning, planning, language) and determines the inclusion or exclusion from conscious processing of salient information, being also engaged in information retrieval processes. The amygdala is important in evaluating the emotional significance of events, especially for those related to anger, aggression, and fear. It acts with the hippocampus, actively modulating the stored information. The hippocampus is involved in learning and memory processes and is fundamental for the recovery of episodic memories and the ability to remember when and where an event occurred (temporal and spatial components). All these cerebral structures play a key role in assessing a situation, selecting an appropriate response (prefrontal cortex) in reference to emotional content (amygdala) and content of past experiences (hippocampus and prefrontal cortex).

Studies on humans showed a volume of hippocampus shrinkage not only in mild cognitive impairment and Alzheimer's disease (AD) [54] but also in type 2 diabetes [55], chronic major depression [56], Cushing disease [57], and PTSD [58]. Also, in non-pathological conditions such as chronic stress [59], chronic inflammation [60], lack of physical activity [61], and jet lag [62], smaller hippocampal or temporal lobe volumes were reported [17].

Several neuroimaging studies were performed with functional magnetic resonance imaging (fMRI) on subjects with PTSD and healthy control subjects, along with nuclear medicine studies (PET, positron emission tomography) on traumatized subjects [63]. In these studies, hippocampal reduction in subjects with PTSD, such as war veterans, women with history of prolonged sexual abuse, people undergoing protracted physical and psychological abuse and CT, was reported [64, 65]. Neuroimaging studies using proton magnetic resonance spectroscopy, a method to evaluate in vivo the metabolism of brain regions, showed reduction in N-acetyl aspartate (NAA) levels, a marker of neuronal integrity, in the hippocampus of adults with PTSD. Bremner et al. [65] and Stein [66] reported a reduction in left hippocampal volume in adults with CT and a current diagnosis of PTSD or dissociative identity disorder. Further, people with smaller hippocampus are more sensitive and more vulnerable to develop post-traumatic reactions. Reduction of hippocampal volume in traumatized individuals may reflect the accumulated toxic effects after repeated exposure to high levels of glucocorticoids. Studies using functional neuroimaging have also shown that patients with PTSD have a deficit in hippocampal activation during verbal declarative memory and reduced ability to formulate adequate responses to stress, as well as deficit in discriminating between safe and unsafe environmental settings.

The amygdala shows quite different responses to acute and chronic stress compared to hippocampus. It responds to glucocorticoid stimulation in the formation of emotionally charged memories [67, 68]. Acute stress causes delayed formation of dendritic spines in basolateral amygdala neurons, and increase of anxiety after 10 days from exposure [69]. Chronic stress of the same type, which in the hippocampus impairs dentate gyrus neurogenesis and causes dendritic shrinkage and spine loss in Ammon's horn neurons, determines expansion of dendrites in the basolateral amygdala [70], causing spine downregulation in the medial amygdala [71]. The latter is dependent from the tissue plasminogen activator, a key mediator of spine plasticity [71]. Amygdala hyperactivity is reported in major depression, as well as in anxiety disorders, such as PTSD [72], and amygdala enlargement has been reported in acute depression [73]. Functional imaging studies have revealed hyper-reactivity of the amygdala in subjects with PTSD during the presentation of stressful stimuli, specifically memories of traumatic events, and also during emotional stimuli such as faces of sad, happy, and angry people.

A PET study performed on subjects with PTSD showed, during the exposure to traumatic stories, an increased activity only in the right hemisphere, in areas most involved in emotional activation and more intimately associated with the amygdala. As regard to CT, imaging studies on amygdala volume in abuse survivors reported an 8% reduction in bilateral amygdaloid volume in women with a history of childhood abuse and a diagnosis of borderline personality disorders [74]. Teicher et al. analyzed amygdala volume in young adults with a history of repeated sexual abuse and found an 8.4% reduction in the size of the left amygdala [51]. The activation of these structures was accompanied by increased activity of the right visual cortex, which reflects the visual re-experience of trauma reported by patients. One of the most significant aspects of these studies is the finding that left Broca area, a cerebral region responsible for translating experiences into communicable words, was completely silent, "switched off," without detectable activation during re-experience. This latter finding correlates with what is called "dumb terror" and to the tendency of PTSD subjects to experience emotions in the form of physical states rather than as verbally codified experiences. It has been hypothesized that difficulties of traumatized patients in translating their sensations into words could be linked to structural and functional changes in the activity of brain regions responsible for retrieving emotional memories and for language.

These observations support the hypotheses of LeDoux [75] according to which emotional memories can be established without a conscious evaluation of the encoded information. Specifically, sensory information entering the CNS through the sense organs passes to the thalamus which, in turn, sends this raw sensory information to the amygdala and prefrontal cortex for further evaluation; the amygdala interprets the emotional value of incoming information, codes its emotional meaning, and passes it to brain areas that control behavioral, autonomous, and neurohormonal response systems. In other words, the amygdala transforms sensory stimuli into emotional and hormonal signals, initiating and controlling emotional responses. Because the input from the thalamus reaches the amygdala before the processed information arrives from the neocortex, LeDoux suggests that this first input from the thalamus "prepares" the amygdala to process the information that comes later from the cortex so that the emotional evaluation of sensory input precedes conscious emotional experience. Subjects can, therefore, be activated physiologically and by hormonal pathways even before they are able to consciously evaluate what they are reacting to. Once the amygdala has assigned emotional value to sensory inputs, other brain structures, including the hippocampus and prefrontal cortex, organize the information and integrate it with the preexisting ones. The intensity of hippocampal activation depends on the intensity of the input coming from the amygdala: the greater the value assigned by the amygdala, the more stable input encoding will be, thus giving rise to a more long-lasting memory. However, while moderate or high amygdala activation promotes long-term enhancement of explicit hippocampus-mediated memory, excessive stimulation damages hippocampus functioning. When this happens, the sensory impressions of the experience are stored in memory but, since the hippocampus is not able to fulfill its integrative function and to support the spatio-temporal contextualization of the information, these impressions are not organized in a unitary way: the experience is encoded and later recovered as affective states, sensorimotor modalities, physical sensations, and visual images, perceived as unrelated, and separated from other life experiences.

In the chronic stress models that lead to amygdala neuronal hypertrophy and shrinkage of dendrites in hippocampus, shrinkage of dendrites and loss of spines throughout the medial prefrontal cortex, and dendrites expansion in the orbitofrontal cortex were also hypothesized [76]. Because the orbitofrontal cortex is involved in determining the saliency of reward or punishment [77], this may reinforce the changes observed in the basolateral amygdala. For the medial prefrontal cortex, a stress-induced impairment was linked to poor cognitive flexibility in both animal and human studies [76, 78, 79].

Myelinated regions, such as the corpus callosum, are potentially vulnerable to the impact of early exposure to excessive levels of stress hormones, which suppress the glial cell division critical for myelination [80]. Pioneering studies by Denenberg showed that corpus callosum volume is markedly affected by early experience and that the effects are gender-dependent [81, 82]. Other authors [83] provided the first indication that the corpus callosum may be adversely affected by CT: they noted a marked reduction in the volume of the middle portions of the corpus callosum in child psychiatric patients (in particular boys) with a substantiated history of abuse or neglect versus control subjects. Likewise, De Bellis [84] showed reduced corpus callosum volume in children with a history of abuse and PTSD, with boys more affected than girls. More recently, it has been found that the corpus callosum is particularly vulnerable to the effects of neglect in boys, while seems to be more vulnerable to the adverse effects of sexual abuse in girls [85]. Reduced size of the corpus callosum was associated with reduced communication between left and right hemispheres [86].

Another brain region that is extraordinarily sensitive to the effects of early maltreatment is the cerebellar vermis [7]. The vermis has a protracted period of postnatal development and may produce granule cells postnatally [87]. The vermis has also the highest density of glucocorticoid receptors during development, exceeding that of the hippocampus [88–90], and may be particularly vulnerable to the effects of stress hormones [91, 92]. Anderson and colleagues [93] studied the association between activity in the cerebellar vermis and symptoms of limbic irritability in both healthy young adult control subjects and young adults with a history of repeated sexual abuse. They found a significant correlation between activity in the cerebellar vermis and the degree of limbic irritability in both healthy young adult control subjects and young adults with a history of repeated sexual abuse. However, at any level of limbic symptomatology, there was a marked decrease in perfusion of the vermis in subjects with abuse history, suggesting a functional impairment in the activity of the cerebellar vermis. These findings indicate that early trauma could interfere with the development of the vermis and produce neuropsychiatric symptoms.

Overall, childhood stress is linked to neuronal irritability, electroencephalographic abnormalities, and symptoms suggestive of temporal lobe epilepsy. It is also associated with reduced growth of the left hemisphere (including the neocortex, hippocampus, and amygdala), reduced size of the corpus callosum, and attenuated activity in the cerebellar vermis. There is a close association between the effects of early maltreatments on brain development and the array of psychiatric symptoms observed in abused patients. These modifications in development are designed to adapt individuals to cope with high levels of stress or deprivation that they may expect to encounter throughout the rest of their life. In this way, people, and therefore their brain, select an alternative developmental pathway that will best match their wiring and configuration to the environment in which, based on early experience, they expected to survive and reproduce.

A future research goal should provide a neurobiological framework for understanding positive health, positive effect, self-efficacy, and self-esteem. Moreover, it should necessarily understand how these components are biologically embedded in a nurturing environment by epigenetic influences, including effects on reactive alleles in the genome. Finally, it should clarify how children's experiences during development contribute to such biological processes and how they may affect the individual response to stress later in their life.

### 18.2.4 Trauma and Disturbances Correlated to Neurobiological Alterations

#### 18.2.4.1 CTs and Chronic Diseases in Adulthood

As mentioned in the previous paragraphs, CTs are psychosocial stressors that occur in childhood and have been shown to be associated with significant morbidity and mortality in adulthood. Increased risky behaviors that contribute to poor health have been associated with increased incidence of chronic diseases in adulthood, such as ischemic heart disease, type II diabetes, obesity, autoimmune disease, liver disease, and depression [1, 5, 94–97]. It has been shown that individuals with six or more CTs die nearly 20 years earlier than those with no experience of CT history. One explanation of this mechanism is that CTs are toxic psychosocial stressors that contribute to physiological dysregulation. Such dysregulation occurs in the stress response and influences metabolic hormone levels, inflammatory markers, and immune mediators, thus providing pathways to cardiovascular, metabolic, and immune diseases [98]. More specifically, recent evidence suggests that three CTs stand out as contributing factors of low socioeconomic status, maltreatment, and social isolation [99]. Low socioeconomic status in childhood is a recognized risk factor for age-related disease, such as cardiovascular disease [100]. Childhood socioeconomic disadvantage predicts age-related-disease risks, such as elevated inflammation levels and the clustering of metabolic risk markers in adulthood [14, 101, 102]. Conversely, the impact of low childhood socioeconomic status on later depression risk is still debated [103]. Some studies showed that childhood maltreatment could also contribute to age-related-disease risks. Childhood maltreatment is a recognized predictor of adult depression [104], and emerging evidence suggests that childhood maltreatment could also contribute to the risk of inflammation and metabolic risk in adult life [5, 105, 106]. Finally, it is increasingly recognized that subjects experiencing social isolation are at greater risk for disease [107–109]. Adverse psychosocial experiences, such as social isolation, could be particularly detrimental in the developing child [110], and initial findings suggest that childhood social isolation may have durable effects on the clustering of metabolic risk markers in adult life [111]. Most studies have examined the association between a single CT and a single age-related-disease risk (e.g., childhood maltreatment and adult depression [103] or low childhood socioeconomic status and elevated adult inflammation [101]). Therefore, the following three important issues are still unanswered: (1) Are the effects of different CTs distinct from each other? To the extent that multiple CTs co-occur in the same individuals, is it possible that their effects on adult health are not independent and unique? (2) Are the effects of different CTs pervasive in different biological systems? Could each CT influence a single age-related-disease risk in a single stress-sensitive system or each CT influence multiple age-related disease risks? (3) Are the effects of CTs independent from the influence of other known risk factors for age-related disease? Have CTs to coexist with other developmental risk factors for poor adult health, including low birth weight [112], family history of disease [113], and childhood overweight [114]? Thus, it is important to test whether CTs exert an influence on adult outcomes that is independent of these established risk factors.

#### 18.2.4.2 CTs and Dementia Risk

A number of studies confirm that stressful life events may have adverse effects on cognitive function, especially in old age [115], being associated with increased risk for dementia and AD in particular [116, 117]. Previous studies have reported that older APOE-E4 allele carriers, when exposed to stress, suffered from worse cognitive functioning, compared to older persons with stress, not carriers of the APOE-E4 allele [118, 119].

The relationship between childhood adversity and cognitive decline or dementia in older adults is also emerging across various populations. Some longitudinal cohort studies demonstrated an association between dementia, particularly AD, and adverse childhood factors including death of a parent in childhood [120], or lower socioeconomic status [121] (Swedish and US cohorts, respectively). In the Longitudinal Aging Study Amsterdam, older adults with a history of childhood stress had more rapid cognitive decline over a 10-year period, but only in the context of depressive symptoms [122]. Similarly, in a French cohort, late-life depression and cognitive decline were associated with some early stressful life events [123, 124]. A large cross-sectional study in central Africa [125] also found that a diagnosis of mild cognitive impairment was associated with a higher number of early stressful life events, although here the association was limited to those without depressive symptoms, and was not found for dementia diagnosis. Finally, a study conducted on older Aboriginal Australians clearly showed the impact of childhood stress on emotional health and dementia, in particular AD [126].

The mechanisms underlying relationship between childhood stress, adult mental health problems, and late-life dementia include the following: reduced development of prefrontal cortex and frontostriatal limbic circuitry, with corresponding reduced executive functioning, developmental dysfunctions of the HPA axis, that secretes the stress hormones, and stress response throughout the lifespan [127–129]. As previously discussed, cortisol is able to cross the blood–brain barrier and to bind receptors in various regions of the brain known to be involved in memory and learning, such as the hippocampus, amygdala, and frontal lobes, consequently influencing cognitive function [130]. Then, subsequent chronic inflammation could precipitate the amyloid cascade which, coupled with reduced brain/cognitive reserve and accumulated hippocampal insults/atrophy, may accelerate the onset of AD [127, 131]. Furthermore, epigenetic and intergenerational factors could also contribute to increased susceptibility for some individuals to the neurobiological effects of the exposure to early life stress [132].

There is evidence that strongly suggests potent effects of early life stress at the cellular level, which can be evidenced by network electrophysiology. Most of this work focuses on LTP, aiming at better understanding effects of early experience on memory, and also pathophysiology-related modifications in epilepsy models [133]. Moreover, there are several studies suggesting that experience of CTs can affect neurodevelopment through epigenetic mechanisms. Thus, early adverse environment can lead to changes in gene expression causing functional and structural changes to brain, neuroendocrine, autonomic and immune functions that may affect the way individuals respond to stress later in life [1, 127, 134–137]. Beyond affecting the biological response to adult stressors, childhood adversity increases the likelihood of engaging in health-risk behaviors in adolescence and adulthood [5, 138–140], and may adversely affect adult-life social resources, coping skills and emotional functioning/distress [141]. All such factors may lead to a disabling process [142], generating negative changes in body structure and function (impairments), activity limitations, and participation restrictions. In other words, CTs may be associated with increased rates of disability, which increase with the number of CT categories experienced. Indeed, there is evidence that childhood adversity can cause persistent alterations in the biological stress response, which may affect how individuals react to disabling impairments in adulthood.

Recognition of these facts provides clear opportunity for early interventions [143, 144].

#### 18.2.4.3 CTs, Amnesia for Childhood Events, and Memory Integration in Adult Survivors

Several studies have been conducted in order to investigate the impact of CTs on memory. Disturbances in autobiographical memories [145, 146] related to childhood are constituted by memory disruptions, and characterized by the inability to remember events from this life period. Despite having been identified by Freud nearly 100 years ago [147], childhood autobiographical memory disturbances remain poorly understood. Relationships between childhood physical and sexual abuse and deficits in short-term memory [148] and childhood autobiographical memory disturbances [149] have been observed among adult survivors. Brown and collaborators [150] showed a strong graded relation between the accumulation of CT experiences and autobiographical memory disturbances with no gender differences. These findings are in line with those of Edwards and collaborators [149], and suggest that experiencing multiple forms of CT is associated with increased likelihood of childhood autobiographical memory disturbances.

Current evidence from epidemiological and neurobiological studies suggests that trauma occurring in early life, such as child maltreatment and related traumatic stressors, could alter brain structure and function resulting in possible long-term effects [1, 151, 152]. As a consequence, the normal development of the hippocampus, which plays a role in memory storage and retrieval, may be vulnerable to the upregulation of the HPA axis that occurs in maltreated children [153–158]. In fact, lower hippocampal volume has been observed in women sexually abused during their childhood [9, 13]. The prevalence of childhood autobiographical memory disturbance decreased with increasing age, perhaps because of older respondents' fear of dementia-related memory loss or of stereotypes related to aging [149]. However, in Brown's study, participants were unaware of the concept that autobiographical memory disturbance is related to their childhood experiences, and there is no reason to suspect systematic over-reporting of traumatic experiences by younger persons with autobiographical memory disturbance.

Schiffer et al. [159] used auditory-evoked potentials to study laterality and hemispheric integration of memories in adults with a history of childhood maltreatment who were currently well functioning and who had no active psychiatric diagnosis. Subjects were asked to actively recall neutral or work-related memory, and were then asked to recall a disturbing memory from childhood and the associated affect. In normal subjects, both hemispheres appeared to be equally involved in the recall of these memories while, in adults with a history of CT, there was a dramatic difference. During recall of the neutral memory, there was a marked suppression of the evoked potential response over the left hemisphere indicative of increased left hemisphere processing. During recall of the disturbing memory, there was a robust shift in laterality with the evoked potential response being suppressed over the right hemisphere, which is indicative of enhanced right hemisphere activation. These results suggest that early maltreatment is associated with increased hemispheric laterality and decreased hemispheric integration.

#### 18.2.4.4 CTs and Other Disturbances: Sleep Disorders, Language, and Vasovagal Syncope

One study conducted in the USA showed that adults with experience of CTs could have multiple sleep disorders [160]. Sleep disorders are estimated to affect 50–70 million Americans, and have cumulative effects over time [161]. They do not only contribute to diminished work-related productivity but also lead to severe health outcomes such as obesity, hypertension, diabetes, occupational injuries, depression, and premature mortality [162–166]. The exposure to stress or trauma in childhood causes general circadian dysregulation, thereby disrupting sleep regulation. However, the mechanism through which CT experiences alter circadian regulation is unclear. There is substantial evidence of a key role of the CRH in sleep disorders, such that elevated CRH and HPA axis hyperactivity are associated with reduced sleep [167, 168]. It is therefore possible that childhood adversity increases CRH reactivity, which subsequently affects sleep quality, but researches concerning the association between CT experiences and cortisol are still inconclusive. Several investigators have shown that a history of childhood abuse or maltreatment is associated with increased cortisol reactivity, supporting the proposed hypothesis [169–173], while others have shown that CT experiences are associated with diminished cortisol reactivity [174–176]. Thus, it is unclear whether or not elevated CRH explains the association with poorer sleep.

Another proposed mechanism of this association is the increased neuronal activity, as measured by electroencephalography (EEG), particularly in patients with insomnia. There is evidence that individuals suffering from insomnia have increased β EEG activity [177–179]. Additionally, this mechanism may be linked to the previously discussed one involving elevated CRH and neuronal hyperactivity. Finally, another plausible mechanism that could explain the relationship between experiences of CTs and adult sleep outcomes involves the development of a mood disorder due to traumatic stress experienced during childhood or adolescence. Childhood and adolescence are periods of rapid brain development. Neuroimaging studies have provided evidence that traumatic experiences during this developmental stage interfere with normal brain development and impair the capability to develop life skills necessary for becoming independent adults [180]. This impairment is often associated with psychiatric disorders, such as PTSD and major depressive disorder, of which sleep disturbance is a common symptom [180]. However, each proposed mechanism does not fully explain the observed relationships between CTs and sleep disturbances; therefore, it is more probable that these mechanisms may interact with one another involving both biological and social pathways [160].

In the few published studies, deficits in language development, related to potential abnormalities in the superior temporal gyrus, were also reported in maltreated children [181]. Moreover, it has been demonstrated that CT, in particular childhood sexual and physical abuse and maltreatment, can contribute to a lifelong vasovagal syncope tendency [182–184]. A possible explanation of the relations between CT and vasovagal syncope is that a stress response could be decoupled from the original acute stressor (i.e., such that a stress response occurs despite no ongoing threat—as in panic disorder or PTSD). Thus, a vasovagal syncope tendency may become habituated in youth exposed to CT and persist throughout one's lifetime. Accordingly, vasovagal syncope tendency and psychological responses to CT could therefore have common neural substrates (i.e., CT may lead to alterations in brain regions with relevance to both vasovagal syncope tendency and psychological responses) [185, 186].

## 18.3 Conclusions

Neither youth nor innocence provides protection from the ravages of fate. In fact, exposure to traumatic stress may occur at any point along the developmental continuum. Since it occurs during a time when the brain is undergoing enormous changes, the impact of severe stress may leave an indelible print on the brain structure and function [7] as it is designed to be sculpted into its final configuration by the effects of early experience [187].

The ongoing effects of CT need to be recognized as people grow older, particularly in terms of dementia prevention and care, as well as in populations with greater exposure to childhood adversity. In particular, data from the literature indicate that children exposed to CT experiences are more likely to have age-related-disease risks in adult life, regardless of their familial liability for diseases, birth weight, childhood weight, and adult low socioeconomic status and health behaviors. Thus, promoting healthy psychosocial experiences for children is necessary to improve the quality of longer lives and reduce health-care costs of age-related diseases [99]. Moreover, early detection of history of CT experiences is critical in developing trauma-informed care for subsequent health consequences and in preventing further adverse health outcomes.

## References

1. Anda RF, Felitti VJ, Bremner JD, Walker JD, Whitfield C, Perry BD, et al. The enduring effects of abuse and related adverse experiences in childhood. A convergence of evidence from neurobiology and epidemiology. Eur Arch Psychiatry Clin Neurosci. 2006;256(3):174–86.
2. Brown DW, Anda RF, Tiemeier H, Felitti VJ, Edwards VJ, Croft JB, et al. Adverse childhood experiences and the risk of premature mortality. Am J Prev Med. 2009;37(5):389–96.
3. Chapman DP, Whitfield CL, Felitti VJ, Dube SR, Edwards VJ, Anda RF. Adverse childhood experiences and the risk of depressive disorders in adulthood. J Affect Disord. 2004;82(2):217–25.
4. Adverse Childhood Experiences Study. Centers for Disease Control and Prevention Web site. 2009. http://www.cdc.gov/nccdphp/
5. Felitti VJ, Anda RF, Nordenberg D, Williamson DF, Spitz AM, Edwards V, et al. Relationship of childhood abuse and household dysfunction to many of the leading causes of death in adults. The Adverse Childhood Experiences (ACE) Study. Am J Prev Med. 1998;14(4):245–58.
6. Wordsworth W. My heart leaps up. The complete poetical works. London: Macmillan; 1802; Bartleby.com
7. Teicher MH, Andersen SL, Polcari A, Anderson CM, Navalta CP. Developmental neurobiology of childhood stress and trauma. Psychiatr Clin North Am. 2002;25(2):397–426.
8. Perry BD, Pollard R. Homeostasis, stress, trauma, and adaptation. A neurodevelopmental view of childhood trauma. Child Adolesc Psychiatr Clin N Am. 1998;7(1):33–51.
9. Teicher MH. Wounds that time wouldn't heal: the neurobiology of childhood abuse. Cerebrum. 2000;2:50–67.
10. De Bellis MD, Thomas LA. Biologic findings of post-traumatic stress disorder and child maltreatment. Curr Psychiatry Rep. 2003;5(2):108–17.
11. Gorman JM, Mathew S, Coplan J. Neurobiology of early life stress: nonhuman primate models. Semin Clin Neuropsychiatry. 2002;7(2):96–103.

12. Gutman DA, Nemeroff CB. Neurobiology of early life stress: rodent studies. Semin Clin Neuropsychiatry. 2002;7(2):89–95.
13. Heim C, Nemeroff CB. The role of childhood trauma in the neurobiology of mood and anxiety disorders: preclinical and clinical studies. Biol Psychiatry. 2001;49(12):1023–39.
14. Repetti RL, Taylor SE, Seeman TE. Risky families: family social environments and the mental and physical health of offspring. Psychol Bull. 2002;128(2):330–66.
15. Bremner JD, Vermetten E. Stress and development: behavioral and biological consequences. Dev Psychopathol. 2001;13:473–89.
16. Miller DB, O'Callaghan JP. Do early-life insults contribute to the late-life development of Parkinson and Alzheimer diseases? Metabolism. 2008;57(Suppl 2):S44–S49.
17. McEwen BS. Physiology and neurobiology of stress and adaptation: central role of the brain. Physiol Rev. 2007;87(3):873–904.
18. De Bellis MD, Zisk A. The biological effects of childhood trauma. Child Adolesc Psychiatr Clin N Am. 2014;23(2):185–222.
19. De Bellis MD. Developmental traumatology: a contributory mechanism for alcohol and substance use disorders. Psychoneuroendocrinology. 2002;27(1–2):155–70.
20. Charney DS, Deutch AY, Krystal JH, Southwick SM, Davis M. Psychobiologic mechanisms of posttraumatic stress disorder. Arch Gen Psychiatry. 1993;50(4):295–305.
21. De Bellis MD, Putnam FW. The psychobiology of childhood maltreatment. Child Adolesc Psychiat Clin North Am. 1994;3:663–77.
22. Bremner JD, Innis RB, Ng CK, Staib LH, Salomon RM, Bronen RA, et al. Positron emission tomography measurement of cerebral metabolic correlates of yohimbine administration in combat-related posttraumatic stress disorder. Arch Gen Psychiatry. 1997;54(3):246–54.
23. Bremner JD, Krystal JH, Southwick SM, Charney DS. Noradrenergic mechanisms in stress and anxiety: II. Clin Stud Synapse. 1996;23(1):39–51.
24. Southwick SM, Krystal JH, Morgan CA, Johnson D, Nagy LM, Nicolaou A, et al. Abnormal noradrenergic function in posttraumatic stress disorder. Arch Gen Psychiatry. 1993;50(4):266–74.
25. Martin SJ, Grimwood PD, Morris RGM. Synaptic plasticity and memory: an evaluation of the hypothesis. Annu Rev Neurosci. 2000;23:649–711.
26. Neves G, Cooke SF, Bliss TVP. Synaptic plasticity, memory and the hippocampus: a neural network approach to causality. Nat Rev Neurosci. 2008;9(1):65–75.
27. Pastalkova E, Serrano P, Pinkhasova D, Wallace E, Fenton AA, Sacktor TC. Storage of spatial information by the maintenance mechanism of LTP. Science. 2006;313(5790):1141–4.
28. Whitlock JR, Heynen AJ, Shuler MG, Bear MF. Learning induces long-term potentiation in the hippocampus. Science. 2006;313(5790):1093–7.
29. Lømo T. Frequency potentiation of excitatory synaptic activity in the dentate are of the hippocampal formation. Acta Physiol Scand. 1966;68(277):128.
30. Bliss TV, Lømo T. Long-lasting potentiation of synaptic transmission in the dentate area of the anaesthetized rabbit following stimulation of the perforant path. J Physiol. 1973;232:331–56.
31. Barnes CA, Jung MW, McNaughton BL, Korol DL, Andreasson K, Worley PF. LTP saturation and spatial learning disruption: effects of task variables and saturation levels. J Neurosci. 1994;14(10):5793–806.
32. Morris RG, Anderson E, Lynch GS, Baudry M. Selective impairment of learning and blockade of long-term potentiation by an N-methyl-D-aspartate receptor antagonist, AP5. Nature. 1986;319(6056):774–6.
33. Moore RY, Bloom FE. Central catecholamine neuron systems: anatomy and physiology of the norepinephrine and epinephrine systems. Annu Rev Neurosci. 1979;2:113–68.
34. Segal M. Norepinephrine modulates reactivity of hippocampal cells to chemical stimulation in vitro. Exp Neurol. 1982;77(1):86–93.
35. Madison DV, Nicoll R. Cyclic adenosine 3¢,5¢-monophosphate mediates beta-receptor actions of noradrenaline in rat hippocampal pyramidal cells. J Physiol. 1986;372:245–59.
36. Frey S, Bergado JA, Frey JU. Modulation of late phases of long-term potentiation in rat dentate gyrus by stimulation of the medial septum. Neuroscience. 2003;118(4):1055–62.

37. Thomas SA, Palmiter RD. Disruption of the dopamine beta-hydroxylase gene in mice suggests roles for norepinephrine in motor function, learning, and memory. Behav Neurosci. 1997;111(3):579–89.
38. Czyrack A, Maćkowiak M, Chocyk A, Fijal K, Wedzony K. Role of glucocorticoids in the regulation of dopaminergic neurotransmission. Pol J Pharmacol. 2003;55(6):667–74.
39. Dalmman MF, Scribner KS, Paecoraro N. Chronic stress induced effects of corticosterone on brain: direct and indirect. Ann N Y Acad Sci. 2004;1018:141–50.
40. Marinelli M, Rudick CN, Hu XT, White FJ. Excitability of dopamine neurons: modulation and physiological consequences. CNS Neurol Disord Drug Targets. 2006;5(1):79–97.
41. Piazza PV, Barrot M, Rouge-Pont F, Marinelli M, Maccari S, Abrous DN, et al. Suppression of glucocorticoid secretion and antipsychotic drugs have similar effects on the mesolimbic dopaminergic transmission. Proc Natl Acad Sci U S A. 1996;93(26):15445–50.
42. Cabib S, Puglisi-Allegra S. Stress, depression and the mesolimbic dopamine system. Psychopharmacology. 1996;128(4):331–42.
43. Butts KA, Phillips AG. Glucocorticoid receptors in the prefrontal cortex regulate dopamine efflux to stress via descending glutamatergic feedback to the ventral tegmental area. Int J Neuropsychopharmacol. 2013;16(8):1799–807.
44. Reinhard JF Jr, Bannon MJ, Roth RH. Acceleration by stress of dopamine synthesis and metabolism in prefrontal cortex: antagonism by diazepam. Naunyn Schmiedeberg's Arch Pharmacol. 1982;318(4):374–7.
45. Weele CMV, Siciliano CA, Tye KM. Dopamine tunes prefrontal outputs to orchestrate aversive processing. Brain Res. 1713;2018:16–31.
46. Moore H, Rose HJ, Grace AA. Chronic cold stress reduces the spontaneous activity of ventral tegmental dopamine neurons. Neuropsychopharmacology. 2001;24(4):410–9.
47. Engler D, Redei E, Kola I. The corticotropin-release inhibitory factor hypothesis: a review of the evidence for the existence of inhibitory as well as stimulatory hypophysiotropic regulation of adrenocorticotropin secretion and biosynthesis. Endocr Rev. 1999;20(4):460–500.
48. Goodman M, New A, Siever L. Trauma, genes, and the neurobiology of personality disorders. Ann N Y Acad Sci. 2004;1032:104–16.
49. Ressler KJ, Nemeroff CB. Role of serotonergic and noradrenergic systems in the pathophysiology of depression and anxiety disorders. Depress Anxiety. 2000;12(Suppl 1):2–19.
50. Kim YK, Amidfar M, Won E. A review on inflammatory cytokine-induced alterations of the brain as potential neural biomarkers in post-traumatic stress disorder. Prog Neuropsychopharmacol Biol Psychiatry. 2019;91:103–12.
51. Teicher MH, Andersen SL, Polcari A, Anderson CM, Navalta CP, Kim DM. The neurobiological consequences of early stress and childhood maltreatment. Neurosci Biobehav Rev. 2003;27(1–2):33–44.
52. Squire LR, Zola-Morgan S. The medial temporal lobe memory system. Science. 1991;253(5026):1380–6.
53. LeDoux JE. Emotion and amygdala. New York: Wiley Liss; 1992.
54. de Leon MJ, George AE, Golomb J, Tarshish C, Convit A, Kluger A, et al. Frequency of hippocampus atrophy in normal elderly and Alzheimer's disease patients. Neurobiol Aging. 1997;18(1):1–11.
55. Gold SM, Dziobek I, Sweat V, Tirsi A, Rogers K, Bruehl H, et al. Hippocampal damage and memory impairments as possible early brain complications of type 2 diabetes. Diabetologia. 2007;50(4):711–19.
56. Sheline YI. Neuroimaging studies of mood disorder effects on the brain. Biol Psychiatry. 2003;54(3):338–52.
57. Starkman MN, Giordani B, Gebarski SS, Berent S, Schork MA, Schteingart DE. Decrease in cortisol reverses human hippocampal atrophy following treatment of Cushing's disease. Biol Psychiatry. 1999;46(12):1595–602.
58. Gurvits TV, Shenton ME, Hokama H, Ohta H, Lasko NB, Gilbertson MW, et al. Magnetic resonance imaging study of hippocampal volume in chronic, combat-related posttraumatic stress disorder. Biol Psychiatry. 1996;40(11):1091–9.

59. Gianaros PJ, Jennings JR, Sheu LK, Greer PJ, Kuller LH, Matthews KA. Prospective reports of chronic life stress predict decreased grey matter volume in the hippocampus. NeuroImage. 2007;35(2):795–803.
60. Marsland AL, Gianaros PJ, Abramowitch SM, Manuck SB, Hariri AR. Interleukin-6 covaries inversely with hippocampal grey matter volume in middle-aged adults. Biol Psychiatry. 2008;64(6):484–90.
61. Erickson KI, Prakash RS, Voss MW, Chaddock L, Hu L, Morris KS, et al. Aerobic fitness is associated with hippocampal volume in elderly humans. Hippocampus. 2009;19(10):1030–9.
62. Cho K. Chronic 'jet lag' produces temporal lobe atrophy and spatial cognitive deficits. Nat Neurosci. 2001;4(6):567–8.
63. Bremner JD, Vythilingam M, Vermetten E, Southwick SM, McGlashan T, Nazeer A, et al. MRI and PET study of deficits in hippocampal structure and function in women with childhood sexual abuse and posttraumatic stress disorder. Am J Psychiatry. 2003;160(5):924–32.
64. Bremner JD, Randall P, Scott TM, Bronen RA, Seibyl JP, Southwick SM, et al. MRI-based measurement of hippocampal volume in patients with combat-related posttraumatic stress disorder. Am J Psychiatry. 1995;152(7):973–81.
65. Bremner JD, Randall P, Vermetten E, Staib L, Bronen RA, Mazure C, et al. Magnetic resonance imaging-based measurement of hippocampal volume in posttraumatic stress disorder related to childhood physical and sexual abuse--a preliminary report. Biol Psychiatry. 1997;41(1):23–32.
66. Stein MB. Hippocampal volume in women victimized by childhood sexual abuse. Psychol Med. 1997;27:951–9.
67. Roozendaal B, Hahn EL, Nathan SV, de Quervain DJ, McGaugh JL. Glucocorticoid effects on memory retrieval require concurrent noradrenergic activity in the hippocampus and basolateral amygdala. J Neurosci. 2004;24(37):8161–9.
68. McEwen BS. Brain on stress: how the social environment gets under the skin. Proc Natl Acad Sci U S A. 2012;109(Suppl 2):17180–5.
69. Mitra R, Jadhav S, McEwen BS, Vyas A, Chattarji S. Stress duration modulates the spatiotemporal patterns of spine formation in the basolateral amygdala. Proc Natl Acad Sci U S A. 2005;102(26):9371–6.
70. Vyas A, Mitra R, Shankaranarayana Rao BS, Chattarji S. Chronic stress induces contrasting patterns of dendritic remodeling in hippocampal and amygdaloid neurons. J Neurosci. 2002;22(15):6810–8.
71. Bennur S, Shankaranarayana Rao BS, Pawlak R, Strickland S, McEwen BS, Chattarji S. Stress-induced spine loss in the medial amygdala is mediated by tissue-plasminogen activator. Neuroscience. 2007;144(1):8–16.
72. Drevets WC. Neuroimaging studies of mood disorders. Biol Psychiatry. 2000;48(8):813–29.
73. Frodl T, Meisenzahl EM, Zetzsche T, Born C, Jager M, Groll C, et al. Larger amygdala volumes in first depressive episode as compared to recurrent major depression and healthy control subjects. Biol Psychiatry. 2003;53(4):338–44.
74. Driessen M, Herrmann J, Stahl K, Zwaan M, Meier S, Hill A, et al. Magnetic resonance imaging volumes of the hippocampus and the amygdala in women with borderline personality disorder and early traumatization. Arch Gen Psychiatry. 2000;57(12):1115–22.
75. LeDoux JE. The emotional brain: the mysterious underpinnings of emotional life. New York, NY: Simon & Schuster; 1996.
76. Liston C, Miller MM, Goldwater DS, Radley JJ, Rocher AB, Hof PR, et al. Stress-induced alterations in prefrontal cortical dendritic morphology predict selective impairments in perceptual attentional set-shifting. J Neurosci. 2006;26(30):7870–4.
77. Schoenbaum G, Roesch M. Orbitofrontal cortex, associative learning, and expectancies. Neuron. 2005;47(5):633–6.
78. Dias-Ferreira E, Sousa JC, Melo I, Morgado P, Mesquita AR, Cerqueira JJ, et al. Chronic stress causes frontostriatal reorganization and affects decision-making. Science. 2009;325(5940):621–5.
79. Liston C, McEwen BS, Casey BJ. Psychosocial stress reversibly disrupts prefrontal processing and attentional control. Proc Natl Acad Sci U S A. 2009;106(3):912–7.

80. Lauder JM. Hormonal and humoral influences on brain development. Psychoneuroendocrinology. 1983;8(2):121–55.
81. Berrebi AS, Fitch RH, Ralphe DL, Denenberg JO, Friedrich VL Jr, Denenberg VH. Corpus callosum: region-specific effects of sex, early experience and age. Brain Res. 1988;438(1–2):216–24.
82. Denenberg VH. In: Glick SD, editor. Hemispheric laterality, behavioral asymmetry, and the effects of early experience in rats. Orlando, FL: Academic Press; 1985.
83. Teicher MH, Ito Y, Glod CA, Andersen SL, Dumont N, Ackerman E. Preliminary evidence for abnormal cortical development in physically and sexually abused children using EEG coherence and MRI. Ann N Y Acad Sci. 1997;821:160–75.
84. De Bellis MD, Keshavan MS, Clark DB, Casey BJ, Giedd JN, Boring AM, et al. A.E. Bennett Research Award. Developmental traumatology. Part II: Brain development. Biol Psychiatry. 1999;45(10):1271–84.
85. Teicher MH, Andersen SL, Dumont NL, et al. Childhood neglect attentuates development of the corpus callosum. Soc Neurosci Abstr. 2000;26:549.
86. Yazgan MY, Wexler BE, Kinsbourne M, Peterson B, Leckman JF. Functional significance of individual variations in callosal area. Neuropsychologia. 1995;33(6):769–79.
87. Altman J, Bayer SA. Development of the cerebellar system in relation to its evolution, structure, and functions. Boca Raton, FL: CRC Press; 1997.
88. Sanchez MM, Young LJ, Plotsky PM, Insel TR. Distribution of corticosteroid receptors in the rhesus brain: relative absence of glucocorticoid receptors in the hippocampal formation. J Neurosci. 2000;20(12):4657–68.
89. Lawson A, Ahima RS, Krozowski Z, Harlan RE. Postnatal development of corticosteroid receptor immunoreactivity in the rat cerebellum and brain stem. Neuroendocrinology. 1992;55(6):695–707.
90. Pavlik A, Buresova M. The neonatal cerebellum: the highest level of glucocorticoid receptors in the brain. Brain Res. 1984;314(1):13–20.
91. Schapiro S. Hormonal and environmental influences on rat brain and behavior. In: Sterman MB, McGinty DJ, editors. Brain development and behavior. NY: Academic Press; 1971. p. 307–34.
92. Ferguson SA, Holson RR. Neonatal dexamethasone on day 7 causes mild hyperactivity and cerebellar stunting. Neurotoxicol Teratol. 1999;21(1):71–6.
93. Anderson CM, Teicher MH, Polcari A, Renshaw PF. Abnormal T2 relaxation time in the cerebellar vermis of adults sexually abused in childhood: potential role of the vermis in stress-enhanced risk for drug abuse. Psychoneuroendocrinology. 2002;27(1–2):231–44.
94. Dong M, Giles WH, Felitti VJ, Dube SR, Williams JE, Chapman DP, et al. Insights into causal pathways for ischemic heart disease: adverse childhood experiences study. Circulation. 2004;110(13):1761–6.
95. Williamson DF, Thompson TJ, Anda RF, Dietz WH, Felitti V. Body weight and obesity in adults and self-reported abuse in childhood. Int J Obes Relat Metab Disord. 2002;26(8):1075–82.
96. Wright RJ, Hanrahan JP, Tager I, Speizer FE. Effect of the exposure to violence on the occurrence and severity of childhood asthma in an inner-city population. Am J Respir Crit Care Med. 1997;155:A972.
97. Chapman DP, Anda RF, Felitti VJ, Dube SR, Edwards VJ, Whitfield CL. Epidemiology of adverse childhood experiences and depressive disorders in a large health maintenance organization population. J Affect Disord. 2004;82:217–25.
98. Van Niel C, Pachter LM, Wade R Jr, Felitti VJ, Stein MT. Adverse events in children: predictors of adult physical and mental conditions. J Dev Behav Pediatr. 2014;35(8):549–51.
99. Danese A, Moffitt TE, Harrington H, Milne BJ, Polanczyk G, Pariante CM, et al. Adverse childhood experiences and adult risk factors for age-related disease: depression, inflammation, and clustering of metabolic risk markers. Arch Pediatr Adolesc Med. 2009;163(12):1135–43.
100. Galobardes B, Smith GD, Lynch JW. Systematic review of the influence of childhood socioeconomic circumstances on risk for cardiovascular disease in adulthood. Ann Epidemiol. 2006;16(2):91–104.

101. Brunner E, Davey Smith G, Marmot M, Canner R, Beksinska M, O'Brien J. Childhood social circumstances and psychosocial and behavioural factors as determinants of plasma fibrinogen. Lancet. 1996;347(9007):1008–13.
102. Poulton R, Caspi A, Milne BJ, Thomson WM, Taylor A, Sears MR, et al. Association between children's experience of socioeconomic disadvantage and adult health: a life-course study. Lancet. 2002;360(9346):1640–5.
103. Costello EJ, Compton SN, Keeler G, Angold A. Relationships between poverty and psychopathology: a natural experiment. JAMA. 2003;290(15):2023–9.
104. Widom CS, DuMont K, Czaja SJ. A prospective investigation of major depressive disorder and comorbidity in abused and neglected children grown up. Arch Gen Psychiatry. 2007;64(1):49–56.
105. Danese A, Pariante CM, Caspi A, Taylor A, Poulton R. Childhood maltreatment predicts adult inflammation in a life-course study. Proc Natl Acad Sci U S A. 2007;104(4):1319–24.
106. Thomas C, Hypponen E, Power C. Obesity and type 2 diabetes risk in midadult life: the role of childhood adversity. Pediatrics. 2008;121(5):e1240–9.
107. Baumeister RF, Leary MR. The need to belong: desire for interpersonal attachments as a fundamental human motivation. Psychol Bull. 1995;117(3):497–529.
108. Cacioppo JT, Patrick W. Loneliness: human nature and the need for social connection. New York: W Norton & Co Inc.; 2008.
109. House JS, Landis KR, Umberson D. Social relationships and health. Science. 1988;241(4865):540–5.
110. Shonkoff JP, Phillips D. From neurons to neighborhoods: the science of early childhood development. Washington, DC: National Academy Press; 2000.
111. Caspi A, Harrington H, Moffitt TE, Milne BJ, Poulton R. Socially isolated children 20 years later: risk of cardiovascular disease. Arch Pediatr Adolesc Med. 2006;160(8):805–11.
112. Barker DJ, Winter PD, Osmond C, Margetts B, Simmonds SJ. Weight in infancy and death from ischaemic heart disease. Lancet. 1989;2(8663):577–80.
113. Guttmacher AE, Collins FS, Carmona RH. The family history—more important than ever. N Engl J Med. 2004;351(22):2333–6.
114. Baker JL, Olsen LW, Sorensen TI. Childhood body-mass index and the risk of coronary heart disease in adulthood. N Engl J Med. 2007;357(23):2329–37.
115. Comijs HC, van den Kommer TN, Minnaar RW, Penninx BW, Deeg DJ. Accumulated and differential effects of life events on cognitive decline in older persons: depending on depression, baseline cognition, or ApoE epsilon4 status? J Gerontol B Psychol Sci Soc Sci. 2011;66(Suppl 1):i111–20.
116. Andel R, Crowe M, Hahn EA, Mortimer JA, Pedersen NL, Fratiglioni L, et al. Work-related stress may increase the risk of vascular dementia. J Am Geriatr Soc. 2012;60(1):60–7.
117. Wang HX, Wahlberg M, Karp A, Winblad B, Fratiglioni L. Psychosocial stress at work is associated with increased dementia risk in late life. Alzheimers Dement. 2012;8(2):114–20.
118. Lee BK, Glass TA, Wand GS, McAtee MJ, Bandeen-Roche K, Bolla KI, et al. Apolipoprotein e genotype, cortisol, and cognitive function in community-dwelling older adults. Am J Psychiatry. 2008;165(11):1456–64.
119. Peavy GM, Lange KL, Salmon DP, Patterson TL, Goldman S, Gamst AC, et al. The effects of prolonged stress and APOE genotype on memory and cortisol in older adults. Biol Psychiatry. 2007;62(5):472–8.
120. Persson G, Skoog I. A prospective population study of psychosocial risk factors for late onset dementia. Int J Geriatr Psychiatry. 1996;11:15–22.
121. Moceri VM, Kukull WA, Emanual I, van Belle G, Starr JR, Schellenberg GD, et al. Using census data and birth certificates to reconstruct the early-life socioeconomic environment and the relation to the development of Alzheimer's disease. Epidemiology. 2001;12(4):383–9.
122. Korten NC, Penninx BW, Pot AM, Deeg DJ, Comijs HC. Adverse childhood and recent negative life events: contrasting associations with cognitive decline in older persons. J Geriatr Psychiatry Neurol. 2014;27(2):128–38.

123. Ritchie K, Jaussent I, Stewart R, Dupuy AM, Courtet P, Ancelin ML, et al. Association of adverse childhood environment and 5-HTTLPR genotype with late-life depression. J Clin Psychiatry. 2009;70(9):1281–8.
124. Ritchie K, Jaussent I, Stewart R, Dupuy AM, Courtet P, Malafosse A, et al. Adverse childhood environment and late-life cognitive functioning. Int J Geriatr Psychiatry. 2011;26(5):503–10.
125. Pilleron S, Guerchet M, Ndamba-Bandzouzi B, Mbelesso P, Dartigues JF, Preux PM, et al. Association between stressful life events and cognitive disorders in Central Africa: results from the EPIDEMCA program. Neuroepidemiology. 2015;44(2):99–107.
126. Radford K, Delbaere K, Draper B, Mack HA, Daylight G, Cumming R, et al. Childhood stress and adversity is associated with late-life dementia in aboriginal Australians. Am J Geriatr Psychiatry. 2017;25(10):1097–106.
127. Danese A, McEwen BS. Adverse childhood experiences, allostasis, allostatic load, and age-related disease. Physiol Behav. 2012;106(1):29–39.
128. Noble KG, Houston SM, Brito NH, Bartsch H, Kan E, Kuperman JM, et al. Family income, parental education and brain structure in children and adolescents. Nat Neurosci. 2015;18(5):773–8.
129. Teicher MH, Samson JA. Annual research review: enduring neurobiological effects of childhood abuse and neglect. J Child Psychol Psychiatry. 2016;57(3):241–66.
130. Lupien SJ, Maheu F, Tu M, Fiocco A, Schramek TE. The effects of stress and stress hormones on human cognition: implications for the field of brain and cognition. Brain Cogn. 2007;65(3):209–37.
131. Ritchie K, Ritchie CW, Yaffe K, Skoog I, Scarmeas N. Is late-onset Alzheimer's disease really a disease of midlife? Alzheimers Dement (N Y). 2015;1(2):122–30.
132. Weder N, Zhang H, Jensen K, Yang BZ, Simen A, Jackowski A, et al. Child abuse, depression, and methylation in genes involved with stress, neural plasticity, and brain circuitry. J Am Acad Child Adolesc Psychiatry. 2014;53(4):417–24. e5
133. Ali I, Salzberg MR, French C, Jones NC. Electrophysiological insights into the enduring effects of early life stress on the brain. Psychopharmacology. 2011;214(1):155–73.
134. McGowan PO, Sasaki A, D'Alessio AC, Dymov S, Labonte B, Szyf M, et al. Epigenetic regulation of the glucocorticoid receptor in human brain associates with childhood abuse. Nat Neurosci. 2009;12(3):342–8.
135. Szyf M. The early life social environment and DNA methylation: DNA methylation mediating the long-term impact of social environments early in life. Epigenetics. 2011;6(8):971–8.
136. Taylor SE, Way BM, Seeman TE. Early adversity and adult health outcomes. Dev Psychopathol. 2011;23(3):939–54.
137. Toyokawa S, Uddin M, Koenen KC, Galea S. How does the social environment 'get into the mind'? Epigenetics at the intersection of social and psychiatric epidemiology. Soc Sci Med. 2012;74(1):67–74.
138. Anda RF, Croft JB, Felitti VJ, Nordenberg D, Giles WH, Williamson DF, et al. Adverse childhood experiences and smoking during adolescence and adulthood. JAMA. 1999;282(17):1652–8.
139. Dube SR, Anda RF, Felitti VJ, Edwards VJ, Croft JB. Adverse childhood experiences and personal alcohol abuse as an adult. Addict Behav. 2002;27(5):713–25.
140. Dube SR, Felitti VJ, Dong M, Chapman DP, Giles WH, Anda RF. Childhood abuse, neglect, and household dysfunction and the risk of illicit drug use: the adverse childhood experiences study. Pediatrics. 2003;111(3):564–72.
141. McLaughlin KA, Conron KJ, Koenen KC, Gilman SE. Childhood adversity, adult stressful life events, and risk of past-year psychiatric disorder: a test of the stress sensitization hypothesis in a population-based sample of adults. Psychol Med. 2010;40(10):1647–58.
142. Schussler-Fiorenza Rose SM, Eslinger JG, Zimmerman L, Scaccia J, Lai BS, Lewis C, et al. Adverse childhood experiences, support, and the perception of ability to work in adults with disability. PLoS One. 2016;11(7):e0157726.
143. Felitti VJ. Adverse childhood experiences and adult health. Acad Pediatr. 2009;9(3):131–2.

144. Felitti VJ, Anda RF. The relationship of adverse childhood experiences to adult health, well-being, social function, and healthcare. Cambridge, UK: Cambridge University Press; 2009.
145. Howe ML, Courage ML. On resolving the enigma of infantile amnesia. Psychol Bull. 1993;113(2):305–26.
146. Nelson K, Fivush R. The emergence of autobiographical memory: a social cultural developmental theory. Psychol Rev. 2004;111(2):486–511.
147. Pillemer DB. What is remembered about early childhood events? Clin Psychol Rev. 1998;18(8):895–913.
148. Bremner JD, Randall P, Scott TM, Capelli S, Delaney R, McCarthy G, et al. Deficits in short-term memory in adult survivors of childhood abuse. Psychiatry Res. 1995;59(1–2):97–107.
149. Edwards VJ, Fivush R, Anda RF, Felitti VJ, Nordenberg DF. Autobiographical memory disturbances in childhood abuse survivors. In: Freyd JJ, DePrince AP, editors. TacsAmom, science, experience. Binghamton, NY: Haworth Press; 2001.
150. Brown DW, Anda RF, Edwards VJ, Felitti VJ, Dube SR, Giles WH. Adverse childhood experiences and childhood autobiographical memory disturbance. Child Abuse Negl. 2007;31(9):961–9.
151. De Bellis MD, Baum AS, Birmaher B, Keshavan MS, Eccard CH, Boring AM, et al. A.E. Bennett research award. Developmental traumatology. Part I: Biological stress systems. Biol Psychiatry. 1999;45(10):1259–70.
152. Thomas LA, De Bellis MD. Pituitary volumes in pediatric maltreatment-related posttraumatic stress disorder. Biol Psychiatry. 2004;55(7):752–8.
153. Gould E, Tanapat P. Stress and hippocampal neurogenesis. Biol Psychiatry. 1999;46(11):1472–9.
154. Ladd CO, Owens MJ, Nemeroff CB. Persistent changes in corticotropin-releasing factor neuronal systems induced by maternal deprivation. Endocrinology. 1996;137(4):1212–8.
155. Liu D, Diorio J, Tannenbaum B, Caldji C, Francis D, Freedman A, et al. Maternal care, hippocampal glucocorticoid receptors, and hypothalamic-pituitary-adrenal responses to stress. Science. 1997;277(5332):1659–62.
156. McEwen BS, Angulo J, Cameron H, Chao HM, Daniels D, Gannon MN, et al. Paradoxical effects of adrenal steroids on the brain: protection versus degeneration. Biol Psychiatry. 1992;31(2):177–99.
157. Sapolsky RM. Why stress is bad for your brain. Science. 1996;273(5276):749–50.
158. Sapolsky RM, Uno H, Rebert CS, Finch CE. Hippocampal damage associated with prolonged glucocorticoid exposure in primates. J Neurosci. 1990;10(9):2897–902.
159. Schiffer F, Teicher MH, Papanicolaou AC. Evoked potential evidence for right brain activity during the recall of traumatic memories. J Neuropsychiatry Clin Neurosci. 1995;7(2):169–75.
160. Kajeepeta S, Gelaye B, Jackson CL, Williams MA. Adverse childhood experiences are associated with adult sleep disorders: a systematic review. Sleep Med. 2015;16(3):320–30.
161. Colten HR, Altevogt BM. Sleep disorders and sleep deprivation: an unmet public health problem. Press; IoMNA, editor. Washington, DC: Institute of Medicine Committee on Sleep Medicine and Research; 2006.
162. Cappuccio FP, D'Elia L, Strazzullo P, Miller MA. Sleep duration and all-cause mortality: a systematic review and meta-analysis of prospective studies. Sleep. 2010;33(5):585–92.
163. Combs K, Smith PJ, Sherwood A, Hoffman B, Carney RM, Freedland K, et al. Impact of sleep complaints and depression outcomes among participants in the standard medical intervention and long-term exercise study of exercise and pharmacotherapy for depression. J Nerv Ment Dis. 2014;202(2):167–71.
164. Davies SK, Ang JE, Revell VL, Holmes B, Mann A, Robertson FP, et al. Effect of sleep deprivation on the human metabolome. Proc Natl Acad Sci U S A. 2014;111(29):10761–6.
165. Gangwisch JE, Heymsfield SB, Boden-Albala B, Buijs RM, Kreier F, Pickering TG, et al. Short sleep duration as a risk factor for hypertension: analyses of the first National Health and nutrition examination survey. Hypertension. 2006;47(5):833–9.

166. Uehli K, Miedinger D, Bingisser R, Durr S, Holsboer-Trachsler E, Maier S, et al. Sleep quality and the risk of work injury: a Swiss case-control study. J Sleep Res. 2014;23(5):545–53.
167. Buckley TM, Schatzberg AF. On the interactions of the hypothalamic-pituitary-adrenal (HPA) axis and sleep: normal HPA axis activity and circadian rhythm, exemplary sleep disorders. J Clin Endocrinol Metab. 2005;90(5):3106–14.
168. Vgontzas AN, Bixler EO, Lin HM, Prolo P, Mastorakos G, Vela-Bueno A, et al. Chronic insomnia is associated with nyctohemeral activation of the hypothalamic-pituitary-adrenal axis: clinical implications. J Clin Endocrinol Metab. 2001;86(8):3787–94.
169. Arborelius L, Owens MJ, Plotsky PM, Nemeroff CB. The role of corticotropin-releasing factor in depression and anxiety disorders. J Endocrinol. 1999;160(1):1–12.
170. Elzinga BM, Spinhoven P, Berretty E, de Jong P, Roelofs K. The role of childhood abuse in HPA-axis reactivity in social anxiety disorder: a pilot study. Biol Psychol. 2010;83(1):1–6.
171. Heim C, Newport DJ, Heit S, Graham YP, Wilcox M, Bonsall R, et al. Pituitary-adrenal and autonomic responses to stress in women after sexual and physical abuse in childhood. JAMA. 2000;284(5):592–7.
172. Nicolson NA, Davis MC, Kruszewski D, Zautra AJ. Childhood maltreatment and diurnal cortisol patterns in women with chronic pain. Psychosom Med. 2010;72(5):471–80.
173. Videlock EJ, Adeyemo M, Licudine A, Hirano M, Ohning G, Mayer M, et al. Childhood trauma is associated with hypothalamic-pituitary-adrenal axis responsiveness in irritable bowel syndrome. Gastroenterology. 2009;137:1954–62.
174. Bicanic IA, Postma RM, Sinnema G, De Roos C, Olff M, Van Wesel F, et al. Salivary cortisol and dehydroepiandrosterone sulfate in adolescent rape victims with post traumatic stress disorder. Psychoneuroendocrinology. 2013;38(3):408–15.
175. Carpenter LL, Tyrka AR, Ross NS, Khoury L, Anderson GM, Price LH. Effect of childhood emotional abuse and age on cortisol responsivity in adulthood. Biol Psychiatry. 2009;66(1):69–75.
176. MacMillan HL, Georgiades K, Duku EK, Shea A, Steiner M, Niec A, et al. Cortisol response to stress in female youths exposed to childhood maltreatment: results of the youth mood project. Biol Psychiatry. 2009;66(1):62–8.
177. Bader K, Schafer V, Schenkel M, Nissen L, Schwander J. Adverse childhood experiences associated with sleep in primary insomnia. J Sleep Res. 2007;16(3):285–96.
178. Perlis ML, Giles DE, Mendelson WB, Bootzin RR, Wyatt JK. Psychophysiological insomnia: the behavioural model and a neurocognitive perspective. J Sleep Res. 1997;6(3):179–88.
179. Riemann D, Spiegelhalder K, Feige B, Voderholzer U, Berger M, Perlis M, et al. The hyperarousal model of insomnia: a review of the concept and its evidence. Sleep Med Rev. 2010;14(1):19–31.
180. Sher L. The concept of post-traumatic mood disorder and its implications for adolescent suicidal behavior. Minerva Pediatr. 2008;60(6):1393–9.
181. Cahill LT, Kaminer RK, Johnson PG. Developmental, cognitive, and behavioral sequelae of child abuse. Child Adolesc Psychiatr Clin N Am. 1999;8(4):827–43.
182. Emiroglu FN, Kurul S, Akay A, Miral S, Dirik E. Assessment of child neurology outpatients with headache, dizziness, and fainting. J Child Neurol. 2004;19(5):332–6.
183. O'Hare C, McCrory C, O'Leary N, O'Brien H, Kenny RA. Childhood trauma and lifetime syncope burden among older adults. J Psychosom Res. 2017;97:63–9.
184. Tannemaat MR, van Niekerk J, Reijntjes RH, Thijs RD, Sutton R, van Dijk JG. The semiology of tilt-induced psychogenic pseudosyncope. Neurology. 2013;81(8):752–8.
185. Beacher FD, Gray MA, Mathias CJ, Critchley HD. Vulnerability to simple faints is predicted by regional differences in brain anatomy. NeuroImage. 2009;47(3):937–45.
186. Saleh A, Potter GG, McQuoid DR, Boyd B, Turner R, MacFall JR, et al. Effects of early life stress on depression, cognitive performance and brain morphology. Psychol Med. 2017;47(1):171–81.
187. Jacobson M. Developmental neurobiology. New York: Plenum Press; 1991.

# Part IV

# Social and Therapeutic Implications

# Childhood Trauma and Stigma

# 19

Nerisa Banaj and Clelia Pellicano

## 19.1 The Hurtful and Subverting Experience of Childhood Trauma

Childhood trauma (CT) experiences [1] include physical, emotional, and sexual abuse; physical and emotional neglect; growing up in a home with divorced parents, domestic violence, substance abuse, or mentally ill or incarcerated household members. These experiences negatively affect childhood development, leading to physical and mental health problems throughout life. Indeed, CT is associated with multiple adverse outcomes in adulthood, such as cardiovascular disease, liver disease, chronic obstructive pulmonary disease, suicide attempts, alcohol dependence, marital problems, intravenous drug use, and many more [1, 2]. Moreover, trauma survivors are often found in psychiatric and psychological outpatient clinics and practices as well as in inpatient psychiatric settings; it has been reported that 50–70% of patients in these settings are trauma survivors [3, 4]. Prevalence of CT exposure in borderline personality disorder patients has been evidenced to be as high as 92% [5] and it reaches 82% in individuals diagnosed with psychotic or affective disorders [6]. According to van der Kolk's definition [7], complex trauma in childhood and the potential for negative outcomes in adult life differ from "simple trauma" or a single traumatic event, such as natural disaster. Symptoms of post-traumatic stress disorders (PTSD) in chronically traumatized children are usually not prominent and tend to be obscured by cognitive, affective, social, and physical problems that range from learning disabilities to aggression against the self and others [8–12]. Lack of capacity for emotional self-regulation is probably the most striking feature of chronically traumatized children [13]. Indeed, chronic maltreatments have pervasive effects on the development of mind and brain. Chronic trauma interferes with development of

N. Banaj (✉) · C. Pellicano
Laboratory of Neuropsychiatry, IRCCS Santa Lucia Foundation, Rome, Italy
e-mail: n.banaj@hsantalucia.it; pellicanoc@gmail.com



G. Spalletta et al. (eds.), *Childhood Trauma in Mental Disorders*,
https://doi.org/10.1007/978-3-030-49414-8_19

several brain areas [14] and therefore with the capability to integrate sensory, emotional, and cognitive information into a cohesive whole. Indeed, chronic trauma corrupts the normal maturation of brain areas involved not only in the regulation of body homeostasis, such as the brain stem and locus coeruleus, but also of areas related to the brain memory systems (including hippocampus, amygdala, frontal cortex) and of areas involved in executive functioning (including the orbitofrontal cortex, the cingulate, and the dorsolateral prefrontal cortex) [15]. Further, trauma may also affect the neuroendocrine system, including the hypothalamic–pituitary–adrenal (HPA) axis, and every conceivable neurotransmitter system [16, 17].

## 19.2 Defining the Concept of Stigma

Variations in conceptualization, definition, type, and degree of stigma exist and derive from both the situational and multidisciplinary application of the term [18]. Although using separate and often different terminology and measures, a conspicuous body of research exists on mental illness stigma [19, 20], minority sexual identity stigma [21, 22], HIV+/AIDS stigma [23–25], epilepsy stigma [26], and domestic violence [27, 28], among others. The landmark publication in the field is indubitably "Stigma: Notes on the Management of Spoiled Identity" of Erving Goffman [29]. His pioneering treatise goes beyond the realm of sociology—from where it arose—to reach also other fields, including medicine, health sciences, criminology, and psychology. Goffman [29] viewed stigma as a process based on the social construction of identity. Persons who become associated with a stigmatized condition thus pass from a "normal" to a "discredited" or "discreditable" social status [29]. The concept of stigma has undergone important shifts in definition and characterization since this initial articulation and has been investigated both at the individual and social level [30]. Corrigan and colleagues [31] postulate that the impact of stigma is twofold. Public stigma comprises reactions of the general public toward a group based on stigma about that group [32]. Perceived social stigma may become internalized and may result in self-stigma (i.e., the personal endorsement of stereotypes about oneself and the resulting prejudice and self-discrimination) [31, 33, 34]. Both public and self-stigma may be understood in terms of three components: stereotypes, prejudice, and discrimination [31, 32] (see Table 19.1).

Although stereotypes are considered an efficient way to categorize information on different social groups [35, 36], when linked to a negative label (i.e., strange, dangerousness, incompetence) they change into negative beliefs (negative stereotypes). People who endorse negative stereotypes develop negative feelings and emotional reactions (i.e., anger, fear, shame): this is known as prejudice. From this emotional reaction comes discrimination or the behavioral response to having negative thoughts and feelings about a person (or the self) in a stigmatized out-group.

Within the area of social psychology, several functions are attributed to stigmatization: to maintain inequalities between groups (*keeping people down*), to encourage deviants to conform to in-group norms (*keeping people in*), and to exclude dangerous deviants to preserve survival of the group (*keeping people away*) [37, 38].

**Table 19.1** Components of public and self-stigma (based on Rusch and collaborators [32])

| Public stigma | Self-stigma |
|---|---|
| **Stereotype:** | **Stereotype:** |
| Negative belief about a group such as | Negative belief about the self such as |
| • Incompetence | • Incompetence |
| • Character weakness | • Character weakness |
| • Dangerousness | • Dangerousness |
| **Prejudice:** | **Prejudice:** |
| Agreement with belief and/or negative emotional reaction such as | Agreement with belief and/or negative emotional reaction such as |
| • Anger or | • Low self-esteem or |
| • Fear | • Low self-efficacy |
| **Discrimination:** | **Discrimination:** |
| Behavior response to prejudice such as: | Behavior response to prejudice such as |
| • Avoidance of work and housing opportunities | • Fails to pursue work and housing opportunities |
| • Withholding help | • Does not seek help |

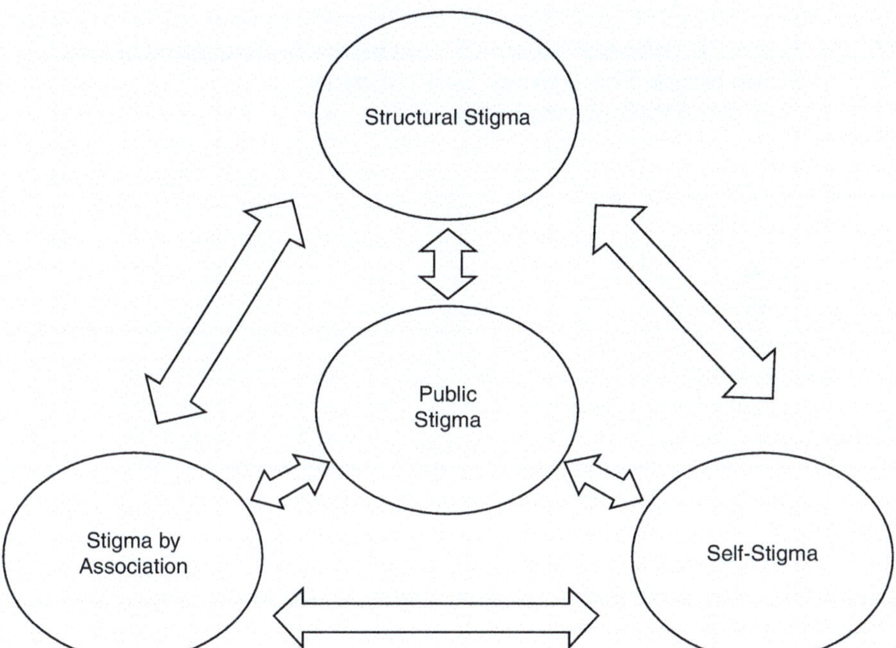


**Fig. 19.1** Four types of stigma (adapted from Pryor & Reeder, 2011 and Arjan E. R, 2013 [18, 39])

More recently, Pryor and Reeder [39] proposed a conceptual model that, building on previous theories [33, 40], depicts four dynamically interrelated manifestations of stigma (see Fig. 19.1).

Public stigma has been conceptualized as the core of this system. Ultimately, these authors refer to public stigma as people's social and psychological reactions to someone they perceive as having a stigmatized condition. Stigma by association involves social and psychological reactions to people associated with a stigmatized person and people's reactions to being associated with a stigmatized person. Structural stigma is defined as the "legitimatization and perpetuation of a stigmatized status by society's institutions and ideological systems" [39]. Finally, self-stigma reflects the social and psychological impact of possessing a stigma and the potential internalization of the negative beliefs, stereotypes, and feelings associated with the stigmatized condition. Self-stigma has cognitive, affective, and behavioral components [41] and operates at both the explicit and implicit level [42].

Transversally, the burden of stigmatization oppresses the individual both on the social and personal sphere. On a societal level, it implies discriminatory behaviors, victimization, isolation, and alienation and fosters myths and stereotypes [43]. At the individual level, it determines unnecessary suffering since stigmatized identities have higher levels of psychological distress [44], it limits help-seeking behaviors [45], and impedes positive treatment effects [46].

## 19.3 From Childhood Trauma Experience to Stigmatization Experience: The Young "Self" Shame and the Attributional Style

Despite the very large effort placed in fighting the stigma surrounding mental illness, the importance of the stigma related to CT still remains in dim light. Stigma is rooted in the collective processes of shaming, social control, and maintenance of in-group boundaries and can silence the actors in their attempts to give voice to their distress and to identify sources of trauma [47]. Healing from the negative effects of CT is anything but easy, particularly when the truth of the traumatic experience itself is denied or nullified by the victim or by her/his inner circle. Adult survivors often feel ashamed about and stigmatized for their childhood adversity. This makes it difficult to even recognize that these events occurred. Further, in the case, the abuse is committed within the context of trust violation and by a significant person (or institution) on whom the child depends, recovery is enormously complicated [48].

CT negatively affect the child's developing sense of self, impairing psychological growth, allocation of blame, emotion regulation, and modeling of constructive coping skills [49, 50]. The victims of CT struggle to form a coherent sense of self: their representations of self, others, and the world are deeply compromised [51]. Chronic traumatic experiences in childhood are internalized as dysfunctional allocation of blame and shame resulting in self-blame and self-loathing [7]. When victims feel bad and blameworthy, the experience of stigmatization arises. Feiring and colleagues [52] define stigmatization as an attributional style of self-blame for negative events and the emotional experience of shame. In the self-evaluation process, two dimensions are crucial: the locus of cause, e.g. internal versus external, and the specificity, e.g. global versus specific [53]. In particular, internal attributions occur when the self is blamed for the abuse and external attributions when another person is blamed for the negative event. Global attributions involve the whole self, whereas specific attributions refer to particular aspects of the self. These two dimensions of attribution are central to determining shame, guilt, and pride. Shame results from attributing failures to an internal, global cause. Guilt is determined by the attribution of failures to internal, specific aspects of the self. When success is attributed to internal, specific factors, the corresponding emotion is pride [53, 54]. A third dimension in attribution processes is the stability as opposed to instability, regarding the perceived reason for an event to having occurred [55]. Internal, stable global attributions for negative events are most likely to lead to shame [52, 53].

Several works suggest that guilt and shame are phenomenologically distinct and involve different cognitive processes [56–58]. Both involve negative affect, but they differ on focus. Guilt focuses on the specific features or actions of the self that led to failures. Conversely, shame has the total or global self as the focus. Because the cognitive-attributional process is focalized on actualized actions, rather than on the global self, the feeling of guilt is not as intensely harmful as shame. The phenomenological experience of shame is a desire to hide, disappear, or die: shame is a dejected, humiliation-based emotional state in which the individual desires to shrink and hide the exposed and spoiled self [53]. In the dynamic of stigmatization [59], shame is the emotional component and a self-blaming attributional style is the cognitive component. Both the clinical and empirical literature link shame with affective disorders and trauma sequelae [53, 60–63]. Indeed, how the victim evaluates the abuse and, in particular, a self-blaming attributional style, has been found to be related to psychological distress [64–68].

## 19.4 Neuroscience of Stigma

The modern social neuroscience approach integrates ideas from multiple research areas in psychology and neuroscience to address questions about possible neurobiological correlates of social processes. Ultimately, stigma, prejudice, and discrimination, although deeply influenced by our sociocultural environment [69], depend upon complex interactions among brain areas and counterbalance of neurotransmitters. Exploring the role of specific neurobiological systems and brain structures underlying complex social psychological phenomena, such as person perception and racial bias, will improve our understanding and consolidate our knowledge on the mechanisms underlying commonly observed behavioral effects of stigma [70]. Within this field, functional imaging techniques, such as functional MRI (fMRI), give the opportunity to closely investigate the relationship between environment and brain. Although available studies refer to the effect of stigma in the general population more than in childhood, it is worth to quickly overview their findings. To date, most work in social neuroscience has analyzed mental processes and brain regions that are activated in the perpetrator when people employ stereotypes to perceive and interpret their social world [71–78]. Emerging social neuroscience research demonstrated the pivotal role of amygdala and insula in coordinating rapid responses to emotionally salient or threatening situations [76, 79, 80]. For example, brain-imaging studies have demonstrated that when viewing black versus white faces, responses in the amygdala powerfully predict unconscious measures of racial prejudice [74]. Beyond the amygdala and the insula, a vast network of complementary neural regions is engaged during implicit or explicit evaluations of stigmatized targets [73, 77, 80–82]. They include the anterior cingulate cortex (ACC) [81, 83, 84], the lateral prefrontal cortex (PFC), the ventrolateral prefrontal cortex (VLPFC), and the dorsolateral prefrontal cortex (DLPFC) [73, 77, 82, 85–88]. Automatic activation of phylogenetically ancient emotion-processing mechanisms within the brain may trigger biases toward stigmatized individuals and prejudicial thoughts, whereas evolutionarily more recent frontal brain regions manage conscious and regulatory control. Although these areas are the same generally engaged in emotional control, more recent studies have demonstrated that stigma regulation has a unique activity profile distinct from general emotion regulation. Krendl and colleagues [79] in an fMRI study found that participants had higher activity in the ACC, and in the lateral and medial PFC during attempts to regulate their negative affect to images of stigmatized individuals, as compared to images of non-stigmatized individuals.

Less effort has been allocated to understanding the perspective of stigmatized individuals with respect to what it means and how it feels to belong to a stigmatized group, how they deal with the effects of stereotype threats, and how they use coping mechanisms to protect their social identity. Stereotype threat occurs when individuals worry about conforming to a negative stereotype regarding their group, resulting in the feeling of being judged or treated in terms of a negative stereotype [89], and leading to underperformance on evaluative or other tasks. Examples of stereotype threat are applicable to African Americans, who are well aware of the negative stereotypes questioning their intellectual ability [89], as well as to women affected by the negative stereotype regarding their gender and math performance [90], or the negative stereotype related to women's inferior spatial reasoning abilities [91]. When compared with situations requiring them to display that specific ability, women fear confirming the stereotype, and because of that, they end up activating neural processes that impede their optimal performance. In particular, Krendl and colleagues [90], in an fMRI study, compared a control group of women with a group of women in the threat condition where they have been reminded of the negative stereotype about women and math performance. Their findings indicated that in performing a math test, the control group properly activated neural regions associated with math learning, such as the inferior prefrontal cortex, left inferior parietal cortex, and bilateral angular gyrus [92, 93]. Conversely, women under the stereotype threat condition did not show this enhanced activation. Instead, they activated brain areas involved in emotional self-regulation and in processing affective information and social feedback like the ventral anterior cingulate cortex [94, 95]. These results suggest that stereotype threat afflicts the normal brain functioning through a twofold mechanism: by disengaging executive and cognitive control necessary for achieving a specific task and by recruiting areas responsible for affective regulations, hyperarousal, emotional salience, and for managing the physiological stress response. These studies demonstrate the detrimental effect of a threatening stereotype, potentially suggesting a risk for internalization of negative beliefs and feelings associated with a stigmatized condition. The main point is that such conditions can affect behaviors and decisions in domains unrelated to the threatening stereotype, determining a threat to the whole self-concept even at the neural level.

## 19.5 Childhood Sexual Abuse and Stigma: A Paradigmatic Case

CT, childhood sexual abuse (SA) in particular, is significantly associated with higher levels of stigmatization and to the risk of developing psychopathology and interpersonal problems [96]. SA is defined as experiencing sexual molestation by an adult or older adolescent who is a family member, relative, familiar person, or stranger [97] and occurs in any setting, from home to aggregation places e.g. school or church. By definition, both physical contact (e.g., fondling, oral, anal, or vaginal penetration) and nonphysical contact (e.g., exhibitionism or exposure to pornography) qualify for SA. Given the peculiar nature of SA linked to the obtrusive attack to the self and to the social condemnation of those who violate the rules of sexual conduct, victims of childhood SA are particularly at risk for self-blame attributions and experiencing shame. Indeed, the concealed circumstances in which childhood SA occurs, the devaluation of the victim by the perpetrator, and the social taboos against this kind of criminal behaviors increase the chance that children will experience shame and self-blame and, in turn, consider themselves as responsible for the abuse [96]. Moreover, because females are more likely to experience shame when they perceive to have transgressed a social rule, abused girls are more likely to implement poor adjustment strategies and to develop depression compared to boys [98]. A cross-sectional study [99] found that college women with a history of childhood SA reported more stigma, including some aspects related to shame, after adulthood sexual assault, compared to women without history of childhood SA. The developmental period during which the abuse occurs is a further determinant in the stigmatization processes: the prepubescent period seems to be more at risk compared to early childhood or adolescence [100]. Therefore, the victim's attributions about the event, the victim's sex, the developmental period, and the presence of social support determine the differential likeness that shame will follow childhood SA. The negative emotion of stigmatization may occur during each phase of the abuse experience: during abuse perpetration or when the abuse is revealed and continues once the abuse and its discovery have ended [52]. Individuals who report more stigmatization and lack of coping resources, such as social support [101], have higher levels of psychological distress [44]. In particular, the level of shame and self-blame intensity can help identifying those SA victims who are most at risk for psychological and interpersonal problems. Several research focuses on long-term outcomes of childhood SA [96, 100, 102]. Berkowski and colleagues [102] investigated possible coping strategies in victims of childhood SA and, in particular, the potential development of paranormal beliefs. Authors speculated that the latter might arise as coping mechanisms and means of control in order to deal with the added psychological stress of stigmatization. Therefore, authors predict a more strict correlation between trauma and the paranormal beliefs considered to involve control (e.g., witchcraft, psychic abilities, superstition, precognition). Their results demonstrated that stigmatization was strongly related to trauma experience and also to increased paranormal beliefs and that the interaction between stigma and trauma may result in increased paranormal beliefs [102].

Subsequent sexual difficulties may occur in childhood SA victims [59]. Distorted feelings and beliefs about the self during nonconsensual sex may corrupt sexual feelings and behaviors during consensual sex. The development of a positive sexual self-schema can be disrupted by self-blame and shame [103]. When the sexual self–schema is damaged, the elaboration of a coherent sexuality becomes significantly more challenging, and the development of intimacy is substituted by fear of self-exposure, anxiety about partner's disapproval, and expectations of reprobation from others [104]. Individuals with a self-blaming attribution style believe they deserve disrespectful and aggressive behaviors from their partners [105, 106]. This style determines frustration in engaging in close relationships, poor relationship quality, and risky relationship behavior [97, 107, 108]. Recently, Feiring and colleagues [96], while investigating the development of romantic intimacy problems after childhood SA, found that the degree of stigmatization, more than abuse severity, predicted sexual difficulties. Their findings emphasize the observation that by targeting youths' cognitive and emotional processing of self-blame attributions and shame, concurrent distress and subsequent sexual problems may be alleviated and prevented. Further, promoting healthier sexuality not only reduces the risk of infection and sexual dysfunctions but also encourages more satisfying and intimate romantic relationships.

## 19.6 Stigma Strikes Twice: The Rough Reconstruction of the Self

As already stated, stigmatization following CT experiences fragments the sense of the self. The victims experience the disruption of affect regulation, attachment patterns, autonomy, a balanced worldview, and a delay in the development of other competencies which are supposed to mature in childhood [7]. Their identity as individuals has been seriously thwarted impacting their ability to form reciprocal and solid relationships with others and in turn making them difficult to reach healthy adulthood [109]. Traumatized children have to grow with a plethora of emotional and psychological turmoil and to accept them as emanating from the self. Unable to justify their struggle, they sometimes conceal their trauma for many years, rather than making it visible. Early adult life accelerates a search for meaning, relief, and professional help. Help-seeking provides a route to externalizing shame and blame and to bring an enduring sense to adult life. Once this path has been undertaken, these individuals are forced to deal with new frustrations. Indeed, there is the concrete possibility for many who have experienced CT to be revictimized by having their distress labeled with a mental health diagnosis [109]. Up to 70% of psychiatric inpatients are reported to have experienced some form of childhood abuse [110]. Initially, a psychiatric diagnostic label is the essential key that gives meaning to years of unexplainable distress, and allows the victims to embrace self-value, and adult identity as separate from that of a traumatized child. The diagnostic label provides a context for their distress and creates a new narrative frame for their trauma history allocating it as only a part of their life story, not their entire identity. The psychiatric diagnosis is the instrument that separates the "self" as a worthy adult from the "self" as a traumatized child and facilitates positive changes. Nevertheless, despite the beneficial effect of receiving a diagnosis, being categorized under a psychiatric label conceptually separate these individuals from others who do not have a label. They experience the fear of being labeled as "a crazy person" and are concerned about social judgment and once again, about the stigma surrounding mental illness [109], thus potentially putting the self under attack again. These internalized stereotypes influence how individuals with CT experiences expect to be treated [111], which is mirrored by self-judgments based on the social assumption that those with mental illness are somehow "less than". They are often so deeply influenced by stereotypes concerning mental illness to become perpetrators themselves. Schomerus and colleagues [112] reported that persons who attribute their mental health problems to CT have less tolerant attitudes toward other persons with mental health problems and experience more stigma-related stress. Comparisons of groups with and without a history of sexual trauma revealed that those who experienced it exhibit a stronger endorsement of negative stereotypes about persons with mental illness, but also report more experiences of discrimination, social withdrawal, and alienation [113]. Overall, these data suggest that trauma history may lead to an increased vulnerability to stigmatizing beliefs. A potential underlying mechanism leading to negative and stigmatizing beliefs is that traumatized persons can be prone to unconscious scanning of the environment for threat [114], based on their personal experience that humans are able to be violent [115], which may also serve as a confirmation of negative self-beliefs [113]. Their personal experience might make them particularly susceptible to the prevalent public stereotype and may represent a bias resulting from rejection sensitivity. Finally, according to the progressive model of self-stigma [116], this negative evaluation of others would translate into more self-stigma.

## 19.7 Stigma: The Key for Therapeutic Intervention

We explored the concept of stigma, in its different components and types, and the relationship with CT. We highlighted the possible role that stigma has in the processes of acceptance and elaboration of the traumatic event. Hence, intervention programs on CT survivors should focus on stigma and self-blame considering the treatment of stigmatized experiences as an additional target, besides symptom reduction. In a recent work, Shneider and colleagues [117] observed that the experience of stigmatization not only increases the risk of developing PTSD but also influences treatment effectiveness and the likelihood of spontaneous remission. Further, stigma could weaken the ability to react and hinder the processes of narration and care-seeking after trauma [47]. Stigmatized individuals showed a reduced frequency of spontaneous remission, and higher symptoms scores before and after 6 weeks of treatment, demonstrating the negative powerful influence that stigmatization has on mental health recovery. Gilligan and Akhtar [118] showed that stigma is a key factor in the underreporting of abuse and related psychological disturbance. Specifically, they showed that mothers of abused children felt intense anxiety about the potential criticism that might arise in the community, and form local social services. Thus, they concealed the identities of the perpetrators in the community, rather than putting the emotional needs of their children first, thereby endorsing a collective ethos of shame and stigmatization. It can be inferred that various types of stigma act as nonsystematic barriers (i.e., fear of others' opinion, denial of the problem, lack of knowledge of treatment options due to noncommunication of the traumatic experience), thus dissuading the request for help and treatment. Therefore, it is clear that identifying through appropriate measures, the specific type of implicated stigma (public stigma, self-stigma, stigma by association, and/or structural stigma) is crucial to lay the groundwork for a constructive change and to direct the psychological intervention. In fact, while public stigma can discourage people with CT experiences from seeking help, due to feelings of embarrassment or shame [119, 120], self-stigma can also disincentives treatment when resulting in loss of self-esteem, poor self-efficacy, and questioning the point of trying to get better [121].

To allow a complete overview of the stigmatization processes, besides individual or group interviews, a series of different measurements suitable to assess diverse components and levels of stigma have been provided. In the past decades, a growing body of research has produced self-report and behavioral measures of implicit and explicit stigma. Internalized stigma scales assess to what extent the group's negative perceptions about a person determine that individual's feeling of shame or potential changes in behavior, in a way that exacerbates feelings of being different; a scale of internalized stigma might include items of self-blame or guilt [122]. Perceived stigma measures ask respondents to report, from their own perspective, how others perceive them or behave toward them [123]. Enacted stigma measures assess actual acts of discrimination [124, 125], for instance, when someone refuses to interact with or to provide assistance to someone with a stigmatized condition [126]. Perceived stigma measures generally capture subjective thoughts and feelings of the survivor, while enacted stigma measures involve reports as to whether or not a particular discriminatory act has occurred. Based on the type of the event and the narration of the same by the actors involved, it is possible to select the appropriate measures that best explore the levels of stigmatization encountered. Once identified, they can become an important part in structuring a personalized treatment path on the individual's needs.

In fact, interventions that are carefully planned, specific, and based on current theoretical knowledge and empirical evidence are more likely to be effective [127, 128]. Among the various approaches aiming to decrease self-stigma (or "internalized stigma"), the "narrative enhancement and cognitive therapy" (NECT) has been developed and experimentally evaluated [129, 130]. In the NECT, the process of externalizing establishes a context where people experience themselves as separate from the traumatic experiences allowing the new individual narrative to take form and be expressed [131]. Further, the trauma-focused cognitive-behavioral therapy has shown to be more effective in reducing self-blame attributions and shame than other forms of child-centered therapy [132]. Another approach that intends to reduce the impact of stigma is based on metacognitive strategies, a spectrum of mental activities involving thinking about thinking. A greater metacognitive capacity, better self-esteem, and less negative symptoms are shown to be associated with stigma resistance [133]. Mashiach-Eizenberg and colleagues [134] found that self-esteem mediated the association between internalized stigma and hope, and hope partially mediated the relationship between self-esteem and quality of life. Results of a study by Hasson-Ohayon and colleagues [135] point to a mediating role of shame-proneness between insight and self-stigma, suggesting that the former can be conceptualized as a vulnerability factor for self-stigma experiences. Experiencing the self as diminished or unproductive may put one at risk for internalizing public stigma, as one may have no alternative internal experiences to reject stigma [136].

Overall, interventions targeting stigmatization and developing stigma-reducing programs might lead to better therapy outcome and long-term therapy success. Moreover, this could sensitize the population regarding negative impacts of stigmatization, highlighting the necessity to reduce stigmatization and discrimination in order to allow survivors to recover and reintegrate into the community.

## 19.8 Conclusion

CT experiences are associated with multiple, detrimental, outcomes in adulthood. Indeed, CT survivors have a variety of disorders or diseases with a higher prevalence compared to the general population, not only in the psychiatric field but also regarding common organic diseases. Traumatic experiences in childhood have pervasive effects on the neurobiological development, leading to problems with self-regulation, aggression against the self and others, problems with attention and dissociation, physical problems, and difficulties in self-concept and the capacity to negotiate satisfactory interpersonal relationships. These adverse experiences are internalized as a dysfunctional allocation of blame and shame, resulting in self-blame and self-loathing since victims tend to feel themselves responsible for the abuses and to experience stigmatization and self-blame for negative events. Stigmatized individuals are regarded as flawed, damaged, and identifiable among others by the presence of their label. These internal and social dynamics undermine the development and the conceptualization of the self. Given the damage to the self and the potential detrimental long-life effects, understanding the stigmatizing mechanisms that follow CT experiences and their role as nonsystematic barriers is crucial to laying the foundations for a constructive change and for directing psychosocial intervention. As a matter of fact, stigma does not only imply discriminatory behaviors against oneself and others but also limits help-seeking and treatment outcomes. Therefore, interventions on CT survivors should consider stigma within the various phenomena linked to the traumatic experience in order to develop and offer a more effective and personalized clinical treatment.

## References


1. Felitti VJ, Anda RF, Nordenberg D, Williamson DF, Spitz AM, Edwards V, et al. Relationship of childhood abuse and household dysfunction to many of the leading causes of death in adults: the adverse childhood experiences (ACE) study. Am J Prev Med. 1998;14:245–58. https://doi.org/10.1016/s0749-3797(98)00017-8.
2. Fergusson DM, McLeod GFH, Horwood LJ. Childhood sexual abuse and adult developmental outcomes: findings from a 30-year longitudinal study in New Zealand. Child Abus Negl. 2013;37(9):664–74. https://doi.org/10.1016/j.chiabu.2013.03.013.
3. Chapman DP, Whitfield CL, Felitti VJ, Dube SR, Edwards VJ, Anda RF. Adverse childhood experiences and the risk of depressive disorders in adulthood. J Affect Disord. 2004;82(2):217–25. https://doi.org/10.1016/j.jad.2003.12.013.
4. Whitfield CL. Adverse childhood experiences and trauma. Am J Prev Med. 1998;14(4):361–4. https://doi.org/10.1016/S0749-3797(98)00013-0.
5. Yen S, Shea MT, Battle CL, et al. Traumatic exposure and posttraumatic stress disorder in borderline, schizotypal, avoidant, and obsessive-compulsive personality disorders: findings from the collaborative longitudinal personality disorders study. J Nerv Ment Dis. 2002;190(8):510–8. https://doi.org/10.1097/00005053-200208000-00003.
6. Larsson S, Andreassen OA, Aas M, et al. High prevalence of childhood trauma in patients with schizophrenia spectrum and affective disorder. Compr Psychiatry. 2013;54(2):123–7. https://doi.org/10.1016/j.comppsych.2012.06.009.
7. van der Kolk BA. Developmental trauma disorder: toward a rational diagnosis for children with complex trauma histories. Psychiatr Ann. 2005;35(5):401–8.
8. Cicchetti D, White J. Emotion and developmental psychopathology. In: Leventhal B, Trebasso T, editors. Stein N. Psychological and biological approaches to emotion. Lawrence Erlbaum Associates; 1990. p. 359–82.
9. Pynoos RS, Frederick C, Nader K, Arroyo W, Steinberg A, Eth S, et al. Life threat and posttraumatic stress in school-age children. Arch Gen Psychiatry. 1987;44(12):1057–63. https://doi.org/10.1001/archpsyc.1987.01800240031005.
10. Van der Kolk BA, Perry JC, Herman JL. Childhood origins of self-destructive behavior. Am J Psychiatry. 1991;148(12):1665–71. https://doi.org/10.1176/ajp.148.12.1665.
11. Van der Kolk BA, Fisler RE. Childhood abuse and neglect and loss of self-regulation. Bull. Menninger Clin. 1994;58(2):145.
12. Streeck-Fischer A, van der Kolk BA. Down will come baby, cradle and all: diagnostic and therapeutic implications of chronic trauma on child development. Aust N Z J Psychiatry. 2000;34(6):903–18. https://doi.org/10.1046/j.1440-1614.2000.00827.x.
13. Cicchetti D, Toth SL. A developmental psychopathology perspective on child abuse and neglect. J Am Acad Child Adolesc Psychiatry. 1995;34:541–65. https://doi.org/10.1097/00004583-199505000-00008.
14. Ford JD, Kidd P. Early childhood trauma and disorders of extreme stress as predictors of treatment outcome with chronic posttraumatic stress disorder. J Trauma Stress. 1998;11(4):743–61. https://doi.org/10.1023/A:1024497400891.
15. Van der Kolk BA. The neurobiology of childhood trauma and abuse. Child Adolesc Psychiatr Clin N Am. 2003;12(2):293–317. https://doi.org/10.1016/S1056-4993(03)00003-8.
16. Wiener SG, Lowe EL, Levine S. Pituitary-adrenal response to weaning in infant squirrel monkeys. Psychobiology. 1992;20(1):65–70. https://doi.org/10.3758/BF03327163.
17. Stanton ME, Gutierrez YR, Levine S. Maternal deprivation potentiates pituitary-adrenal stress responses in infant rats. Behav Neurosci. 1988;102(5):692. https://doi.org/10.1037/0735-7044.102.5.692.
18. Bos AER, Pryor JB, Reeder GD, Stutterheim SE. Stigma: advances in theory and research. Basic Appl Soc Psych. 2013;35(1):1–9. https://doi.org/10.1080/01973533.2012.746147.
19. Livingston JD, Boyd JE. Correlates and consequences of internalized stigma for people living with mental illness: a systematic review and meta-analysis. Soc Sci Med. 2010;71(12):2150–61. https://doi.org/10.1016/j.socscimed.2010.09.030.

20. Corrigan P. On the stigma of mental illness: practical strategies for research and social change. Washington, DC: American Psychological Association; 2005.
21. Meyer IH. Prejudice, social stress, and mental health in lesbian, gay, and bisexual populations: conceptual issues and research evidence. Psychol Bull. 2003;129(5):674. https://doi.org/10.1037/0033-2909.129.5.674.
22. Hatzenbuehler ML. How does sexual minority stigma "get under the skin"? A psychological mediation framework. Psychol Bull. 2009;135(5):707. https://doi.org/10.1037/a0016441.
23. Earnshaw VA, Chaudoir SR. From conceptualizing to measuring HIV stigma: a review of HIV stigma mechanism measures. AIDS Behav. 2009;13(6):1160. https://doi.org/10.1007/s10461-009-9593-3.
24. Kalichman SC, Simbayi LC, Cloete A, Mthembu PP, Mkhonta RN, Ginindza T. Measuring AIDS stigmas in people living with HIV/AIDS: the internalized AIDS-related stigma scale. AIDS Care. 2009;21(1):87–93. https://doi.org/10.1080/09540120802032627.
25. Zhao G, Li X, Zhao J, Zhang L, Stanton B. Relative importance of various measures of HIV-related stigma in predicting psychological outcomes among children affected by HIV. Community Ment Health J. 2012;48(3):275–83. https://doi.org/10.1007/s10597-011-9424-7.
26. Jacoby A, Austin JK. Social stigma for adults and children with epilepsy. Epilepsia. 2007;48:6–9. https://doi.org/10.1111/j.1528-1167.2007.01391.x.
27. Fischbach RL, Herbert B. Domestic violence and mental health: correlates and conundrums within and across cultures. Soc Sci Med. 1997;45(8):1161–76. https://doi.org/10.1016/S0277-9536(97)00022-1.
28. Overstreet NM, Quinn DM. The intimate partner violence stigmatization model and barriers to help seeking. Basic Appl Soc Psych. 2013;35(1):109–22. https://doi.org/10.1080/01973533.2012.746599.
29. Goffman I. Stigma: notes on the management of spoiled identity. NJ: Prentice-Hall; 1963.
30. Brohan E, Gauci D, Sartorius N, Thornicroft G. Self-stigma, empowerment and perceived discrimination among people with bipolar disorder or depression in 13 European countries: the GAMIAN–Europe study. J Affect Disord. 2011;129:56–63. https://doi.org/10.1016/j.jad.2010.09.001.
31. Corrigan PW, Watson AC. Understanding the impact of stigma on people with mental illness. World Psychiatry. 2002;1(1):16. https://doi.org/10.1016/S0013-4686(01)00816-7.
32. Rüsch N, Angermeyer MC, Corrigan PW. Mental illness stigma: concepts, consequences, and initiatives to reduce stigma. Eur Psychiatry. 2005;20:529–39. https://doi.org/10.1016/j.eurpsy.2005.04.004.
33. Corrigan P. How stigma interferes with mental health care. Am Psychol. 2004;59(7):614. https://doi.org/10.1037/0003-066X.59.7.614.
34. Corrigan PW, Rao D. On the self-stigma of mental illness: stages, disclosure, and strategies for change. Can J Psychiatr. 2012;57(8):464–9. https://doi.org/10.1177/070674371205700804.
35. Hilton JL, von Hippel W. Stereotypes. Annu Rev Psychol. 1996;47:237–71.
36. Hyler S, Gabbard G, Schneider I. Homicidal maniacs and narcissistic parasites: stigmatization of mentally ill persons in the movies. Hosp Community Psychiatry. 1991;42:1044–8. https://doi.org/10.1176/ps.42.10.1044.
37. Kurzban R, Leary MR. Evolutionary origins of stigmatization: the functions of social exclusion. Psychol Bull. 2001;127(2):187. https://doi.org/10.1037/0033-2909.127.2.187.
38. Phelan JC, Link BG, Dovidio JF. Stigma and prejudice: one animal or two? Soc Sci Med. 2008;67(3):358–67. https://doi.org/10.1016/j.socscimed.2008.03.022.
39. Pryor JB, Reeder GD. HIV/AIDS in the post-HAART era: manifestations, treatment, and epidemiology. Hall J. C. Shelton, CT: PMPH-USA; 2011. p. 790–806.
40. Herek GM. Confronting sexual stigma and prejudice: theory and practice. J Soc Issues. 2007;63(4):905–25. https://doi.org/10.1111/j.1540-4560.2007.00544.x.
41. Mak WWS, Cheung RYM. Affiliate stigma among caregivers of people with intellectual disability or mental illness. J Appl Res Intellect Disabil. 2008;21(6):532–45. https://doi.org/10.1111/j.1468-3148.2008.00426.x.

42. Rüsch N, Corrigan PW, Todd AR, Bodenhausen GV. Automatic stereotyping against people with schizophrenia, schizoaffective and affective disorders. Psychiatry Res. 2011;186(1):34–9. https://doi.org/10.1016/j.psychres.2010.08.024.
43. Quinn DM, Williams MK, Quintana F, Gaskins JL, Overstreet NM, Pishori A, et al. Examining effects of anticipated stigma, centrality, salience, internalization, and outness on psychological distress for people with concealable stigmatized identities. PLoS One. 2014;9(5):e96977. https://doi.org/10.1371/journal.pone.0096977.
44. Quinn DM, Chaudoir SR. Living with a concealable stigmatized identity: the impact of anticipated stigma, centrality, salience, and cultural stigma on psychological distress and health. J Pers Soc Psychol. 2009;97(4):634. https://doi.org/10.1037/a0015815.
45. Starr S, Campbell LR, Herrick CA. Factors affecting use of the mental health system by rural children. Issues Ment Health Nurs. 2002;23(3):291–304. https://doi.org/10.1080/016128402753543027.
46. Sirey JA, Bruce ML, Alexopoulos GS, Perlick DA, Friedman SJ, Meyers BS. Perceived stigma and patient-rated severity of illness as predictors of antidepressant drug adherence. Psychiatr Serv. 2001;52(12):1615–20. https://doi.org/10.1176/appi.ps.52.12.1615.
47. Budden A. The role of shame in posttraumatic stress disorder: a proposal for a socio-emotional model for DSM-V. Soc Sci Med. 2009;69:1032–9. https://doi.org/10.1016/j.socscimed.2009.07.032.
48. Freyd JJ, Klest B, Allard CB. Betrayal trauma: relationship to physical health, psychological distress, and a written disclosure intervention. J Trauma Dissociation. 2005;6(3):83–104. https://doi.org/10.1300/j229v06n03_04.
49. Kinniburgh K, Jentoft Kinniburgh K, Blaustein M, Spinazzola J. Attachment, self-regulation, and competency: a comprehensive intervention framework for children with complex trauma. Psychiatr Ann. 2017;35(5):424–30.
50. McCormack L, White S, Cuenca J. A fractured journey of growth: making meaning of a 'broken' childhood and parental mental ill-health. Community Work Fam. 2017;20(3):327–45. https://doi.org/10.1080/13668803.2015.1117418.
51. Cook A, Spinazzola J, Ford J, et al. Complex trauma in children and a adolescents. Psychiatr Ann. 2003;38(3):515–27.
52. Feiring C, Taska L, Lewis M. A process model for understanding adaptation to sexual abuse: the role of shame in defining stigmatization. Child Abus Negl. 1996;20:767–82.
53. Lewis M. Shame: the exposed self; 1992.
54. Feiring C, Taska L, Chen K. Trying to understand why horrible things happen: attribution, shame, and symptom development following sexual abuse. Child Maltreat. 2002;7(1):25–39. https://doi.org/10.1177/1077559502007001003.
55. Weiner B. Attribution in personality psychology. In: Pervin LA, editor. Handbook of personal. New York: Guilford Press; 1990. p. 465–85.
56. Ferguson TJ, Stegge H, Damhuis I. children's understanding of guild and shame. Child Dev. 1991;62(4):827–39. https://doi.org/10.1111/j.1467-8624.1991.tb01572.x.
57. Lewis HB. Shame: the "sleeper" in psychopathology. In: Lewis HB, editor. Role shame symptom form. Hillsdale, NJ: Erlbaum; 1987. p. 1–28.
58. Wicker FW, Payne GC, Morgan RD. Participant descriptions of guilt and shame. Motiv Emot. 1983;7(1):25–39. https://doi.org/10.1007/BF00992963.
59. Feiring C, Taska LS. The persistence of shame following sexual abuse: a longitudinal look at risk and recovery. Child Maltreat. 2005;10(4):337–49. https://doi.org/10.1177/1077559505276686.
60. Andrews B, Brewin CR, Rose S, Kirk M. Predicting PTSD symptoms in victims of violent crime: the role of shame, anger, and childhood abuse. J Abnorm Psychol. 2000;109(1):69. https://doi.org/10.1037/0021-843X.109.1.69.
61. Coffey P, Leitenberg H, Henning K, Turner T, Bennett RT. Mediators of the long-term impact of child sexual abuse: perceived stigma, betrayal, powerlessness, and self-blame. Child Abus Negl. 1996;20(5):447–55. https://doi.org/10.1016/0145-2134(96)00019-1.
62. Hoblitzelle W. Differentiating and measuring shame and guilt: the relation between shame and depression. In: Lewis HB, editor. Role shame symptom form. Hillsdale, NJ: Lawrence Erlbaum; 1987. p. 207–36.

63. Tangey JP, Burggraf SA, Wagner PE. Shame-prone ness, guilt-proneness and psychological symptoms. In: Tangey JP, Fischer KW, editors. Self-conscious emotions psychology shame, guilt, embarrassment pride. New York: Guilford; 1995. p. 343–67.
64. Mannarino AP, Cohen JA. Abuse-related attributions and perceptions, general attributions, and locus of control in sexually abused girls. J Interpers Violence. 1996;11(2):162–80. https://doi.org/10.1177/088626096011002002.
65. Mannarino AP, Cohen JA. A follow-up study of factors that mediate the development of psychological symptomatology in sexually abused girls. Child Maltreat. 1996;1(3):246–60. https://doi.org/10.1177/1077559596001003007.
66. Morrow KB. Attributions of female adolescent incest victims regarding their molestation. Child Abus Negl. 1991;15(4):477–83. https://doi.org/10.1016/0145-2134(91)90031-8.
67. Wolfe VV, Gentile C, Wolfe DA. The impact of sexual abuse on children: a PTSD formulation. Behav Ther. 1989;20(2):215–28. https://doi.org/10.1016/S0005-7894(89)80070-X.
68. Wolfe D, McGee R. Assessment of emotional status among maltreated children. In: Starr R, Wolfe DA, editors. The effects of child abuse and neglect: issues and research. New York: Guilford; 1991. p. 257–77.
69. Amodio DM. Can neuroscience advance social psychological theory? Social neuroscience for the behavioral social psychologist. Soc Cogn. 2011;28(6):695–716. https://doi.org/10.1521/soco.2010.28.6.695.
70. Derks B, Inzlicht M, Kang S. The neuroscience of stigma and stereotype threat. Gr Process Intergr Relations. 2008;11(2):163–81. https://doi.org/10.1177/1368430207088036.
71. Amodio DM, Devine PG, Harmon-Jones E. Mechanisms for the regulation of intergroup responses: insights from a social neuroscience approach. In: Winkielman P, Harmon-Jones E, editors. Social neuroscience: integrating biological and psychological explanations of social behavior. New York: Guilford; 2007. p. 353–75.
72. Bartholow BD, Dickter CL, Sestir MA. Stereotype activation and control of race bias: cognitive control of inhibition and its impairment by alcohol. J Pers Soc Psychol. 2006;90(2):27. https://doi.org/10.1037/0022-3514.90.2.272.
73. Cunningham WA, Johnson MK, Raye CL, Gatenby JC, Gore JC, Banaji MR. Separable neural components in the processing of black and white faces. Psychol Sci. 2004;15(12):806–13. https://doi.org/10.1111/j.0956-7976.2004.00760.x.
74. Hart AJ, Whalen PJ, Shin LM, McInerney SC, Fischer H, Rauch SL. Differential response in the human amygdala to racial outgroup vs ingroup face stimuli. Neuroreport. 2000;11(11):2351–4.
75. Ito TA, Willadsen Jensen E, Correll J. Social neuroscience and social perception: new perspectives on categorization, prejudice, and stereotyping. In: Winkielman P, Harmon-Jones E, editors. Social neuroscience: integrating biological and psychological explanations of social behavior. New York: Guilford; 2007. p. 401–21.
76. Phelps EA, O'Connor KJ, Cunningham WA, Funayama ES, Gatenby JC, Gore JC, et al. Performance on indirect measures of race evaluation predicts amygdala activation. J Cogn Neurosci. 2000;12(5):729–38. https://doi.org/10.1162/089892900562552.
77. Richeson JA, Baird AA, Gordon HL, Heatherton TF, Wyland CL, Trawalter S, et al. An fMRI investigation of the impact of interracial contact on executive function. Nat Neurosci. 2003;6(12):1323–8. https://doi.org/10.1038/nn1156.
78. Wheeler ME, Fiske ST. Controlling racial prejudice social-cognitive goals affect amygdala and stereotype activation. Psychol Sci. 2005;16(1):56–63. https://doi.org/10.1111/j.0956-7976.2005.00780.x.
79. Krendl AC, Kensinger EA, Ambady N. How does the brain regulate negative bias to stigma? Soc Cogn Affect Neurosci. 2012;7(6):715–26. https://doi.org/10.1093/scan/nsr046.
80. Harris LT, Fiske ST. Dehumanizing the lowest of the the low. Psychol Sci. 2006;17(10):847–53. https://doi.org/10.1111/j.1467-9280.2006.01793.x.
81. Amodio DM, Harmon-Jones E, Devine PG, Curtin JJ, Hartley SL, Covert AE. Neural signals for the detection of unintentional race bias. Psychol Sci. 2004;15(2):88–93. https://doi.org/10.1111/j.0963-7214.2004.01502003.x.

82. Krendl AC, Macrae CN, Kelley WM, Fugelsang JA, Heatherton TF. The good, the bad, and the ugly: an fMRI investigation of the functional anatomic correlates of stigma. Soc Neurosci. 2006;1(1):5–15. https://doi.org/10.1080/17470910600670579.
83. MacDonald AW, Cohen JD, Andrew Stenger V, Carter CS. Dissociating the role of the dorsolateral prefrontal and anterior cingulate cortex in cognitive control. Science. 2000;288(5472):1835–8. https://doi.org/10.1126/science.288.5472.1835.
84. Amodio DM, Kubota JT, Harmon-Jones E, Devine PG. Alternative mechanisms for regulating racial responses according to internal vs external cues. Soc Cogn Affect Neurosci. 2006;1(1):26–36. https://doi.org/10.1093/scan/nsl002.
85. Lieberman Gaunt Gilbert DT, Trope Y, MDR. Reflection and reflexion: a social cognitive neuroscience approach to attributional inference. Adv Exp Soc Psychol. 2002;34:199–24.
86. Vogeley K, Bussfeld P, Newen A, Herrmann S, Happé F, Falkai P, et al. Mind reading: neural mechanisms of theory of mind and self-perspective. NeuroImage. 2001;14(1):170–81. https://doi.org/10.1006/nimg.2001.0789.
87. Samson D, Apperly IA, Kathirgamanathan U, Humphreys GW. Seeing it my way: a case of a selective deficit in inhibiting self-perspective. Brain. 2005;128(5):1102–11. https://doi.org/10.1093/brain/awh464.
88. Satpute AB, Lieberman MD. Integrating automatic and controlled processes into neurocognitive models of social cognition. Brain Res. 2006;1079(1):86–97. https://doi.org/10.1016/j.brainres.2006.01.005.
89. Steele CM, Aronson J. Stereotype threat and the intellectual test performance of African Americans. J Pers Soc Psychol. 1995;69(5):797. https://doi.org/10.1037/0022-3514.69.5.797.
90. Krendl AC, Richeson JA, Kelley WM, Heatherton TF. The negative consequences of threat: a functional magnetic resonance imaging investigation of the neural mechanisms underlying women's underperformance in math: research article. Psychol Sci. 2008;19(2):168–75. https://doi.org/10.1111/j.1467-9280.2008.02063.x.
91. Wraga M, Helt M, Jacobs E, Sullivan K. Neural basis of stereotype-induced shifts in women's mental rotation performance. Soc Cogn Affect Neurosci. 2007;2(1):12–9. https://doi.org/10.1093/scan/nsl041.
92. Dehaene S, Spelke E, Pinel P, Stanescu R, Tsivkin S. Sources of mathematical thinking: behavioral and brain-imaging evidence. Science. 1999;284(5416):970–4. https://doi.org/10.1126/science.284.5416.970.
93. Delazer M, Domahs F, Bartha L, Brenneis C, Lochy A, Trieb T, et al. Learning complex arithmetic—an fMRI study. Cogn Brain Res. 2003;18(1):76–88. https://doi.org/10.1016/j.cogbrainres.2003.09.005.
94. Moran JM, Macrae CN, Heatherton TF, Wyland CL, Kelley WM. Neuroanatomical evidence for distinct cognitive and affective components of self. J Cogn Neurosci. 2006;18(9):1586–94. https://doi.org/10.1162/jocn.2006.18.9.1586.
95. Somerville LH, Heatherton TF, Kelley WM. Anterior cingulate cortex responds differentially to expectancy violation and social rejection. Nat Neurosci. 2006;9(8):1007–8. https://doi.org/10.1038/nn1728.
96. Feiring C, Simon VA, Cleland CM. Childhood sexual abuse, stigmatization, internalizing symptoms, and the development of sexual difficulties and dating aggression. J Consult Clin Psychol. 2009;77(1):127. https://doi.org/10.1037/a0013475.
97. Feiring C, Rosenthal S, Taska L. Stigmatization and the development of friendship and romantic relationships in adolescent victims of sexual abuse. Child Maltreat. 2000;5(4):311–22. https://doi.org/10.1177/1077559500005004003.
98. Cutler SE, Nolen-Hoeksema S. Accounting for sex differences in depression through female victimization: childhood sexual abuse. Sex Roles. 1991;24(7–8):425–38. https://doi.org/10.1007/BF00289332.
99. Gibson LE, Leitenberg H. The impact of child sexual abuse and stigma on methods of coping with sexual assault among undergraduate women. Child Abus Negl. 2001;25(10):1343–61. https://doi.org/10.1016/S0145-2134(01)00279-4.
100. Friedrich WN. Behavior problems in sexually abused children: an adaptational perspective. In: Wyatt GE, Powell GJ, editors. Lasting effects of child sexual abuse. Newbury Park, CA: Sage; 1988. p. 171–91.

101. Markowitz FE, Angell B, Greenberg JS. Stigma, reflected appraisals, and recovery outcomes in mental illness. Soc Psychol Q. 2011;74(2):144–65. https://doi.org/10.1177/0190272511407620.
102. Berkowski M, Macdonald DA. Childhood trauma and the development of paranormal beliefs. J Nerv Ment Dis. 2014;202(4):305–12. https://doi.org/10.1097/NMD.0000000000000123.
103. Meston CM, Rellini AH, Heiman JR. Women's history of sexual abuse, their sexuality, and sexual self-schemas. J Consult Clin Psychol. 2006;74:229. https://doi.org/10.1037/0022-006X.74.2.229.
104. Tangney J, Dearing R. Shame and guilt. New York: Guilford Press; 2002.
105. Covert MV, Tangney JP, Maddux JE, Heleno NM. Shame-proneness, guilt-proneness, and interpersonal problem solving: a social cognitive analysis. J Soc Clin Psychol. 2003;22(1):1–2. https://doi.org/10.1521/jscp.22.1.1.22765.
106. Gilbert P, Allan S, Goss K. Parental representations, shame, interpersonal problems, and vulnerability to psychopathology. Clin Psychol Psychother. 2002;3(1):23–34. https://doi.org/10.1002/(sici)1099-0879(199603)3:1<23::aid-cpp66>3.3.co;2-f.
107. Fletcher GJO, Fitness J, Blampied NM. The link between attributions and happiness in close relationships: the roles of depression and explanatory style. J Soc Clin Psychol. 2011;9(2):243–55. https://doi.org/10.1521/jscp.1990.9.2.243.
108. Liem JH, Boudewyn AC. Contextualizing the effects of childhood sexual abuse on adult self- and social functioning: an attachment theory perspective. Child Abus Negl. 1999;23(11):1141–57. https://doi.org/10.1016/S0145-2134(99)00081-2.
109. McCormack L, Thomson S. Complex trauma in childhood, a psychiatric diagnosis in adulthood: making meaning of a double-edged phenomenon. Psychol Trauma Theory Res Pract Policy. 2017;9:156–65. https://doi.org/10.1037/tra0000193.
110. Herman JL. Trauma and recovery. New York: Basic Books; 1992.
111. Link B. Understanding labeling effects in the area of mental disorders: an assessment of the effects of expectations of rejection. Am Sociol Rev. 1987;52:96–112.
112. Schomerus G, Matschinger H, Angermeyer MC. Causal beliefs of the public and social acceptance of persons with mental illness: a comparative analysis of schizophrenia, depression and alcohol dependence. Psychol Med. 2014;44(2):303–14. https://doi.org/10.1017/S003329171300072X.
113. Outcalt SD, Lysaker PH. The relationships between trauma history, trait anger, and stigma in persons diagnosed with schizophrenia spectrum disorders. Psychosis. 2012;4(1):32–41. https://doi.org/10.1080/17522439.2011.591422.
114. Pineles SL, Shipherd JC, Welch LP, Yovel I. The role of attentional biases in PTSD: is it interference or facilitation? Behav Res Ther. 2007;45(8):1903–13. https://doi.org/10.1016/j.brat.2006.08.021.
115. Stolzenburg S, Freitag S, Schmidt S, Schomerus G. Associations between causal attributions and personal stigmatizing attitudes in untreated persons with current mental health problems. Psychiatry Res. 2018;260:24–9. https://doi.org/10.1016/j.psychres.2017.11.014.
116. Corrigan PW, Rafacz J, Rüsch N. Examining a progressive model of self-stigma and its impact on people with serious mental illness. Psychiatry Res. 2011;189(3):339–43. https://doi.org/10.1016/j.psychres.2011.05.024.
117. Schneider A, Conrad D, Pfeiffer A, Elbert T, Kolassa IT, Wilker S. Stigmatization is associated with increased PTSD risk after traumatic stress and diminished likelihood of spontaneous remission: a study with East African conflict survivors. Front Psych. 2018;9:1–10. https://doi.org/10.3389/fpsyt.2018.00423.
118. Gilligan P, Akhtar S. Cultural barriers to the disclosure of child sexual abuse in Asian communities: listening to what women say. Br J Soc Work doi. 2006;36(8):1361–77. https://doi.org/10.1093/bjsw/bch309.
119. Keyes KM, Hatzenbuehler ML, McLaughlin KA, Link B, Olfson M, Grant BF, et al. Stigma and treatment for alcohol disorders in the United States. Am J Epidemiol. 2010;172(12):1364–72. https://doi.org/10.1093/aje/kwq304.
120. Cumming C, Troeung L, Young JT, Kelty E, Preen DB. Barriers to accessing methamphetamine treatment: a systematic review and meta-analysis. Drug Alcohol Depend. 2016;168:263. https://doi.org/10.1016/j.drugalcdep.2016.10.001.

121. Corrigan PW, Bink AB, Schmidt A, Jones N, Rüsch N. What is the impact of self-stigma? Loss of self-respect and the "why try" effect. J Ment Health. 2016;25(1):10–5. https://doi.org/10.3109/09638237.2015.1021902.
122. Simbayi LC, Kalichman S, Strebel A, Cloete A, Henda N, Mqeketo A. Internalized stigma, discrimination, and depression among men and women living with HIV/AIDS in Cape Town, South Africa. Soc Sci Med. 2007;64(9):1823–31. https://doi.org/10.1016/j.socscimed.2007.01.006.
123. Liu SH, Srikrishnan AK, Zelaya CE, Solomon S, Celentano DD, Sherman SG. Measuring perceived stigma in female sex workers in Chennai, India. AIDS Care - Psychol Socio-Medical Asp AIDS/HIV. 2011;23(5):619–27. https://doi.org/10.1080/09540121.2010.525606.
124. Lasalvia A, Zoppei S, Van Bortel T, et al. Global pattern of experienced and anticipated discrimination reported by people with major depressive disorder: a cross-sectional survey. Lancet. 2013;381(9860):55–62. https://doi.org/10.1016/S0140-6736(12)61379-8.
125. Thornicroft G, Brohan E, Rose D, Sartorius N, Leese M. Global pattern of experienced and anticipated discrimination against people with schizophrenia: a cross-sectional survey. Lancet. 2009;373(9661):408–15. https://doi.org/10.1016/S0140-6736(08)61817-6.
126. Link BG, Phelan JC. Stigma and its public health implications. Lancet. 2006;30(3):511–41. https://doi.org/10.1016/S0140-6736(06)68184-1.
127. Bartholomew LK, Parcel GS, Kok G, Gottlieb NH, Fernández ME. Planning health promotion programs: an intervention mapping approach. 3rd ed. San Francisco, CA: Jossey Bass; 2011.
128. Bos AER, Schaalma HP, Pryor JB. Reducing AIDS-related stigma in developing countries: the importance of theory- and evidence-based interventions. Psychol Heal Med. 2008;13(4):450–60. https://doi.org/10.1080/13548500701687171.
129. Roe D, Hasson-Ohayon I, Derhi O, Yanos PT, Lysaker PH. Talking about life and finding solutions to different hardships: a qualitative study on the impact of narrative enhancement and cognitive therapy on persons with serious mental illness. J Nerv Ment Dis. 2010;198:807. https://doi.org/10.1097/NMD.0b013e3181f97c50.
130. Yanos PT, Roe D, West ML, Smith SM, Lysaker PH. Group-based treatment for internalized stigma among persons with severe mental illness: findings from a randomized controlled trial. Psychol Serv. 2012;9(3):248. https://doi.org/10.1037/a0028048.
131. Besley AC. Foucault and the turn to narrative therapy. Br J Guid Couns. 2002;30:125. https://doi.org/10.1080/03069880220128010.
132. Cohen J, Mannarino A, Deblinger E. Treating trauma and traumatic grief in children and adolescents. New York: Guilford Press; 2006.
133. Nabors LM, Yanos PT, Roe D, Hasson-Ohayon I, Leonhardt BL, Buck KD, et al. Stereotype endorsement, metacognitive capacity, and self-esteem as predictors of stigma resistance in persons with schizophrenia. Compr Psychiatry. 2014;55(4):792–8. https://doi.org/10.1016/j.comppsych.2014.01.011.
134. Mashiach-Eizenberg M, Hasson-Ohayon I, Yanos PT, Lysaker PH, Roe D. Internalized stigma and quality of life among persons with severe mental illness: the mediating roles of self-esteem and hope. Psychiatry Res. 2013;208(1):15–20. https://doi.org/10.1016/j.psychres.2013.03.013.
135. Hasson-Ohayon I, Ehrlich-Ben Or S, Vahab K, Amiaz R, Weiser M, Roe D. Insight into mental illness and self-stigma: the mediating role of shame proneness. Psychiatry Res. 2012;200(2–3):802–6. https://doi.org/10.1016/j.psychres.2012.07.038.
136. Hasson-Ohayon I, Mashiach-Eizenberg M, Elhasid N, Yanos PT, Lysaker PH, Roe D. Between self-clarity and recovery in schizophrenia: reducing the self-stigma and finding meaning. Compr Psychiatry. 2014;55(3):675–80. https://doi.org/10.1016/j.comppsych.2013.11.009.

# Treatment of Childhood Trauma: Pharmacological Approach

# 20

Isabella Panaccione, Gianluigi Di Cesare, Gabriele Sani, and Giuseppe Ducci

## 20.1 Introduction

Early stressful or traumatic experiences may lead to the development of anxiety, dissociative, somatoform, and affective symptoms both during childhood and later in life.

Childhood trauma (CT) has been recognized as a significant factor in the pathogenesis of mental and physical disorders, including mood disorders (i.e., major depressive and bipolar disorders), psychotic disorders, anxiety disorders, post-traumatic stress disorder (PTSD), substance abuse, chronic fatigue or chronic pain disorders, and functional gastrointestinal conditions [1–6]. Increased stress and emotional reactivity appears to play a crucial role in the pathogenesis of these disorders, and this reactivity is likely enhanced by the exposure to CT. Moreover, psychiatric patients with a history of CT usually show earlier onset, increased severity of symptoms, delayed access to treatment, worse course of illness, increased rate of comorbidity, proneness to substance misuse, and detrimental outcomes [7–14].

Treatment refractoriness in psychiatric patients with CT seems to be differently mediated by specific mechanisms. CT-related cognitive impairments have been postulated to underlie worse response to treatment in psychotic patients. In PTSD patients,

I. Panaccione (✉)
Mental Health Department, ASL Roma 1, Rome, Italy

Centro Lucio Bini, Rome, Italy
e-mail: isabella.panaccione@gmail.com

G. Di Cesare · G. Ducci
Mental Health Department, ASL Roma 1, Rome, Italy
e-mail: gianluigi.dicesare@aslroma1.it; giuseppe.ducci@aslroma1.it

G. Sani
Centro Lucio Bini, Rome, Italy

Department of Neuroscience, Section of Psychiatry, Università Cattolica del Sacro Cuore, Rome, Italy
e-mail: gabriele.sani@unicatt.it



G. Spalletta et al. (eds.), *Childhood Trauma in Mental Disorders*,
https://doi.org/10.1007/978-3-030-49414-8_20

CT seems to make the brain more vulnerable to subsequent stress, therefore producing more severe and difficult-to-treat symptoms. In bipolar disorder, the relationship between CT and treatment resistance appears to be mediated by greater severity of manic and depressive episodes and earlier onset of the disease, while in major depressive disorder the mediating factors are persistent, chronic symptoms, and frequent recurrences [15]. Also, studies focusing on the occurrence of depressive symptoms in CT patients postulated the existence of distinct subtypes of depression as a function of CT, which are responsive to different kinds of treatment [4].

These peculiar characteristics might be mediated by neurobiological changes occurring in the brain as a result of traumatic events taking place at an early age. In fact, the experience of CT, intended both as neglect and physical/psychological abuse, has been related to structural and functional alterations in several brain regions involved in emotional regulation, such as the amygdala and the hippocampus [16–18]. Other neurobiological mechanisms that appear to be modified by CT include dysregulation of the hypothalamic–pituitary–adrenal (HPA) axis and the inflammatory and metabolic systems [19–22]. This is particularly relevant since increasing evidence has been linking the abnormal functioning/activation of the immunoendocrine system to the development of several major psychiatric diseases, including major depression, bipolar disorder, and schizophrenia [4, 23–25].

CT might also interact with several genes (i.e., the serotonin transporter gene) possibly via epigenetic modifications, eventually leading to pathological manifestations in adulthood [26–29]. Furthermore, evidence suggests that the presence of CT significantly influences the response to treatment in psychiatric patients, resulting in pharmacological resistance to several types of drugs, such as antidepressants and mood stabilizers [30, 31], as well as to combined psychotherapy–pharmacotherapy treatment [32]. This observation is likely consequent to the aforementioned multiple effects that CT exerts on neurobiological mechanisms involved in the pathogenesis of psychiatric disorders. Therefore, the efficacy of pharmacotherapy in these conditions might vary with the presence –or the severity – of CT. Finally, an important point to keep in mind is that CT patients frequently have poor premorbid functioning and are more likely to discontinue treatment (Fig. 20.1) [33].

The management of CT is mainly based on psychotherapy, aimed at elaborating the traumatic experiences and offering coping strategies, and has been extensively addressed elsewhere in this book. However, pharmacological intervention is required when the presence of CT leads – or is associated to – the onset of relevant psychiatric symptoms or diseases. Pharmacotherapy is typically used as one component of a more comprehensive multiple modality treatment approach. The pharmacological therapy of these conditions should be carried out following the current guidelines for treating the specific symptom or disease; nevertheless, as described earlier, literature data suggest that psychiatric patients exposed to CT may represent a particular clinical population, with peculiar response to treatment. In general, psychiatric patients with severe CT tend to be prescribed higher doses of medications, especially antipsychotics and mood stabilizers [34], have significantly longer time to remit, and need more frequently a combined treatment with an additional class of medication [32, 35, 36].

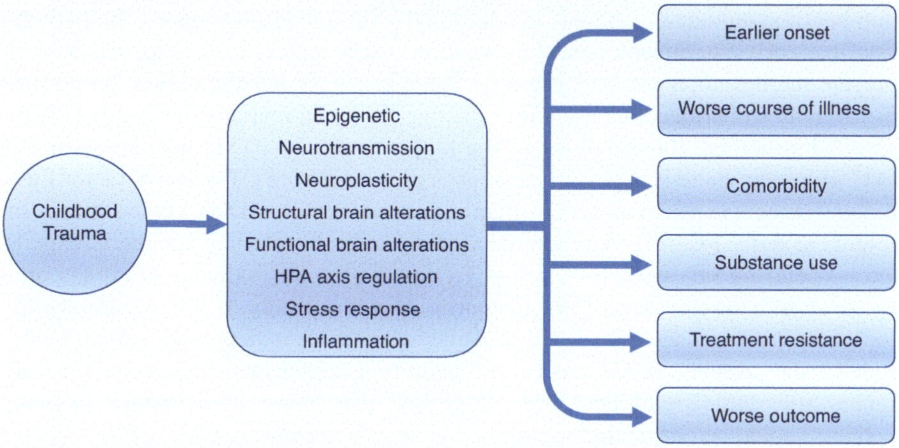


**Fig. 20.1** The exposure to childhood trauma is related to several neurobiological alterations and negatively influences clinical presentation, response to treatment, and outcome of psychiatric disorders. HPA: hypothalamic–pituitary–adrenal

In this chapter, we review the most relevant evidence investigating the effects of psychotropic medications on psychiatric patients with a history of CT.

## 20.2 Childhood Trauma and Different Classes of Medications

### 20.2.1 Antidepressants

In patients with mood disorders, the presence of CT is associated with differences in the clinical response to antidepressants. Such differential outcomes might result from variability in neurobiological pathways underlying mood alterations, which might be differentially affected by drugs or psychotherapy [37]. Nemeroff et al. observed that the atypical antidepressant nefazodone was less effective than psychotherapy in depressed patients with a history of CT, while showing similar efficacy in patients without CT [38]. Preliminary data found that depressed patients with CT show a better clinical response when treated with antidepressants with higher affinity for the serotonin transporter, such as citalopram, fluoxetine, paroxetine, and venlafaxine [39]. This might be due to the observed alterations in the serotonin transporter gene associated with CT [26], but clinical trials are required before drawing conclusions. Patients with a history of relevant CT are usually prescribed higher doses of antidepressants, with a prevalence of more sedating substances (i.e., mirtazapine). These compounds are chosen because of their effectiveness in treating agitation and hyperarousal-related symptoms. On the other hand, more activating substances (i.e., fluoxetine) tend to be used in patients exposed to less severe traumatic experiences [34]. A preclinical study found that treatment with fluoxetine in rats exposed to juvenile stress protocols significantly reduced the number of symptomatic animals in adulthood [40]; this result supports the hypothesis that early pharmacological intervention might represent an effective way to prevent the development of psychiatric disorders. A systematic review found that 3–12 months pharmacotherapy with selective serotonin reuptake inhibitors (SSRI) reverted structural abnormalities in the hippocampus of PTSD patients with CT [41]. However, there is a general consensus on considering CT as a risk factor for a much worse response to all types of antidepressants, primarily as a consequence of trauma severity [42–44]. A randomized study also showed that traumatic experiences occurring very early (i.e., before 7 years of age) are predictive of even poorer response to antidepressants [30]. Comorbidity with anxiety or dissociative symptoms has also been associated with treatment resistance in depressed patients [45]. Finally, the presence of CT has been related to a higher rate of depressive recurrences after initial remission [46], suggesting that longer term therapy in these patients should be considered in order to achieve better outcomes.

### 20.2.2 Mood Stabilizers

The experience of CT is associated with inadequate response to mood stabilizers, both lithium and anticonvulsant agents. Poor responders to long-term anticonvulsant treatment showed higher scores of emotional and physical abuse at the Childhood Trauma Questionnaire (CTQ) [47], although subjects exposed to more severe traumatic experiences usually receive higher doses of mood stabilizers [34]. Regarding bipolar disorder, subjects exposed to CT were more likely to be nonresponders to prophylactic lithium treatment, independently from clinical features and course of illness [48]. Patients exposed to at least two traumatic experiences (emotional, physical, or sexual abuse) were nearly fivefold more at risk to be lithium nonresponders than patients not reporting any form of abuse. In these subjects, augmentation with antipsychotics (i.e., risperidone) showed some efficacy in managing mood episodes, particularly in younger individuals [49].

### 20.2.3 Antipsychotics

Although only few studies investigated the relationship between CT and treatment outcomes in psychotic diseases, findings suggest that the efficacy of antipsychotic treatment is reduced in CT patients [50]. Subjects with severe CT are usually prescribed a higher dosage of antipsychotics, with partial response [34]. Misiak et al. found a significantly more frequent history of CT in first-episode schizophrenia patients not responding to a 12-weeks treatment with second-generation antipsychotics. They also observed that emotional abuse was a predictor of poor response to treatment [51].

Some evidence suggests that clinical response to antipsychotics is mediated by the interaction between predisposing factors and the neurobiological changes correlated to CT. For example, polymorphisms within matrix metallopeptidase 9

(MMP9), a gene involved in synaptic plasticity, seem to play a role in determining differences in the clinical efficacy of the injectable flupenthixol decanoate, monitored over 12 months [52]. However, results are still controversial: a recent genome-wide association study could not demonstrate a relationship among 1,178,234 single-nucleotide polymorphisms and treatment resistance in CT schizophrenia patients [53].

Psychotic patients with a history of CT show several abnormalities in the activity of the HPA and the immune system that, in turn, mediate and are predictive of poorer response to second-generation antipsychotic treatment [54, 55]. Another study found that a 2-weeks treatment with atypical antipsychotics can revert some of these neuroendocrine alterations, although no significant correlations with clinical outcomes were observed [56].

Interestingly, a preclinical study reported that the atypical antipsychotic quetiapine improves depressive-like symptoms in animals subjected to an experimental model of CT (i.e., maternal deprivation) via an epigenetic mechanism, thus suggesting a potential role for this compound in treating depression in CT patients [57].

### 20.2.4 Benzodiazepines

So far, evidence from a systematic review suggests that benzodiazepines should be considered relatively contraindicated for patients with recent trauma [58], possibly because these compounds interfere with the normal HPA–stress response and increase vulnerability to subsequent stress [59]. Moreover, it is important to keep in mind that traumatic experiences represent a risk factor for the development of drug misuse and addictive behaviors, including prolonged use of benzodiazepines [2, 60]. Therefore, medications with a high potential for abuse should be prescribed with caution in these patients.

## 20.3 Novel Strategies: Targeting the Immune System

CT and early-life immune activation are associated with a large range of psychiatric disorders, including depression, bipolar disorder, schizophrenia, and PTSD. Both preclinical and clinical studies have shown that early-life stress is associated with epigenetic changes leading to alterations in the glucocorticoid signaling that could produce high inflammation levels. Greater baseline levels of inflammation, in turn, seem to induce higher stress-related responses in animal models and humans. Moreover, both CT and hyperactivation of the immune system appear to be associated not only with higher risk of developing psychiatric disorders but also with unfavorable longitudinal course of illness and treatment response [61–64].

Evidence from the literature shows that subjects with a history of CT have increased peripheral inflammation markers, such as C-reactive protein (CRP), fibrinogen, and white blood cell counts. Interestingly, these markers are elevated both in CT subjects that later developed depressive symptoms and subjects who did not, whereas depressed individuals without a history of childhood maltreatment did not show such alterations [21, 65–67]. A recent meta-analysis on 53 studies described increased levels of Interleukin-1 beta (IL-1b) in depressed patients with early exposure to CT, and more frequent polymorphisms in the IL-1b gene in subjects with a history of CT [68]. A longitudinal study found that adolescents exposed to early-life stressors showed greater increases in both IL-6 and CRP after developing depressive symptoms than subjects without a history of CT [69]. Similar results have been observed in schizophrenia and bipolar disorder patients [31, 70, 71], further supporting the hypothesis that inflammation may contribute to psychopathology after CT. Taken together, these findings suggest that the increased inflammation could be considered as a "biological scar" of the early exposure to high levels of stress, associated with abnormalities in both mental and physical health. Interestingly, intrinsic mild anti-inflammatory effects were described in animal and human studies for antidepressants (e.g., fluoxetine and clomipramine), mood stabilizers (e.g., sodium valproate and lithium), and antipsychotics (e.g., clozapine and risperidone), proposing that appeasing inflammation may play a role in relieving psychiatric symptoms [72]. However, this conclusion is not consensual, as inconsistent findings were frequently reported.

Moving from these considerations, several lines of research are investigating whether peripheral inflammatory biomarkers may be used as a strategy to screen psychiatric patients with a history of childhood stressful experiences, who may benefit from treatments targeting the immune system, and whether anti-inflammatory strategies could prevent the onset of clinical outcomes in these subjects [73]. Preclinical studies found that the administration of anti-inflammatory agents could prevent the expression of long-term consequences of CT. For example, rodent pups exposed to maternal separation protocols showed a proinflammatory phenotype and more depressive-like symptoms compared to unchallenged animals and these differences were reduced by the administration of an anti-inflammatory cyclo-oxygenase 2 (COX-2) inhibitor [74] or the anti-inflammatory IL-10 [75]. In humans, clinical phenotypes associated with elevated CRP included childhood adversity and specific depressive and anxious symptoms. Thus, patients with major depressive disorder (MDD) stratified for proinflammatory biomarkers, like CRP, could have a distinctive clinical presentation that might be responsive to second-line treatment with anti-inflammatory drugs [76].

Other studies suggest that non-pharmacological strategies with certain anti-inflammatory properties, such as psychotherapy, psychosocial interventions, and physical exercise, might be beneficial in reducing the effects of trauma exposure [64].

This evidence is of particular interest in the field of stratified (or precision) medicine, which aims to identify subgroups of psychiatric patients with specific mechanisms of disease or peculiar responses to treatment, in order to develop patient-tailored strategies of intervention. Because of its distinctive impact on neurobiological, neuroendocrine, and inflammatory functioning, CT might identify one of such subgroups.

Since treatment-resistant patients with elevated baseline inflammation levels may benefit from anti-inflammatory compounds, such as the tumor necrosis factor (TNF)-α inhibitor, infliximab [77], the stratified biological findings in maltreated patients suggest potential treatment targets.

Future studies should test if immunomodulatory interventions could improve the treatment of psychiatric patients with a history of CT. For example, further research should investigate whether the clinical subpopulation of depressed patients with CT and increased levels of IL-1b may benefit from IL-1b-targeted therapies and whether blocking IL-1b in patients with a CT history and elevated IL-1b levels may help to prevent MDD onset or relapses.

## 20.4 Conclusions

Traumatic experiences during childhood are related to a higher risk of developing psychiatric symptoms and diseases, which also show increased clinical severity, treatment resistance, and worse outcome. Carefully investigating the presence of CT could add significant prognostic information to patients' assessment and can also be important for planning both pharmacological and psychotherapeutic treatment. Especially when managing subjects with mood disorders, clinicians should keep in mind that CT can make patients more resistant to treatment and, consequently, should consider more intensive and alternative therapeutic options for this clinical subgroup.

Data from the literature describe several neurobiological, neuroendocrine, and inflammatory alterations, induced by traumatic experiences, which appear to mediate the onset of psychiatric diseases and decreased response to treatment. Within this framework, if inflammation is part of the pathophysiological pathway linking predisposing and environmental risk factors, such as CT, to the development of psychiatric disorders and their comorbidities, modulating the immune response in these subjects could provide an opportunity to reduce the risk of developing mental disorders. Similar strategies, such as anti-inflammatory interventions and potentiation of adaptive immunity, could be used to improve the unfavorable treatment response in psychiatric patients with CT.

## References

1. Agid O, Shapira B, Zislin J, Ritsner M, Hanin B, Murad H, et al. Environment and vulnerability to major psychiatric illness: a case control study of early parental loss in major depression, bipolar disorders and schizophrenia. Mol Psychiatry. 1999;4(2):163–72.
2. De Bellis MD. Developmental traumatology: a contributory mechanism for alcohol and substance use disorders. Psychoneuroendocrinology. 2002;27(1–2):155–70.
3. Read J, van Os J, Morrison AP, Ross CA. Childhood trauma, psychosis and schizophrenia: a literature review with theoretical and clinical implications. Acta Psychiatr Scand. 2005;112(5):330–50.

4. Heim C, Newport DJ, Mletzko T, Miller AH, Nemeroff CB. The link between childhood trauma and depression: insights from HPA axis studies in humans. Psychoneuroendocrinology. 2008;33(6):693–710. https://doi.org/10.1016/j.psyneuen.2008.03.008.
5. Mandelli L, Petrelli C, Serretti A. The role of specific early trauma in adult depression: a meta-analysis of published literature. Childhood trauma and adult depression. Eur Psychiatry. 2015;30(6):665–80.
6. de Codt A, Monhonval P, Bongaerts X, Belkacemi I, Tecco JM. Bipolar disorder and early affective trauma. Psychiatr Danub. 2016;28(Suppl.1):4–8.
7. Garno J, Goldberg J, Ramirez PM, Ritzler BA. Impact of childhood abuse on the clinical course of bipolar disorder. Br J Psychiatry. 2005;186:121–5.
8. Daruy-Filho L, Brietzke E, Lafer B, Grassi-Oliveira R. Childhood maltreatment and clinical outcomes of bipolar disorder. Acta Psychiatr Scand. 2011;124(6):427–34.
9. Etain B, Aas M, Andreassen OA, Lorentzen S, Dieset I, Gard S, et al. Childhood trauma is associated with severe clinical characteristics of bipolar disorder. J Clin Psychiatry. 2013;74(10):991–8.
10. Post R, Altshuler L, Leverich GS, Frye MA, Suppes T, McElroy SL, et al. Role of childhood adversity in the development of medical co-morbidities associated with bipolar disorder. J Affect Disord. 2013;147(1–3):288–94.
11. Aas M, Henry C, Andreassen OA, Bellivier F, Melle I, Etain B. The role of childhood trauma in bipolar disorder. Int J Bipolar Disord. 2016;4(1):2.
12. Agnew-Blais J, Danese A. Childhood maltreatment and unfavourable clinical outcomes in bipolar disorder: a systematic review and meta-analysis. Lancet Psychiatry. 2016;3(4):342–9.
13. Crosta ML, De Simone C, Di Pietro S, Acanfora M, Caldarola G, Moccia L, et al. Childhood trauma and resilience in psoriatic patients: a preliminary report. J Psychosom Res. 2018;106:25–8. https://doi.org/10.1016/j.jpsychores.2018.01.002.
14. Garami J, Valikhani A, Parkes D, Haber P, Mahlberg J, Misiak B, et al. Examining perceived stress, childhood trauma and interpersonal trauma in individuals with drug addiction. Psychol Rep. 2018;33294118764918:433. https://doi.org/10.1177/0033294118764918.
15. Kim JS, Lee SH. Influence of interactions between genes and childhood trauma on refractoriness in psychiatric disorders. Prog Neuro-Psychopharmacol Biol Psychiatry. 2016;70:162–9. https://doi.org/10.1016/j.pnpbp.2016.01.013.
16. Paquola C, Bennett MR, Lagopoulos J. Understanding heterogeneity in grey matter research of adults with childhood maltreatment-a meta-analysis and review. Neurosci Biobehav Rev. 2016;69:299–312.
17. Aas M, Kauppi K, Brandt C, Tesli M, Kaufmann T, Steen NE, et al. Childhood trauma is associated with increased brain responses to emotionally negative as compared with positive faces in patients with psychotic disorders. Psychol Med. 2017;47(4):669–79.
18. Janiri D, Sani G, Rossi P, Piras F, Iorio M, Banaj N, et al. Amygdala and hippocampus volumes are differently affected by childhood trauma in patients with bipolar disorders and healthy controls. Bipolar Disord. 2017;19(5):353–62. https://doi.org/10.1111/bdi.12516.
19. McIntyre RS, Soczynska JK, Liauw SS, Woldeyohannes HO, Brietzke E, Nathanson J, et al. The association between childhood adversity and components of metabolic syndrome in adults with mood disorders: results from the international mood disorders collaborative project. Int J Psychiatry Med. 2012;43(2):165–77.
20. Chen Y, Baram TZ. Toward understanding how early-life stress reprograms cognitive and emotional brain networks. Neuropsychopharmacology. 2015;41(1):197–206.
21. Baumeister D, Akhtar R, Ciufolini S, Pariante CM, Mondelli V. Childhood trauma and adulthood inflammation: a meta-analysis of peripheral C-reactive protein, interleukin-6 and tumour necrosis factor-α. Mol Psychiatry. 2016;21(5):642–9.
22. Wielaard I, Schaakxs R, Comijs HC, Stek ML, Rhebergen D. The influence of childhood abuse on cortisol levels and the cortisol awakening response in depressed and nondepressed older adults. World J Biol Psychiatry. 2017;25:1–10. https://doi.org/10.1080/15622975.2016.1274829.

23. Di Nicola M, Cattaneo A, Hepgul N, Di Forti M, Aitchison KJ, Janiri L, et al. Serum and gene expression profile of cytokines in first-episode psychosis. Brain Behav Immun. 2013;31:90–5. https://doi.org/10.1016/j.bbi.2012.06.010.
24. Pariante CM. Why are depressed patients inflamed? A reflection on 20 years of research on depression, glucocorticoid resistance and inflammation. Eur Neuropsychopharmacol. 2017;27(6):554–9. https://doi.org/10.1016/j.euroneuro.2017.04.001.
25. Rosenblat JD, McIntyre RS. Bipolar Disorder and Immune Dysfunction: Epidemiological Findings, Proposed Pathophysiology and Clinical Implications. Brain Sci. 2017;7(11):E144. https://doi.org/10.3390/brainsci7110144.
26. Caspi A, Sugden K, Moffitt T, Taylor A, Craig IW, Harrington H, et al. Influence of life stress on depression: moderation by a polymorphism in the 5-HTT gene. Science. 2003;301(5631):386–9.
27. Aas M, Djurovic S, Athanasiu L, Steen NE, Agartz I, Lorentzen S, et al. Serotonin transporter gene polymorphism, childhood trauma, and cognition in patients with psychotic disorders. Schizophr Bull. 2012;38(1):15–22.
28. Yang BZ, Zhang H, Ge W, Weder N, Douglas-Palumberi H, Perepletchikova F, et al. Child abuse and epigenetic mechanisms of disease risk. Am J Prev Med. 2013;44(2):101–7.
29. Suderman M, Borghol N, Pappas JJ, Pinto Pereira SM, Pembrey M, Hertzman C, et al. Childhood abuse is associated with methylation of multiple loci in adult DNA. BMC Med Genet. 2014;7:13.
30. Williams LM, Debattista C, Duchemin AM, Schatzberg AF, Nemeroff CB. Childhood trauma predicts antidepressant response in adults with major depression: data from the randomized international study to predict optimized treatment for depression. Transl Psychiatry 2016; 6: e799.
31. Jaworska-Andryszewska P, Rybakowski JK. Childhood trauma in mood disorders: neurobiological mechanisms and implications for treatment. Pharmacol Rep. 2018;71(1):112–20. https://doi.org/10.1016/j.pharep.2018.10.004.
32. Nanni V, Uher R, Danese A. Childhood maltreatment predicts unfavorable course of illness and treatment outcome in depression: a meta-analysis. Am J Psychiatry. 2012;169(2): 141–51.
33. Conus P, Cotton S, Schimmelmann BG, Berk M, Daglas R, McGorry PD. Pretreatment and outcome correlates of past sexual and physical trauma in 118 bipolar I disorder patients with a first episode of psychotic mania. Bipolar Disord. 2010;12(3):244–52.
34. Schneeberger AR, Muenzenmaier K, Castille D, Battaglia J, Link B. Use of psychotropic medication groups in people with severe mental illness and stressful childhood experiences. J Trauma Dissociation. 2014;15(4):494–511. https://doi.org/10.1080/15299732.2014.903550.
35. Putnam FW, Hulsmann JE. Pharmacotherapy for survivors of childhood trauma. Semin Clin Neuropsychiatry. 2002;7(2):129–36.
36. Miniati M, Rucci P, Benvenuti A, Frank E, Buttenfield J, Giorgi G, et al. Clinical characteristics and treatment outcome of depression in patients with and without a history of emotional and physical abuse. J Psychiatr Res. 2009;44(5):302–9.
37. Mayberg HS. Modulating dysfunctional limbic–cortical circuits in depression: towards development of brain-based algorithms for diagnosis and optimised treatment. Br Med Bull. 2003;65:193–207.
38. Nemeroff CB, Heim CM, Thase ME, Klein DN, Rush AJ, Schatzberg AF, et al. Differential responses to psychotherapy versus pharmacotherapy in patients with chronic forms of major depression and childhood trauma. Proc Natl Acad Sci U S A. 2003;100(24):14293–6.
39. Quilty L, Marshe V, Lobo D, Harkness KL, Müller DJ, Bagby RM. Childhood abuse history in depression predicts better response to antidepressants with higher serotonin transporter affinity: a pilot investigation. Neuropsychobiology. 2016;74(2):78–83.
40. Ariel L, Inbar S, Edut S, Richter-Levin G. Fluoxetine treatment is effective in a rat model of childhood-induced post-traumatic stress disorder. Transl Psychiatry. 2017;7(11):1260. https://doi.org/10.1038/s41398-017-0014-5.

41. Thomaes K, Dorrepaal E, Draijer N, Jansma EP, Veltman DJ, van Balkom AJ. Can pharmacological and psychological treatment change brain structure and function in PTSD? A systematic review. J Psychiatr Res. 2014;50:1–15. https://doi.org/10.1016/j.jpsychires.2013.11.002.
42. Johnstone JM, Luty SE, Carter JD, Mulder RT, Frampton CM, Joyce PR. Childhood neglect and abuse as predictors of antidepressant response in adult depression. Depress Anxiety. 2009;26(8):711–7. https://doi.org/10.1002/da.20590.
43. Lewis CC, Simons AD, Nguyen LJ, Murakami JL, Reid MW, Silva SG, et al. Impact of childhood trauma on treatment outcome in the treatment for adolescents with depression study (TADS). J Am Acad Child Adolesc Psychiatry. 2010;49(2):132–40.
44. Douglas KM, Porter RJ. The effect of childhood trauma on pharmacological treatment response in depressed inpatients. Psychiatry Res. 2012;200(2–3):1058–61. https://doi.org/10.1016/j.psychres.2012.06.015.
45. Kaplan MJ, Klinetob NA. Childhood emotional trauma and chronic posttraumatic stress disorder in adult outpatients with treatment-resistant depression. J Nerv Ment Dis. 2000;188(9):596–601.
46. Lara ME, Klein DN, Kasch KL. Psychosocial predictors of the short-term course and outcome of major depression: a longitudinal study of a nonclinical sample with recent-onset episodes. J Abnorm Psychol. 2000;109(4):644–50.
47. Cakir S, Tasdelen Durak R, Ozyildirim IE, Sar V. Childhood trauma and treatment outcome in bipolar disorder. J Trauma Dissociation. 2016;17(4):397–409.
48. Etain B, Lajnef M, Brichant-Petitjean C, Geoffroy PA, Henry C, Gard S, et al. Childhood trauma and mixed episodes are associated with poor response to lithium in bipolar disorder. Acta Psychiatr Scand. 2017;135(4):319–27.
49. Pavuluri MN, Henry DB, Carbray JA, Sampson GA, Naylor MW, Janicak PG. A one-year open-label trial of risperidone augmentation in lithium nonresponder youth with preschool-onset bipolar disorder. J Child Adolesc Psychopharmacol. 2006;16(3):336–50.
50. Hassan AN, De Luca V. The effect of lifetime adversities on resistance to antipsychotic treatment in schizophrenia patients. Schizophr Res. 2015;161(2–3):496–500.
51. Misiak B, Frydecka D. A history of childhood trauma and response to treatment with antipsychotics in first-episode schizophrenia patients: preliminary results. J Nerv Ment Dis. 2016;204(10):787–92.
52. McGregor N, Thompson N, O'Connell KS, Emsley R, van der Merwe L, Warnich L. Modification of the association between antipsychotic treatment response and childhood adversity by MMP9 gene variants in a first-episode schizophrenia cohort. Psychiatry Res. 2018;262:141–8.
53. Koga A, Bani-Fatemi A, Hettige N, Borlido C, Zai C, Strauss J, et al. GWAS analysis of treatment resistant schizophrenia: interaction effect of childhood trauma. Pharmacogenomics. 2017;18(7):663–71.
54. Mondelli V, Ciufolini S, Belvederi Murri M, Bonaccorso S, Di Forti M, Giordano A, et al. Cortisol and inflammatory biomarkers predict poor treatment response in first episode psychosis. Schizophr Bull. 2015;41(5):1162–70. https://doi.org/10.1093/schbul/sbv028.
55. Ciufolini S, Gayer-Anderson C, Fisher HL, Marques TR, Taylor H, Di Forti M, et al. Cortisol awakening response is decreased in patients with first-episode psychosis and increased in healthy controls with a history of severe childhood abuse. Schizophr Res. 2018. S0920-9964(18)30260-3; https://doi.org/10.1016/j.schres.2018.05.002.
56. Mondelli V, Dazzan P, Hepgul N, Di Forti M, Aas M, D'Albenzio A, et al. Abnormal cortisol levels during the day and cortisol awakening response in first-episode psychosis: the role of stress and of antipsychotic treatment. Schizophr Res. 2010;116(2–3):234–42. https://doi.org/10.1016/j.schres.2009.08.013.
57. Ignácio ZM, Réus GZ, Abelaira HM, Maciel AL, de Moura AB, Matos D, et al. Quetiapine treatment reverses depressive-like behavior and reduces DNA methyltransferase activity induced by maternal deprivation. Behav Brain Res. 2017;320:225–32. https://doi.org/10.1016/j.bbr.2016.11.044.

58. Guina J, Rossetter SR, DeRhodes BJ, Nahhas RW, Welton RS. Benzodiazepines for PTSD: a systematic review and meta-analysis. J Psychiatr Pract. 2015;21(4):281–303. https://doi.org/10.1097/PRA.0000000000000091.
59. Matar MA, Zohar J, Kaplan Z, Cohen H. Alprazolam treatment immediately after stress exposure interferes with the normal HPA-stress response and increases vulnerability to subsequent stress in an animal model of PTSD. Eur Neuropsychopharmacol. 2009;19(4):283–95. https://doi.org/10.1016/j.euroneuro.2008.12.004.
60. Vogel M, Dürsteler-Macfarland KM, Walter M, Strasser J, Fehr S, Prieto L, et al. Prolonged use of benzodiazepines is associated with childhood trauma in opioid-maintained patients. Drug Alcohol Depend. 2011;119(1–2):93–8. https://doi.org/10.1016/j.drugalcdep.2011.05.037.
61. Eraly SA, Nievergelt CM, Maihofer AX, Barkauskas DA, Biswas N, Agorastos A, et al. Assessment of plasma C-reactive protein as a biomarker of posttraumatic stress disorder risk. JAMA Psychiat. 2014;71:423–31.
62. Hodes GE, Pfau ML, Leboeuf M, Golden SA, Christoffel DJ, Bregman D, et al. Individual differences in the peripheral immune system promote resilience versus susceptibility to social stress. Proc Natl Acad Sci U S A. 2014;111:16136–41.
63. Cattaneo A, Gennarelli M, Uher R, Breen G, Farmer A, Aitchison KJ, et al. Candidate genes expression profile associated with antidepressants response in the GENDEP study: differentiating between baseline 'predictors' and longitudinal 'targets'. Neuropsychopharmacology. 2013;38(3):377–85.
64. Danese A, Lewis S. Psychoneuroimmunology of early-life stress: the hidden wounds of childhood trauma? Neuropsychopharmacology. 2017;42(1):99–114. https://doi.org/10.1038/npp.2016.198.
65. Danese A, Pariante CM, Caspi A, Taylor A, Poulton R. Childhood maltreatment predicts adult inflammation in a life-course study. Proc Natl Acad Sci U S A. 2007;104(4):1319–24.
66. Danese A, Moffitt TE, Pariante CM, Ambler A, Poulton R, Caspi A. Elevated inflammation levels in depressed adults with a history of childhood maltreatment. Arch Gen Psychiatry. 2008;65(4):409–15. https://doi.org/10.1001/archpsyc.65.4.409.
67. Danese A, Caspi A, Williams B, Ambler A, Sugden K, Mika J, et al. Biological embedding of stress through inflammation processes in childhood. Mol Psychiatry. 2011;16(3):244–6. https://doi.org/10.1038/mp.2010.5.
68. Ellul P, Boyer L, Groc L, Leboyer M, Fond G. Interleukin-1 β-targeted treatment strategies in inflammatory depression: toward personalized care. Acta Psychiatr Scand. 2016;134(6):469–84. https://doi.org/10.1111/acps.12656.
69. Miller GE, Cole SW. Clustering of depression and inflammation in adolescents previously exposed to childhood adversity. Biol Psychiatry. 2011;72:34–40.
70. Hepgul N, Pariante CM, Di Pasquale S, Di Forti M, Taylor H, Marques TR, et al. Childhood maltreatment is associated with increased body mass index and increased C-reactive protein levels in first-episode psychosis patients. Psychol Med. 2012;42(9):1893–901.
71. Fond G, Godin O, Boyer L, Berna F, Andrianarisoa M, Coulon N, et al. Chronic low-grade peripheral inflammation is associated with ultra resistant schizophrenia. Results from the FACE-SZ cohort. Eur Arch Psychiatry Clin Neurosci. 2019;269(8):985–92. https://doi.org/10.1007/s00406-018-0908-0.
72. Fond G, Hamdani N, Kapczinski F, Boukouaci W, Drancourt N, Dargel A, et al. Effectiveness and tolerance of anti-inflammatory drugs' add-on therapy in major mental disorders: a systematic qualitative review. Acta Psychiatr Scand. 2014;129:163–79.
73. Cattaneo A, Macchi F, Plazzotta G, Veronica B, Bocchio-Chiavetto L, Riva MA, et al. Inflammation and neuronal plasticity: a link between childhood trauma and depression pathogenesis. Front Cell Neurosci. 2015;9:40. https://doi.org/10.3389/fncel.2015.00040.
74. Brenhouse HC, Andersen SL. Nonsteroidal anti-inflammatory treatment prevents delayed effects of early life stress in rats. Biol Psychiatry. 2011;70:434–40.
75. Wieck A, Andersen SL, Brenhouse HC. Evidence for a neuroinflammatory mechanism in delayed effects of early life adversity in rats: relationship to cortical NMDA receptor expression. Brain Behav Immun. 2013;28:218–26.

76. Chamberlain SR, Cavanagh J, de Boer P, Mondelli V, Jones DNC, Drevets WC, et al. Treatment-resistant depression and peripheral C-reactive protein. Br J Psychiatry. 2019;214(1):11–9. https://doi.org/10.1192/bjp.2018.66.
77. Raison CL, Rutherford RE, Woolwine BJ, Shuo C, Schettler P, Drake DF, et al. A randomized controlled trial of the tumor necrosis factor antagonist infliximab for treatment-resistant depression: the role of baseline inflammatory biomarkers. JAMA Psychiat. 2013;70(1):31–41.

# Childhood Trauma-Related Interventions: Treatment at Different Stages Across the Life Span

# 21

Mayra A. Gomez and Lisa M. Brown

## 21.1 Introduction

Childhood development is a sensitive period where exposure to adversity can have deleterious and enduring effects into adulthood. Potentially traumatic events can range from large, public catastrophes such as school violence, disasters, and warfare, to individual trauma such as abuse, death of a parent, serious accident, or life-threatening illness. Moreover, it is not uncommon for children to experience a combination of chronic, episodic, collective, and/or individual traumas that are cumulative and adversely influence mental health. For example, chronic exposure can include exposure to domestic violence and discrimination. The cumulative effect of these life adversities can and do occur in combination with other traumas.

It is estimated that more than half of children and adolescents experience at least one potentially traumatic event before reaching adulthood [1, 2]. The most commonly reported events by youth are violence (physical, domestic), motor vehicle accidents, natural disasters, life-threatening illness, and neglect (U.S. Department of Health & Human Services, Administration for Children and Families, Administration on Children, Youth and Families, Children's Bureau, 2018). Epidemiological research suggests that prevalence rates of some childhood traumatic experiences (e.g., maltreatment, sexual assault) are underestimated due to undetected and/or unreported events [3]. Nonetheless, existing literature reveals that adverse experiences can interrupt developmental trajectories and increase the risk for psychopathology later in life [1, 4, 5]. Post-traumatic stress disorder (PTSD),

M. A. Gomez
Clinical Psychology, Palo Alto University, Palo Alto, CA, USA
e-mail: mgomez@paloaltou.edu

L. M. Brown (✉)
Trauma Program, Risk and Resilience Research Lab, Palo Alto University, Palo Alto, CA, USA
e-mail: lbrown@paloaltou.edu



G. Spalletta et al. (eds.), *Childhood Trauma in Mental Disorders*,
https://doi.org/10.1007/978-3-030-49414-8_21

depression, anxiety disorders, borderline personality disorder (BPD), and substance abuse problems are common mental health disorders associated with early trauma exposure [6–8].

A dose–effect response of childhood trauma frequency has been identified and believed to gravely impact later functionality and quality of life [9]. A study conducted by the Center for Diseases and Control Prevention and Kaiser-Permanente examined the association between adverse childhood experiences (ACE) and negative health outcomes in more than 17,000 adults [1]. It was found that about 40% of adults reported at least two or more ACEs in childhood. Further, a higher ACEs score was associated with greater negative mental health outcomes (depression, substance abuse, increased suicidality), negative physical health outcomes (heart, lung, and liver disease), and even increased risk for early-life mortality [1, 10]. Additionally, a child's age when the trauma occurred, event type, intensity of traumatic event, and lack of emotional support was found to further influence the manifestation of mental illness. Since children and adolescent populations are particularly vulnerable, accurate detection and early intervention are needed to minimize psychological damage. Childhood trauma can instigate a series of disruptions starting with altered neurodevelopmental mechanisms and ending with negative health management behaviors that result in poor quality of life [1, 11]. Treatment should include a thorough evaluation using normed screening and assessment measures, a trauma-informed case conceptualization, and care with evidence-based treatments.

In this chapter, we review assessments, interventions, and therapeutic strategies for people who have experienced trauma during their childhood or adolescence. This chapter follows a phase model of intervention comprised of three levels of response: (1) Psychological First Aid (PFA), a Level 1 intervention implemented during the first few hours, days, or weeks of mass trauma exposure; (2) Crisis Intervention/Skills for Psychological Recovery (SPR), a Level 2 intervention used weeks or months after a traumatic event to restore psychosocial functioning and minimize lasting psychological sequalae; and (3) Psychotherapy, a Level 3 response that includes evidence-based treatments. This chapter reviews issues and therapeutic strategies that can be considered across different developmental periods. This chapter provides a framework for managing crises and providing psychotherapy to those who would benefit from more intensive treatment.

## 21.2 Brief Historical Perspective of Trauma and Treatment in Childhood

Historically, the effects of trauma exposure in childhood and adolescence have not been a focus of research. Early perspectives posited that children lacked perception, cognition, and social maturity to understand traumatic events [10]. In 1952, the Diagnostic and Statistical Manual of Mental Disorders, First Edition (DSM-I) included "gross stress reaction" based on Freud's theory of traumata. Freud believed that "pre-morbid psychic functioning" and family history of mental illness were determinants of long-term "mental disturbances" [12, 13]. In the 1970s, the US

Congress passed the Child Abuse Prevention and Treatment Act to protect child victims of maltreatment and abuse. This Act catalyzed research funding that would reveal that adverse emotional and psychological consequences resulted from trauma experienced by minors. Simultaneously, a paradigm shift was occurring in the recognition and conceptualization of psychological trauma in adults, specifically among active-duty members and war veterans. Similarly, victims of sexual violence (e.g., rape) were experiencing outward reactions to this life-threatening experience, but these interpersonal traumas were pathologized as "hysteria." Despite these advances in research, DSM-II did not include trauma as a diagnostic category.

By 1980, a third iteration of the DSM (DSM-III) named psychological trauma "delayed stress syndrome" [12] and reclassified it to extend beyond war-related trauma called "post-traumatic stress disorder" or PTSD. This assisted in validating and legitimizing trauma experiences from a broader range of traumatic events to include "invisible" interpersonal traumas. Shortly after the term "Rape Trauma Syndrome" was added, the definition of trauma in the DSM-III expanded to include a wider scope of traumatic events [14].

A focus on ameliorating child maltreatment, physical and sexual abuse, spurred trauma-related interventions for youth populations. By the late 1980s, significant advancements occurred in the development of trauma interventions for both adults and children. As theoretical and etiological perspectives of trauma emerged, models of interventions and treatments followed suit.

In the 1980s, crisis interventions, such as Psychological Debriefing (PD), were introduced as short-term crisis interventions for adults, followed by adolescents [15]. Though PD was originally deemed effective, a growing body of research revealed that to a significant number of people, PD was of little benefit and for a smaller number of trauma survivors it actually resulted in harm [16]. PD for children was considered as potentially harmful as it exacerbated behavioral problems and provoked onset of psychopathology [17]. To address the need for a crisis intervention that could be delivered by trained laypeople and used to aid large number of those affected by disasters and terrorism, Psychological First Aid (PFA) was developed and introduced in 2006. Although PFA is an evidence-informed intervention and not an evidence-based intervention, it is now considered by many as the gold standard for crisis intervention for children and adults alike [18].

Psychotherapy is a treatment provided by highly trained professionals and has existed for decades. Adult psychotherapy evolved earlier than therapy for young people. During the era of contemporary psychology in the 1900s, a new approach to psychotherapy evolved called psychoanalysis. Because Sigmund Freud's conceptualized that childhood processes continued into adulthood, the field was starting to note the method usefulness for young people. Shortly after, other theories developed that facilitated the progression of psychotherapeutic methods rising from behaviorism, cognitive, and humanistic perspectives. The child psychotherapy primarily focused on behaviorism (modeling and conditioning) and became one of the most popular forms of treatment for children by the end of the twentieth century [19]. Trauma-related therapeutic approaches in children remained conspicuously absent, until an increased attention and recognition of the impact of

PTSD. Organizations such as The National Center for Trauma-Informed Care, U.S. Department of Health and Human Services, The National Child Traumatic Stress Network (NCTSN), and The National Center for Children in Poverty encouraged researchers to examine the impact of trauma among children and adolescents. As psychologists and researchers verified the value of trauma-focused treatment modalities, the notion of trauma-informed care was born [20].

Trauma-informed care is defined as the integration of safety, trustworthiness, choice, peer support, collaboration, resilience/empowerment, and cultural considerations as critical components in promoting resilience and stress management [20, 21] Agencies providing trauma-informed care strive to identify trauma-related needs and strengths as means to enhance provision of individual and family services. A trauma-informed approach must be reliable, valid, and facilitated by trained clinicians to effectively assist those processing their trauma emotions with minimal harm. It is in response to these core values that interventions and treatments such as Crisis Counseling and Trauma-Focused Cognitive Behavioral Therapy (TF-CBT) [2, 20] were developed.

## 21.3 Trauma-Focused Diagnostics: Screeners and Assessments

Use of age-appropriate post-trauma assessment measures is critical in taking the first step toward designing and planning suitable psychological care. Due to developmental differences, children and adolescents have different reactions to potential traumatic events than adults. Prior to the DSM-5, PTSD in children age 6 and younger had not been differentiated from the diagnostic criteria of PTSD in adults [22]. The diagnostic implications in the taxonomy of PTSD in children highlighted the need to include developmental differences and variations in stress-reactions among children and adolescents. Screening and assessment instruments developed for this age group are geared toward diagnostic interviews and self-report informant measures completed by their adult caregivers.

Before describing the use of screening tools and assessments, it is worthwhile to discuss differences between these types of measures. *Screeners* are quick and easy to use in diverse settings to detect trauma-related symptoms and determine if further evaluation is warranted [23, 24]. Typically, screening tools can be used by any properly trained individual (e.g., case worker, nurse, or first responder). In contrast, *assessments* are more comprehensive tools for determining the possible etiology of the problem, severity of symptoms, and level of functioning [23]. Because assessments inform the process of case conceptualization, treatment planning, and making a formal diagnosis, a trained mental health professional is needed to administer, score, and interpret the results.

To select the right assessments to evaluate a person who has experienced a traumatic event, the NCTSN recommends considering the following three properties: *validity*; *reliability*; and *standardization of norms* [25]. All measurement tools that are used for treatment and diagnostic purposes should have these properties, but the clinician is responsible for selecting the appropriate measure for each specific patient.

Methods of administration are another factor to consider when using screening measures or assessments. Two types of administration tools exist: questionnaires/self-report instruments and structured/semi-structured interviews. Questionnaires are typically self-report tools that can be administered verbally, via paper-pencil, or on a computer by any layperson that has been trained to conduct testing. However, interpretation of results and referrals for further evaluation should be done by a trained professional. Length of the questionnaire and available time can vary and is contingent upon the setting and the timing of when it is used after the traumatic event (crisis counseling as the first response vs. psychotherapy in a clinic setting). Necessary assessment considerations for children and adolescents involve developmental factors and appropriateness of questions. Literacy issues, such as language fluidity and comprehension between children, adolescents, and adults, are a primary factor to consider. Therefore, collecting collateral reports and assessments from parents or guardians is frequently used to strengthen the clinical picture of young patients.

The second type of administrative tools is structured and semi-structured interviews that require administration by a trained mental health professional who possesses a higher level of clinical skill. Structured interviews often include prompts designed to identify specific problem areas. Specifically, trauma-focused interviews aim to understand etiology, severity, and frequency of trauma symptoms, while capturing level of functionality through subjective reports and clinical observation of respondent's behaviors. Semi-structured interviews function similarly but differ in their leniency of prompt questions. That is, the administrator may deviate from the prompt if it results in adding clarity to the interview. These types of interviews obtain more information than self-report questionnaires but take longer to administer. In younger children, these interviews are typically administered to primary caregivers while child behaviors are closely observed.

## 21.4 Introduction to Interventions and Treatments

Before we review a list of trauma-focused interventions and treatments for childhood traumas, it is helpful to understand the difference between interventions and treatments. In psychology literature, these terms are often used interchangeably but some distinctions between the two are warranted [26]. An *intervention* is defined as an "*act or method of interfering with the outcome or course… to prevent harm or improve functioning*" [26]. The PFA intervention for disaster related-events uses psychosocial techniques (*act*) in promoting safety, connectedness, calm, and hope with a goal to restore cognitive and emotional well-being (change in *outcome*). Conversely, the term *treatment* is defined as the formal process of using psychological methods by a trained professional in providing services for an extended duration [26]. The goal in treatment is to decrease, manage, or reverse illness. Treatment typically involves a form of psychotherapy with a specialized treatment approach selected on the basis of symptom presentation, age, and developmental capacity [27].

Similar to PFA, crisis counseling is a trauma-informed intervention that has been widely used to aid children and adolescents. In the days and weeks after a trauma, young people have elevated risks for trauma sequelae. Because crisis counseling is adaptable, it is frequently used in school settings as means to encourage normal day-to-day functioning and to strengthen social connections. This coping skills approach aims to strengthen individual and group resiliency, while decreasing symptomatology. Similarly, crisis counseling for adults provides intermediate strength-based skill-building techniques with the goal to reduce distress. The difference lies in the topics discussed and resources provided. For example, adults may engage in problem-solving techniques to lessen their financial burden, improve their living situations, use effective parenting skills, and enhance their stress management. Adults may engage in crisis counseling at home, church, local shelter, or community center.

When selecting psychological treatments, the age, type and duration of trauma, time since traumatic event, symptom type, and current level of functioning inform the selection of the psychotherapy modality. Mental health professionals should consider use of evidence-based treatments when a person's stress reactions have overwhelmed their natural ability to respond and are adversely affecting their day-to-day functioning. Those whose symptoms are left untreated run the risk of worsening symptoms and a negative prognosis [28].

Adult psychotherapy involves some form of talk therapy and is typically provided in a clinical setting. In children under age 7, therapy usually involves some type of play combined with observations. Unlike adults and adolescents, talk therapy is not very useful for very young children. While older children [6, 23, 29–32] and adolescents [7, 33–36] can benefit from some traditional components of talk therapy, in children, this involves teaching skills (emotional awareness, stress management, healthy communication, and behavioral activation) in a concrete manner through play, modeling, and exercises [15, 37]. Psychotherapy for adolescents can involve more complex approaches than those used with younger children and can incorporate "take-home assignments" to promote use of newly learned skills at home. Some parental or caregiver involvement for children and adolescents is required in order to support change at home. The following paragraphs describe the uses of trauma-informed care that range from short-term interventions to recovery-based psychotherapy.

### 21.4.1 Psychological First Aid

During World War II, the first PFA model was developed to return soldiers without physical wounds to the battlefield as soon as possible. The model included four key principles: (1) support of victim's emotional experience; (2) understand injurious psychological limitations; (3) observe individuals' level of capacities; and (4) awareness of one's own limitations as providers [38, 39]. Later, PFA was reintroduced by the American Psychiatric Association (APA) during the Cold War era to help lay workers prepare for a nuclear attack [38]. Many versions of PFA have recently been developed for specific populations that have sparked questions regarding the benefits of the intervention. The Red Cross attempted to tackle this problem by standardizing the PFA model and creating training manuals [40]. However, little is known about how trained laypeople deliver the PFA in real-life settings and about the outcomes for those who have received the intervention.

PFA is now used worldwide as the primary intervention of choice in the hours or days after a mass trauma (e.g. terrorism, natural disaster, or large-scale accidents). PFA was designed for children, families, adults, first responders, and other disaster relief workers. PFA is malleable in nature and designed to be offered in diverse settings from hospitals to public shelters [41]. The goal is to promote resilience and adaptive coping through practical care by providing social support, comfort, and referral to resources [42]. PFA is not intended to help people to process the traumatic experience. The National Center for PTSD (NCPTSD) and National Center of Traumatic Stress Network (NCTSN) developed a field operations guide designed to train healthcare professionals and lay people to systematically deliver services after a disaster or terrorist attack. However, many organizations are encouraging the use of PFA for any mass crisis, regardless of size, where a person is distressed and would benefit from coping, managing stress, and referral to resources if needed. For example, PFA may prove useful to a person of any age who has experienced a car accident or loss of a pet.

Most training resources for PFA are available for free online and with consultation services. Despite global endorsement by large organizations, questions still remain regarding PFA's lack of systematic research showing efficacy of the intervention. Attempts to increase empirical evidence have propelled qualitative and quantitative researcher to assess the outcomes of PFA. Studies conducted internationally have identified PFA as effective in providing emotional and social support without causing any harm [43]. Others have reported that PFA had some cultural limitations because more women than men were receptive to receiving psychosocial support, while others felt that PFA was too limited in providing basic care such as food, water, and shelters [43]. Nonetheless, PFA remains the treatment of choice to treat people of all ages who have experienced a traumatic event.

### 21.4.2 Psychological First Aid for Children and Adolescents

PFA intervention for children and adolescents uses age and culturally appropriate skills to assist this subgroup of the population in feeling safe while meeting their needs. The unique vulnerabilities this subgroup experiences, such as developmental disruptions and age-related challenges in verbalizing their needs, require extra attention to fully support young people after a potentially traumatic event [44]. An important first step in proving PFA is to identify those who are at risk for adverse outcomes. Risk factors can range from separation from their caregiver; death of a loved one; disruption of current foster care systems;

medical, mental, or developmental problems; trauma-related physical injuries; risk-taking behaviors; and substance use [42]. If risk factors are identified, PFA trained staff attends to the person's needs and makes age-appropriate recommendations for care to promote safety and feelings of trust. Typically, in public shelters, there should be a designated secure child-friendly space that fosters social engagement.

PFA psychosocial support for young people is intended to facilitate collaboration with caregivers and family members. PFA-trained volunteers are encouraged to teach practical skills, assist in action planning, facilitate connection, and support social cohesion. In cases where a child or adolescent has been separated from family or if the death of a caregiver has occurred, qualified volunteers can provide temporary guardianship and emotional support until a legal guardian can be identified. Specialized PFA volunteers support recovery by engaging trauma-exposed children and adolescents in enjoyable physical or social activities to reduce their stress and increase adaptive coping.

### 21.4.3 Crisis Counseling

Level 1 interventions enable recovery by encouraging psychosocial resilience and access to basic resources. For some people, PFA may not be sufficient in addressing their trauma symptoms due to unresolved internal or external issues. *Crisis counseling* is a Level 2 intervention that includes some components of PFA in that it involves practical skills, promotes resilience, and connects survivors to resources [2]. In general, crisis counseling has substantial variability in its application. It is used with individuals and groups of all ages and who have experienced some type of trauma. Crisis counseling is delivered in short sessions, usually less than an hour, that are flexible and delivered in single or multiple sessions in clients' homes, community settings, or via telephone.

Despite its widespread use during the past several decades, the efficacy of crisis counseling still remains unknown mainly because protocols and training are not standardized, and data collection is often difficult. Yet, the aftermath of 9/11 helped advance crisis counseling programs to where certified trauma-focused interventions sponsored by The Federal Emergency Management Agency (FEMA) now exist [45]. More recent evaluations of crisis counseling techniques have revealed benefits in reducing suicidal ideation among trauma survivors and improvement in stress management post-disaster [41, 46].

After a mass disaster, crisis counseling services are typically federally funded and offered free of charge to affected populations. Crisis counselors usually have a college degree or previous mental health experience. All receive some type of specialized training from The Substance Abuse and Mental Health Services Administration (SAMHSA) or FEMA that covers stress management and skill-building techniques. One type of crisis counseling that is now widely offered is Skills for Psychology Recovery.

### 21.4.4 Skills for Psychological Recovery

Skills for Psychological Recovery (SPR) are intended to disrupt the developmental course of a mental illness for people of all ages [47, 48]. For those with subclinical mild to moderate distress, SPR is designed to help people engage and learn healthy coping skills. Typically, SPR is used after mass disasters and delivered in a variety of settings (e.g., schools, clinics, assisted-living facilities, community centers, and homes) [47]. To promote recovery, SPR teaches skill building to facilitate problem solving, encourage engagement in positive activities, manage physical and emotional reactions to upsetting situations, support healthy thinking, and rebuild healthy social connections [47].

### 21.4.5 Skills for Psychological Recovery for Children and Adolescents

When SPR is offered to children and adolescents, the counselor collaborates with the parent or caregiver to restore functionality and prevent worsening of existing mental health symptoms. SPR in older children (ages 7–12) and adolescents (ages 13–17) involves use of reflective listening, noninvasive language, and encouragement for social support between peers [49]. With younger children, SPR counselors may assume the role of a secondary, formal caretaker to help support parents or, if child is displaced from their family or caregiver, SPR will care for children until they can be reunited with their family [50]. Intermediate support is provided until more stable resources arrive (i.e., psychotherapy or social services).

Overall, SPR has been found to be more effective than supportive counseling alone [47, 51]. Field studies evaluating use of SPR in national and international settings reported that it was useful in alleviating moderate mental health problems [48, 52, 53]. In children and adolescents, SPR has shown to be beneficial when implemented in school settings [49]. The NC-PTSD and NCTSN offer online training to providers, resources, manuals, and in-person workshops. Prior basic credentials in mental health and experience working with trauma populations are required to learn and deliver SPR [47].

## 21.5 Introduction to Psychotherapy

Over time, the concept of psychotherapy has evolved in meaning and practice from psychological practices grounded in religious beliefs to present-day influences of science and sophisticated research methods that informed the rise of trauma-focused therapies (TFTs) [49, 54]. TFTs focus on those with significantly impaired symptoms and low functioning [20]. The goal of these trauma-informed approaches is to reduce negative psychological outcomes that can be manifested as depression, anxiety, borderline personality disorder, substance use disorders, and PTSD. The majority of TFTs feature a mix of cognitive, behavioral, and exposure techniques that facilitate the processing of the trauma memories [49]. Before treatment implementation, proper evaluation and the use of good diagnostic tools are needed to inform case conceptualization. The following section introduces evidenced-based treatments for childhood related traumas.

### 21.5.1 Trauma-Focused Cognitive Behavioral Therapy

Trauma-Focused Cognitive-Behavioral Therapy (TF-CBT) was developed within a CBT framework to address a variety of negative mental health outcomes resulting from trauma. Originally, TF-CBT was used with victims of sexual abuse but was later broadened to target other types of trauma [35, 55]. TF-CBT in children is geared toward emotion and behavioral-focused treatment, family engagement, and social problems that may present at school or home [32, 33]. For adolescents, the most benefit occurs when they are able to successfully challenge their false cognitions and beliefs because it promotes different perceptions and greater adaptability to their trauma [30, 56]. Involvement of parents or caregivers in therapy sessions is common. The goal is to increase parenting and communication skills as a means to positively influence and support their child's emotional regulation and strengthen coping skills [30, 34]. Research also suggests that TF-CBT is effective for teens in group settings to help destigmatize participants' experiences and reduce their feelings of shame [36].

### 21.5.2 Prolonged Exposure

Prolonged exposure (PE) is another widely used evidenced-based treatment for adult trauma, specifically PTSD. For children (12 and under) research is scant, but the existing literature presents mixed findings [57, 58]. Among adolescents, PE shows promising results [58, 59]. Prolonged Exposure for Adolescents (PE-A) is comprised of 14 weekly sessions that includes a developmentally appropriate module focused on personal safety while encouraging open dialogue to support adolescents' in confronting their traumas [59, 60]. Most studies have reported that PE-A and PE only work about 50%, but failure is mostly attributable to patients' high attrition rate [59, 61, 62].

Most recently, patients can now download a mobile-based app called *PE coach* to use in conjunction with treatment to enhance and facilitate compliance of take-home assignments [49, 63]. Clinicians using PE or PE-A must have a masters- or doctoral-level degree in mental health. Training workshops for PE-A and PE vary in cost. Settings for training include universities, mental health settings, and Veteran Affairs facilities. PE-A training focuses on differentiating how PE for adults differs from PE-A, including eligibility criteria, motivational techniques, sensitive skills development of exposure techniques, and tools to reduce treatment barriers.

### 21.5.3 Eye Movement Desensitization and Reprocessing

Eye Movement Desensitization and Reprocessing (EMDR) is based on the adaptive processing model that posits that PTSD occurs as a result of maladaptive processing of the trauma memory [64–66]. EMDR uses rhythmic left–right bilateral stimulation (eye movement or taps) to help induce REM (rapid eye movement) that in turn, facilitates the release of unaddressed cognitive, emotional, and physiological trauma reactions. Unlike PE, EMDR aims to process the trauma in a relaxed state [31, 43, 67]. Treatment is typically broken down into eight phases where the first three phases involve prep work, psychoeducation, and history gathering, while the later phases include reprocessing and desensitization of trauma, installation, body scans, and self-evaluation of the one's past, present, and future [67].

When EMDR was first introduced in the late 1980s, this treatment method was controversial. However, a growing number of studies, including randomized controlled trials (RTCs), comparative studies, and meta-analysis, indicate that EMDR is an effective type of psychotherapy [43, 55]. The literature suggests that EMDR can treat an array of different traumas, across ages, including adults with a history of childhood and complex traumas [23, 55]. Some researchers have evaluated the application of EMDR in group settings with refugees and victims of mass disasters and reported that it appeared helpful [2, 5, 68].

Similar to other psychotherapies, EMDR trainees must have an advanced degree in mental health. EMDR has private training institutes that provide certification upon completing the program. The minimal training requirements are two weekend in-person classes, assigned readings, practice between trainings, 10 hours of case conceptualization, and consultation [46].

### 21.5.4 Dyadic Psychotherapies

Recent evidenced-based approaches for trauma with children and adolescents are treatments that involve the parent or caregiver in the therapeutic process. Dyadic psychotherapies are based on a developmental model of attachment with integrated psychodynamic, social learning, and cognitive-behavioral techniques. The dyadic therapies described below require training and an advanced degree in mental health.

*Child-parent psychotherapy* (CPP) is appropriate for infants to children age 6 and younger. The focus of treatment is to appropriately strengthen attachment between the caregiver and the child [69, 70]. CPP treatment protocols are intensive and often include up to 50 treatment sessions. Research shows that CPP has been used to successfully address anxious attachment, children living with depressed parents, and children exposed to domestic violence, neglect, or maltreatment [12, 29]. Sessions focus on maladaptive behaviors, rebuilding dyadic emotional bond, and developing skills for appropriate interactions between caregiver and child as a means to increase child's functionality [29, 71].

CPP is one of the few psychodynamic and attachment-based psychotherapy that has been empirically validated to treat trauma among very young children. RTCs have shown CPP to be as effective as TF-CBT and CPP to be a flexible framework in treating dyads with diverse cultural values and in a wide range of settings [24, 43, 72]. CPP training is offered by CPP trainers who provide live or video demonstrations, case conceptualization, consultation, and the treatment manual.

*Parent–child interaction therapy* (PCIT) is another evidence-based treatment model that is tailored for parents of children who are 2 to 8 years of age with externalizing behavioral problems resulting from abuse [29, 73]. PCIT is helpful for children with severe disruptive and impulsive behaviors that are problematic at home and school. The goal is to reduce the child's negative behaviors and improve the relationship with their parent(s) by restructuring parent–child interactions through parental coaching. PCIT has two phases of treatment. Phase 1 focuses on relationship enhancement, while phase 2 targets effective discipline skills and limit setting [29]. Treatment ranges between 12 and 20 weekly, 1-hour sessions. The number of sessions is determined by the child's behavioral changes and caregivers' mastery of parenting skills.

Overall, PCIT is recognized in the United States and internationally as an empirically supported treatment that decreases negative externalizing behaviors in children with a trauma history of domestic violence, neglect, maltreatment, and abuse [74, 75]. PCIT can be conducted at home, foster care settings, and clinics [76, 77]. Training for clinicians is offered at private PCIT institutions nationwide [78]. PCIT training is structured and involves a minimum of 40 h of face-to-face training with a certified master trainer or level II trainer, attendance at in-person workshops, completion of online training modules, and monthly consultation via phone, web, or in-person [8, 76].

## 21.6 The Function of Resiliency and Post-Traumatic Growth

Positive psychology is focused on strength-based factors that promote recovery in the face of adversity [79, 80]. When confronted with trauma, most people experience emotional distress. Through the dynamic process of positive adaptation, recovery is possible [58, 61, 81]. This phenomenon, referred to as *resilience*, has been described in the scientific literature for the past several decades. Studies on resilience have identified traits such as greater cognitive flexibility, self-efficacy, active coping, internal locus of control, emotional regulation, optimism, and hopefulness as contributing factors that buffer the negative outcomes of trauma [80, 82].

A newer concept that gained traction in the 1990s is posttraumatic growth (PTG). PTG is defined as a transformative process of positive psychological changes in response to adversity. Recent literature distinguishes between PTG and resilience, while initially, these terms were used interchangeably. Although both constructs share some variance, they reflect two distinct processes [7, 72, 77]. PTG refers to the positive transformation that one undergoes when struggling with trauma, whereas resilience is defined by how well one adapts to an adversity.

Resilience research during the past decade has focused on examining these two mechanisms in children and adolescents. Resilience in youth populations includes environmental stability because young people are more vulnerable to environmental disruptions. Environmental stability is central to resilience [83, 84]. For example, resilience factors in maltreated children have been associated with stable caregiving, family coherence, and good parental relationships [74, 85]. Domhardt and colleagues [86] identified that an intense trauma (e.g., sexual abuse) caused both individual and environmental disruptions to children as well as adolescents [86].

Individuals who experience PTG may not necessarily be resilient. From the process of recovering from trauma, it is believed that some individuals become stronger, gain appreciation and confidence in their own capabilities, and are better equipped to cope with future unforeseeable events [56]. PTG research investigates how PTG in children differs or is similar to that identified in adults. Some factors of PTG are common among people of all ages such as quality of support and rumination ("brooding and reflective pondering") [80, 87, 88]. In children, PTG is strongly associated with competence beliefs. That is, a child believes that they can achieve success after a negative experience [34]. Another study on reactions post natural disaster found greater PTG among younger children, females, and higher positive reappraisal coping [21].

## 21.7 Conclusions and Key Points

This chapter reviewed current interventions and treatment approaches that can be implemented at different phases of recovery from traumatic events occurring in childhood and adolescence. It is important to address presenting concerns among children and adolescents because this developmental period marks greater susceptibility to long-term negative consequences. The development of trauma-informed approaches ensures the availability of appropriate, evidenced-based interventions and treatments that can be used during the acute-, intermediate-, and long-term phases of recovery.

Throughout this chapter, core interventions and treatments were discussed to highlight their role, function, and limitations in relieving distress. PFA is a Level 1 intervention delivered shortly after a mass trauma. Children and adolescents who receive support through education, stabilization, and assistance with resources to bring their lives back to pre-event baseline functioning fare well. When symptoms persist after receiving PFA, phase 2 interventions can help alleviate a child or adolescent's mild-to-moderate trauma-related symptoms. During this "at risk" period, SPR and crisis counseling are interventions that can be used by trained crisis counselors to support healthy coping skills, manage symptom reactions, promote healthy social connections, and enhance problem-solving skills. Although little research exists evaluating the efficacy of crisis counseling techniques, it is believed that children who receive PFA or SPR fare better than those who do not receive some type of evidence-informed intervention.

The third phase of care includes traditional psychotherapy. For some individuals, childhood trauma has long-term implications and causes severe disruption that increases the likelihood of mental illness. TFTs provide more intensive care and are delivered by licensed professionals who use evidence-based treatments to target specific symptoms and disorders. Finally, this chapter also reviewed resilience and posttraumatic growth. Children and adolescents who have experienced a traumatic event benefit from adequate resources, trained volunteers and professionals, and caring parents and caregivers who can support their recovery process.

## References

1. Felitti VJ, Anda RF, Nordenberg D, Williamson DF, Spitz AM, Edwards V, et al. Relationship of childhood abuse and household dysfunction to many of the leading causes of death in adults: The Adverse Childhood Experiences (ACE) Study. Am J Prev Med. 1998;14(4):245–58.
2. North CS, Pfefferbaum B. Mental health response to community disasters: a systematic review. JAMA. 2013;310(5):507–18.
3. Marchette LK, Weisz JR. Practitioner review: empirical evolution of youth psychotherapy toward transdiagnostic approaches. J Child Psychol Psychiatry. 2017;58(9):970–84.
4. McEwen BS. Physiology and neurobiology of stress and adaptation: central role of the brain. Physiol Rev. 2007;87(3):873–904.
5. Van der Kolk BA. Developmental trauma disorder: toward a rational diagnosis for children with complex trauma histories. Psychiatr Ann. 2017;35(5):401–8.
6. Chrisman AK, Dougherty JG. Mass trauma: disasters, terrorism, and war. Child Adolesc Psychiatr Clin. 2014;23(2):257–79.
7. Cougle JR, Timpano KR, Sachs-Ericsson N, Keough ME, Riccardi CJ. Examining the unique relationships between anxiety disorders and childhood physical and sexual abuse in the National Comorbidity Survey-Replication. Psychiatry Res. 2010;177(1–2):150–5.
8. Goodwin RD, Stein MB. Association between childhood trauma and physical disorders among adults in the United States. Psychol Med. 2004;34(3):509–20.
9. Fonzo GA, Ramsawh HJ, Flagan TM, Simmons AN, Sullivan SG, Allard CB, et al. Early life stress and the anxious brain: evidence for a neural mechanism linking childhood emotional maltreatment to anxiety in adulthood. Psychol Med. 2016;46(5):1037–54.
10. Wolitzky-Taylor K, Sewart A, Vrshek-Schallhorn S, Zinbarg R, Mineka S, Hammen C, et al. The effects of childhood and adolescent adversity on substance use disorders and poor health in early adulthood. J Youth Adolesc. 2017;46(1):15–27.
11. McLaughlin KA, Koenen KC, Hill ED, Petukhova M, Sampson NA, Zaslavsky AM, et al. Trauma exposure and posttraumatic stress disorder in a national sample of adolescents. J Am Acad Child Adolesc Psychiatry. 2013;52(8):815–30.
12. Jones E, Wessely S. Psychological trauma: a historical perspective. Psychiatry. 2006;5(7):217–20.
13. Wilson JP. The historical evolution of PTSD diagnostic criteria. In: Psychotraumatology. Boston, MA: Springer; 1995. p. 9–26.
14. DiMauro J, Carter S, Folk JB, Kashdan TB. A historical review of trauma-related diagnoses to reconsider the heterogeneity of PTSD. J Anxiety Disord. 2014;28(8):774–86.
15. Kirk AB, Madden LL. Trauma related critical incident debriefing for adolescents. Child Adolesc Soc Work J. 2003;20(2):123–34.
16. Foa EB, Keane TM, Friedman MJ, Cohen JA. Effective treatments for PTSD: practice guidelines from the International Society for Traumatic Stress Studies: Guilford Press; 2008.
17. Rose S, Bisson J, Churchill R, Wessely S. Psychological debriefing for preventing post traumatic stress disorder (PTSD). Cochrane Database Syst Rev. 2002;2(2)

18. Najavits LM. The problem of dropout from "gold standard" PTSD therapies. F1000prime Rep. 2015:7.
19. Weisz JR. Psychotherapy for children and adolescents: evidence-based treatments and case examples: Cambridge University Press; 2004.
20. Wilson C, Pence DM, Conradi L. Trauma-informed care: Encyclopedia of Social Work; 2013. p. 1–22.
21. Fallot RD, Harris M. Creating cultures of trauma-informed care (CCTIC): a self-assessment and planning protocol. Community Connections. 2009;2(2):1–8.
22. American Psychiatric Association, American Psychiatric Association. Highlights of changes from DSM-IV-TR to DSM-5.
23. Center for Substance Abuse Treatment. SAMHSA/CSAT treatment improvement protocols. Substance Abuse and Mental Health Services Administration. Rockville, MD: Author; 1993.
24. Weiner DA, Schneider A, Lyons JS. Evidence-based treatments for trauma among culturally diverse foster care youth: treatment retention and outcomes. Child Youth Serv Rev. 2009;31(11):1199–205.
25. Screening and Assessment Considerations. (2018). https://www.nctsn.org/sites/default/files/resources/fact-sheet/screening_and_assessment_considerations_for_implementation.pdf. Accessed 17 Aug 2018
26. England MJ, Butler AS, Gonzalez ML. Psychosocial interventions for mental and substance use disorders: A framework for establishing evidence-based standards. National Academy Press; 2015.
27. Barlow DH. Psychological treatments. Am Psychol. 2004;59(9):869.
28. Fink DS, Galea S. Life course epidemiology of trauma and related psychopathology in civilian populations. Curr Psychiatry Rep. 2015;17(5):31.
29. Brymer M, Layne C, Jacobs A, Pynoos R, Ruzek J, Steinberg A, et al. Psychological first aid field operations guide: National Child Traumatic Stress Network; 2006.
30. Buss KE, Warren JM, Horton E. Trauma and treatment in early childhood: a Review of the historical and emerging literature for counselors. Professional Counselor: Research and Practice. 2015;5(2).
31. Chen R, Gillespie A, Zhao Y, Xi Y, Ren Y, McLean L. The efficacy of eye movement desensitization and reprocessing in children and adults who have experienced complex childhood trauma: a systematic review of randomized controlled trials. Front Psychol. 2018; 9:534.
32. Cohen JA, Mannarino AP. Trauma-focused cognitive behavior therapy for traumatized children and families. Child Adolesc Psychiatr Clin. 2015;24(3):557–70.
33. Cohen JA, Mannarino AP, Kliethermes M, Murray LA. Trauma-focused CBT for youth with complex trauma. Child Abuse Negl. 2012;36(6):528–41.
34. Cryder CH, Kilmer RP, Tedeschi RG, Calhoun LG. An exploratory study of posttraumatic growth in children following a natural disaster. Am J Orthopsychiatry. 2006;76(1): 65–9.
35. De Arellano MA, Lyman DR, Jobe-Shields L, George P, Dougherty RH, Daniels AS, et al. Trauma-focused cognitive-behavioral therapy for children and adolescents: assessing the evidence. Psychiatr Serv. 2014;65(5):591–602.
36. Deblinger E, Pollio E, Dorsey S. Applying trauma-focused cognitive–behavioral therapy in group format. Child Maltreat. 2016;21(1):59–73.
37. Weisz JR, Ugueto AM, Cheron DM, Herren J. Evidence-based youth psychotherapy in the mental health ecosystem. J Clin Child Adolesc Psychol. 2013;42(2):274–86.
38. Jacobs GA, Meyer DL. Psychological first aid. Psychological interventions in times of crisis; 2006. p. 57–71.
39. Vernberg EM, Steinberg AM, Jacobs AK, Brymer MJ, Watson PJ, Osofsky JD, et al. Innovations in disaster mental health: psychological first aid. Prof Psychol Res Pract. 2008;39(4):381.
40. Dieltjens T, Moonens I, Van Praet K, De Buck E, Vandekerckhove P. A systematic literature search on psychological first aid: lack of evidence to develop guidelines. PLoS One. 2014;9(12):e114714.

41. Brown LM, Framingham JL, Frahm KA, Wolf LD. Crisis counselors' perceptions and assessment of suicidal behavior among hurricane survivors receiving crisis counseling services. Disaster Med Public Health Prep. 2015;9(3):291–300.
42. World Health Organization. Psychological first aid: guide for field workers. Geneva: World Health Organization; 2011.
43. Schafer A, Snider L, Sammour R. A reflective learning report about the implementation and impacts of psychological first aid (PFA) in Gaza. Disaster Health 2016;3(1):1–10.
44. Ramirez M, Harland K, Frederick M, Shepherd R, Wong M, Cavanaugh JE. Listen protect connect for traumatized schoolchildren: a pilot study of psychological first aid. BMC Psychol. 2013;1(1):26.
45. Donahue SA, Lanzara CB, Felton CJ, Essock SM, Carpinello S. Project liberty: New York's crisis counseling program created in the aftermath of September 11, 2001. Psychiatr Serv. 2006;57(9):1253–8.
46. Norris FH, Rosen CS. Innovations in disaster mental health services and evaluation: national, state, and local responses to Hurricane Katrina (introduction to the special issue). Adm Policy Ment Health Ment Health Serv Res. 2009;36(3):159–64.
47. Berkowitz S, Bryant R, Brymer M, Hamblen J, Jacobs A, Layne C, et al. Skills for psychological recovery: field operations guide. Washington (DC): National Center for PTSD (US Department of Veterans Affairs) and National Child Traumatic Stress Network (funded by US Department of Health and Human Services and jointly coordinated by University of California, Los Angeles, and Duke University); 2010.
48. Forbes D, Creamer M, Wade D. Psychological support and recovery in the aftermath of natural disaster. Int Psychiatr. 2012;9(1):15–7.
49. Marks S. Psychotherapy in historical perspective; 2017.
50. Pynoos RS, Steinberg AM, Brymer MJ. Children and disasters: public mental health. Textbook of disaster psychiatry; 2007. p. 48.
51. Fukkink R, Hermanns J. Counseling children at a helpline: chatting or calling? J Community Psychol. 2009;37(8):939–48.
52. Forbes D, Creamer M, Bisson JI, Cohen JA, Crow BE, Foa EB, et al. A guide to guidelines for the treatment of PTSD and related conditions. J Trauma Stress. 2010;23(5):537–52.
53. Williams JL, Rheingold AA. Novel application of skills for psychological recovery as an early intervention for violent loss: rationale and case examples. OMEGA J Death Dying. 2018;0030222818766138
54. Wagenmans A, Van Minnen A, Sleijpen M, De Jongh A. The impact of childhood sexual abuse on the outcome of intensive trauma-focused treatment for PTSD. Eur J Psychotraumatol. 2018;9(1):1430962.
55. Jensen TK, Holt T, Ormhaug SM. A follow-up study from a multisite, randomized controlled trial for traumatized children receiving TF-CBT. J Abnorm Child Psychol. 2017;45(8): 1587–97.
56. Ogińska-Bulik N, Zadworna-Cieślak M. The role of resiliency and coping strategies in occurrence of positive changes in medical rescue workers. Int Emerg Nurs. 2018;39:40.
57. Forbes D, Fletcher S, Wolfgang B, Varker T, Creamer M, Brymer MJ, et al. Practitioner perceptions of skills for psychological recovery: a training programme for health practitioners in the aftermath of the Victorian bushfires. Aust N Z J Psychiatr. 2010;44(12):1105–11.
58. Luthar SS. Resilience and vulnerability: adaptation in the context of childhood adversities. Cambridge University Press; 2003.
59. Aderka IM, Appelbaum-Namdar E, Shafran N, Gilboa-Schechtman E. Sudden gains in prolonged exposure for children and adolescents with posttraumatic stress disorder. J Consult Clin Psychol. 2011;79(4):441.
60. Gilboa-Schechtman E, Foa EB, Shafran N, Aderka IM, Powers MB, Rachamim L, et al. Prolonged exposure versus dynamic therapy for adolescent PTSD: a pilot randomized controlled trial. J Am Acad Child Adolesc Psychiatry. 2010;49(10):1034–42.
61. Masten AS. Ordinary magic: resilience processes in development. Am Psychol. 2001; 56(3):227.

62. McLean CP, Yeh R, Rosenfield D, Foa EB. Changes in negative cognitions mediate PTSD symptom reductions during client-centered therapy and prolonged exposure for adolescents. Behav Res Ther. 2015;68:64–9.
63. Reger GM, Hoffman J, Riggs D, Rothbaum BO, Ruzek J, Holloway KM, et al. The "PE coach" smartphone application: an innovative approach to improving implementation, fidelity, and homework adherence during prolonged exposure. Psychol Serv. 2013;10(3):342.
64. Shapiro F. Eye movement desensitization and reprocessing (EMDR): basic principles, protocols, and procedures: Guilford Press; 2001.
65. Yurtsever A, Konuk E, Akyüz T, Zat Z, Tükel F, Çetinkaya M, et al. An eye movement desensitization and reprocessing (EMDR) group intervention for Syrian refugees with post-traumatic stress symptoms: results of a randomized controlled trial. Front Psychol. 2018;9
66. Yurtsever A, Konuk E, Akyuz T, Tukel F, Zat Z, Acarturk C, et al. Early EMDR Interventions with Syrian refugees in Turkey R-TEP and G-TEP. In EMDR European conference, Edinburgh, Scotland. Three case studies. J EMDR Pract Res. 2014;5(3):95–110.
67. Shapiro E, Laub B. Early EMDR intervention (EEI): a summary, a theoretical model, and the recent traumatic episode protocol (R-TEP). J EMDR Pract Res. 2008;2(2):79–96.
68. US Department of Health & Human Services Administration for Children and Families Administration on Children, Youth and Families Children's Bureau.
69. Love JR, Fox RA. Home-based parent child therapy for young traumatized children living in poverty: a randomized controlled trial. J Child Adolesc Trauma. 2017:1–1.
70. McCabe K, Yeh M. Parent–child interaction therapy for Mexican Americans: a randomized clinical trial. J Clin Child Adolesc Psychol. 2009;38(5):753–9.
71. Lieberman AF, Ippen CG, Van Horn P. Child-parent psychotherapy: 6-month follow-up of a randomized controlled trial. J Am Acad Child Adolesc Psychiatry. 2006;45(8):913–8.
72. Kilmer RP, Gil-Rivas V, Griese B, Hardy SJ, Hafstad GS, Alisic E. Posttraumatic growth in children and youth: clinical implications of an emerging research literature. Am J Orthopsychiatry. 2014;84(5):506.
73. Hembree-Kigin TL, McNeil CB. Parent—child interaction therapy. Springer Science & Business Media; 2013.
74. Haskett ME, Scott SS, Nears K, Grimmett MA. Lessons from Katrina: disaster mental health service in the Gulf coast region. Prof Psychol Res Pract. 2008;39(1):93.
75. Lewey JH, Smith CL, Burcham B, Saunders NL, Elfallal D, O'Toole SK. Comparing the effectiveness of EMDR and TF-CBT for children and adolescents: a meta-analysis. J Child Adolesc Trauma. 2018:1–6.
76. Herschell AD, Scudder AB, Schaffner KF, Slagel LA. Feasibility and effectiveness of parent–child interaction therapy with victims of domestic violence: a pilot study. J Child Fam Stud. 2017;26(1):271–83.
77. Strasser A. Trauma-focused cognitive behavioral therapy: an evidence based practice applicable with minority children. Gallaud Chronic Psychol. 2015;3(1):38–42.
78. Wilsie C, Campbell C, Chaffin M, Funderburk B. Parent-child interaction therapy in child welfare. In: Parenting and family processes in child maltreatment and intervention. Cham: Springer; 2017. p. 107–25.
79. Moran GS, Nemec PB. Walking on the sunny side: what positive psychology can contribute to psychiatric rehabilitation concepts and practice. Psychiatr Rehabil J. 2013;36(3):202.
80. Vieselmeyer J, Holguin J, Mezulis A. The role of resilience and gratitude in posttraumatic stress and growth following a campus shooting. Psychological Trauma: Theory, Research, Practice, and Policy. 2017;9(1):62.
81. Splevins K, Cohen K, Bowley J, Joseph S. Theories of posttraumatic growth: cross-cultural perspectives. J Loss Trauma. 2010;15(3):259–77.
82. Tedeschi RG, Calhoun LG. Posttraumatic growth: conceptual foundations and empirical evidence. Psychol Inq. 2004;15(1):1–8.
83. Felix E, Afifi T, Kia-Keating M, Brown L, Afifi W, Reyes G. Family functioning and posttraumatic growth among parents and youth following wildfire disasters. Am J Orthopsychiatry. 2015;85(2):191.

84. Grolnick WS, Schonfeld DJ, Schreiber M, Cohen J, Cole V, Jaycox L, et al. Improving adjustment and resilience in children following a disaster: addressing research challenges. Am Psychol. 2018;73(3):215.
85. Afifi TO, MacMillan HL. Resilience following child maltreatment: a review of protective factors. Can J Psychiatry. 2011;56(5):266–72.
86. Domhardt M, Münzer A, Fegert JM, Goldbeck L. Resilience in survivors of child sexual abuse: a systematic review of the literature. Trauma Violence Abuse. 2015;16(4):476–93.
87. Tedeschi RG, Calhoun LG. The posttraumatic growth inventory: measuring the positive legacy of trauma. J Trauma Stress. 1996;9(3):455–71.
88. Taku K, Kilmer RP, Cann A, Tedeschi RG, Calhoun LG. Exploring posttraumatic growth in Japanese youth. Psychological Trauma: Theory, Research, Practice, and Policy. 2012;4(4):411.

The manufacturer's authorised representative in the EU is Springer Nature Customer Service Centre GmbH, Europaplatz 3, 69115 Heidelberg, Germany. If you have any concerns regarding our products, please contact ProductSafety@springernature.com

Printed and bound by CPI Group (UK) Ltd, Croydon, CR0 4YY

30/09/2026

02225299-0009